I0819434

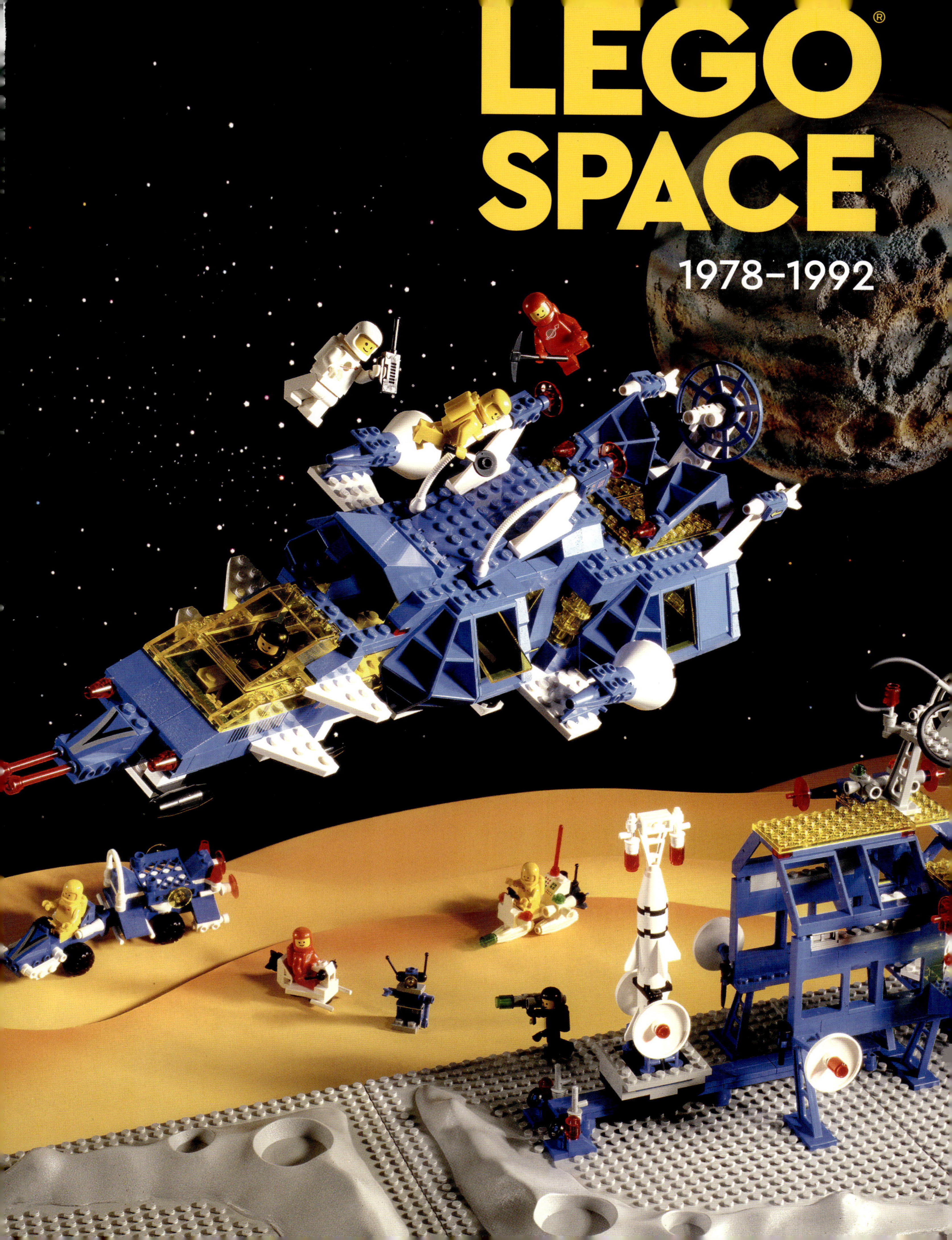
LEGO®
SPACE
1978–1992

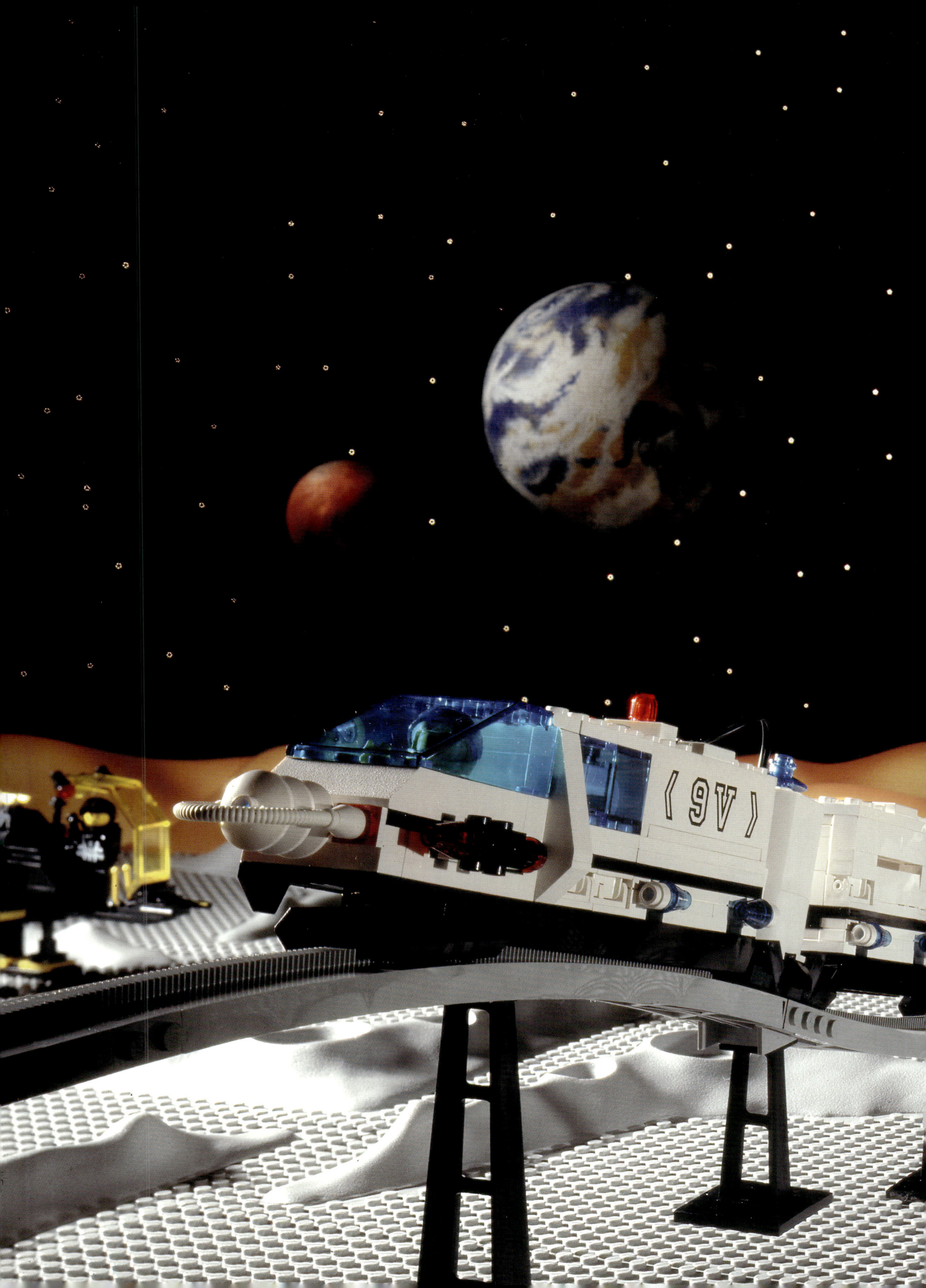
9V

1978–1992

Written by
TIM JOHNSON

Dark Horse Books

President & Publisher
Mike Richardson

Editor
Ian Tucker

Assistant Editor
Anastacia Ferry

Designer
David Nestelle

Digital Art Technicians
Adam Pruett and Josie Weaver

Special thanks to Simon Beecroft and Magdalena Czyżewska, Aleksandra Michalak, and Dan Shepherd at AMEET, and to Randi K. Sørensen, Søren Borup, Hanne Mørk Hede, Paul Hansford, Robin James Pearson, Martin Leighton Lindhardt, Jette Orduna, Kristian Reimer Hauge, and Jakob Liesenfeld at the LEGO Group.

The LEGO Group, Dark Horse, and the author also wish to thank Frédéric Roland Andre, Connie Bork, Niels Milan Pedersen, Mark John Stafford, Jørn Thomsen, William Thorogood, and Bjarne Panduro Tveskov for their contribution.

Photo of the Kristiansen family on page 9 © Copyright bpk.
Photo credit: Hanns Hubmann.
Photo credit of set 1580 on page 81: Kev Levell.

Additional references:
"The Truth about Space," *BrickJournal* 2, no. 6 (Summer 2009), TwoMorrows Publishing.
"Out of This World," *BrickJournal* 2, no. 2 (Summer 2008), TwoMorrows Publishing.
Brickset.com.
Rebrickable.com.
Peeron.com.
BrickLink.com.
BackOfTheBoxBuilds.com.

AMEET

AMEET Sp. z o.o.
Nowe Sady 6, 94–102, Łódź—Poland
ameet@ameet.eu
www.ameet.eu

Published by Dark Horse Books
A division of Dark Horse Comics LLC
10956 SE Main Street
Milwaukie, OR 97222

DarkHorse.com
LEGO.com

Facebook.com/DarkHorseComics
Twitter.com/DarkHorseComics

First edition: November 2023

Ebook ISBN 978-1-50672-519-2
Hardcover ISBN 978-1-50672-518-5

10 9 8 7 6 5 4 3 2
Printed in Heshan, China

IN MARCH 2023, AS THIS BOOK WENT TO PRESS, we received the sad news that Jørn Thomsen passed away. In his thirty-six years with the LEGO Group, he worked not only on LEGO® Space but also LEGO® Trains, LEGO® MINDSTORMS®, LEGO® Western, LEGO® Creator Inventor Sets, LEGO® Architecture, LEGO® Technic, and many more. His inventive designs have given happiness to generations of children, and his legacy lives on in the brick.

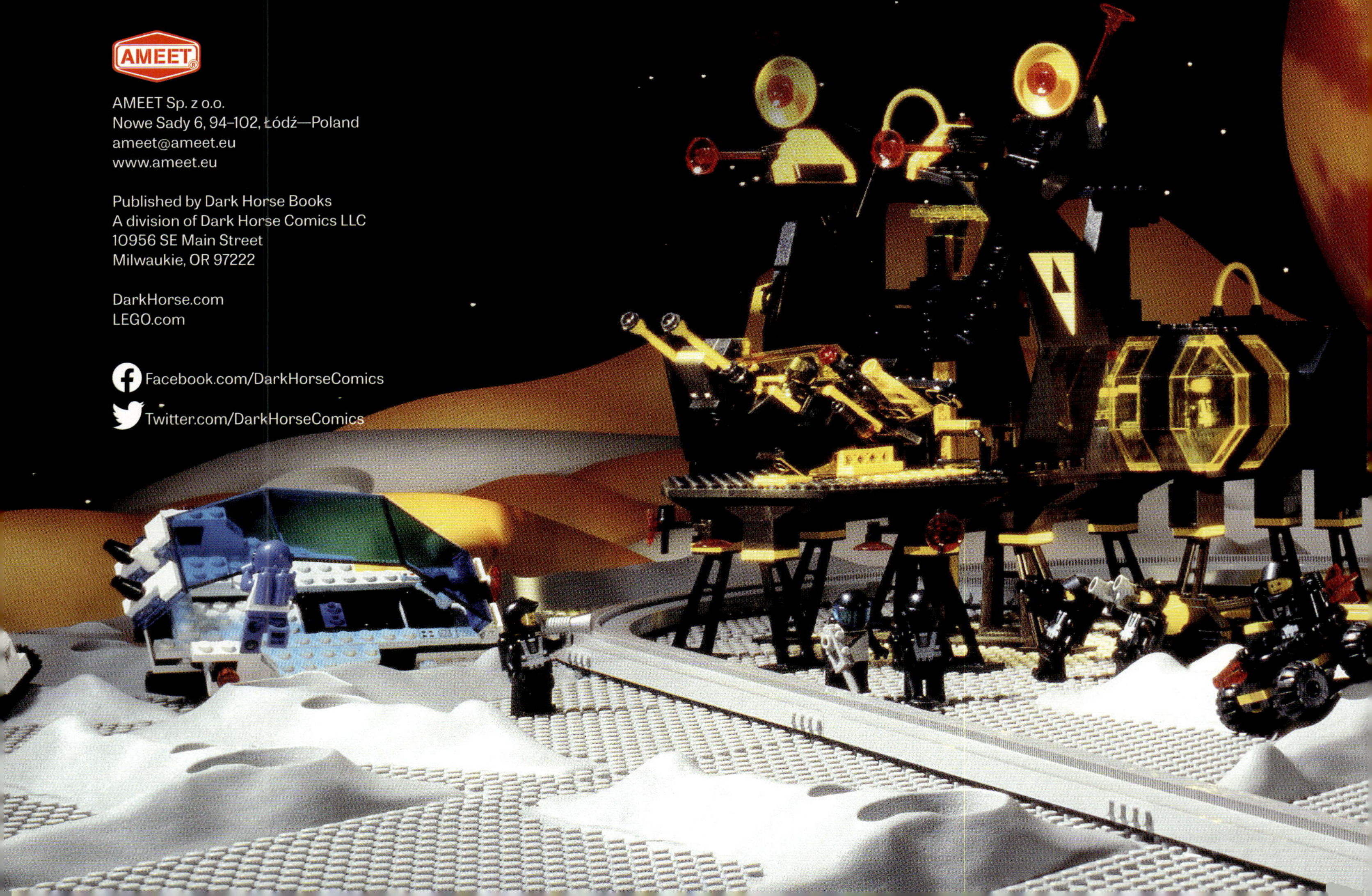

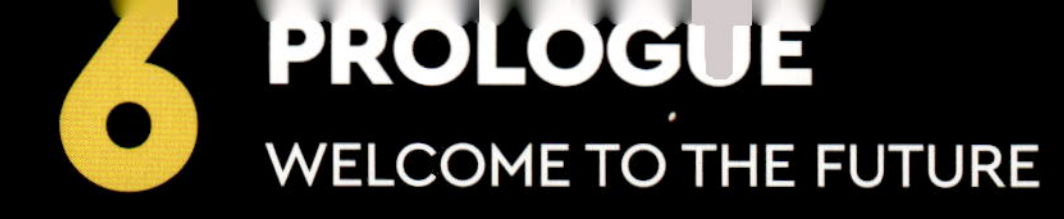

6 **PROLOGUE**
WELCOME TO THE FUTURE

16 **CHAPTER 1**
CLASSIC LEGO® SPACE, 1978–1987

106 **CHAPTER 2**
FUTURON AND BLACKTRON, 1987–1990

150 **CHAPTER 3**
SPACE POLICE, M:TRON, AND BLACKTRON FUTURE GENERATION, 1989–1992

190 **EPILOGUE**
WHAT'S NEXT?

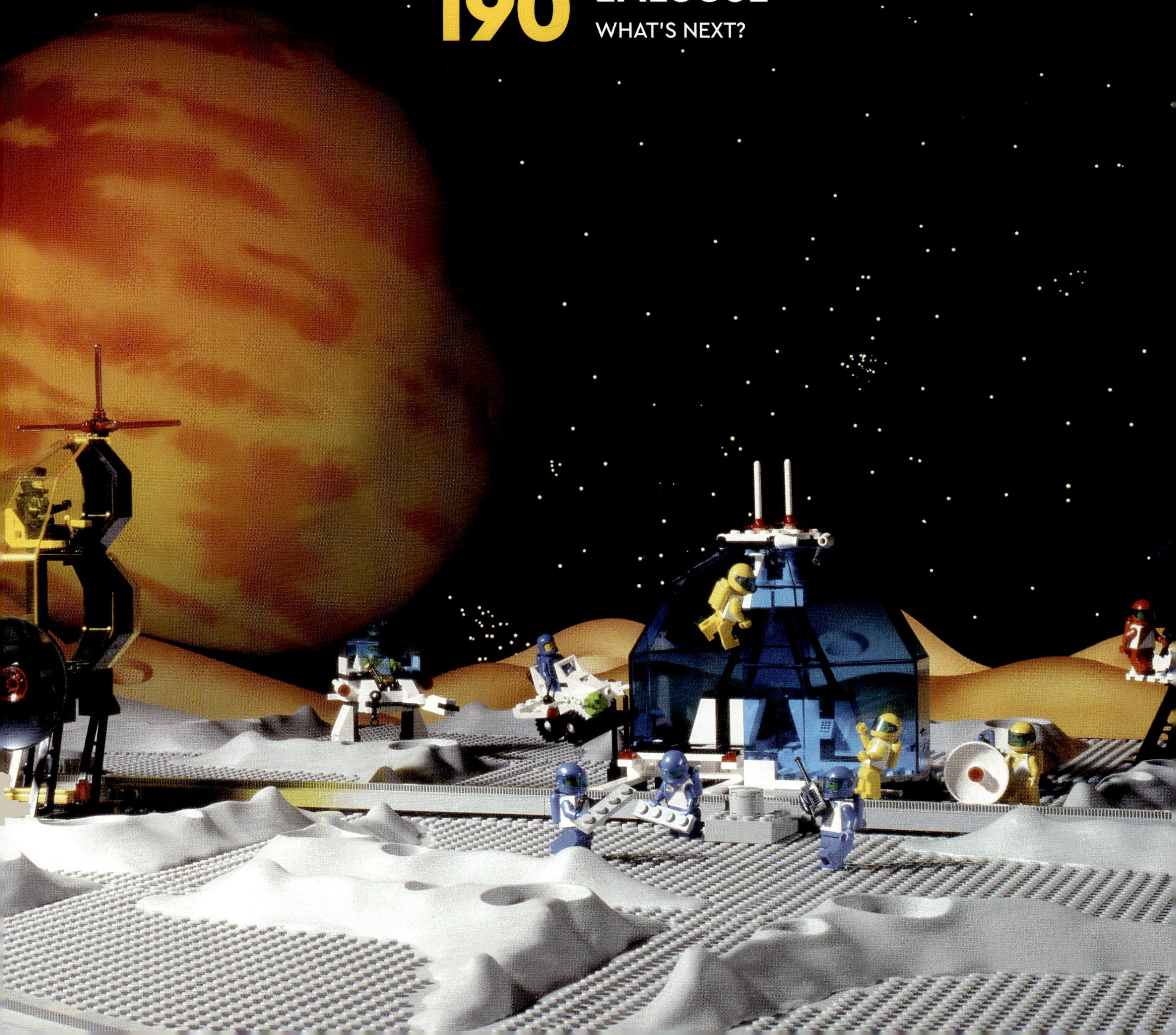

PROLOGUE

Welcome to the future

In 1932, in the small Danish town of Billund, forty-one-year-old carpenter Ole Kirk Kristiansen founded a wooden toy company. At the same time, laboratories around the world were developing new kinds of so-called "plastics"—the first wholly synthetic materials.

After the Second World War, wood supply issues led Ole to eventually favor plastics over wood—a decision that his children, including future owner Godtfred Kirk Christiansen, initially tried to dissuade him from pursuing. Ole, however, was resolute in his belief—especially given the potential he saw in one of their plastic products, a self-locking construction brick.

▼ Some sets in the LEGO® Basic range, such as set 033 from 1968, included images of models of rockets and astronauts to inspire young builders.

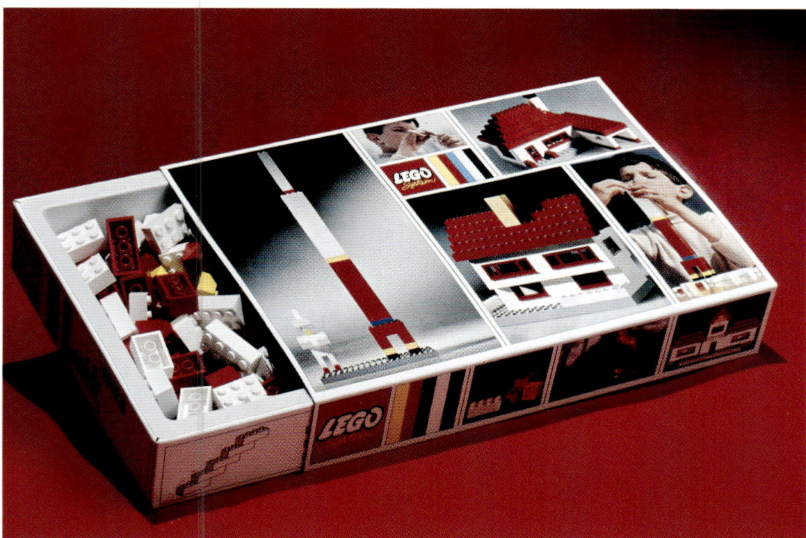

The LEGO® brick soon became the company's primary focus, bolstered by Godtfred's introduction of the LEGO® System in Play in the mid-1950s: a novel concept proposing that a line of toys could provide unlimited play opportunities by being based upon a common underlying system. In 1958, the year of Ole's death, the LEGO Group patented their improved design for the LEGO brick.

Meanwhile, technological advances in supersonic aviation and space flight meant that reality seemed to be catching up with science fiction. The imagination of young minds had been captured by comics and B movies, but increasingly it was humanity's actual steps into space that inspired children's play.

Following the launch of the first successful artificial satellite, the Soviet Union's Sputnik 1 of 1957, space technology progressed rapidly—as did fierce competition between the Soviet Union and the United States, in what became known as the space race. The Soviet Union achieved a series of inspirational firsts: by 1966 they had sent men and women into space, succeeded in the first spacewalk, propelled uncrewed missions to Venus and Mars, and managed the first safe, or "soft," landing of surveying equipment on the moon.

The most notable successes of the US in that same period were of a scientific nature: the discovery of the Van Allen radiation belts that surround the Earth, direct measurements of the solar winds, and the launch of the first space telescope. However, in July of 1969, the US achieved a first that united the world in awe, when the various dramatic successes and disasters of the Apollo missions finally resulted in humanity's first steps on the moon.

Back on Earth, the LEGO product offering had also been evolving. In 1963, the company improved the strength and longevity of LEGO elements by switching plastics from cellulose acetate to acrylonitrile butadiene styrene (ABS). The same year saw the first sets that included building instructions. LEGO Futura ApS was established in 1965: an independent internal product development department tasked with planning the long-term development of models and elements, such as motorization through the LEGO® Trains product line in 1966. One aspect of the lengthy development process was to formally test potential products in focus groups with children.

"IF WE GET THIS RIGHT, WE CAN SELL THESE BRICKS ALL OVER THE WORLD!"

—OLE KIRK KRISTIANSEN

While the LEGO Group had not yet released any space-themed products, it was clearly an attractive prospect. Throughout the 1960s, rockets featured in various LEGO pamphlets, promotional materials, and Idea Books, which provided inspirational building ideas. Even Godtfred's teenage son, Kjeld Kirk Kristiansen, designed his own giant launch rocket using LEGO bricks.

Western Europe was the LEGO Group's target market in the 1950s, but horizons expanded in the 1960s. In 1961 the license to manufacture and sell LEGO products in the US and Canada was given to the Samsonite Corporation, who, in 1964, released the 115-piece set 801 Space Rocket. The red, blue, and white rocket had dozens of curved elements, though the construction used delicate building techniques that would not be considered sufficiently stable today!

In 1968, the first LEGOLAND® theme park opened in Billund to enormous success. At the same time, a new product line was being developed, also called LEGOLAND. Like the current-day LEGO® City theme, it featured housing, public services, and construction and urban vehicles, at a more consistent scale than earlier LEGO product ranges.

Many of the small LEGOLAND cars were designed by a new recruit who began in 1968: Jens Nygaard Knudsen. He soon progressed to creating the larger sets in the range, as well as designing products for the LEGO Trains product line, including some of the newer elements, such as the chassis and wheels.

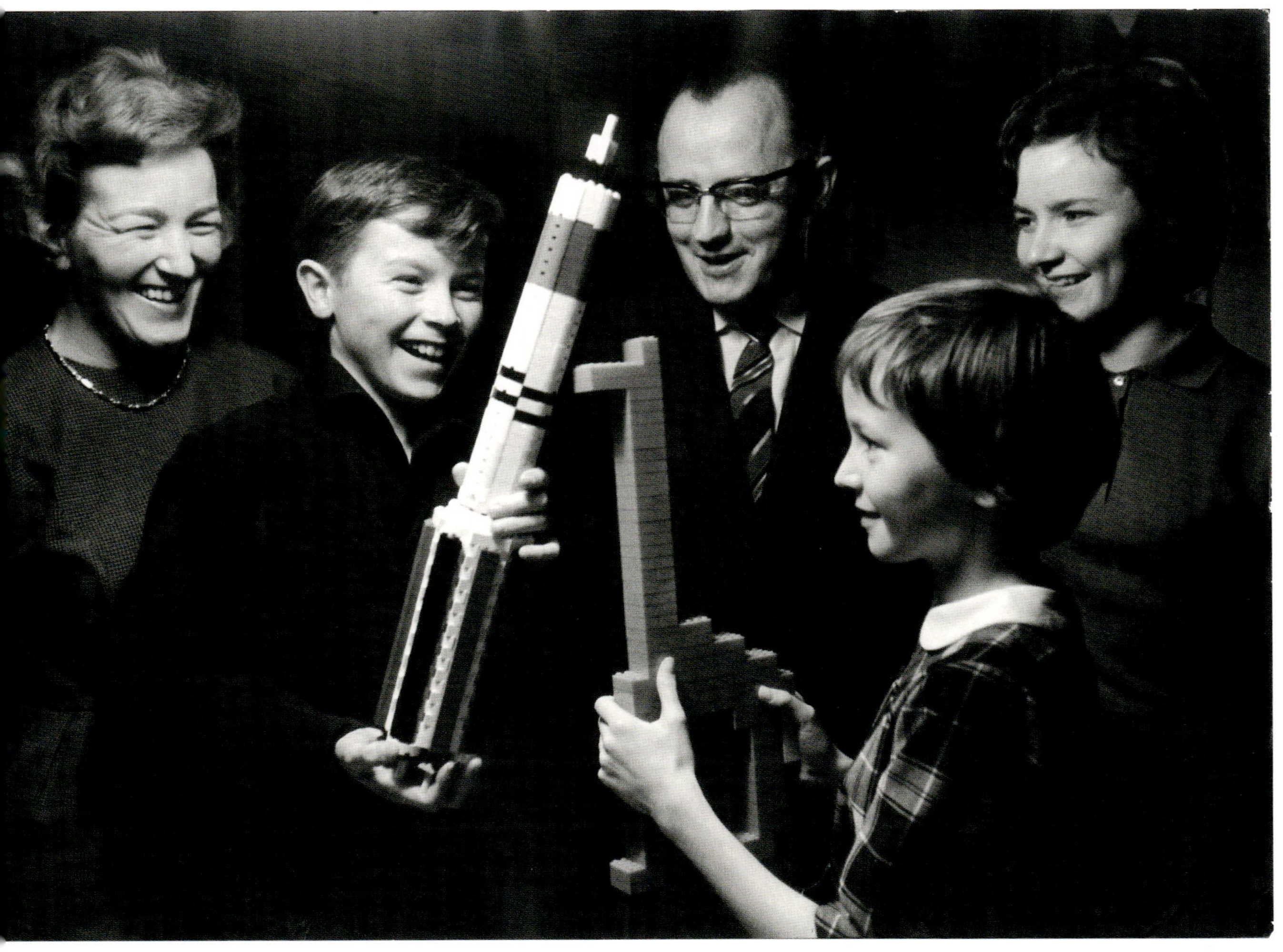

◀ **THE KRISTIANSEN FAMILY**, published in the German-language weekly news magazine *Quick* in 1962. Kjeld Kirk Kristiansen, who became the president and CEO of the LEGO Group in 1979, is holding a custom-built LEGO rocket. *Left to right*: Edith Christiansen, Kjeld Kirk Kristiansen, Godtfred Kirk Christiansen, Hanne Kristiansen, and Gunhild Kristiansen.

▶ In the mid-1970s, a breathtaking model of Cape Kennedy Air Force Station was constructed at the second LEGOLAND park to be built, LEGOLAND Sierksdorf in Germany.

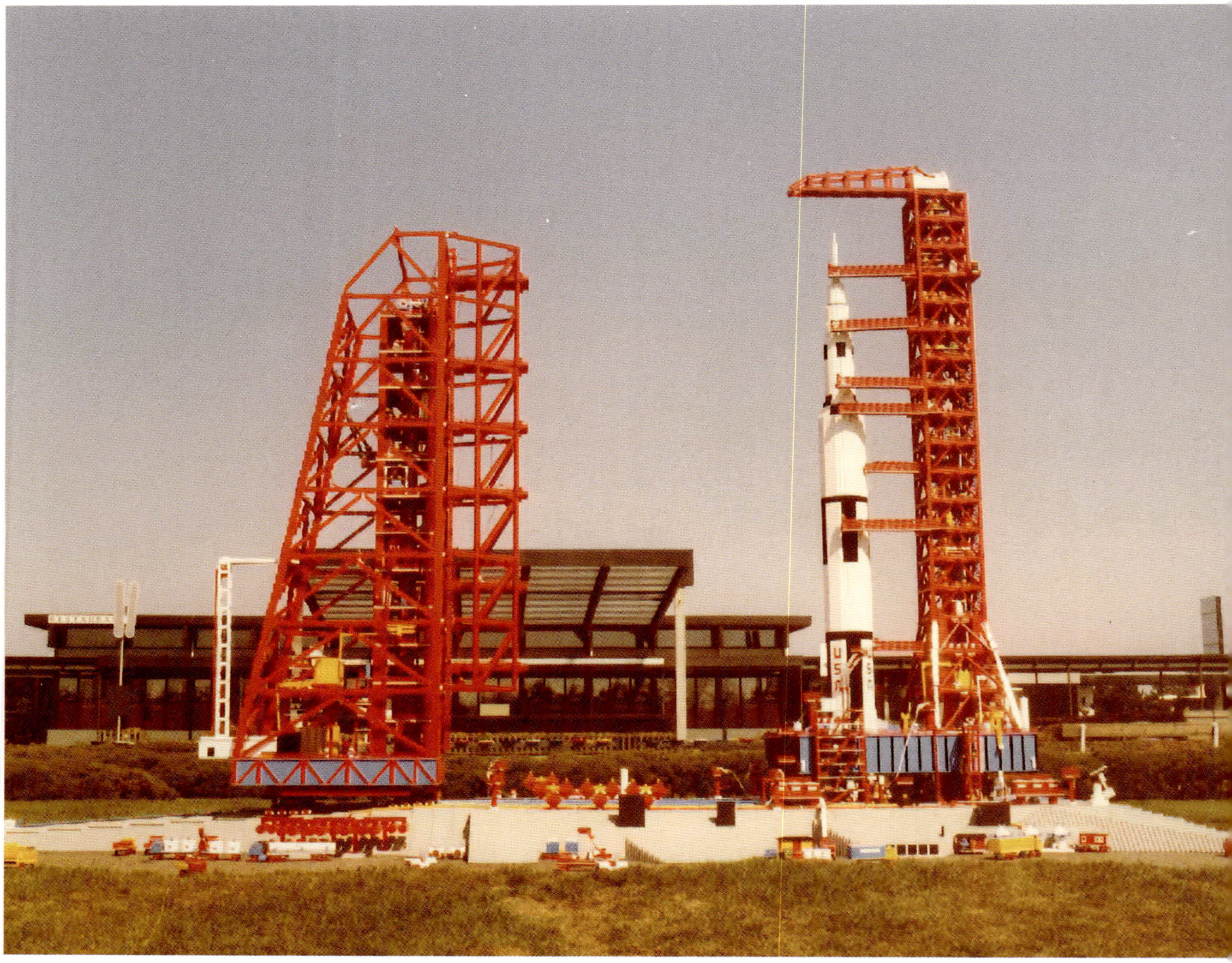

SMALL STEPS INTO SPACE

In 1973, the LEGOLAND® range included a space-themed set for the first time. Although the LEGO Group had now regained the US rights from Samsonite, this set was not released in North America. Designed by Jens Nygaard Knudsen, set 358 Rocket Base contained 275 pieces and depicted a rocket launch pad. A simple carriage resting on two railway track elements allowed the black-and-white rocket to be moved back and forth from the service structure.

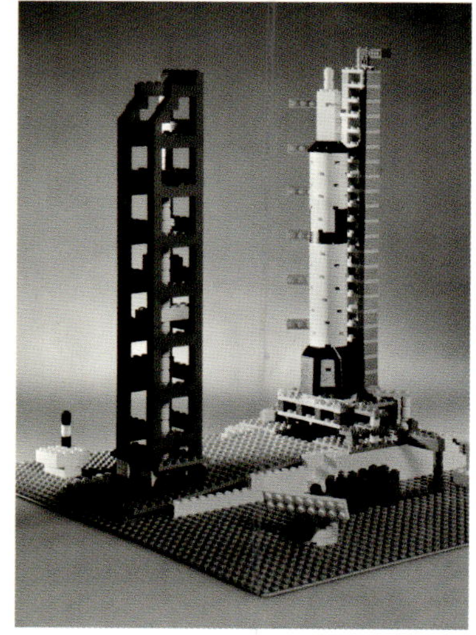

▲ "The rocket itself soars 100 meters into the sky—like a giant toy." In 1970, the 242 Idea Book inspired readers with this model of Cape Kennedy in Florida.

The LEGO® element inventory contained few rounded elements at this time, and so the radar dishes in this set were square, constructed from transparent plates. The rocket, however, was topped by a brand-new rounded element, cone 4 x 4 x 3, but this element was not reused in the LEGO® Space line later on. The whole model was built upon a dramatic red baseplate that was unique to this product and perhaps could have been interpreted by some children as being alien terrain rather than terrestrial.

At this time, the standard brick colors available to the designers were yellow, red, blue, white, and black. Gray, green, and some transparent colors were present but used only sparsely. The company avoided models with a lot of gray bricks, in case children used them to build weapons of war. Furthermore, from a marketing viewpoint, a predominantly gray set may have appeared less attractive than other products on store shelves. Jens settled upon using blue for the structural elements of the building and launch pad. He felt that blue looked sufficiently technical, and the color would remain prominent in the palette of LEGO Space sets for many years to come.

Completing the set was an eight-wheeled vehicle with an open tailgate, which could be used to deliver the astronauts to the rocket. Of course, no astronauts were included, as LEGO sets of this era did not include any figures, an issue which the LEGO Group was addressing at the time. In 1974, large buildable figures were introduced into doll's house sets in the LEGO® Homemaker line. It is interesting to note that although this was the first line to be marketed primarily to girls, the LEGO Group did not consider girls to be the exclusive audience. On the rear of a 1974 brochure titled *Little Girls Think Big*, a note to parents read, "The urge to create is equally strong in all children. . . . A lot of boys like dolls houses. They're more human than spaceships. A lot of girls prefer

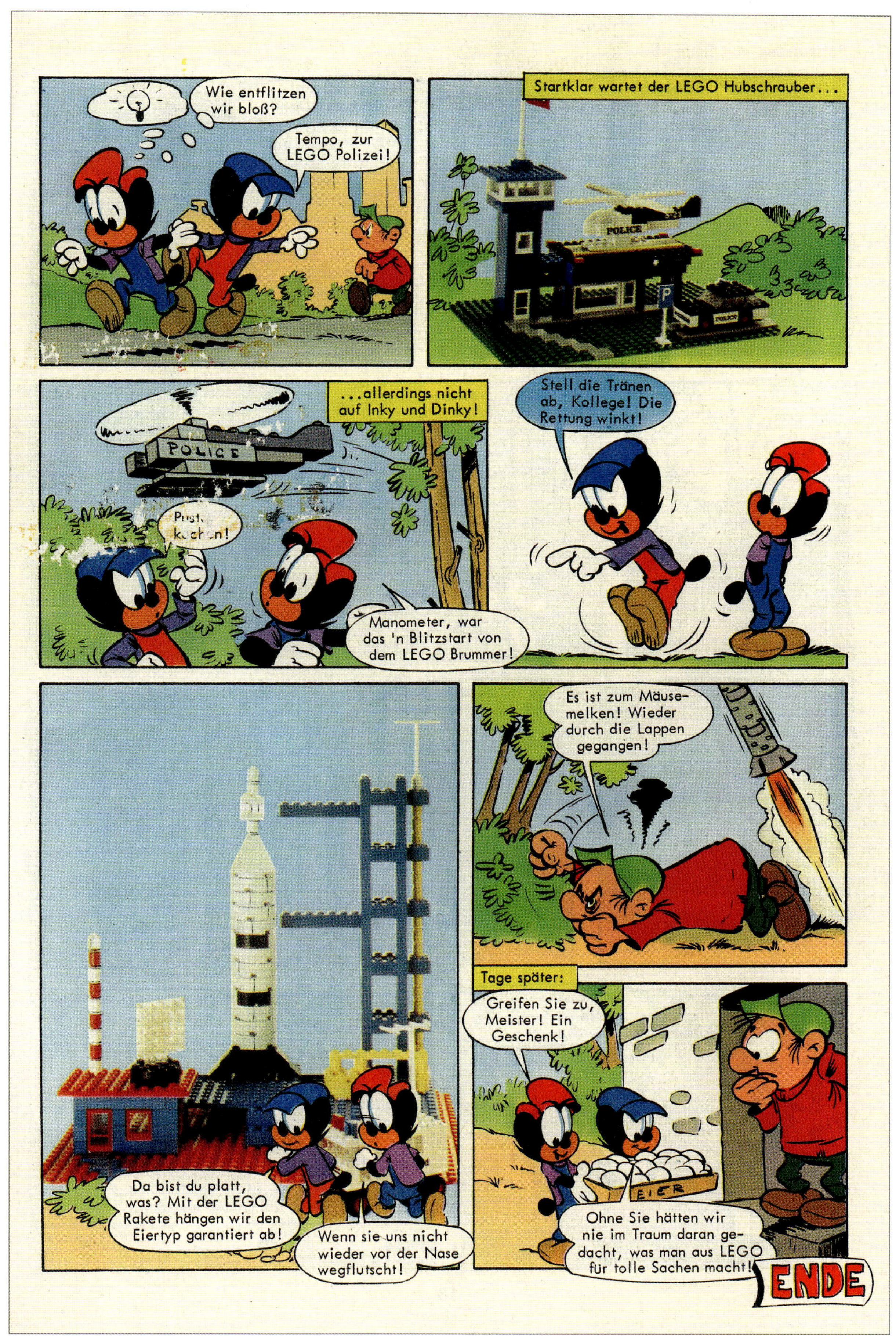

A 1974 German comic following Felix the Cat's nephews Inky and Dinky and featuring several LEGO products. Set 358 Rocket Base provides them with the advantage over their pursuer, but they share their prize with him at the end: "Without you, we would never have dreamed of the amazing things that can be made from LEGO bricks!"

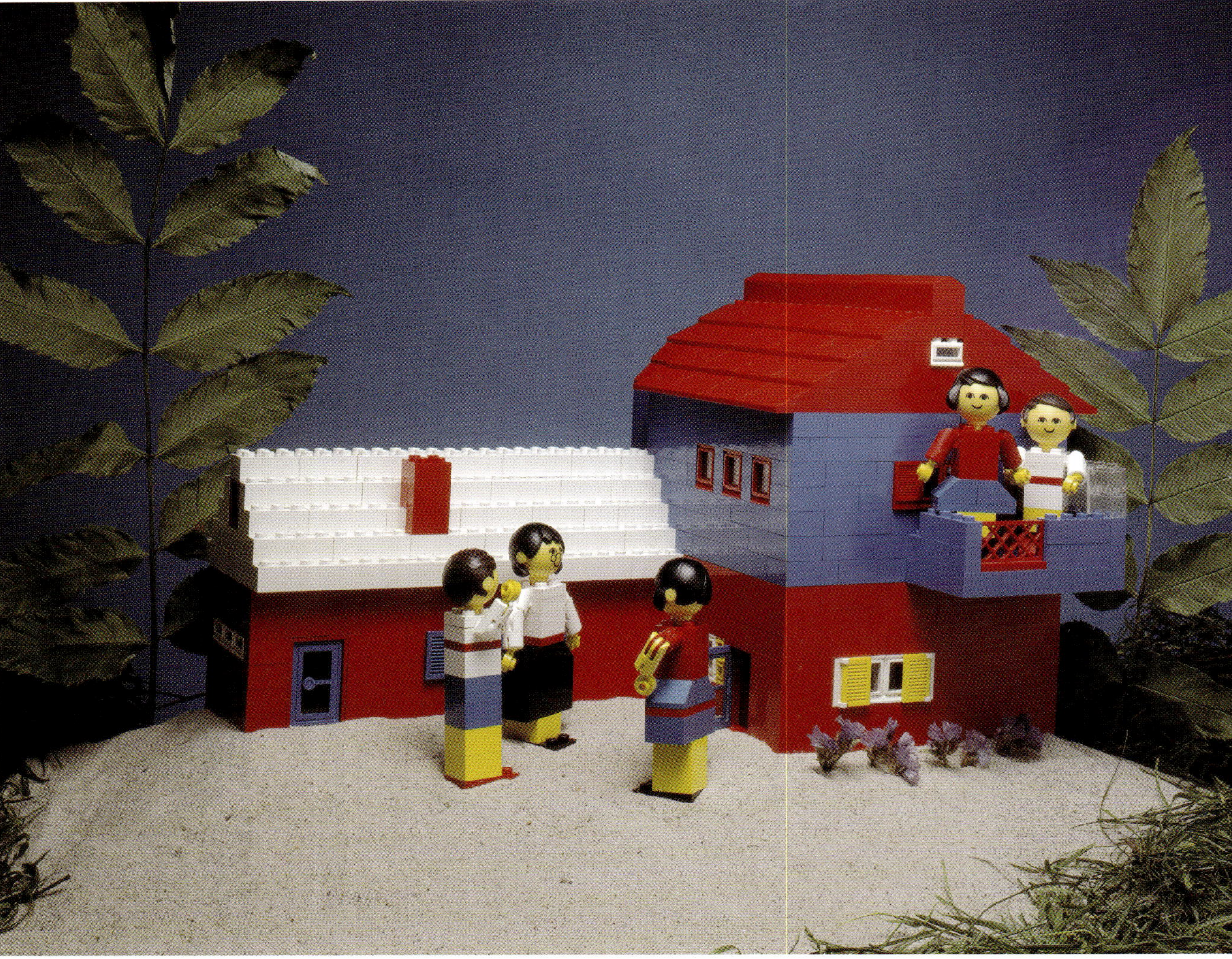

▶ Large-scale buildable figures were introduced in 1974. The torso and arm elements were also used in space-themed sets.

spaceships. They're more exciting than dolls houses. The most important thing is to put the right materials in their hands and let them create whatever appeals to them." Indeed, the very next space-themed product to be released utilized the buildable figures.

> **"WE NEEDED THE PEOPLE IN THERE SO THAT WE HAD THE ROLE-PLAY."**
>
> —CONNIE BORK, BUILDING INSTRUCTIONS SPECIALIST, THE LEGO GROUP, 1979 TO PRESENT

Released in 1975 in Europe as set 367 Moon Landing and in 1976 in the US as set 565, this 364-piece set depicts a blue lunar lander and rover, crewed by three buildable figure astronauts. The central module rests freely on the leg assembly, ready to lift off at a moment's notice, and, although it is hollow, does not accommodate a buildable figure.

White and transparent bricks represented the astronauts' helmets, and buildable figure heads were also provided in the set. Although these astronauts lacked fine detail overall, their arms were highly flexible, meaning they could drive the buggy, look through a handheld telescope, and even salute the American flag. This was

Årets første LEGO® nyheder.

367 - Månelanding kr. 69,50

693 - Brandbil kr. 18,50

694 - Sættevogn kr. 18,50

362 - Hollandsk mølle kr. 48,00

660 - Flytransportør m/fly kr. 12,75

615 - Gaffeltruck m/fører kr. 7,00

LEGO® er nyt legetøj hver dag.

Priserne er vejl. udsalgspriser incl. 15% moms.

Navnene LEGO, LEGOLAND og LEGO DUPLO er registrerede varemærker 1975.

▶ A charming Danish ad of 1975 presents set 367 Moon Landing on an illustrated background.

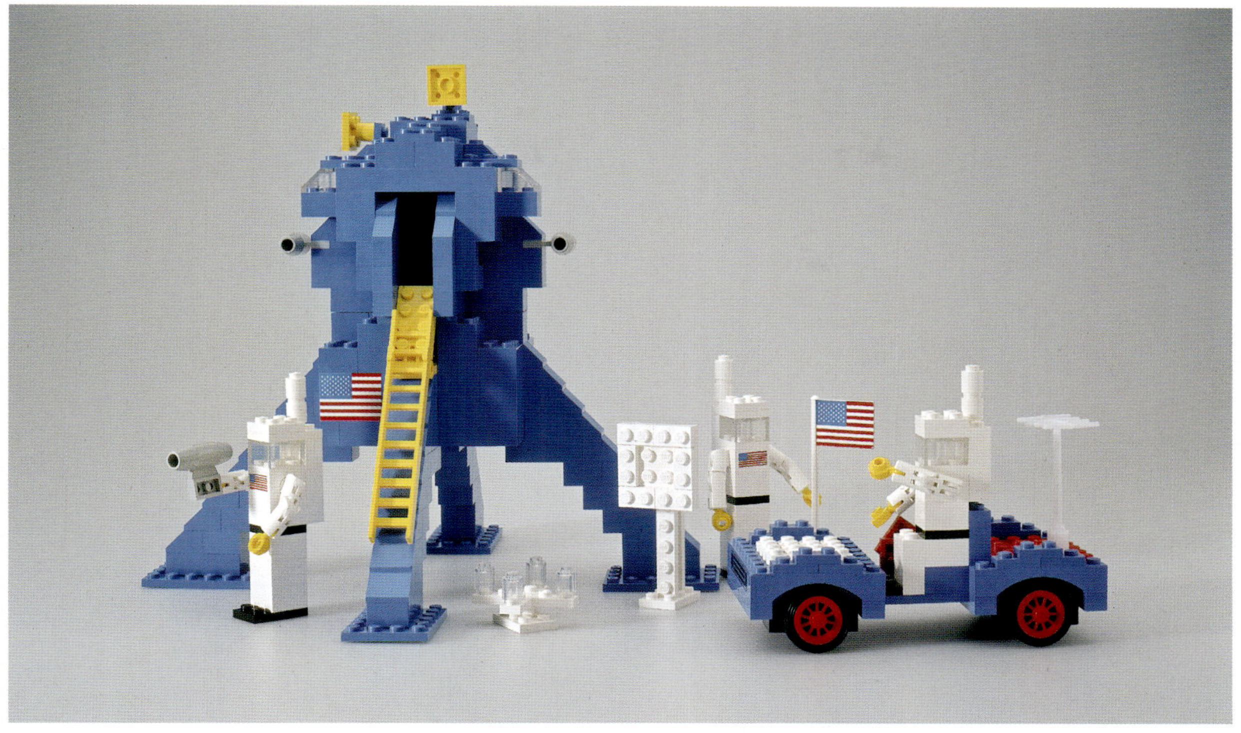

◀ Set 367 Moon Landing (also known as "Space Module with Astronauts").

"THE SPACE RACE CAPTURED CHILDREN'S IMAGINATION AND THAT WAS THE DRIVING FORCE; JENS [NYGAARD KNUDSEN] WANTED TO GIVE A VOICE TO THAT."

—MARK JOHN STAFFORD, SENIOR DESIGNER, THE LEGO GROUP, 2006 TO PRESENT

not the last time these arm elements would be utilized in space sets; however, buildable figures were soon to be replaced by the smaller, more practical minifigure.

The box art was a giant leap forward. Placed on a barren moonscape with the sun rising overhead, the lunar lander was lit in high contrast, emulating the stark, sunlit photographs from the Apollo missions that children were so familiar with. This approach would be further improved with the LEGO Space line.

THE GENESIS OF LEGO® SPACE

In the 1970s, the development of new LEGO® products took several years. Although the LEGO® Space line would not launch until 1978, Jens Nygaard Knudsen began his initial designs in 1973, the same year that 358 Rocket Base was released.

There were only three or four designers at the LEGO Group at this time, and they bore many responsibilities. Their primary areas of work were to develop new concepts, design products, and draw the building instructions. About two years later, some departmental changes were made to LEGO Futura, resulting in three separate teams being established to work specifically on those three areas. Jens was placed on a forward-planning team, and, with his time now primarily focused on developing new ideas, he was able to fully develop his beloved outer-space concept: his designs for the first assortment of LEGO Space products were completed in 1976.

Jens is known to have led the creative design of all the products in the initial LEGO Space line; however, his fellow designers did also contribute to the process. Daniel August Krentz, who was hired as a LEGO designer in 1970, recalls designing one of those initial products, 891 Two Seater Space Scooter. "I made that, but under Jens's close direction; LEGO Space was his favorite, and he wanted to do them all!"

THE SYSTEM WITHIN THE SYSTEM

Godtfred Kirk Christiansen had made a conscious decision in the 1960s to focus the company's output solely on the LEGO® System in Play, but as the product portfolio grew, the need for clearer delineation became more important.

During the 1970s, Godtfred's son, Kjeld Kirk Kristiansen, was taking on more management responsibilities within the company. He was a driving force behind a new development model, broadly segmented into LEGO construction toys, LEGO® DUPLO®, and other associated play materials. Construction toys were then divided

▶ Second- and third-generation owners Godtfred Kirk Christiansen and Kjeld Kirk Kristiansen at home in 1976. The car is a model that Kjeld designed in the 1960s.

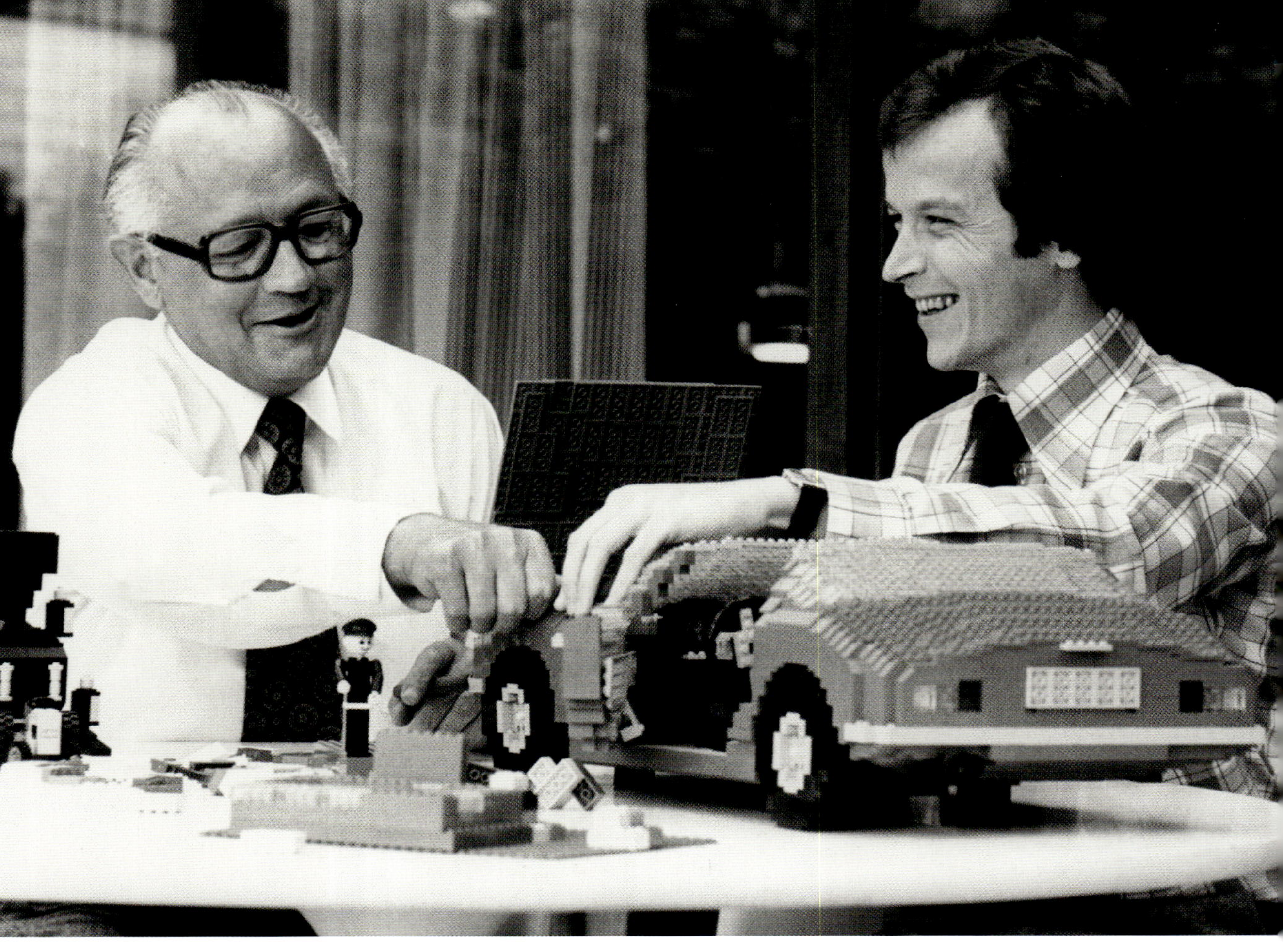

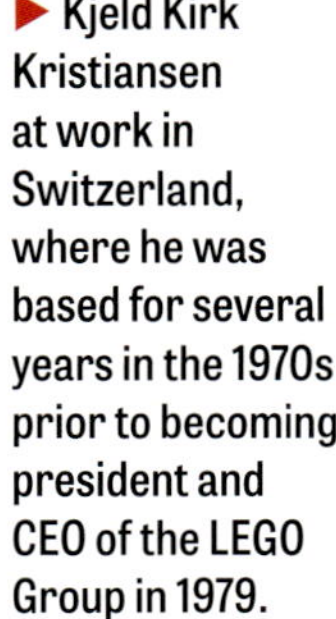

▶ Kjeld Kirk Kristiansen at work in Switzerland, where he was based for several years in the 1970s prior to becoming president and CEO of the LEGO Group in 1979.

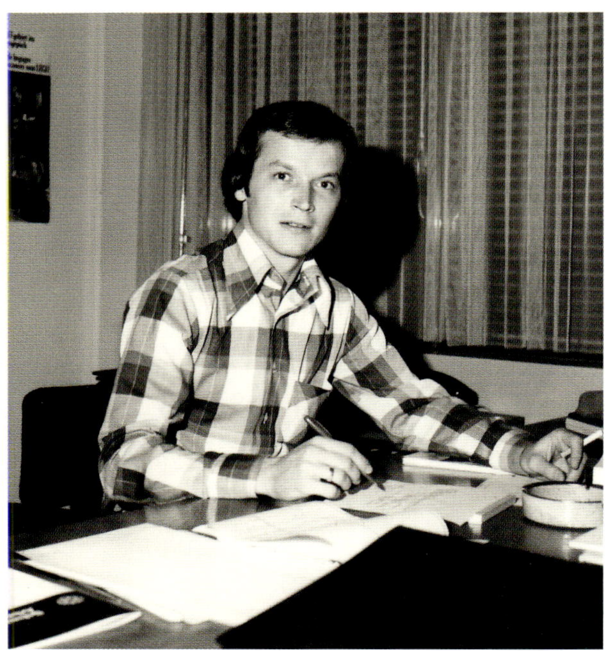

into LEGO® Basic, LEGO Trains, LEGO Homemaker, LEGO® FABULAND®, LEGO® Technic, and of course the LEGOLAND® theme. A significant change, however, was that the LEGOLAND theme was now to be subdivided into three categories: LEGOLAND® Town, LEGOLAND® Castle, and LEGOLAND® Space. This would not only assist consumers with choosing the right model but also help retailers make the differences clear to their customers.

LEGOLAND Town and LEGOLAND Castle were launched in 1978, and LEGOLAND Space was intended to be premiered at the Nuremberg Toy Fair in Germany in 1979. However, concerns about competitor products resulted in some of the sets being released in the American market in 1978.

ONE GIANT LEAP FOR MINIFIGURES

At the same time that Jens Nygaard Knudsen was developing what would eventually be rebranded to the simplified LEGO® Space line in the mid-1970s, he was also working on another new concept: a much smaller figure than the buildable figures from LEGO® Homemaker, which he believed could populate the LEGOLAND® theme. In 1975, the first products containing an early version of what would become known as the LEGO minifigure appeared. The arms and legs of this forerunner to the minifigure did not move, and the head had no face print. Jens and his colleagues continued to develop the new figure, and the final version was ready to be included in the LEGOLAND sets of 1978.

The elements accessorizing the minifigure needed to be usable across all three LEGOLAND lines as much as possible. For example, the air tanks on the back of astronauts first appeared on firefighters in LEGO® Town, and the helmet element was worn by knights in LEGO®

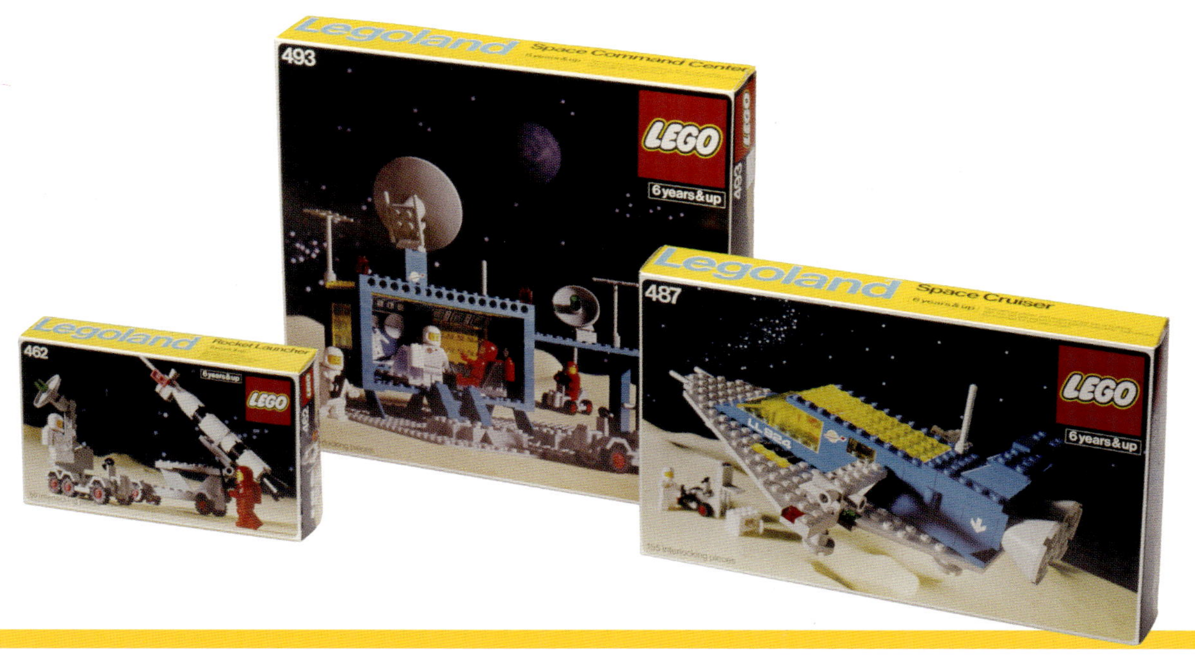

◀ A photograph from 1978 showing three LEGO Space products that were released that year in the United States, prior to the European release.

"JENS WANTED SPACE HELMETS WITH A VISOR, AND ALSO SOME WITH ANTENNAS AND THINGS LIKE THAT. I MADE A LOT OF VERSIONS OF THOSE, BUT THEY WERE NEVER TO BE [AT THAT TIME]."

—NIELS MILAN PEDERSEN, DESIGNER, THE LEGO GROUP, 1980 TO PRESENT

Castle with a special visor attached. With so many new elements needed for LEGO Space, however, visors for the astronauts were left to the imagination until 1987.

Of course, all minifigure head elements were yellow with the same face print. "At that time minifigures had to be smiling and had to be neutral, all of them," recalls LEGO designer Niels Milan Pedersen, who joined Jens's team in 1980. "We tried to change the printing on the faces of the minifigures. I also think that it was Jens who thought we should have some space women, but we couldn't figure out how to show this. I recall that was quite early on. Later, it was actually me who made the first; that was for LEGO® Pirates in 1989."

The distinguishing feature of the astronauts was the dynamic insignia on their chests: a red spaceship performing a gravitational slingshot around a golden moon. This torso decoration was initially supplied on a sticker sheet, rather than being printed directly onto the torso element.

The first astronaut minifigures were provided in red and white. In more recent times, there has been speculation that these two colors may have represented Soviet cosmonauts and American astronauts. Mark John Stafford, a current designer at the LEGO Group who grew up playing with LEGO Space products, discussed the topic with Jens in 2009. "I think that was the original idea, to try and make sure that both sides were represented, but I'm not sure," recalls Mark. "Even if that was the idea at the beginning, they took the conflict out very quickly—probably at the first presentation. Whereas nowadays we'd explore it and see if kids liked it." Niels recalls a similar discussion in the 1980s. "I do remember Jens saying that the white and the yellow guys were the good guys, and the red were the bad guys. But he was mostly joking, of course."

While many ideas and stories constantly flowed through Jens's mind, what is certain is that no nationalities, job roles, or story lines were attributed to minifigures in the original LEGO Space products. Instead, the play theme presented a vast universe of exploration, with models inspired by real-world technology, for children to enjoy open-ended play.

In this book, the short adventures that accompany each set are from the author's imagination, inspired by the original LEGO Space products and their photography and advertising.

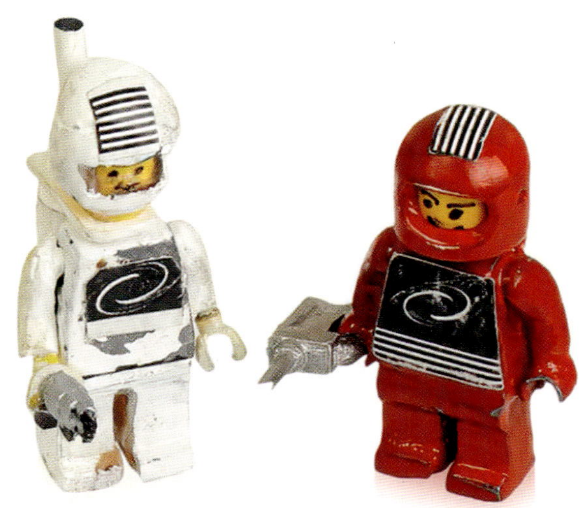

◀ Prototype minifigures developed for the LEGO Space line.

CHAPTER 1

1978–1979

▲ **885**
SPACE SCOOTER

886
SPACE BUGGY

889
RADAR TRUCK

897
MOBILE ROCKET LAUNCHER

891
TWO SEATER SPACE SCOOTER

928
SPACE CRUISER AND MOONBASE

924
SPACE CRUISER

926
COMMAND CENTRE

▶ **918**
ONE MAN SPACE SHIP

894
MOBILE GROUND TRACKING STATION

920
ROCKET LAUNCH PAD

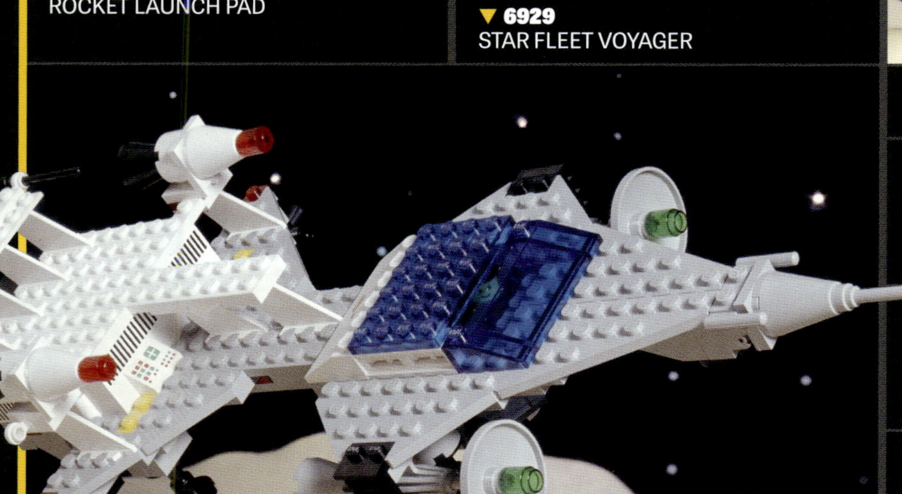

1980

6970
BETA I COMMAND BASE

6901
MOBILE LAB

6841
MINERAL DETECTOR

6821
SHOVEL BUGGY

▼ **6861**
X1 PATROL CRAFT

1981

6870
SPACE PROBE LAUNCHER

6927
ALL-TERRAIN VEHICLE

▶ **6822**
SPACE DIGGER

6801
MOON BUGGY

6842
SMALL SPACE SHUTTLE CRAFT

▼ **6929**
STAR FLEET VOYAGER

1982

6950
MOBILE ROCKET TRANSPORT

▶ **6890**
COSMIC CRUISER

6929
STAR FLEET VOYAGER

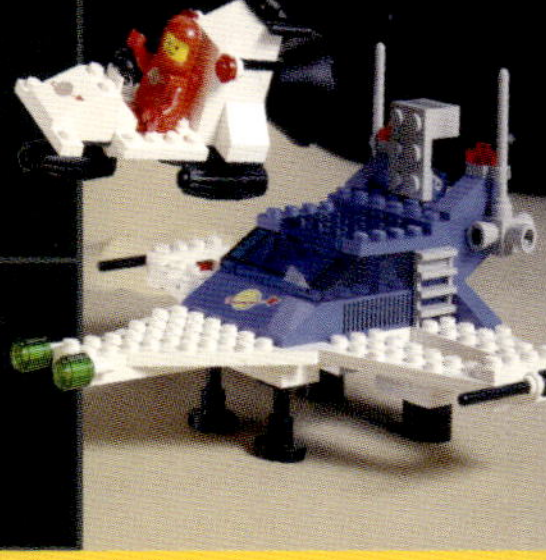

1983

6844
SISMOBILE

6930
SPACE SUPPLY STATION

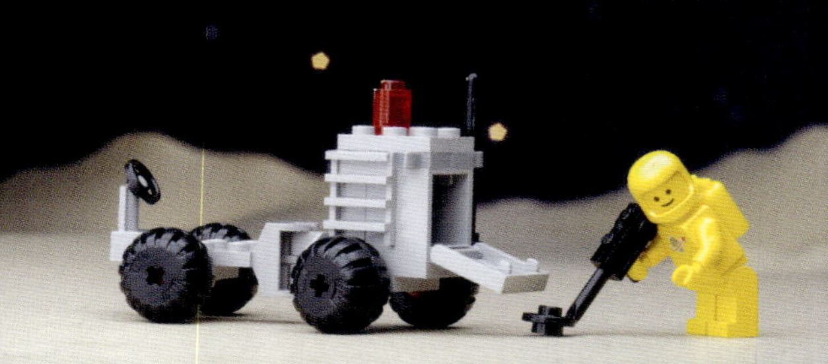

▲ **6823**
SURFACE TRANSPORT

6980
GALAXY COMMANDER

1593
SUPER MODEL BUILDING INSTRUCTIONS

1984

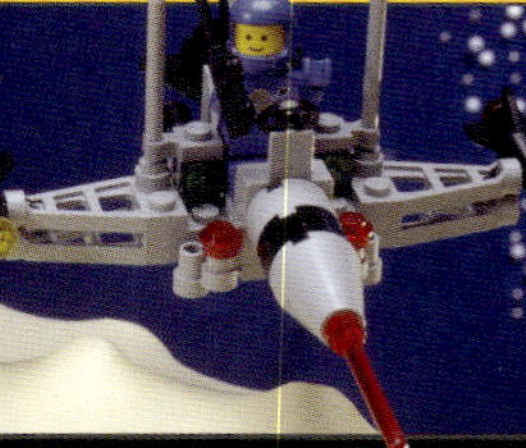

▲ **6824**
SPACE DART I

6881
LUNAR ROCKET LAUNCHER

6951
ROBOT COMMAND CENTER

6871
STAR PATROL LAUNCHER

6971
INTER-GALACTIC COMMAND BASE

▲ **6928**
URANIUM SEARCH VEHICLE

6846
TRI-STAR VOYAGER

6804
SURFACE ROVER

TIMELINE 1978–1987

1985

▶ **6806** SURFACE HOPPER

6847 SPACE DOZER

6807 SPACE SLEDGE WITH ASTRONAUT AND ROBOT

6848 INTER-PLANETARY SHUTTLE

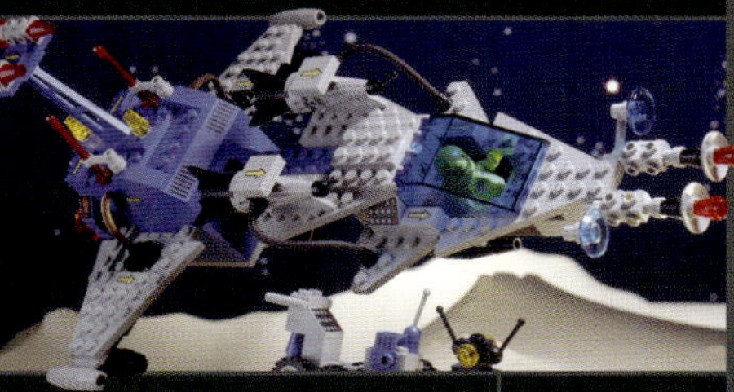

▲ **6931** FX STAR PATROLLER

6872 XENON X-CRAFT

6882 WALKING ASTRO GRAPPLER

6825 COSMIC COMET

6826 CRATER CRAWLER

6891 GAMMA V LASER CRAFT

6805 ASTRO DASHER

▼ **6952** SOLAR POWER TRANSPORTER

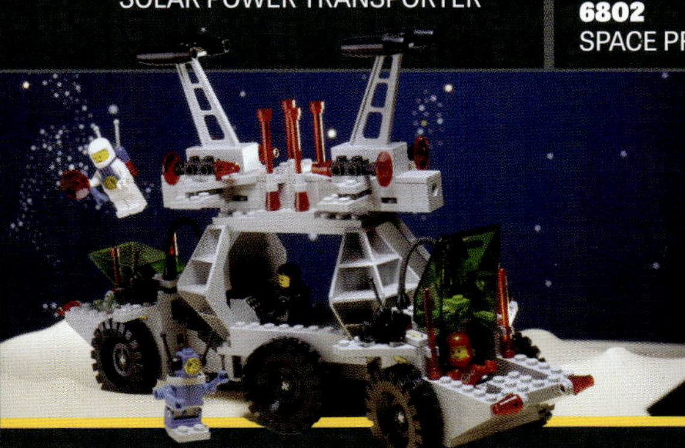

1986

▼ **6750** SONIC ROBOT

1558 MOBILE COMMAND TRAILER

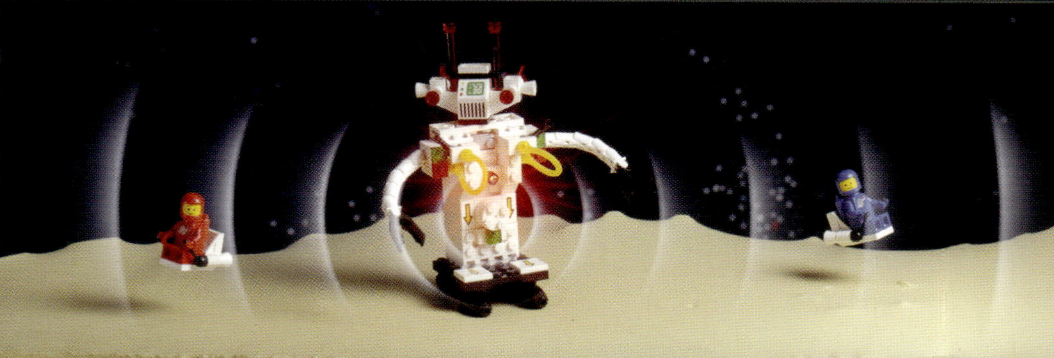

1557 SCOOTER

1580 LUNAR SCOUT

6780 XT STARSHIP

6926 MOBILE RECOVERY VEHICLE

6783 SONAR TRANSMITTING CRUISER

▼ **6892** MODULAR SPACE TRANSPORT

6940 ALIEN MOON STALKER

6985 COSMIC FLEET VOYAGER

6802 SPACE PROBE

6820 STARFIRE I

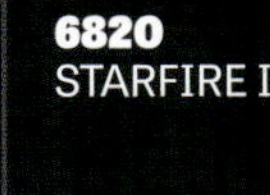

6845 COSMIC CHARGER

6874 MOON ROVER

1987

1498 SPY-BOT

1499 TWIN STARFIRE

6808 GALAXY TREKKOR

▶ **6809** XT-5 AND DROID

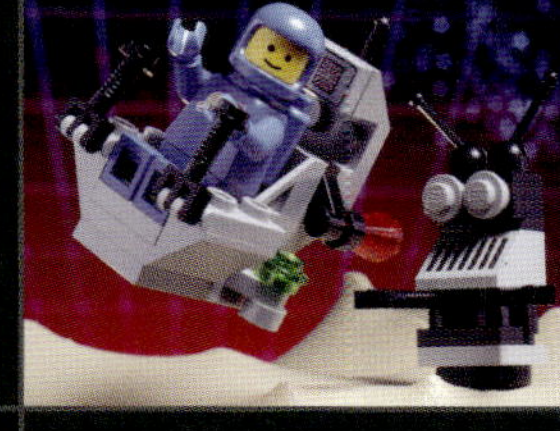

6972 POLARIS I SPACE LAB

6827 STRATA SCOOTER

6883 TERRESTRIAL ROVER

▼ **6849** SATELLITE PATROLLER

CHAPTER 1

CLASSIC LEGO® SPACE

1978–1987

The release of the complete LEGO® Space range in 1979 consisted of eleven models and two packs of baseplates. It was an overwhelming success with both retailers and the buying public. In Europe, LEGO Space accounted for approximately 35 percent of the total turnover of the LEGOLAND® range but did not appear to have had a significant adverse effect on the sales of LEGO® Town. In fact, turnover for the three LEGOLAND lines overall increased by more than 50 percent in 1979. LEGO Space also won coveted industry prizes, including being voted Toy of the Year 1979 by industry experts at the Nuremberg Toy Fair and the National Association of Toy Retailers in the United Kingdom.

Jens Nygaard Knudsen received a promotion, and during this successful period the LEGO Group increased production and staff levels. From 3,150 employees in 1979, the company grew to 3,650 by 1981.

One of the new design recruits to Jens's team was Niels Milan Pedersen, who still works with the LEGO Group to this day. Although Niels is best known for his contributions to the LEGO® Castle and LEGO® Pirates play themes, he has played a crucial role in LEGO Space over the decades.

▼ A range of sets released in 1984 and 1985.

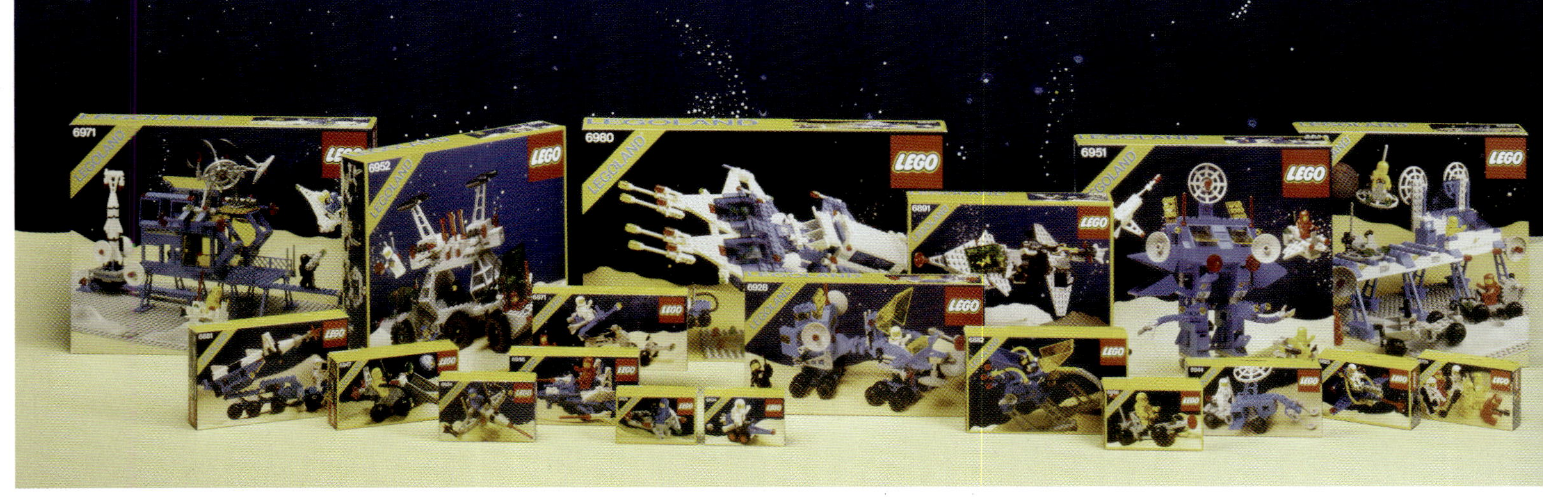

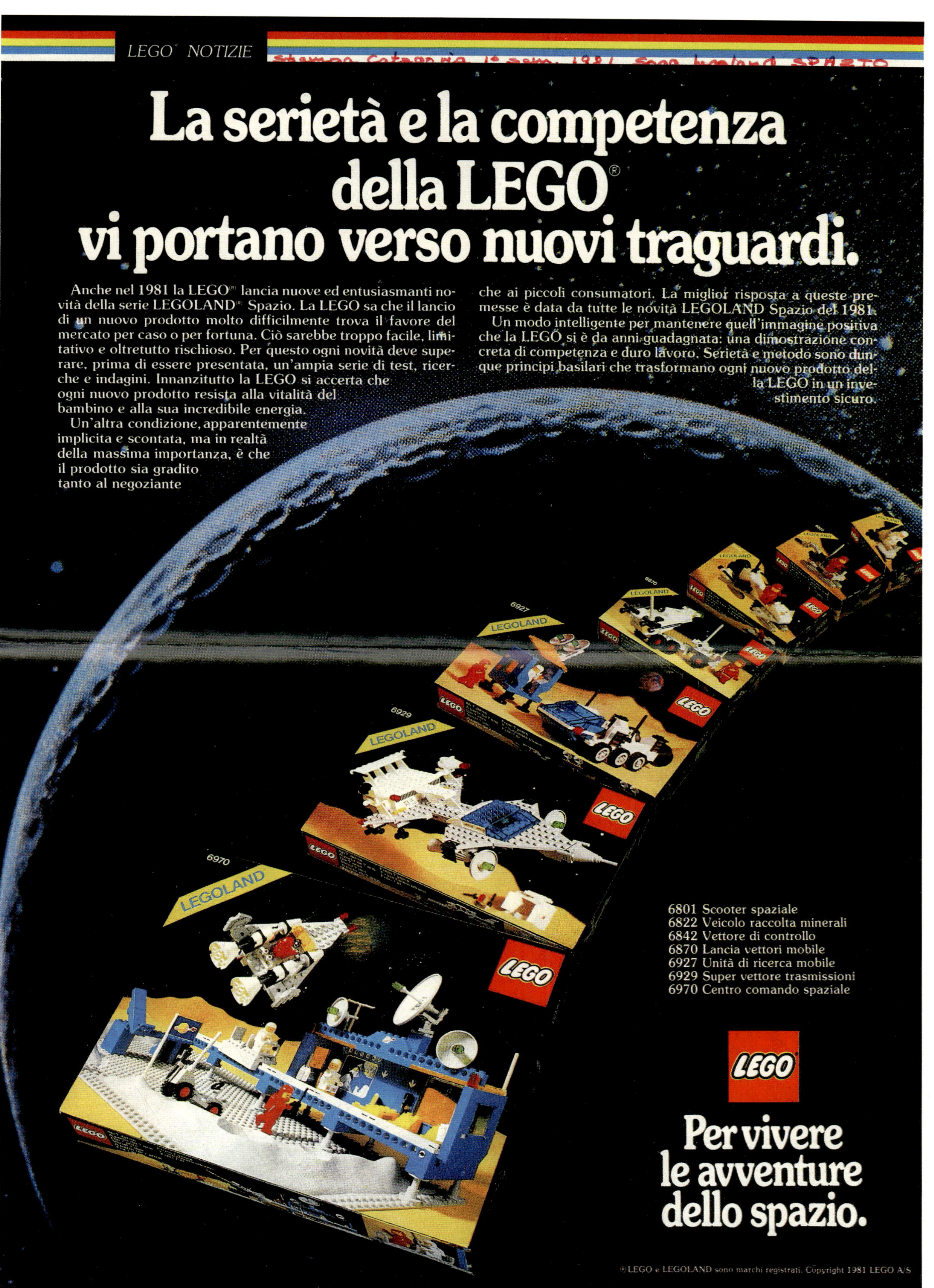

An Italian ad aimed at retailers highlights the virtues of the 1981 LEGO Space range.

NEW WORLDS FOR LEGO® SPACE

Although subthemes were not introduced into LEGO® Space until 1987, the line did undergo various changes in the early 1980s, to keep the products novel and appealing to consumers and retailers year upon year. One notable change was the introduction of new colors for the minifigures' spacesuits: yellow in 1982 and both blue and black in 1984.

New colors also appeared on the models. After the initial wave of sets, with their gray wings, blue fuselage, and transparent yellow cockpits, the palette was slowly expanded with white and transparent blue elements. The large spaceship from 1983, set 6980 Galaxy Commander (see page 56), is an iconic example of this color scheme.

Decisions about color schemes were made by the designers in conjunction with the marketing department. "Jens liked the idea of spaceships being white," designer Mark John Stafford recalls from his conversations with Knudsen, "because real spaceships are white—like the Apollo missions and the space shuttle." However, Knudsen had initially used gray instead, as many elements he needed were already available in that color. "If something is already in production, it means you've already got a box full of those in the factory," explains Stafford, "so you can build from that—you only need to make a few color changes."

Gray elements did not disappear by any means. While there was no clear delineation between color schemes year upon year, in 1985 and 1986 many of the sets sported a distinctive gray with transparent green color scheme. "From what I picked up, Jens didn't mind too much what the colors were," says Stafford. "He wanted to make models that children wanted to play with, in whatever colors there were."

"IF WE HAD A LOT MORE NEW COLORS, THEN THE WINGS IN THOSE EARLY SHIPS WOULD PROBABLY HAVE BEEN WHITE, NOT GRAY."

—JENS NYGAARD KNUDSEN, *BRICKJOURNAL* 2, NO. 6 (SUMMER 2009)

EMERGENT TECHNOLOGIES

Each successive wave of LEGO® Space sets consisted of new spaceships and ground-based vehicles, and some also included command bases. The vehicles ranged from utilitarian rovers and mining vehicles to mobile laboratories and launchers. The emphasis was always on research and exploration, and this was reflected in the advertising.

▼ The limited-release set 1968 included two of the novelties introduced in 1985: a color scheme of gray with transparent green and a helpful droid. This original, uncropped image reveals the techniques that the photographers used to create the alien dunes.

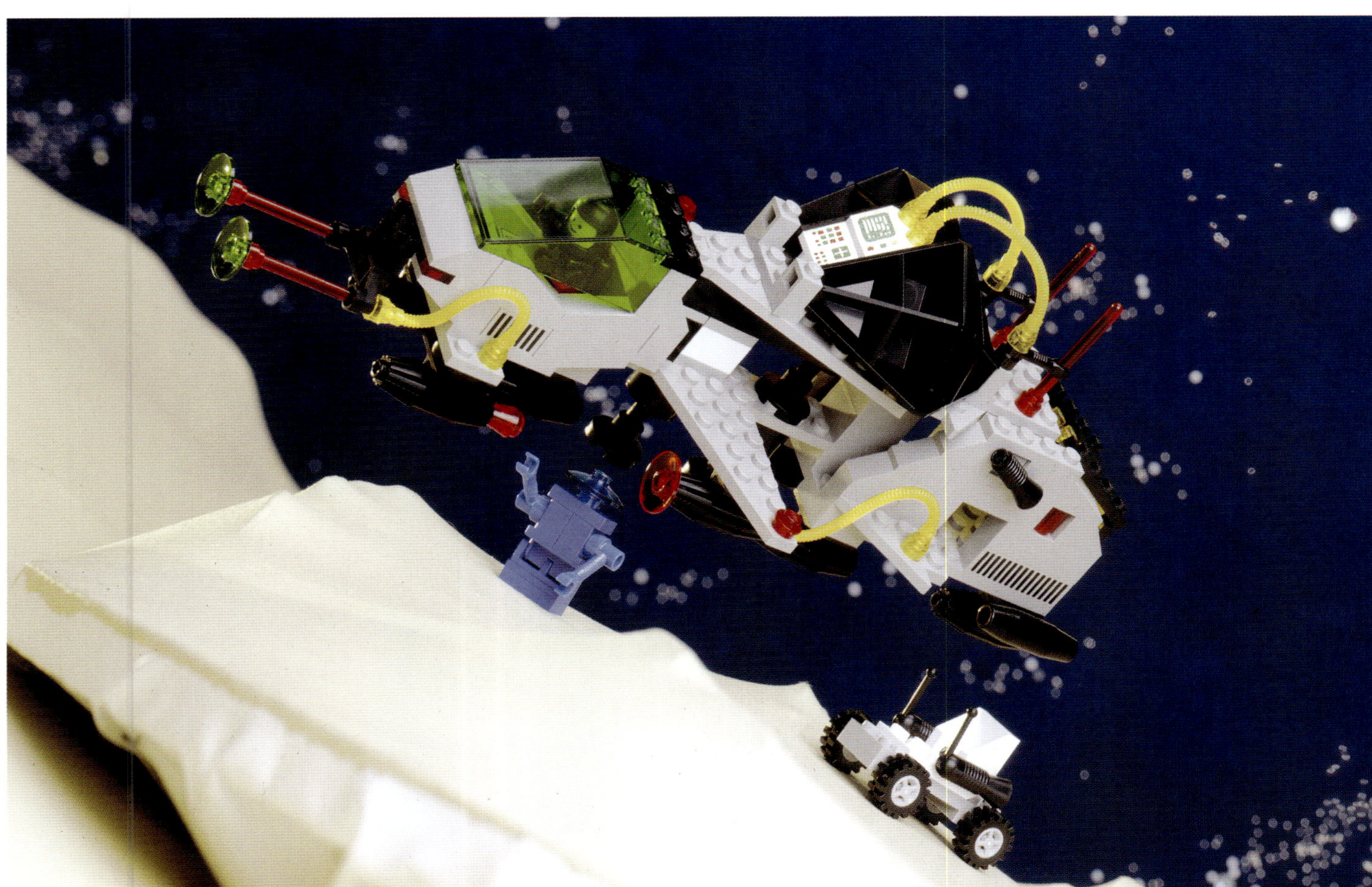

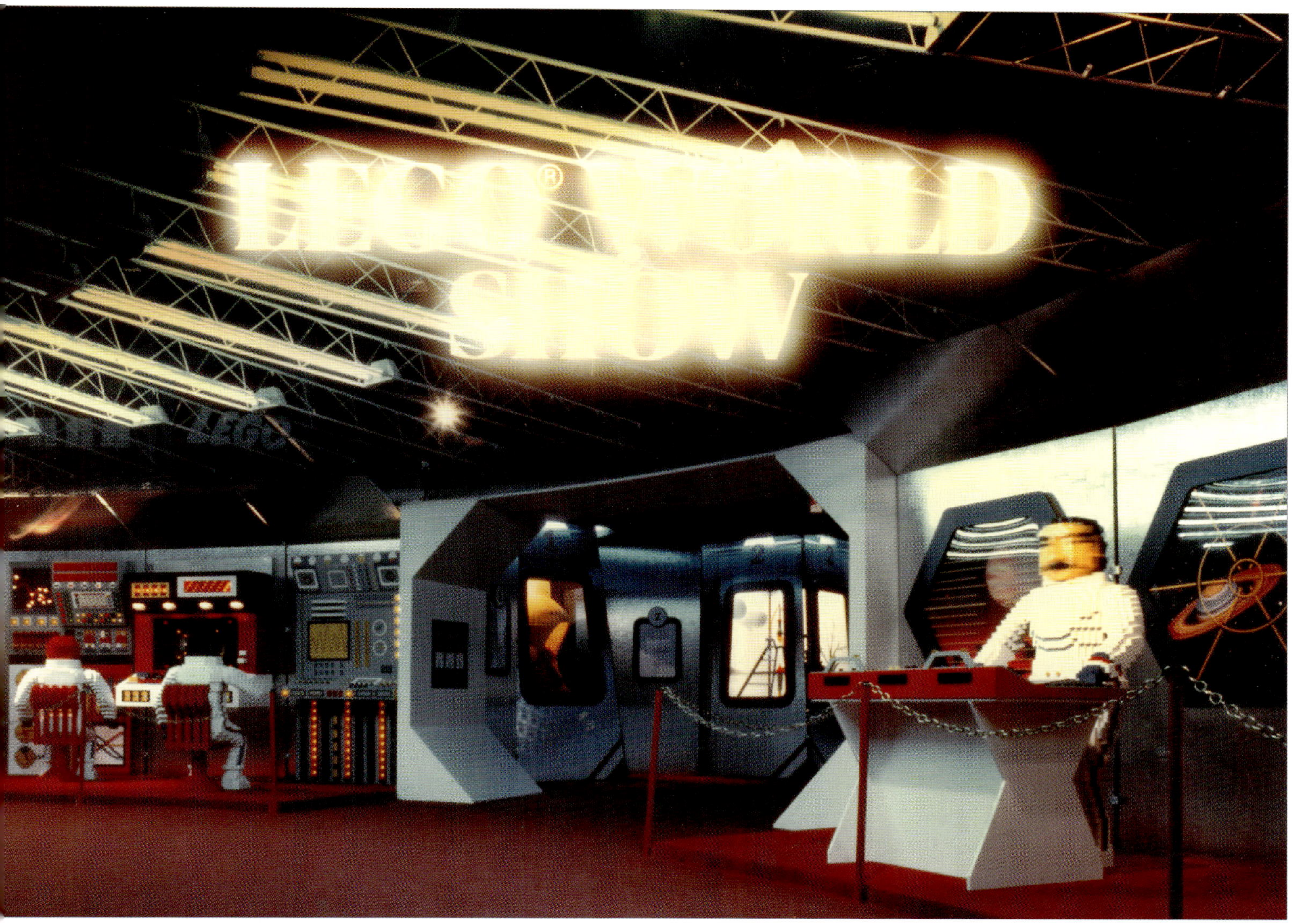

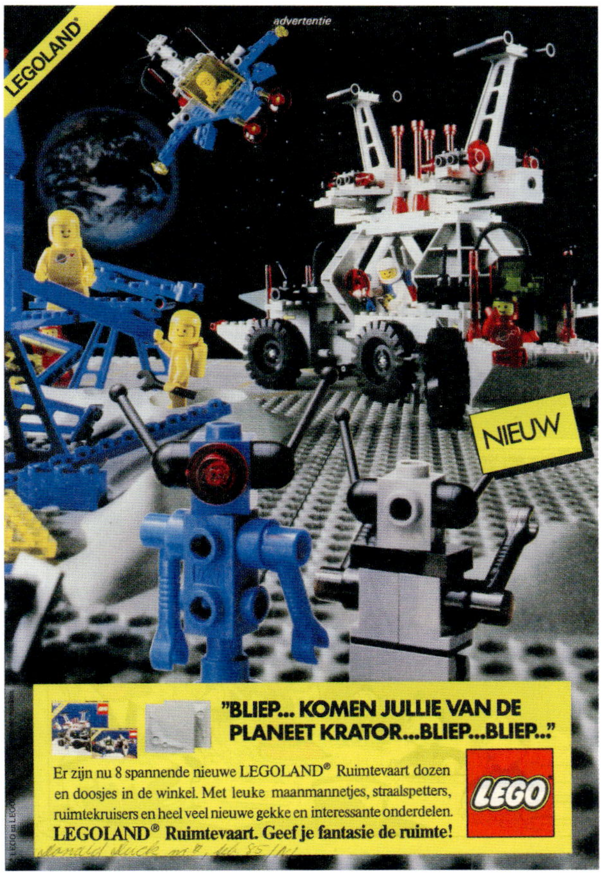

Over time, new concepts began to appear within the models. Science fiction had reached new levels of popularity thanks to movies such as the *Star Wars* franchise and television shows such as *Battlestar Galactica*. LEGO Space sets began to include more fantastical designs and fanciful concepts, such as giant walking robots.

Minifigure-scale, brick-built robots appeared in sets in 1985. They greatly increased the storytelling potential of products and were made from a range of interesting new elements that were also useful for other LEGO building purposes.

In 1986, the Light & Sound System brought an added dimension to LEGO play. Now, spaceships and giant robots could flash colorful lights and emit oscillating sound waves not dissimilar from a car alarm! William Thorogood, who was design manager on the Space Police 3 subtheme in 2009 and is now vice president of design, new business, at the LEGO Group, recalls receiving 6770 Lunar Transporter Patroller as his first-ever LEGO set at the age of six. "The fact that something I had just built had light and sound in it was fascinating," he recalls, "and the way that the bricks conducted electricity from the battery pack to the lights was absolutely incredible." The range of light bricks and sound bricks were connected to the 9 V power box not by wires but

▲ The LEGO® World show was a popular touring exhibition, and for the sixth event, which was held in 1985, model makers spent nearly five thousand hours creating the *Travel in Space* interactive exhibition.

◀ A Dutch ad from 1985 highlights the arrival of "fun moon men" and "lots of crazy and interesting new elements."

▶ The supplementary minifigure pack 6702 from 1986 included two of the new robots and exciting minifigure accessories, such as a jetpack.

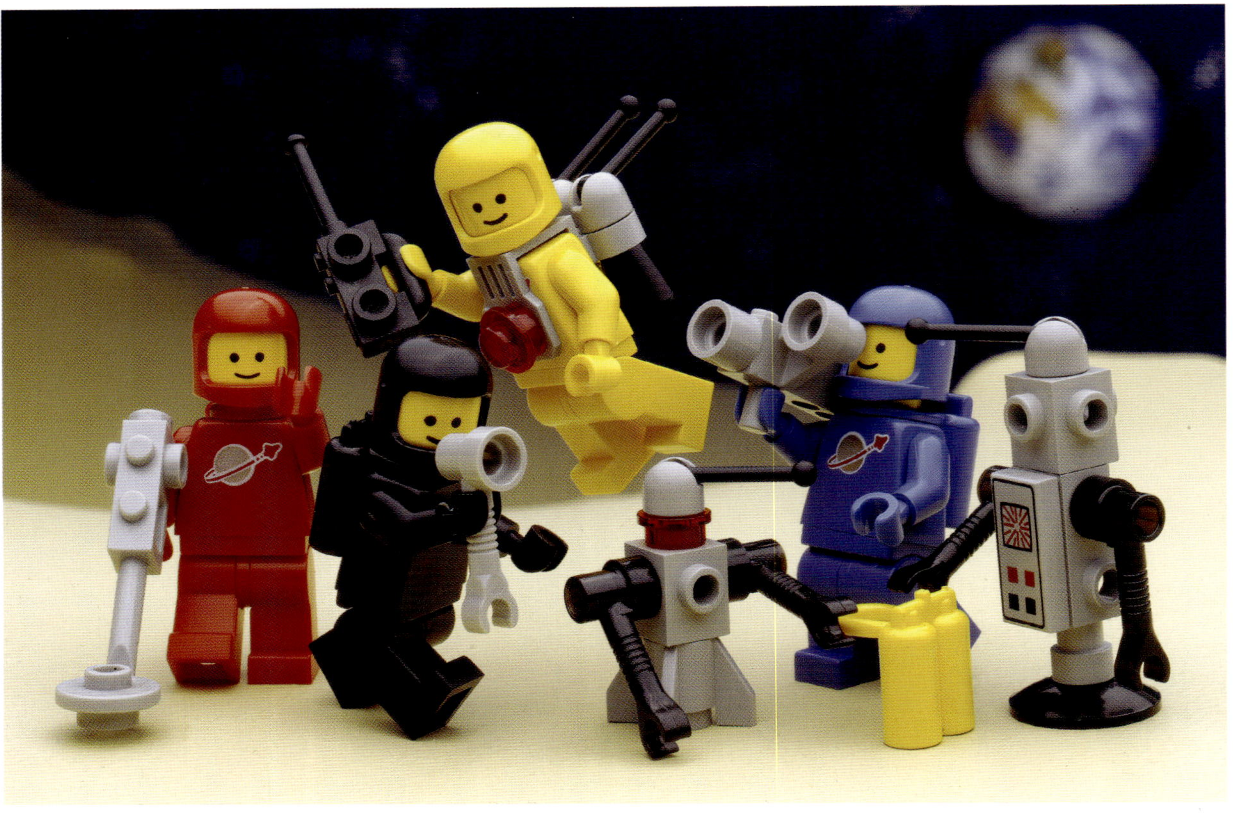

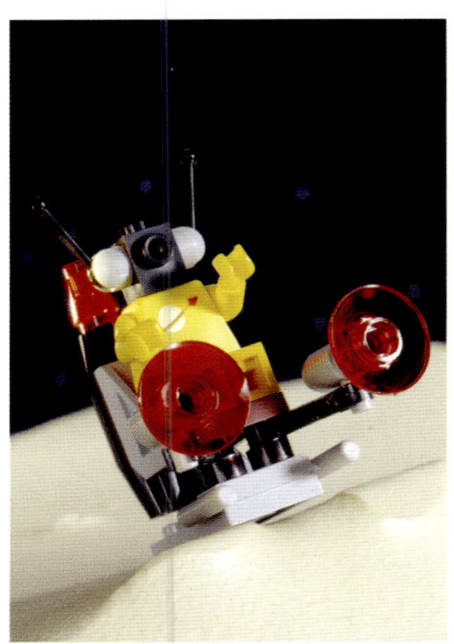

▲ An alternative model from set 6807 Space Sledge with Astronaut and Robot from 1985 proposes constructing an alien head by using the new modified brick 1 x 1 with studs on 4 sides.

by special LEGO plates with strips of metal threaded inside them.

In 1986, three LEGO Space sets included the Light & Sound System, but it was used less frequently after that. Ultimately, the system appeared in thirteen LEGO Space sets, the last four within the Insectoids subtheme released in 1998.

"I remember some new iterations of the Light & Sound System that we worked on a lot," recalls Bjarne Panduro Tveskov, a designer on the LEGO Space team in the 1980s. "It was a smart system where you could have a red switch and put a red motor anywhere on that system, and then those would correspond. It became expensive; even to this day, that's always the challenge for tech stuff."

PARALLEL UNIVERSES

The parameters of LEGO® Space extended beyond the range of construction sets available. For example, there were merchandise, competitions run by local LEGO offices, and many large-scale space models displayed at LEGOLAND® parks and on touring shows.

Building further upon the play theme's success, in 1983 the newly formed LEGO Publishing division investigated the idea of releasing a series of LEGO Space storybooks.

Nowadays, some LEGO themes are based upon existing franchises, and others, like LEGO® NINJAGO®, are developed alongside accompanying narratives of their own. LEGO Space products had no such overarching story; however, the LEGO Group was experimenting with storytelling in this era. For example, the LEGO® FABULAND® theme from 1979 was supplemented by a series of books, a vinyl record, and even a TV series featuring the characters. The 6000 Idea Book, released in 1980, interweaved inspirational models and building instructions with the adventure of two minifigures, who at one point embark on a fantastical journey into outer space.

"THE SOUND BRICK WOULD DRIVE MY PARENTS CRAZY!"

—FRÉDÉRIC ROLAND ANDRE, SENIOR DESIGN MANAGER, THE LEGO GROUP, 2010 TO PRESENT

The first potential storybook, *Trapped in Space*, was written by children's author Douglas Hill and lavishly illustrated with photographs of LEGO models built by the LEGO Space designers, including Niels Milan Pedersen.

"I made all the big interior models," Niels recalls. "There was this central computer which was called MAGS, which had panels with lights. I had to drill holes in all the printed elements we had, wherever the knobs and instruments were, so we could put lights in from the back of the model to create lighting effects. It was hard work, but it was quite fun."

By late 1983, mockups of the completed book were ready, and focus tests were undertaken in Denmark with parents of seven- to nine-year-old boys. The children had found the pictures inspiring, but some parents were confused about the book's intended purpose—was it a picture book, an inspirational building guide, or an advertisement for products? They considered its language too

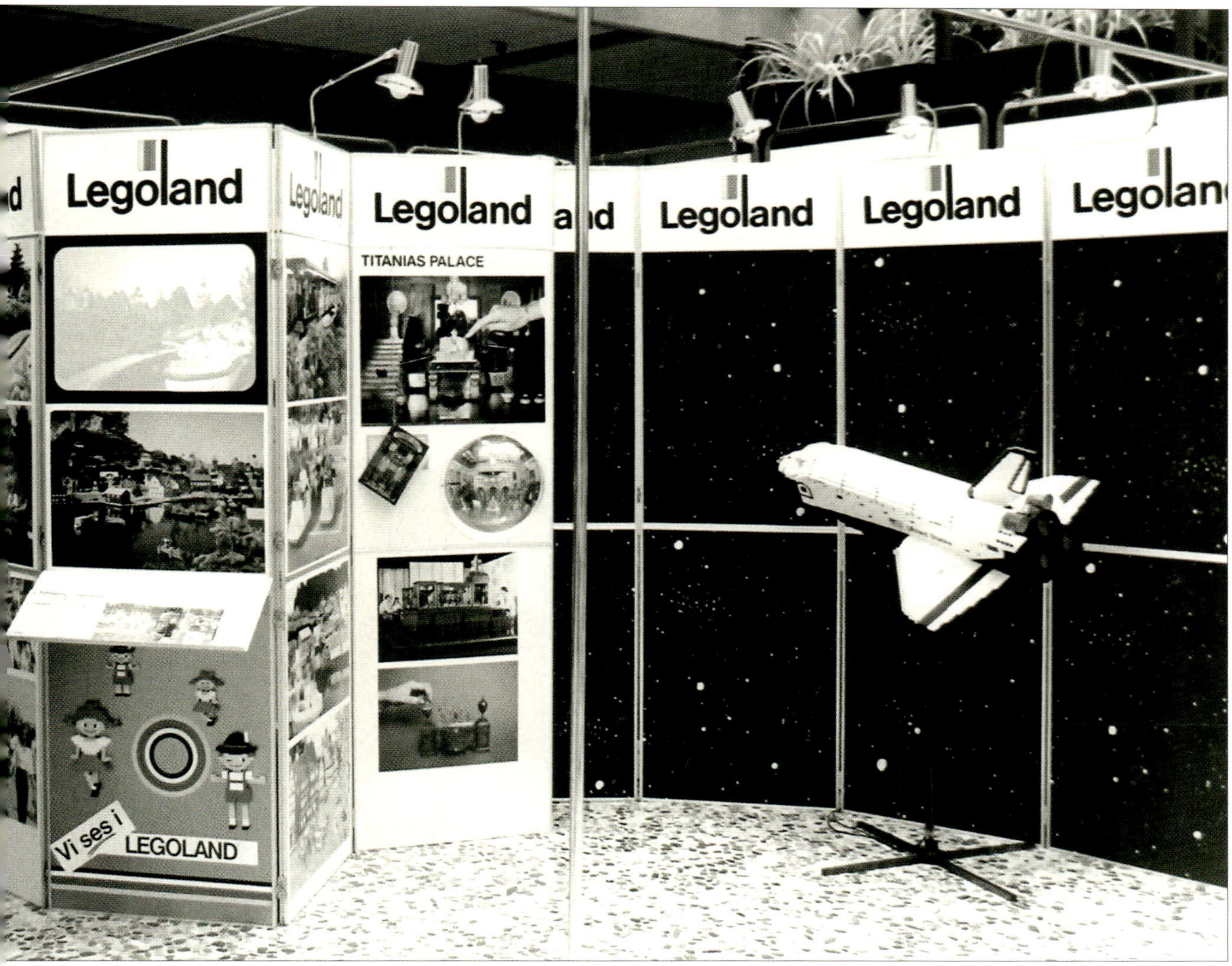

A space shuttle *Columbia* model at a show in Copenhagen in 1983.

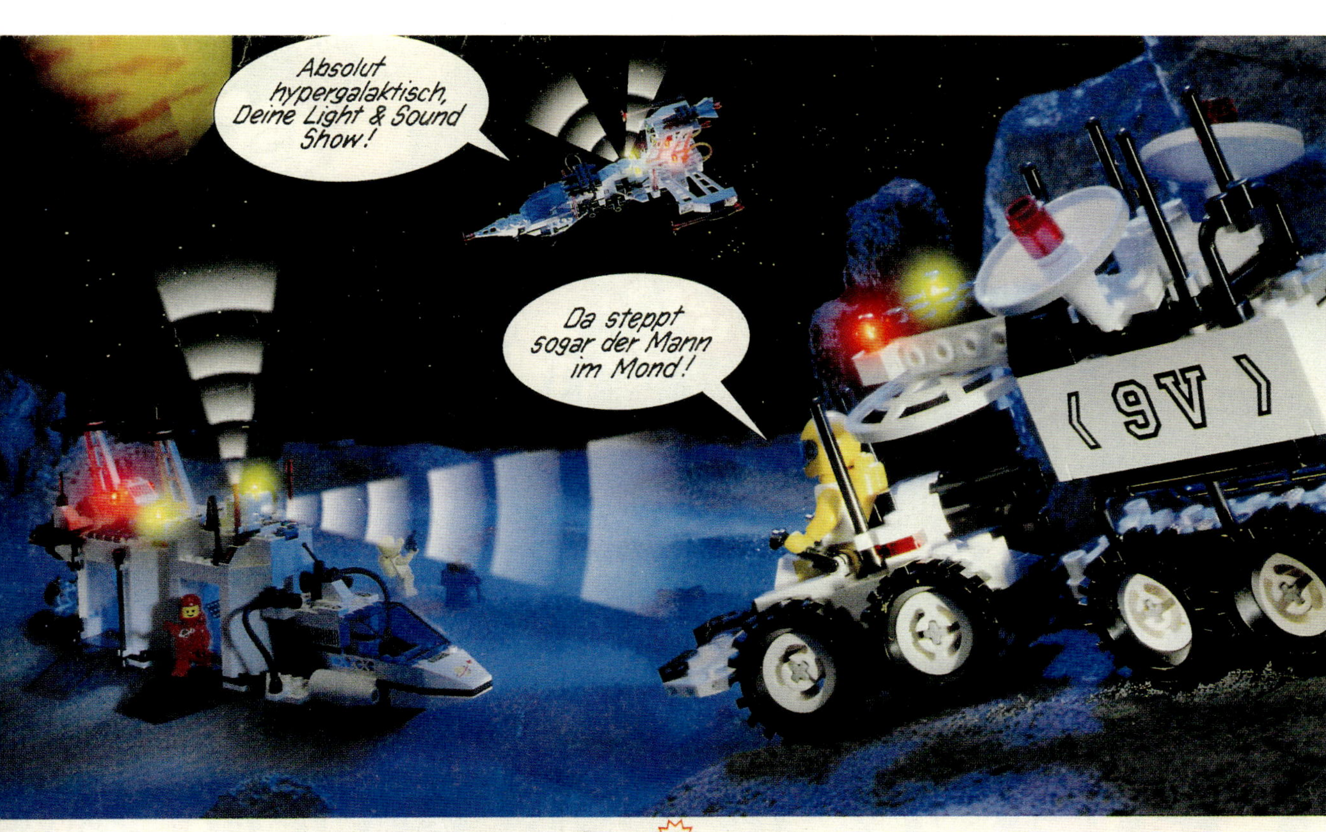

The advertisements for the Light & Sound System, such as this German example from 1988, explained how the elements worked and highlighted the added fun they could bring to storytelling.

▶ A page from the mockup of *Trapped in Space* showing the *Galactic Spearhead* and supercomputer MAGS.

▼ The cover of the unpublished LEGO Space book *Trapped in Space*.

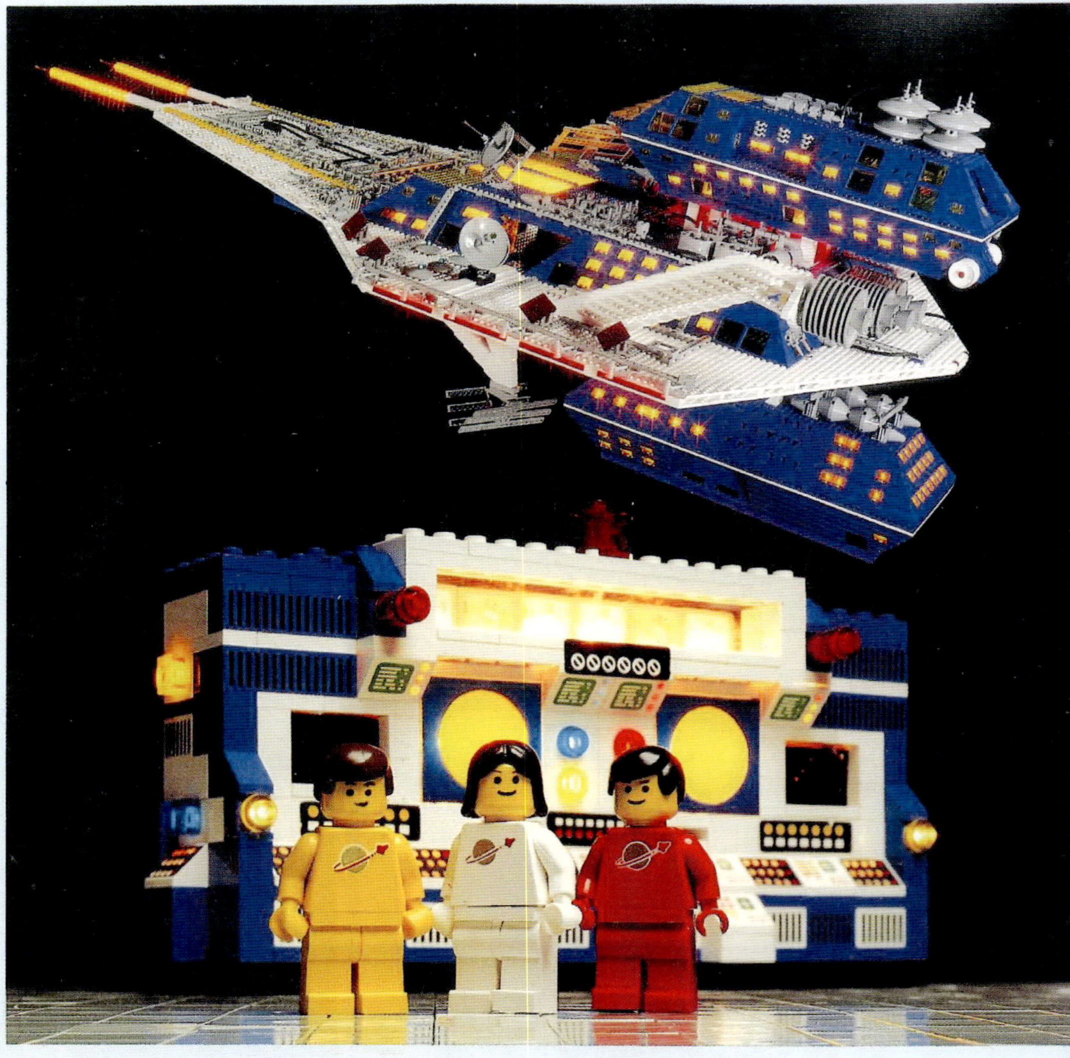

Galactic Spearhead er navnet på en stor anlagt ekspedition, en opdagelsesrejse ud til fjerne og ukendte egne af galakserne.

Spearhead, (Spydspids) er også navnet på det kæmpestore stjerneskib, der bliver brugt til opdagelsesrejsen. Besætningen er på tre hundrede mennesker. De er inddelt i tre hold:

Rødt Hold består af de piloter og navigatører, som flyver Spearhead. De flyver også de mindre fartøjer, som stjerneskibet har med.

Gult Hold består af teknikere og ingeniører. De tager sig af maskiner og mekanik og teknik, og de opfinder også nye ting, når der er brug for det.

Hvidt Hold består af videnskabsfolk, forskere, der undersøger alt det nye, ekspeditionen opdager.

Alle tre hold får hjælp af et vældig stort og temmelig indbildsk elektronisk anlæg, en computer, der bliver kaldt MAGS (Maximised Accel-drive Governing System).

Galactic Spearhead er en opdagelsesrejse med mange formål. Den leder efter beboelige planeter, hvor mennesker kan slå sig ned. Den leder efter planeter med naturrigdomme, der kan blive til gavn for Jorden. Den leder efter andre arter, der ligesom mennesket har en højt udviklet hjerne.

Og hele tiden har den travlt med at søge ny viden om alle de mærkværdige ting, der findes ude i rummet mellem stjernerne.

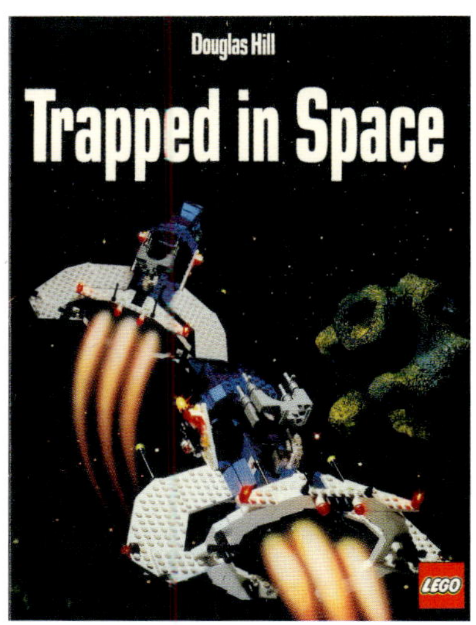

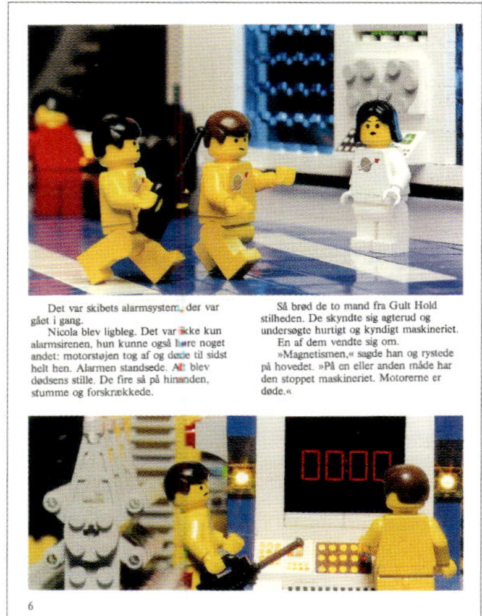

Det var skibets alarmsystem, der var gået i gang.

Nicola blev ligbleg. Det var ikke kun alarmsirenen, hun kunne også høre noget andet: motorstøjen tog af og døde til sidst helt hen. Alarmen standsede. Alt blev dødsens stille. De fire så på hinanden, stumme og forskrækkede.

Så brød de to mand fra Gult Hold stilheden. De skyndte sig agterud og undersøgte hurtigt og kyndigt maskineriet.

En af dem vendte sig om.

»Magnetismen,« sagde han og rystede på hovedet. »På en eller anden måde har den stoppet maskineriet. Motorerne er døde.«

6

▲ The standard smiling minifigure face was replaced by various hand-drawn expressions for *Trapped in Space*.

complicated for the target age range to read on their own, and conversely, it contained too many pictures to be considered the kind of book that the parents would read to children of that age.

Because of this response, *Trapped in Space* was not released, and a more suitable approach was undertaken instead. The LEGO Group already used the comic strip format for advertising products in children's magazines (see pages 9, 71, 77, 85, and 95) and now decided to evolve the *Trapped in Space* concept into a forty-eight-page comic book.

JIM SPACEBORN

The illustrator Frank Madsen was commissioned to write and illustrate a series of comic books which followed the thrilling adventures of a young boy, Jim Spaceborn, who had been found on an abandoned spaceship. Many of the concepts and physical models created for *Trapped in Space* were retained, such as MAGS and their huge spaceship, the *Spearhead*. Notable additions were Jim's friendly giant robot Keko and the villain Kazak, whose henchmen wore black uniforms and flew in black-and-yellow spaceships—an interesting precedent to the

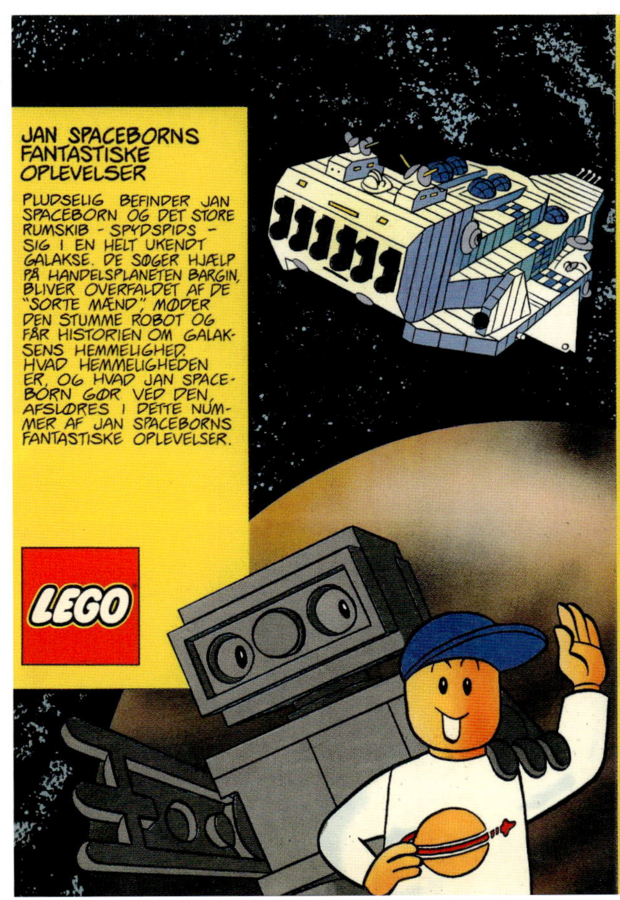

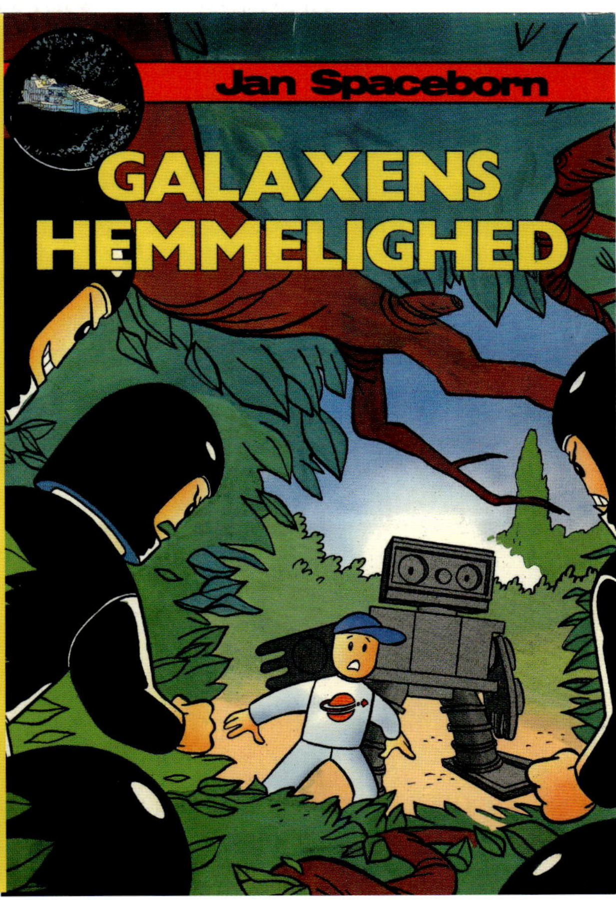

Draft cover art for the first book in the Jim Spaceborn series, *The Unknown Galaxy*. At this stage, the hero's name was Jan.

Sådan blev tegneserien til:

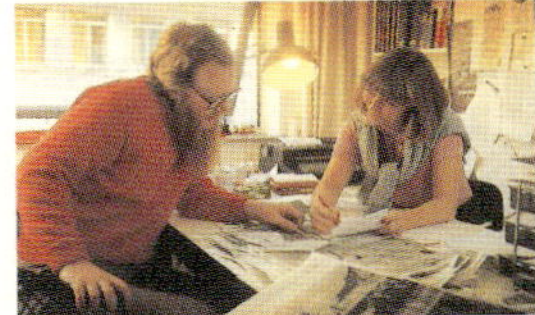

Først fandt forfatterne ud af, hvad historien skulle handle om og skrev det hele ned.

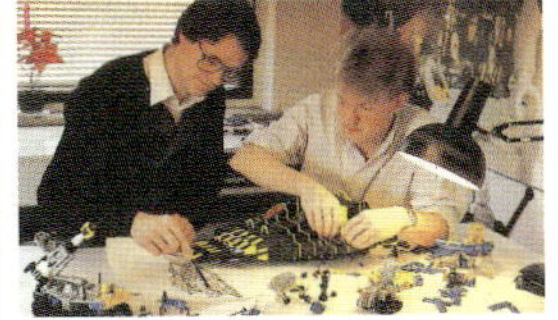

Så byggede en modeldesigner de mange modeller, der skulle bruges i historien.

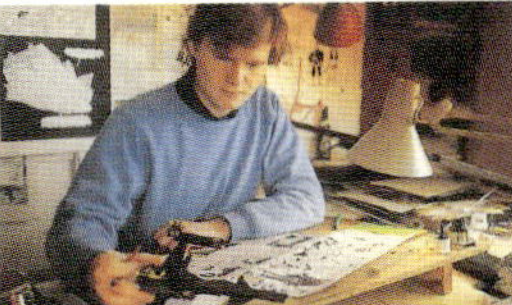

Dem brugte tegneren til at kigge efter, da han tegnede alle tegningerne - først med blyant og bagefter med en sort tuschpen.

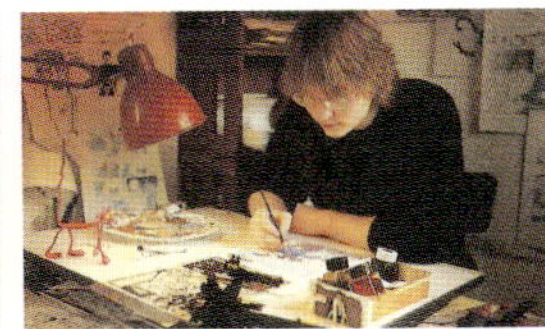

Derefter blev tegningerne farvelagt. Også her måtte farvelæggeren kigge på modellerne for at finde de rigtige farver.

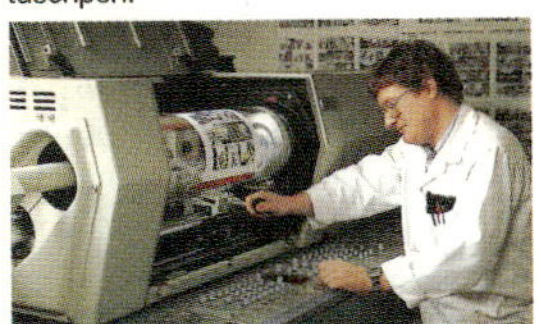

Til sidst blev tegneserien trykt, pakket og sendt ud til den butik, du har købt den i.

At lave en tegneserie tager lang tid. Men det kan også gøres nemmere.

PRØV AT LAVE DIN EGEN TEGNESERIE.

Inddel et stykke papir i felter, som vist her i hæftet og tegn dine egne figurer. Bagefter kan du lave talebobler og lade figurerne sige noget.

Hvis du ikke kan tegne figurerne, kan du prøve at klippe mennesker og figurer ud fra ugebladene og klæbe dem ind i felterne.

God fornøjelse!

In the mockup of the first book in the Jim Spaceborn series, *The Unknown Galaxy*, the final page revealed the creative process, but this was omitted from the final product. The artist, Frank Madsen, is shown in the third picture.

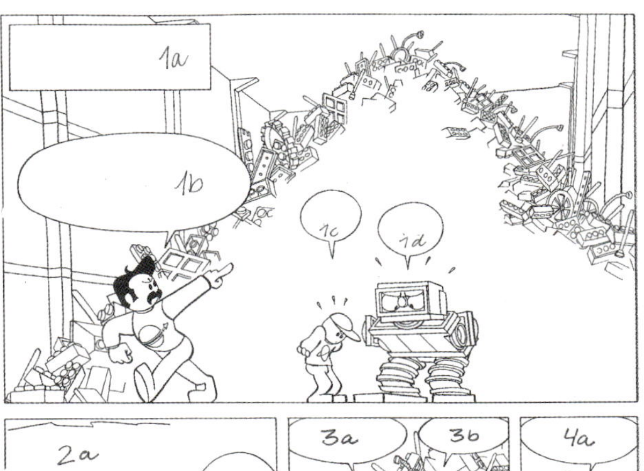

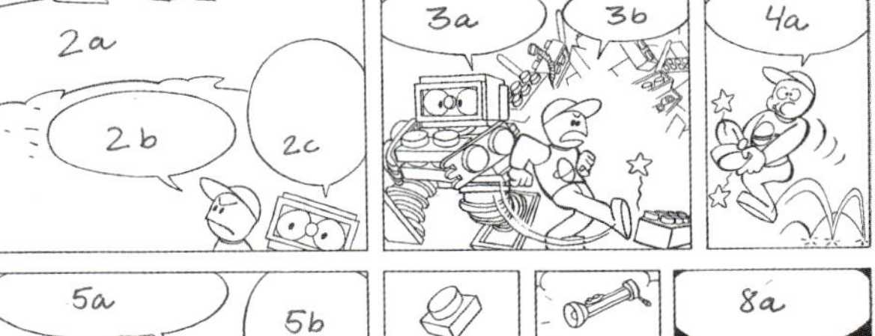

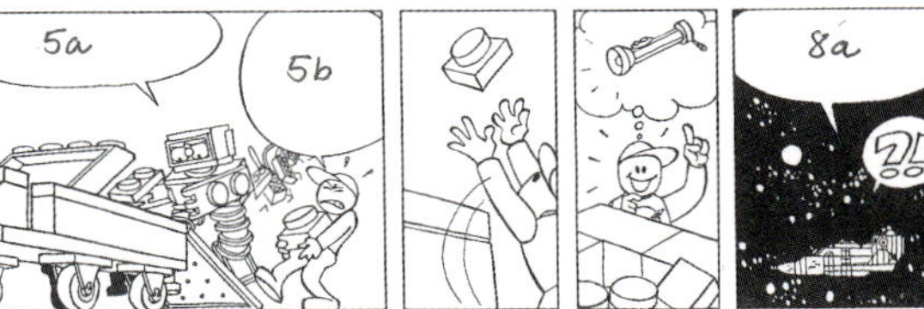

The opening page of *Panic on Board*, the unpublished third installment in the Jim Spaceborn series, at the inking stage of development. Even in space, LEGO bricks need sorting, and Captain Bart Seeker assigns the task to Jim Spaceborn and his robot, Keko.

Blacktron subtheme that was to arrive in a few years' time. However, these were not based upon Blacktron prototypes, and actual LEGO® Space products were never featured in the comic.

Two books, *The Unknown Galaxy* and *The Kidnappers from Swamp Planet*, were published in succession, and a third, *Panic on Board*, was developed, as well as three half-sized comic books, but the LEGO Publishing division closed down in 1987 before they could be completed.

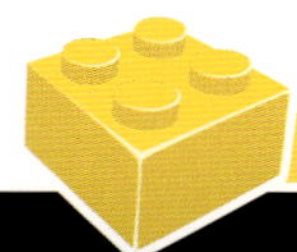

INTERVIEW WITH

NIELS MILAN PEDERSEN

DID YOU PLAY WITH LEGO® BRICKS AS A CHILD?

I didn't own any. My older brother had some of the trains, but he wouldn't let me play with them because he thought I would make them dirty. Which I might have!

I had to make most of my toys myself, and I also made them for the other children. We lived in a very small fishing town on the west coast of Denmark. Nobody had much money, so you learned to make things from scratch from whatever you could find. I would make a modeling pulp out of wood glue and toilet paper, from which I made figures, and I also used a lot of lead from the shipyards.

It was actually my mother, who had seen an advertisement in one of the papers, who thought I should apply to the LEGO Group. It was the spring of 1980 and I was twenty years old. I was trying to get into a course to become an archaeologist, but because we had poor schooling where I came from—we were just supposed to be fishermen—I couldn't get into the university, and so I needed a job. I didn't know how to apply for one, so I sent a letter with a lot of drawings included, and I thought I would never hear from them again!

One day, I came home and in front of my door there was a big box of LEGO® bricks. They asked me to build a space model, so I went down to the nearest toy shop to look at what LEGO® Space models looked like. Then I built a big spaceship that looked quite a lot like those that Jens Nygaard Knudsen had made. The letter also said that I should go to Billund to show what I had built, and that I should bring my hobby stuff.

I remember I had to borrow a jacket from my neighbor because I didn't have any nice-looking clothing! It was one of those 1970s brown velvet jackets and was a little too small for me. I got off the bus outside the LEGOLAND® park—I had assumed that must be where the LEGO office was located—but actually it is somewhat away from there, and it was raining cats and dogs. I was totally soaked, and this jacket was crimping up and starting to smell awful, and I thought, "This is a total catastrophe."

Anyway, I showed Jens Nygaard Knudsen and his colleagues my LEGO model and a lot of drawings of castles, knights, and trolls. I hadn't bothered to unpack my lead figurines of knights, but Jens was throwing himself across the table to unwrap those! I could see that he was really enthusiastic. What I didn't know, because they couldn't yet say it, was that although they had asked me to make a space model, they were actually looking for somebody to work on the LEGO® Castle theme. So I had brought all the right stuff with me without actually knowing it, and that was why I got the job.

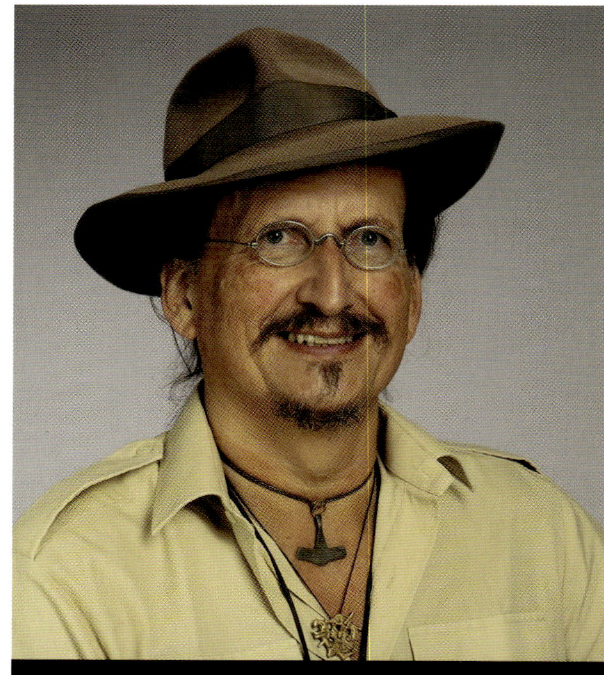

Name: Niels Milan Pedersen
Designation: Designer, the LEGO Group
Years of service: 1980–present
Subthemes: Classic Space, Mars Mission, Space Police 3

I began in the summer of 1980. We were only twelve designers when I started, and apart from Daniel Krentz, who was American, we were all Danes. At that point, it was just Jens, Daniel, and me working on LEGO Space, LEGO Castle, and a Wild West theme that was not released. We also made models for the Idea Books.

Our studio was a long, expansive room, and there were models from the floor to the roof. There were shelves on all the walls filled with models of all kinds, and we actually played with the models after we built them.

From almost the first day, I had to learn how to make technical drawings. I had to teach myself that, but that's how it worked in those days. I didn't have any design schooling as such when I came to the LEGO Group. Actually, nobody had any design training at that point; Jens had been working in a drugstore, and Daniel Krentz had been working in a post office in Chicago!

THE FIRST SPACE CREW

WHAT WAS JENS NYGAARD KNUDSEN LIKE TO WORK WITH?

Jens didn't do much sketching. If you had a brainstorm with Jens, it was a brain hurricane. If you came up with three ideas, he would build and expand on yours with ten other ideas. He was so imaginative. Of course, I came up with a lot of ideas too. And Jens would say, "Yes, that's a good idea," and then he would say, "By the way, look here," and pull out a drawer, and there would be a model of the same idea. He had already built it, because from the start, in the 1970s, he'd already had so many ideas.

Jens would tell stories about his space models. It was not just a design; there was already a story from his side, like a story a child might tell as they are playing. A spaceship flying into an asteroid belt would have some function that could dispel the asteroids, and then suddenly they would see some aliens! He would expand the storytelling around the model and get more and more enthusiastic about it, and you would get into it and become excited too. He was also really interested in what you were building, and he could also see the story in what you built—which would inspire you. They were fun times, that's for sure.

Jens was one of the most creative and wonderful people I ever met. He had a great life and had an impact on so many lives. He and the rest of us could always see things from the child's point of view, and so we more or less put the right things into the model. The marketing department also had to think about what was on the market at the time, selling strategies and so on. We took the approach that we were making what we would have loved to have when we were children. It has to be designed for the child. These are toys.

I think that's one of the reasons why all those who grew up with these models seem to have been very much influenced by them. What they make today is more intricate and so much more professional than our models were. But the relative simplicity of our models was possibly what made such a big impact on people, and why they are considered classics today.

WHAT WERE JENS'S CREATIVE INSPIRATIONS?

I know that one of the reasons Jens loved space so much was because as a child he had been into Buck Rogers and Flash Gordon comics, which would appear in the Sunday paper. In Danish, Flash Gordon is called Jens Lyn—that translates as "Lightning Jens"—so of course Jens had loved that space hero because they had the same name.

He was also into space movies, but there weren't many around then. Jens wasn't that keen on the first *Star Wars* movies. He thought that *Battlestar Galactica* would be much more successful. That was his

▼ Set 6711 from 1983 provided four minifigures, including a yellow astronaut that had been introduced a year earlier. Each carried a different tool; the space camera and the metal detector were elements designed by Niels Milan Pedersen.

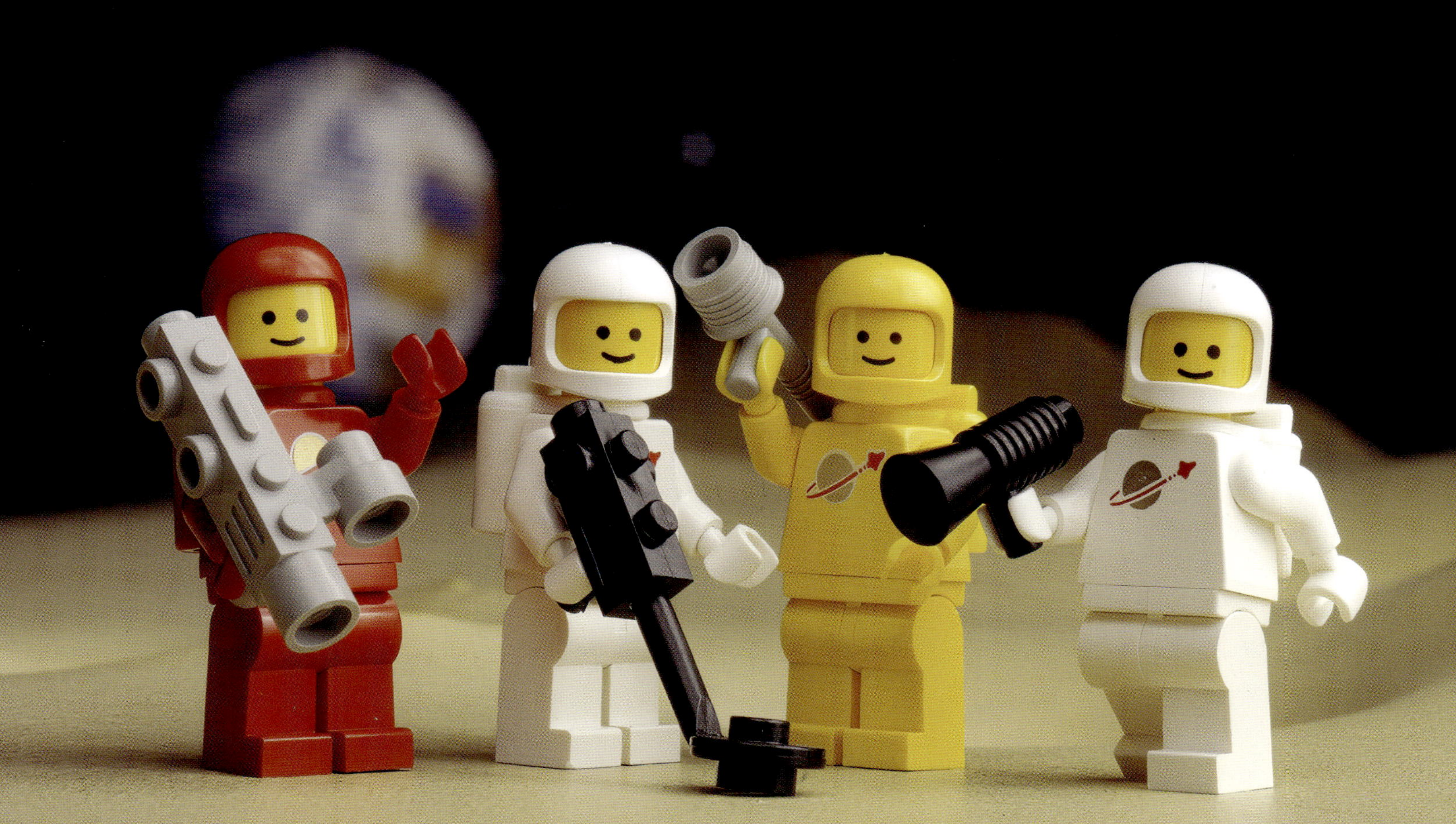

favorite, because it included more of the kind of space vehicles he liked. There was also more action based around the spaceships than there was in *Star Wars*.

DID YOU ENJOY SCIENCE FICTION AS A CHILD?

I didn't know much of it; one of the few things Danish television had ever shown which was a little bit science fiction was *Thunderbirds*, with those puppets. When I was young, that was the only thing I was really inspired by.

However, I was nine years old when the first moon landing occurred, and that made a great impact on me. Everything seemed to be happening back then in the late 1960s and early '70s when I was growing up. I remember we didn't have to go to school that day, because they expected us to watch the moon landing on television. We had a little black-and-white television with a very bad picture, but we sat up all night to see it. I remember standing outside the next evening, looking up at the moon. There were actually people up there. That was amazing. I don't think anybody nowadays would know what a sensational feeling that was, to actually see that happening.

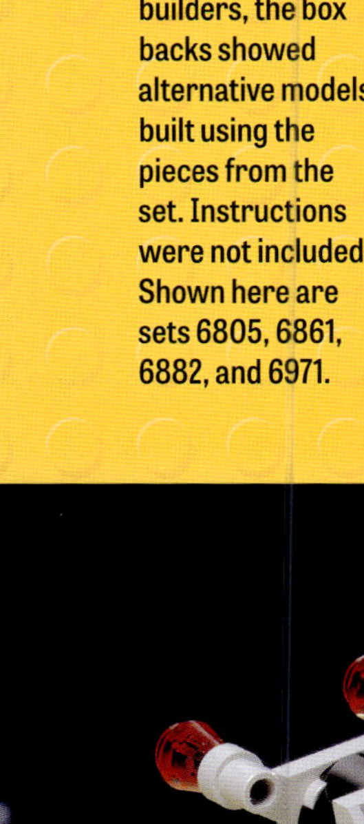

▼ To inspire builders, the box backs showed alternative models built using the pieces from the set. Instructions were not included. Shown here are sets 6805, 6861, 6882, and 6971.

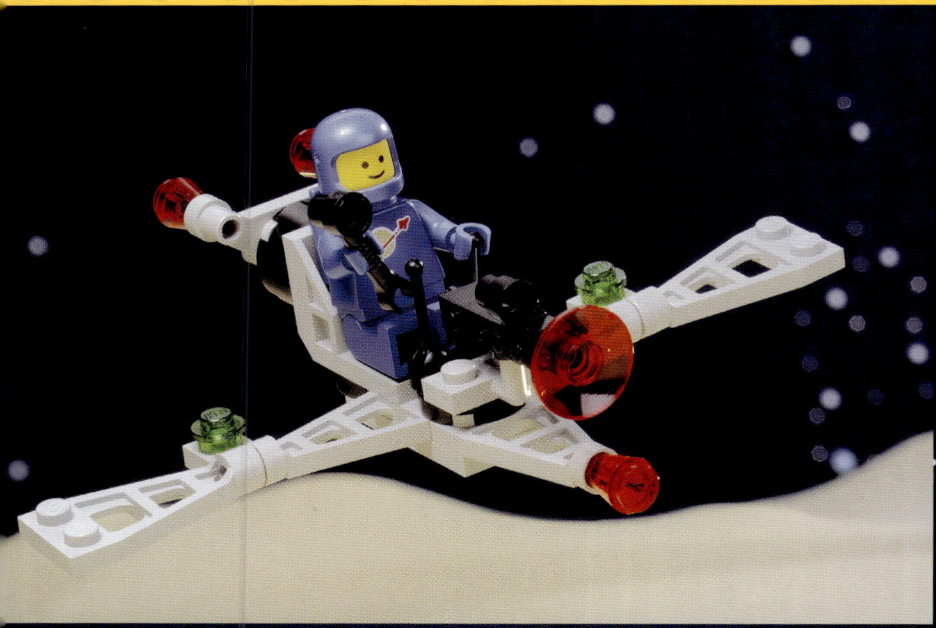

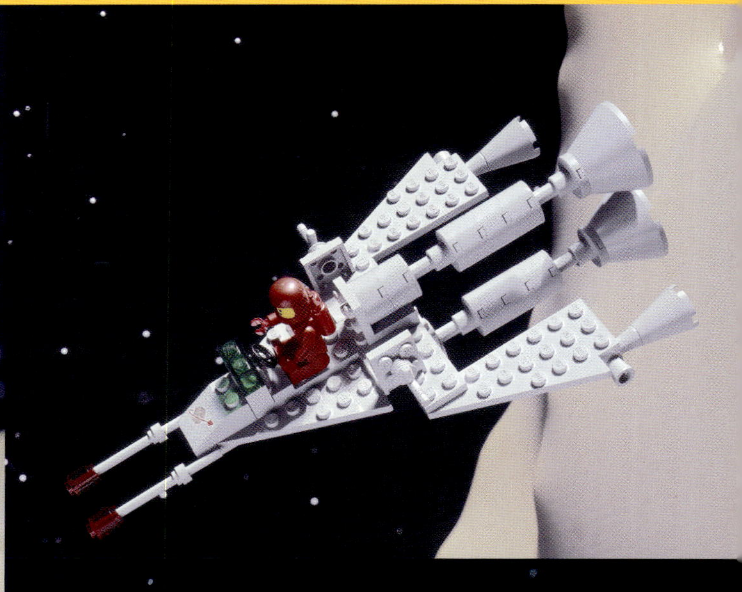

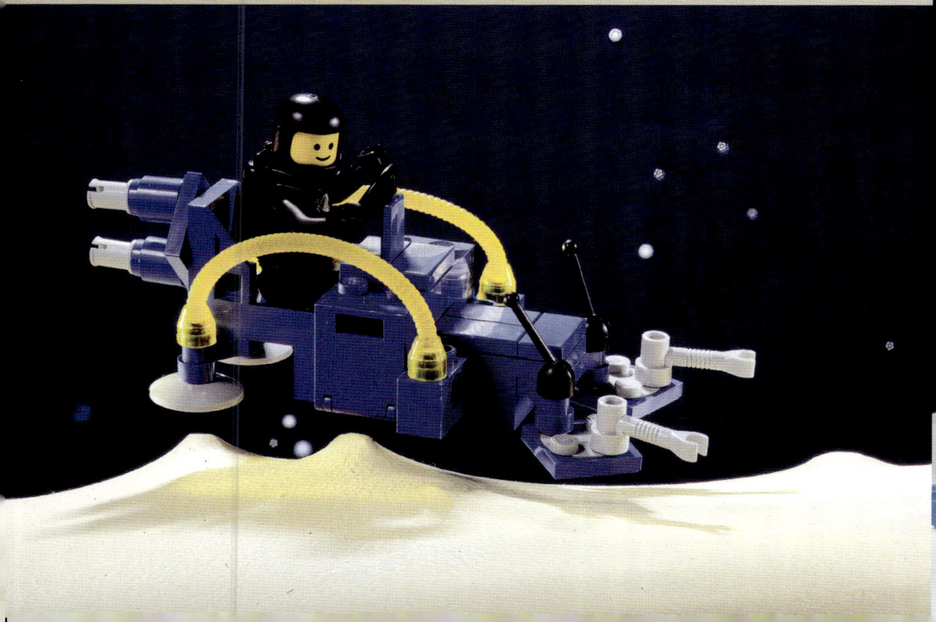

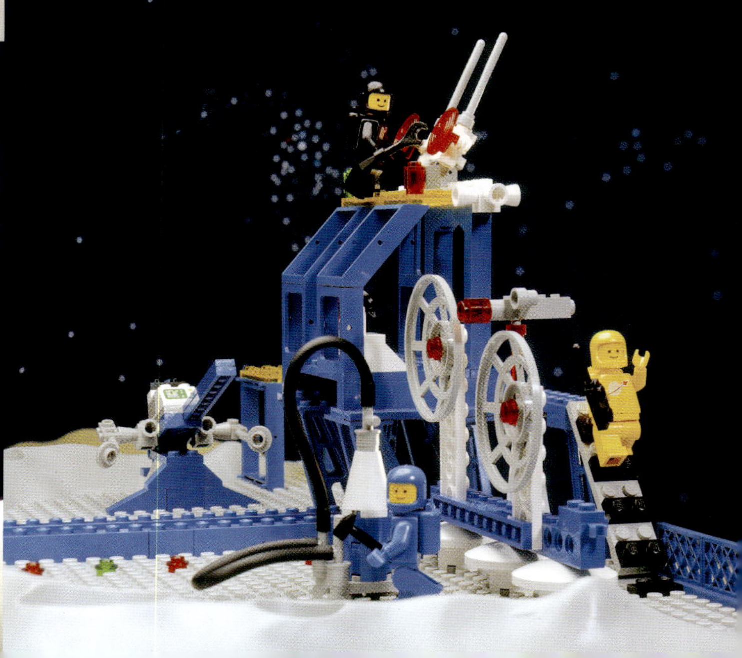

DANIEL AUGUST KRENTZ, LIKE YOURSELF, IS BETTER KNOWN FOR HIS WORK ON LEGO® CASTLE, SO IT WOULD BE INTERESTING TO HEAR ABOUT THE WORK HE DID FOR LEGO® SPACE.

Daniel was very much into science fiction. He was a really funny guy. I remember on the way home from the premiere of the movie *ET: The Extra-Terrestrial*, Daniel was so excited. He said, "We have to prepare—we have to be friends when they come." He really expected that at some point there would be aliens visiting us, and of course we shouldn't start fighting them. We should try to be friendly.

Daniel was very much into building architecture, and he really wanted to create a kind of space architecture. He made some quite odd space buildings, in all sorts of colors. He loved to make them colorful—I remember they would often be in bright yellow or bright red, and in quite a different style to Jens's.

Daniel really loved to build big. He would often begin with the roof—start from the top and then build his way down. That was quite funny. Daniel was very good at mathematics, so he could figure out how you could make angles and build in a more complicated way that was new and fanciful. Jens and I weren't into that—we were more just brick on brick.

After he had been building with LEGO bricks all day, Daniel would build during the evening and nights too, in his spare time. His home was fantastic. There were LEGO models all over the place, even in his restroom! He even made shelves over the doors so he could also put models up there.

I could never tell you how thankful I am for the way Jens and Daniel treated me, from my very first days. I came to Billund from another part of the country as an inexperienced twenty-year-old, feeling quite lonely. But from the first day, they were absolutely like family, and it was so fun to work with them.

Then in 1982 we got Ella Hansen, who was sort of a secretary. Before she came, we didn't have any meeting calendars and there was also quite a mess all over the place—models and stuff everywhere. She was the one who brought order, and she succeeded in that absolutely amazingly. She made us clean up the studio and really kept us in line! She could put a lid on us when we got too carried away. Ella was a really down-to-earth person. Nothing could freak her out; she would take everything in her stride.

At one point, we had one of the earlier types of videotape recorder, so we were making this space movie for fun. We had turned half of the design studio into one big space landscape. In those days, when you were using a video recorder, you couldn't stop and then start [to create frame-by-frame animation], so everything had to be done in one shot. There was only Jens, Daniel, Ella, and me, so models were placed on strings, and we would get everything moving while recording. I remember we had taken one of Jens's classic spaceships and put firecrackers on the back, so it would look like it had a rocket engine. It slid along on a fishing line, so it would look like it was flying. And Ella would have to stand at the end of the line and grab it!

BEING A LEGO® DESIGNER

WHAT WERE YOUR RESPONSIBILITIES AT THAT TIME?

We had to do everything in those days. I had to make new elements, the technical drawings, the graphics, and of course the models. We had to build the steps for the building instructions, make the calculations for the price for the models, and fix all the models for photography for the boxes.

I'd only been at the LEGO Group for eleven days before I had my first element design approved for production! That was the space camera [see page 54]. When we made a new element in those days, we were told that it should be in use for the next ten years at least. If we couldn't foresee that we would be using it for the next ten years, we were not allowed to make it. Well, my space camera is still about!

I later designed the metal detector [see page 61], and the pillar with three holes in it [inclined stanchion support 2 x 4 x 5; see page 55], which was used in a lot of buildings but also for some of the smaller models. It was created for LEGO® Space but also looked a little Victorian, as if it were cast in iron.

The company couldn't introduce too many new elements, because they had to manage the capacity to make the molds and to produce the elements. Also, as the LEGO Space line was already established by the time I joined, we couldn't make many new elements specifically for that line, so we only had maybe three or four new elements per year.

However, we designed many elements, because Jens had so many ideas. He had the ideas, and I made the actual prototypes. I would make the first prototype out of whatever I had to hand, like wood or plastic pieces, but, of course, I was not allowed to use my favorite material—lead.

The perforated radar [inverted webbed dish 6 x 6; see page 53] is an example of an element that I worked on closely with Jens. We did not make weapons in those times, especially not for LEGO Space. We really avoided making any elements that looked too obviously like laser guns or other weapons, so we put radar disks and things like that on top.

WHAT WAS THE PROCESS FOR DESIGNING LEGO® SPACE PRODUCTS?

During the early 1980s Jens made most of the models, and Daniel and I made the alternative models, for which there would only be pictures on the box. Set 6880 from 1982 [see page 53] was my first model: a small space vehicle covered with wheels. It also has my space camera built into the dish. It's the only box that I have kept all these years!

I've always been into making stuff with a lot of detailing, like set 6950 [Mobile Rocket Transport; see page 54], also from 1982. I even made it so that when you drove the vehicle over rough ground, there was some sort of suspension.

We had our different styles, and that shows. When you built a model back in those days, it was pretty much just you, so you can actually see which models Jens built, and Daniel, and so on.

Then, Jens, Daniel, and I—and sometimes some other colleagues—would look at the models, and we would agree which we would show to the marketing team. We also had to narrow it down to products that actually fit into a price point. I can still remember functions and detailing I had to take out, which was a challenge, but the customers never knew.

Jens would include many functions in the models. He made those big spaceships that open up at the back

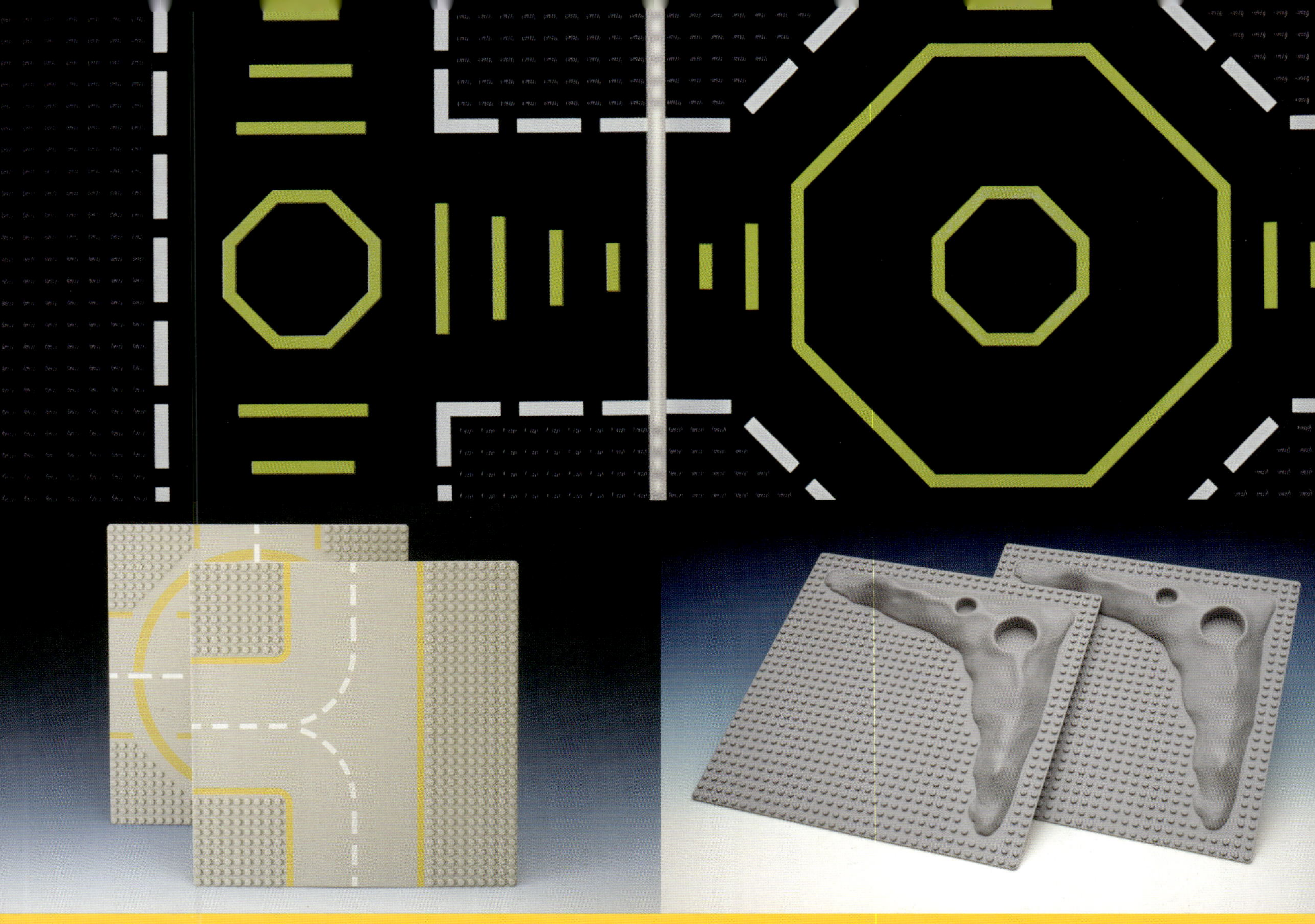

▲ Baseplates were an integral aspect of LEGO play in the 1980s. The LEGO Space line launched them with crater plates and two road plates, which were printed with a T-junction and a landing pad. In 1991, these road plates were updated for the Blacktron Future Generation subtheme (seen at top).

with a model inside the model. Well, originally, there would often be a model in the model in the model! These would disappear, to bring down the price.

Once we had completed a model, another person would build it through to see if improvements could be made. It was our responsibility at this point to be sure that models would be stable as you built them. There weren't many problems with the early models, because they were just built up simply from the table. We didn't have all these fancy angles then.

We always tested models with children, and it became more professional over time. For example, if we made a building with several floors, would they notice if we left out staircases? Ultimately they didn't, so there was no reason to spend time and money building a staircase if the children didn't need one. Of course, in space you don't need staircases, because the figure could just jump to the next level!

In the mid-1980s, there came this idea that children were spending more and more time in front of the television, and therefore some couldn't use the printed building instructions any longer. Indeed, some tests showed that they couldn't extrapolate from the drawings how to build things. So that placed restrictions on what we could build. At some point in the mid to late '80s, the models were given different colors from layer to layer of brick, almost like a checkerboard, so it would be easier to see that you were building a new layer—all because we thought that children couldn't figure out how to build any longer.

LOST IN TIME

DO YOU REMEMBER ANY CONCEPTS THAT DID NOT PROGRESS TO MARKET?

As a child I was really inspired by the Disney movie *20,000 Leagues under the Sea*. For many years, I tried to make a Jules Verne–inspired theme. It was what you might call steampunk, nowadays. Some of the ideas later made it into actual LEGO® models, but they were changed to look like space models. Some of the submarine vehicles in the LEGO® Aquazone subtheme, Aquanauts, are actually based on the Jules Verne models that I made, but mine were much more steampunk.

One of Jens's favorite movies was *2001: A Space Odyssey*. He liked the idea of making a space station that could be built bigger and bigger by attaching new sections and also some of the spaceships from the LEGO® Space range. I remember we actually hung one from the ceiling on fishing lines, but when we came to work one day, there were all these ruined models on the floor. We realized that was not a good idea.

Actually, we did try to find out if we could introduce a way that children could hang models up, because it would be another dimension of play to actually have the model flying. Jens and I worked a lot on it. We tried hooks, magnets, and suction cups; we even tried different pulley systems so you could make things move. But we never came up with a solution.

There were so many elements we never made. At one point, Jens really wanted to make angled bricks so we could also build sideways on the models, but they would make the models too hard for children to build. I also remember at some point we had plastics that change color when you heat them up; Jens was really into that idea, especially for LEGO Space.

We wanted to do a lot more with the vacuum-formed baseplates, and for LEGO Space we designed many more than we ever released. I remember I made a sort of half-moon plate, like a bowl turned upside down. It looked like the surface of a moon, and there were a lot of craters on it. But they were so expensive to make and to print upon that we never came out with very many for the LEGO Space theme. Of course, later we made some for LEGO® Castle and LEGO® Pirates, but not many.

Also, Jens always wanted to make projectile functions, which we have a lot of nowadays. At one point we had a projectile function where you could attach the projectiles with a hook-and-loop fastener—like Velcro—because that might have been cheaper than using magnets. However, tests showed that parents didn't like the idea because it would pick up a lot of dust and be messy. One of the ideas was that you would apply them as stickers, but I think we also worked with seeing how they could actually be molded onto the element. But we couldn't do much with that kind of molding. It wasn't possible to do all the advanced things they can do now. ■

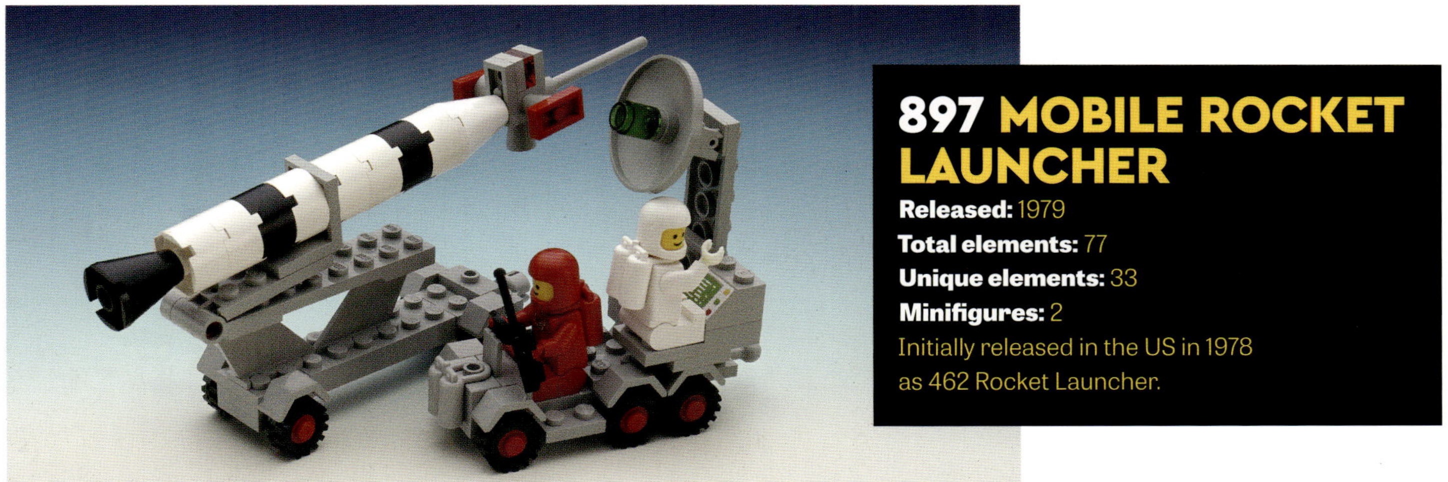

897 MOBILE ROCKET LAUNCHER

Released: 1979
Total elements: 77
Unique elements: 33
Minifigures: 2
Initially released in the US in 1978 as 462 Rocket Launcher.

Deftly referencing the black-and-white patterning of the real-world Saturn rockets, this feature-heavy set placed one of science's most iconic crafts in the hands of eager children. Airplane tails are repurposed as stable and visually arresting stanchions, and a tow ball connection enables easy separation of the control unit and the launcher.

The team had to launch the X7 satellite to search for sources of methane energy on Titan urgently if they were to have any hope of establishing a mining base there. With adverse conditions at Alpha-1, they dispatched this independently operated trailer vehicle with a two-person crew instead. Having located suitable terrain, the mechanic angled the launch arm while the rocket scientist swung into the 360° command post and carefully programmed the payload location, direction of movement, and route to destination into the video terminal.

*Arguably the oddest-looking new element introduced by the original LEGO® Space wave was **modified brick 1 x 1 with 3 space positioning rockets**, clearly inspired by reaction control system thrusters. Having three projecting cones limited the ways in which it could be attached to brick structures, but they did permit sideways building, still uncommon within the LEGO® System in Play at the time. Specialized element designs do not necessarily render an element unusable: it was also utilized as public address loudspeakers in the LEGO® Town line and has appeared in LEGO sets almost constantly ever since.*

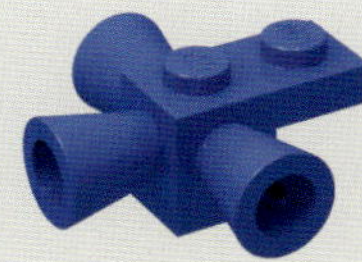

▲ **ELEMENT: 3963**
INTRODUCED: 1979

894 MOBILE GROUND TRACKING STATION

Released: 1979
Total elements: 79
Unique elements: 39
Minifigures: 1
Initially released in the US in 1978 as 452 Mobile Tracking Station.

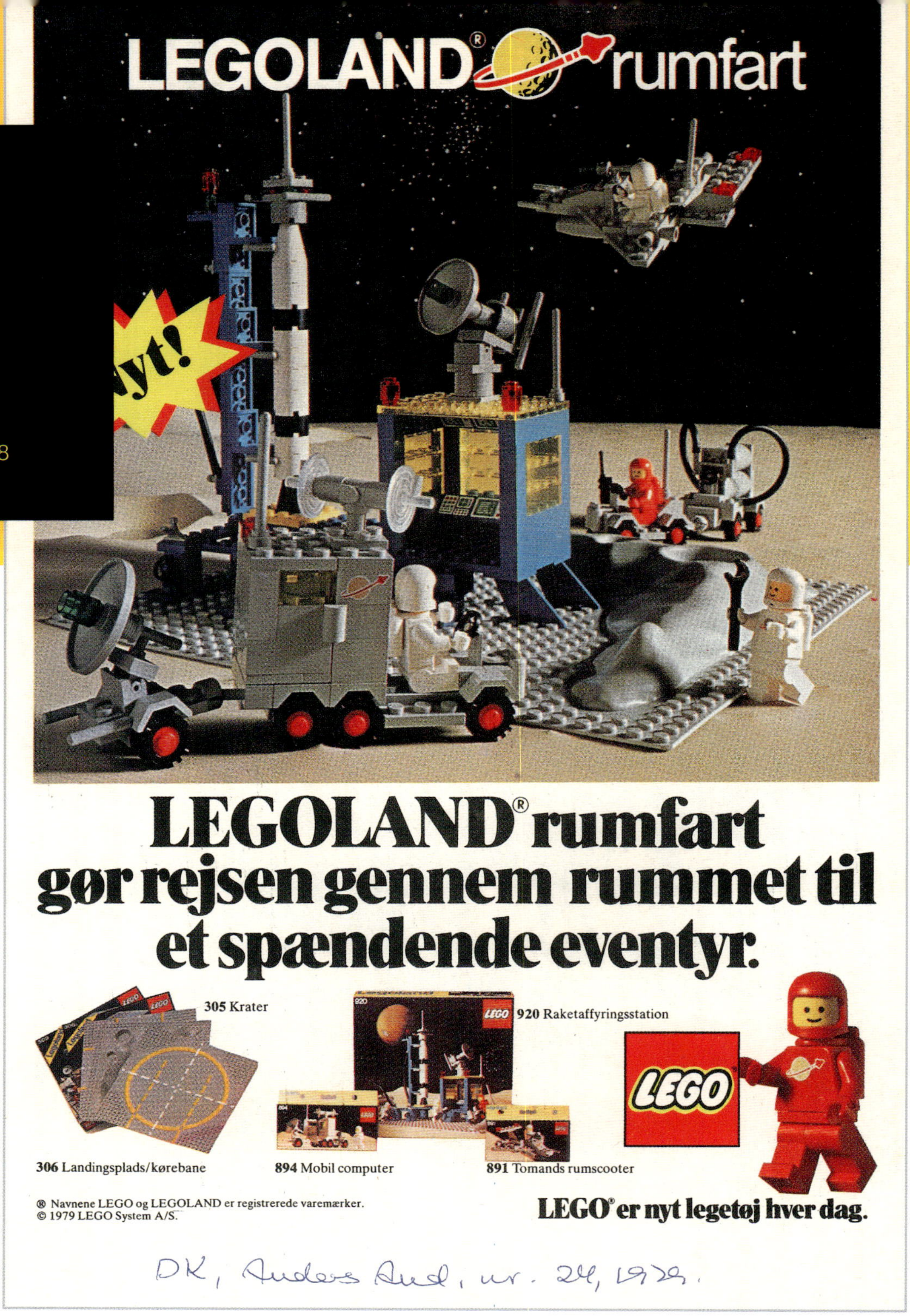

LEGOLAND® rumfart

Nyt!

LEGOLAND® rumfart gør rejsen gennem rummet til et spændende eventyr.

305 Krater
920 Raketaffyringsstation
306 Landingsplads/kørebane
894 Mobil computer
891 Tomands rumscooter

® Navnene LEGO og LEGOLAND er registrerede varemærker.
© 1979 LEGO System A/S.

LEGO® er nyt legetøj hver dag.

DK, Anders And, nr. 24, 1979.

Danish ad from 1979, with set 894 prominently featured in the foreground.

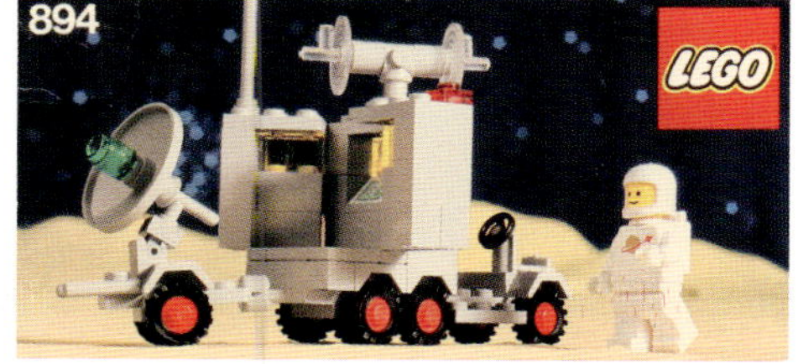

An unusual model with limited interior space and detail, set 894 nevertheless featured several exciting elements that added to play functionality, such as the turntable, hinges, and transparent yellow bricks. On the rotating antenna, a LEGO® Technic element is employed to allow for building sideways.

For observation and analysis on the nitrogen ice fields, this off-road tracking station proved essential. Its thick paneling protected the sensitive equipment from the hostile temperatures, allowing the radiation assessment detector to measure cosmic rays and solar particles in the atmosphere. Being a one-person operation, it had a single central control panel that received, processed, and displayed all radar, satellite data, and multiwavelength infrared images.

*The **round minifigure shield with stud** is a beautiful element, featuring concentric rings that echo the design of the minifigure utensil space gun / torch. This is the only set where it was molded in transparent clear plastic: it returned in 1984 in transparent red, as well as being used in the LEGO® Castle theme in dark gray.*

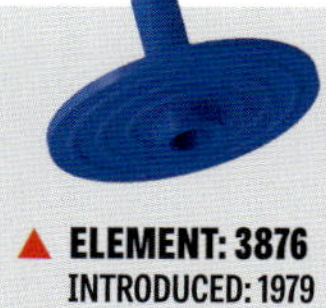

ELEMENT: 3876
INTRODUCED: 1979

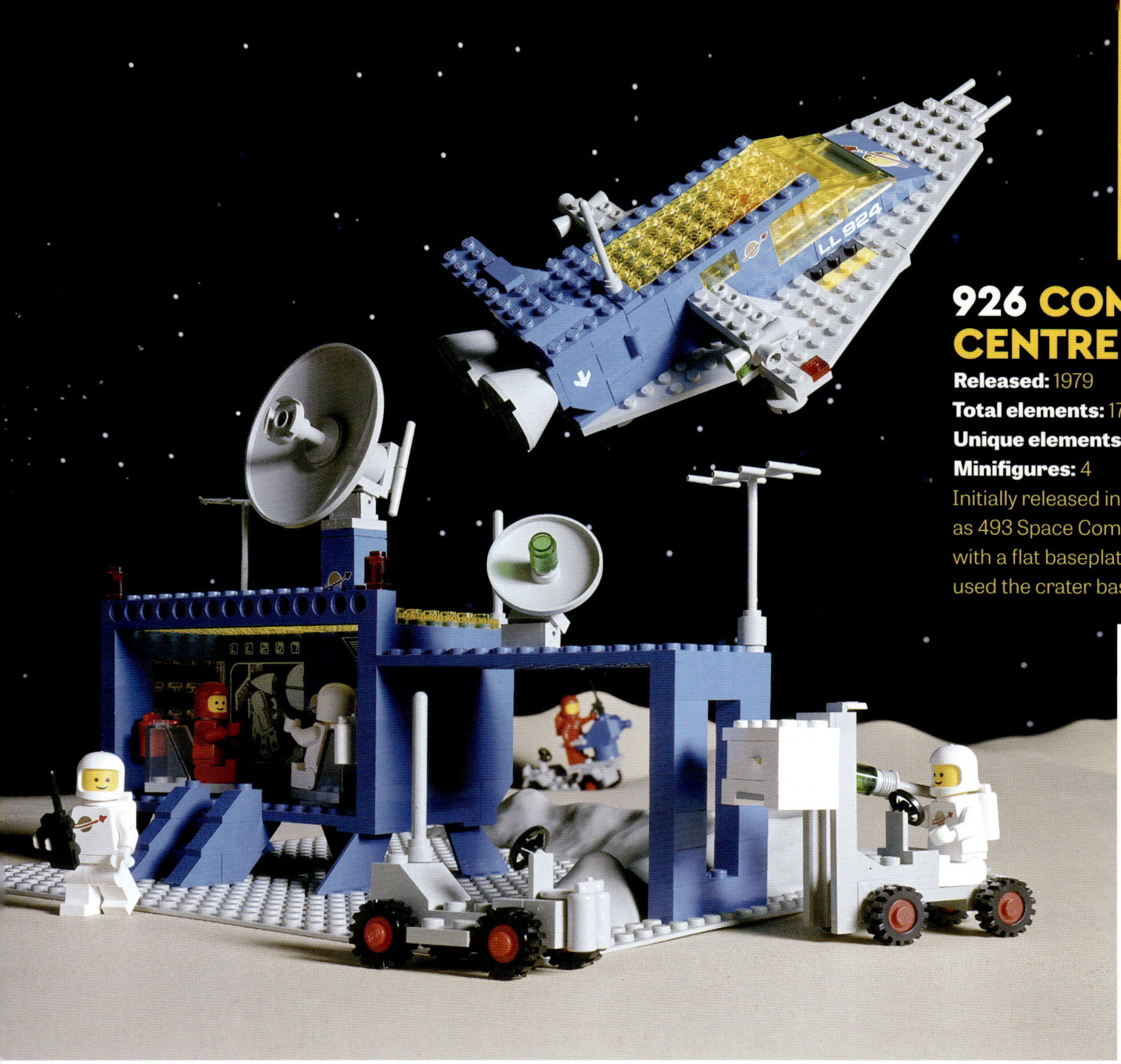

926 COMMAND CENTRE

Released: 1979
Total elements: 177
Unique elements: 58
Minifigures: 4

Initially released in the US in 1978 as 493 Space Command Center, with a flat baseplate. Later versions used the crater baseplate.

◀ Set 926, with set 924 taking off.

Unlike the poky control box provided with 920 Rocket Launch Pad, here two astronauts can lie back in their transparent chairs (with handy hooks for their air tanks) and swivel to view the array of screens and controls. The structure itself, however, is understandably stark. A 1 x 16 LEGO® Technic brick is used purely for its decorative nature—an effective and low-cost solution—while airplane tails supporting the center add some diagonal interest, giving the architecture a modernist feel. However, this set is really about the tech. Eight elements feature printed controls, of four different designs. Both sizes of the new dishes are included, alongside two antennas with side spokes. Two vehicles service the center: a buggy of identical design to set 886, and the other featuring a rear-mounted device cleverly utilizing an element intended for attaching airplane rotors.

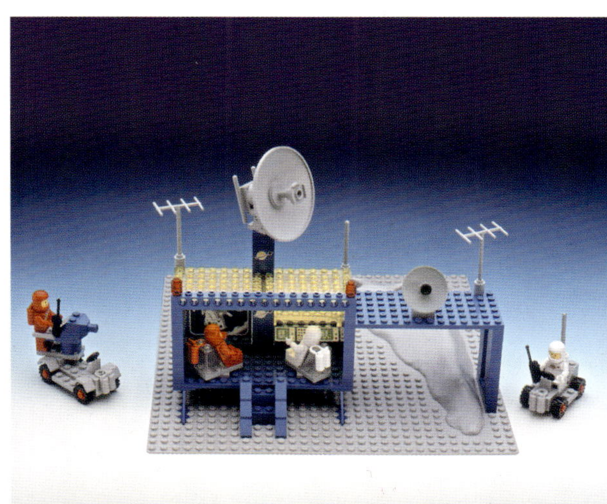

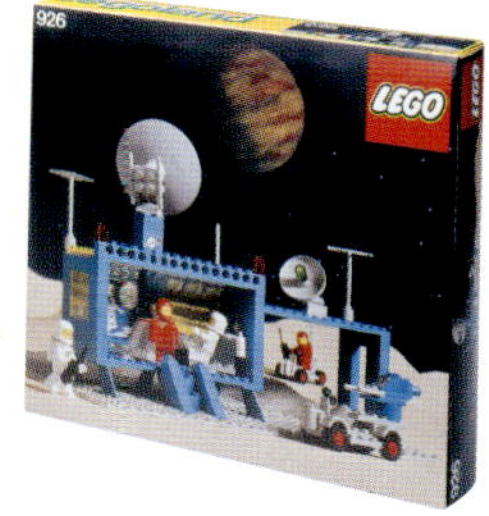

▲ Original mockup of box.

The team were quick to assemble the new command center on Charon, but installing the electronics was a much more involved task. Thankfully, once the gravity wall was activated, they could work much faster. They finished not a moment too soon—with the X11 satellite in urgent need of repair before the rest of the Triton crew could arrive, they were able to remotely guide one of the astronauts on LL 2079 through the complex process.

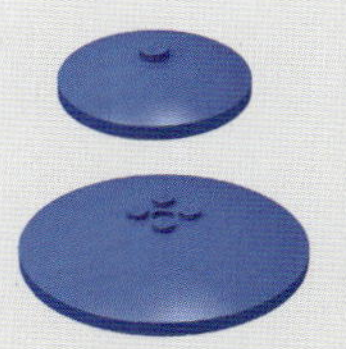

▲ **ELEMENTS: 3960, 3961**
INTRODUCED: 1979

*Dishes were a crucial addition to the LEGO parts inventory and remain in use to this day. For his earlier space-themed sets 358 and 565, Jens had resorted to brick-built dishes which looked more like square signs. For LEGO® Space, **inverted dish 4 x 4** and **inverted dish 8 x 8** were initially introduced, but many more sizes were to follow.*

924 SPACE TRANSPORTER

Released: 1979
Total elements: 172
Unique elements: 69
Minifigures: 2
Initially released in the US in 1978 as 487 Space Cruiser.

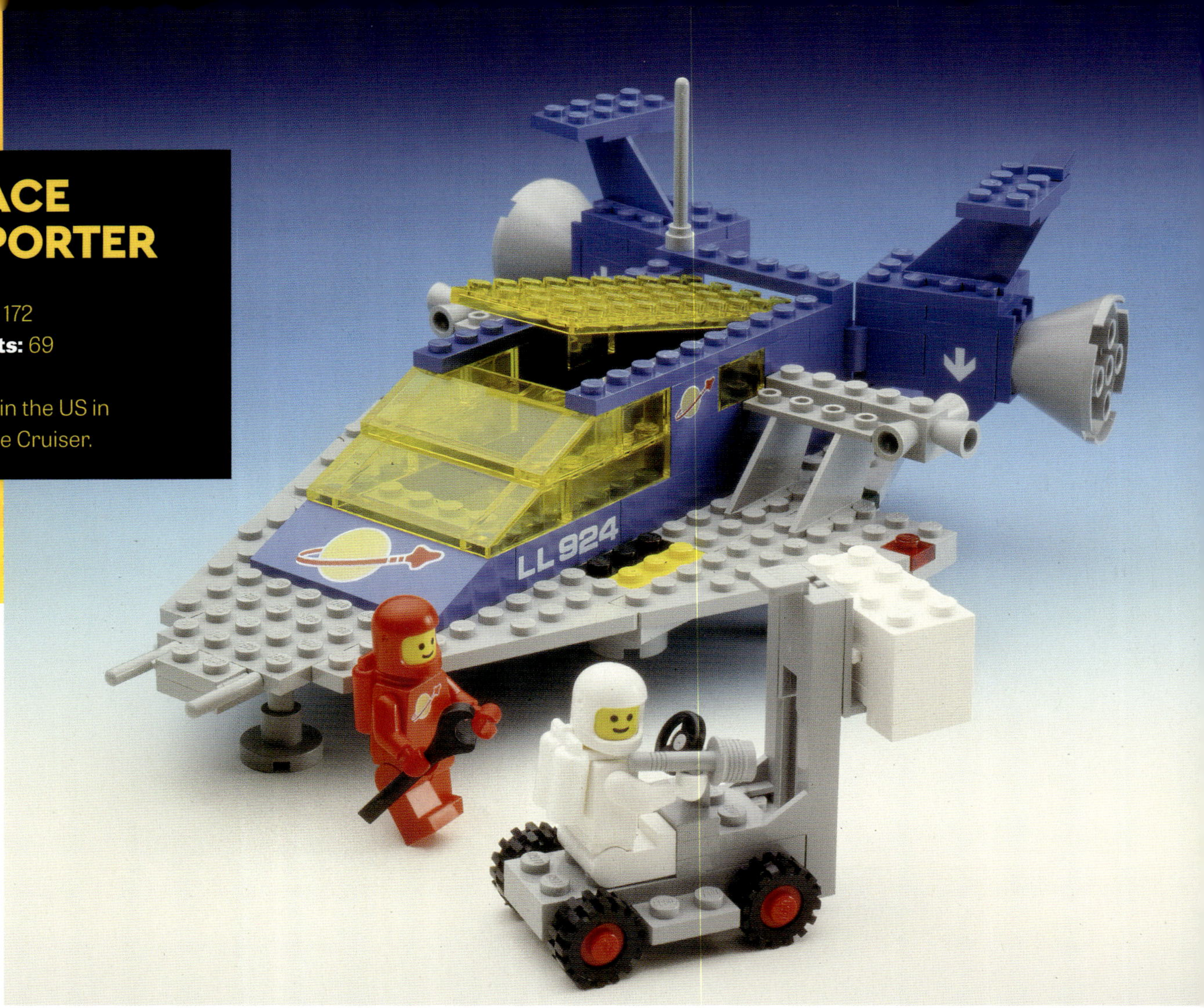

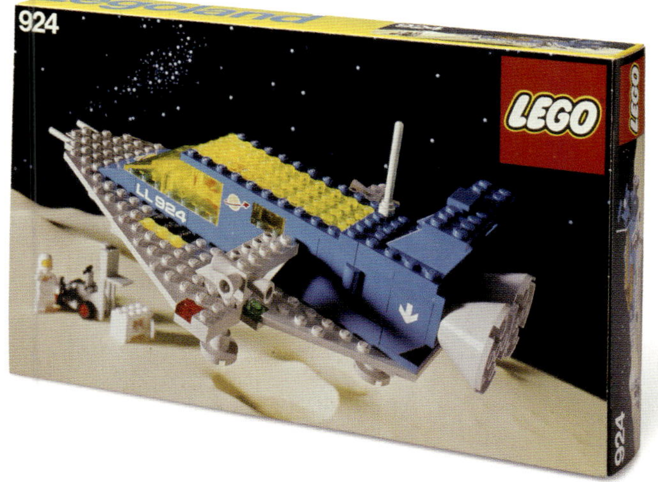

▲ Original box art for set 924 Space Cruiser.

Here, the fun is in the fuselage. The ship is long, tall enough to fit two minifigures, and the rear swings open in two halves to accommodate the crate delivered by the accompanying forklift. As a consequence, the fuselage feels bulky compared to the wings, which have a noticeably snub-nosed planform compared to sets 918 and 928 and are the closest to a diamond shape. The empennage (tail assembly) is more complex than 918: in addition to a twin T-tail there are wing-mounted fins. Hazard markings add a welcome sense of industrial realism.

HAZARD MARKINGS ADD A WELCOME SENSE OF INDUSTRIAL REALISM.

With tons of compressed nitrogen ready for shipment to the new base on Charon, the team readied LL 924 to fly. They knew they could rely upon its sturdy, balanced design to deliver this highly unstable cargo, and the reaction control system thrusters would guide them through any unexpected gravitational flux as they passed close to Pluto. The cargo bay slammed shut, and the scientist clambered in behind the pilot, his hands visibly shaking as he pulled the canopy shut.

Contrary to some people's recollections, the LEGO® Space line did not exclusively introduce specialized elements into the LEGO® System in Play. One of the most universal elements imaginable, ***plate 6 x 6****, appeared in two spaceships in its first year of production and was not used by other LEGO lines until the following year.*

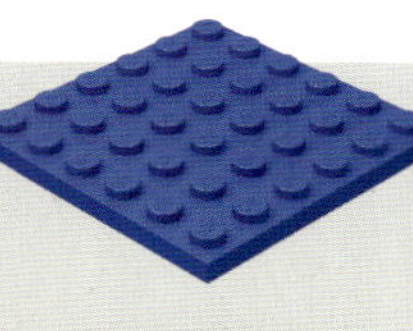

▲ **ELEMENT: 3958**
INTRODUCED: 1979

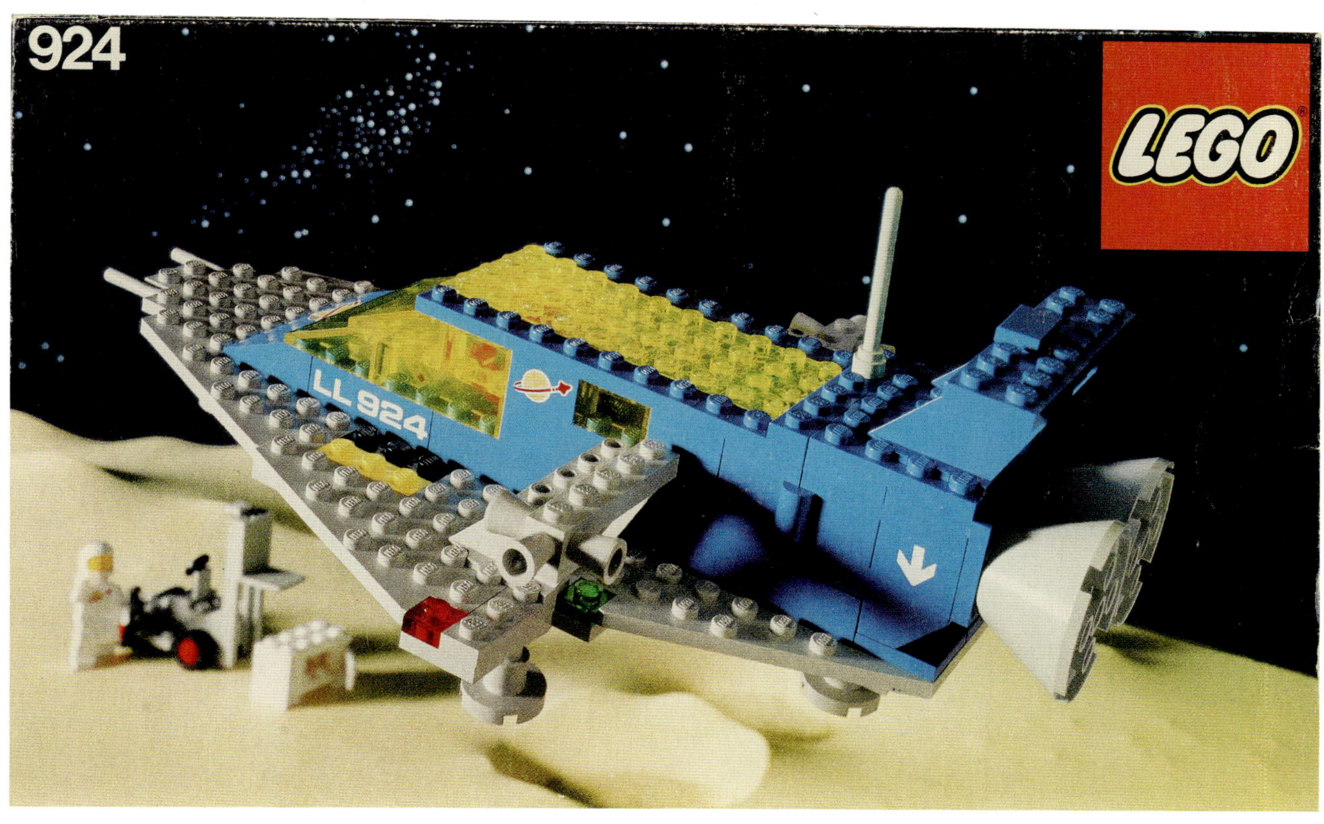
924
LEGO
LL 924

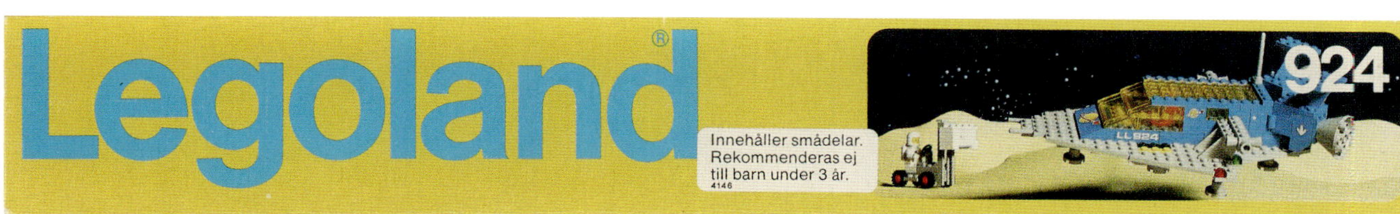
Legoland
924
Innehåller smådelar.
Rekommenderas ej
till barn under 3 år.
4146

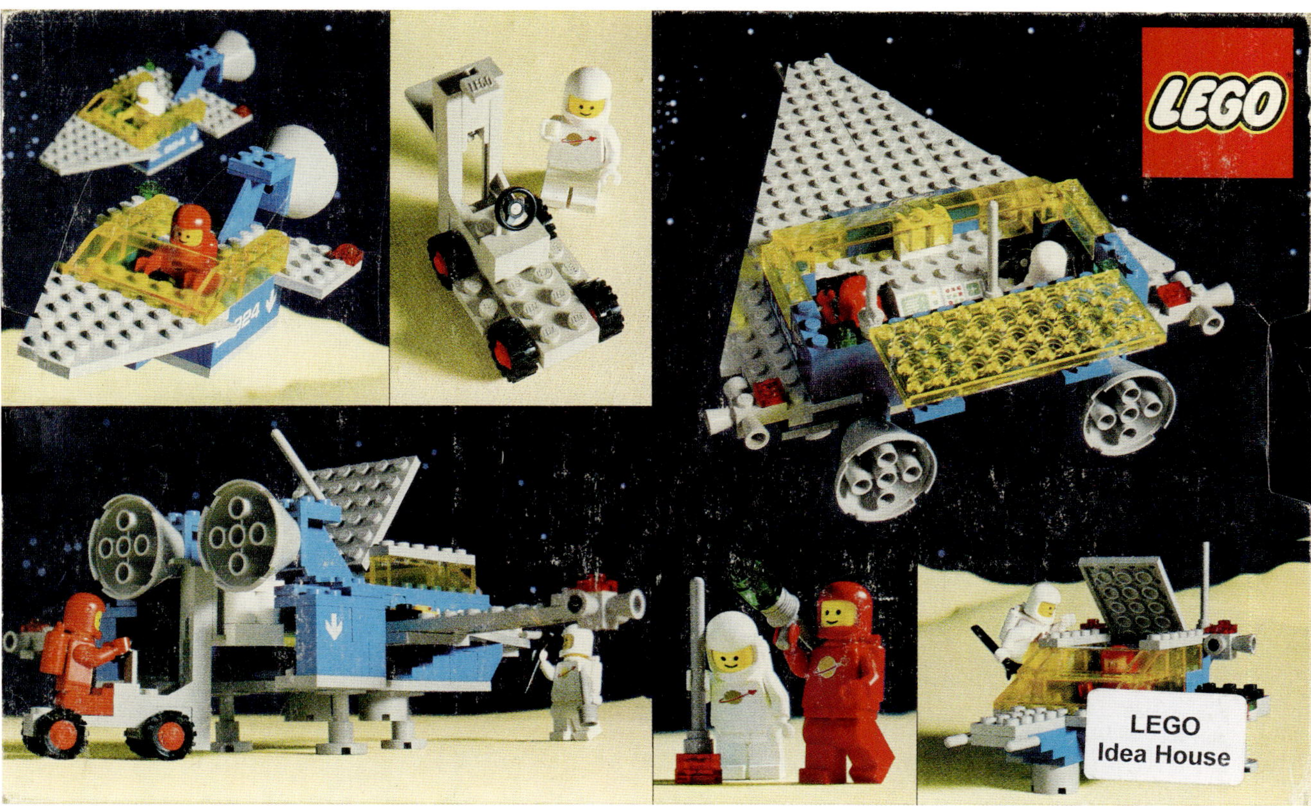
LEGO
LEGO
Idea House

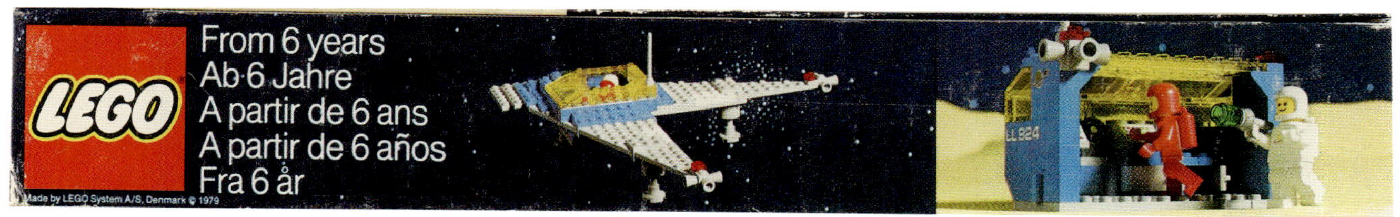
LEGO
From 6 years
Ab 6 Jahre
A partir de 6 ans
A partir de 6 años
Fra 6 år
Made by LEGO System A/S, Denmark © 1979

928 SPACE CRUISER AND MOONBASE

Released: 1979
Total elements: 342
Unique elements: 97
Minifigures: 4
Released in the US as 497 Galaxy Explorer.

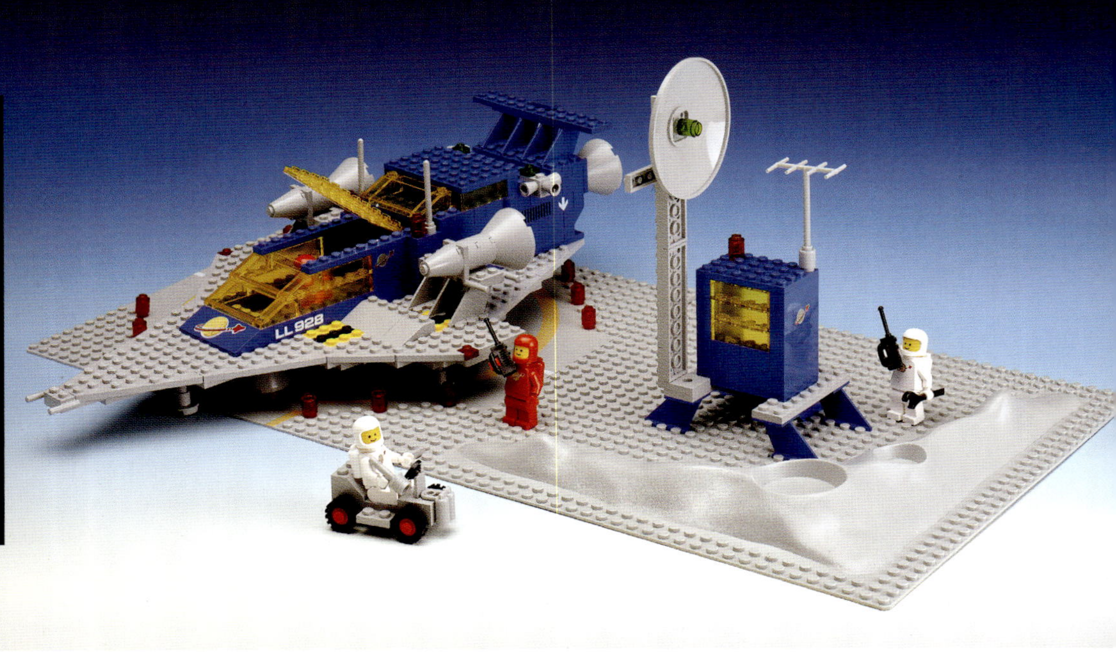

Every novelty of the 1979 LEGO® Space line is present here in one set: baseplates, ground vehicles, a control center, and of course an incredible spaceship. Most are actually relatively basic structures, with the exception of the Galaxy Explorer itself. Most striking is the ship's compound, tapered leading edge, switching from 67° to 45° wing plates. This increased planform allows for additional wing-mounted engines, spoilers, and hazard markings. The fuselage widens to six modules at the rear: generous enough for the buggy to be driven inside!

"Ground crew to the landing pad! LL 928 is cleared for stop-and-go." Knowing their colleague was in a critical condition, the team performed their duties flawlessly. There was only a brief moment for them to take in the incredible sight of the first-ever interstellar craft descending slowly toward them. With a hiss, the cargo bay swung open, the ramp descended, and out shot the medic in his buggy, speeding across the ice dunes toward sickbay! Within moments the doors were resealed and fuel levels were topped up, and with a sigh of relief from its pilot, the Galaxy Explorer lifted off to resume its first mission to Alpha Centauri.

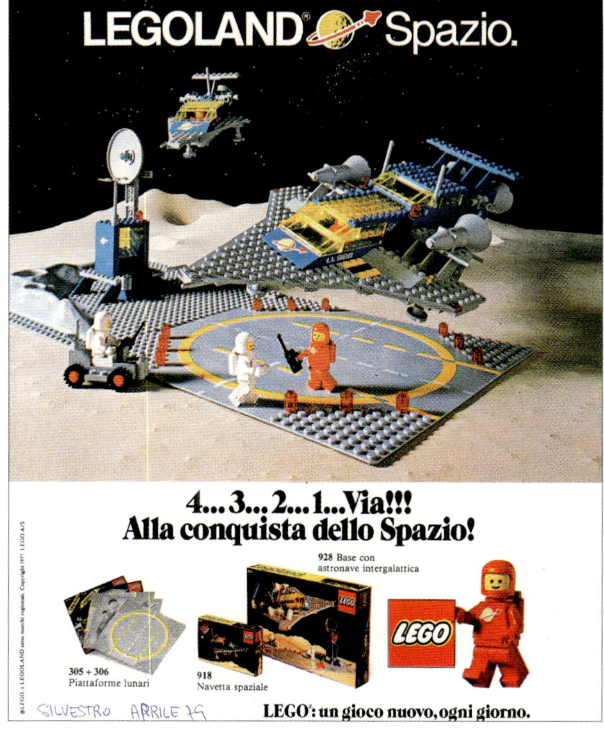

▶ An Italian ad from 1979 invites us to conquer space with sets 928 and 918.

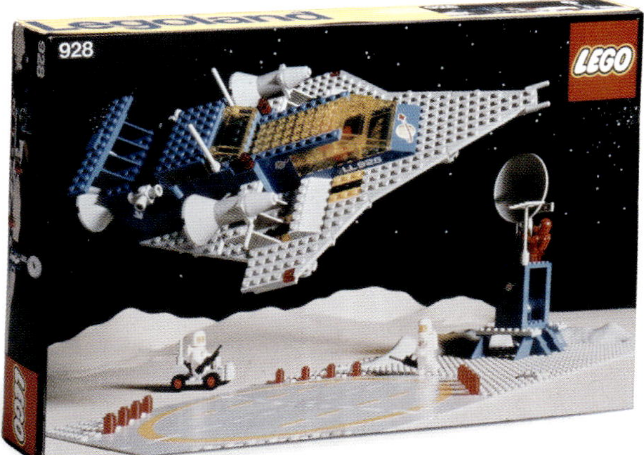

▶ Original box art for set 928 Space Cruiser and Moonbase.

The cone introduced in the 1973 set 358 Rocket Base never appeared in LEGO® Space, perhaps because its tapering—from a four-module diameter at its base to a one-module top—was too restrictive for Jens's requirements and offered insufficient clutch at the tip. Instead, two new rounded elements were introduced: the ***cone 2 x 2 x 2*** *and the truncated* ***cone 4 x 4 x 2****. Usually they represented the nose and the exhaust, respectively, but were utilized in all sorts of ways across many LEGO lines. In the 1980s, both of these cone elements were redesigned to include central holes, for additional connection options.*

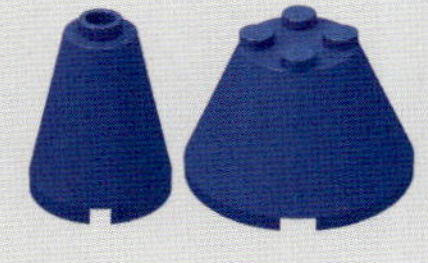

▲ **ELEMENTS: 3942, 3943**
INTRODUCED: 1979

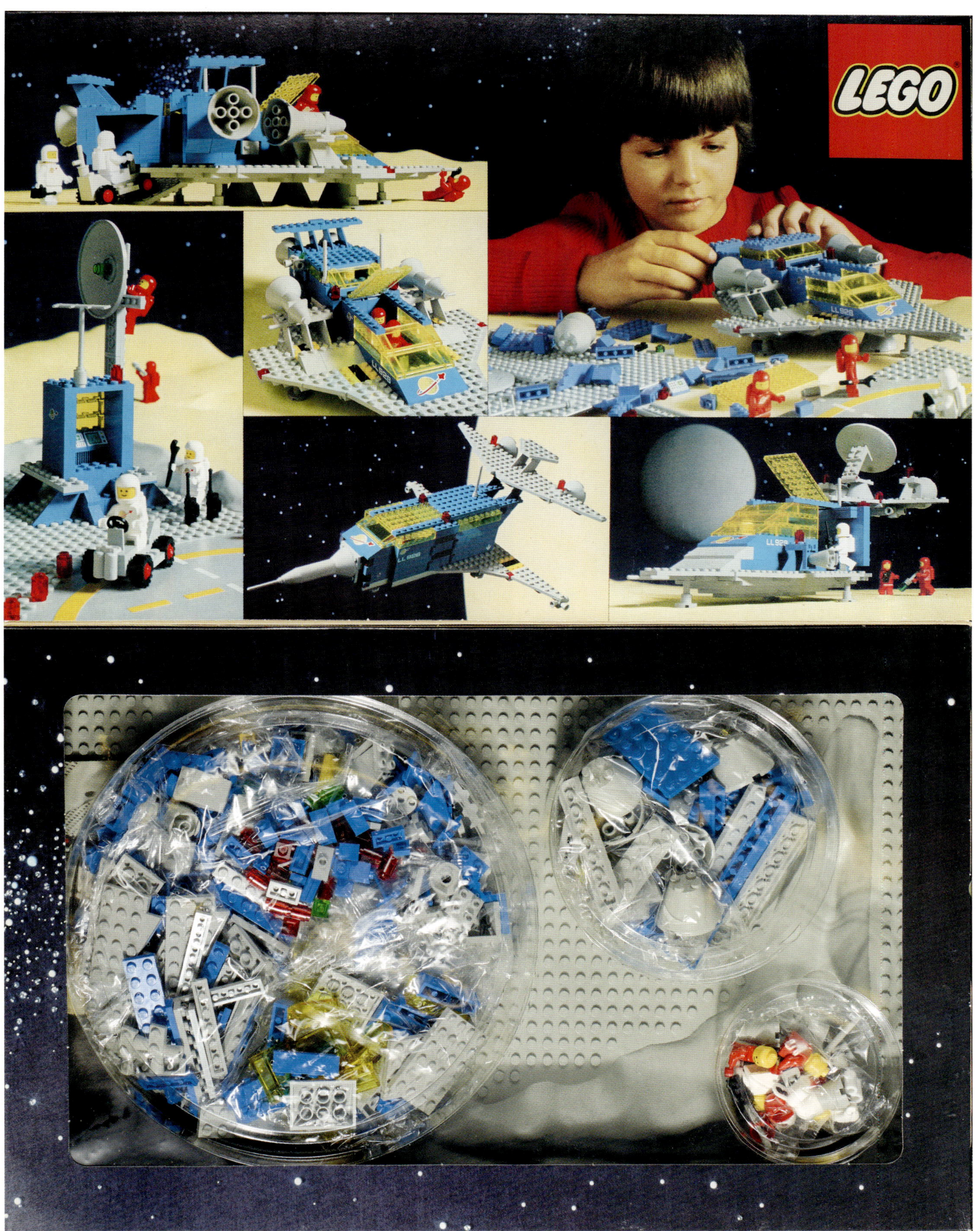

The interior packaging of set 928 was carefully considered. Transparent film displayed the elements, separated into tubs to reveal the crater plate below.

889 RADAR TRUCK

Released: 1979
Total elements: 29
Unique elements: 17
Minifigures: 1
Not released in the US.

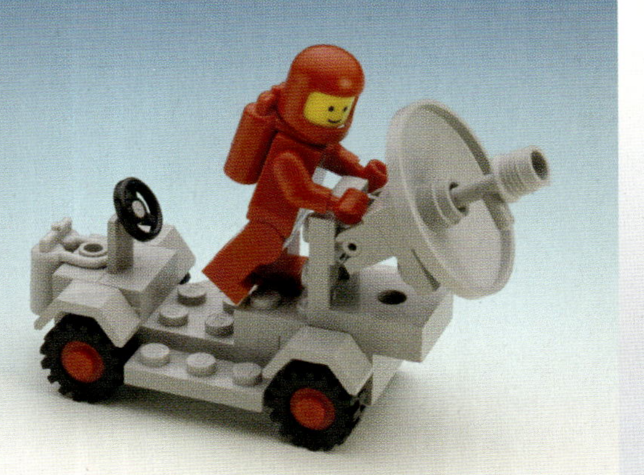

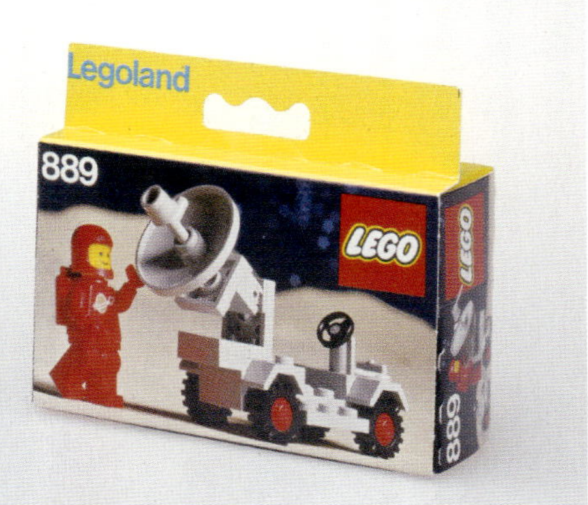

This small set makes clever use of just a few elements: the bracket 2 x 2 - 2 x 2, the hinge brick, and the space torch all combine with the dish to create adjustable radar equipment.

The cosmic meteorologist and his trustworthy truck proved invaluable to the team once again. Not only did he detect the oncoming dust storm, but he also guided the supply ship to a safe landing at Alpha-1 at the height of the tempest. Finding himself lost in the aftermath, he angled the dish up toward Neptune to calculate his location.

*The **bracket 2 x 2 - 2 x 2 with 2 holes** was among a smattering of radical new LEGO elements that permitted construction at 90° to the traditional plane of building. Because the LEGO brick is not based upon a perfect 1:1 ratio cube, a challenge with using brackets is that it can prove difficult to reattach the sideways build to the main build in a second position. In 1987, brackets with a thinner vertical section were introduced, which solved the issue in many situations; however, the thick style of bracket is also used to this day—and as recently as 2019, a new bracket was introduced (41682) that looks like a sibling of this 1979 element.*

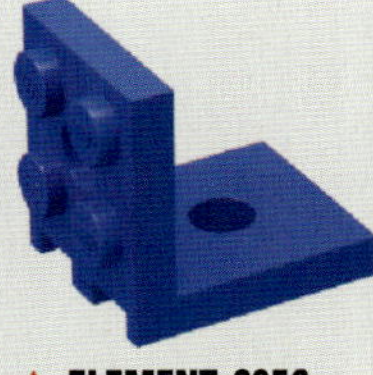

▲ **ELEMENT: 3956**
INTRODUCED: 1979

891 TWO SEATER SPACE SCOOTER

Released: 1979
Total elements: 40
Unique elements: 23
Minifigures: 1
Initially released in the US in 1978 as 442 Space Shuttle.

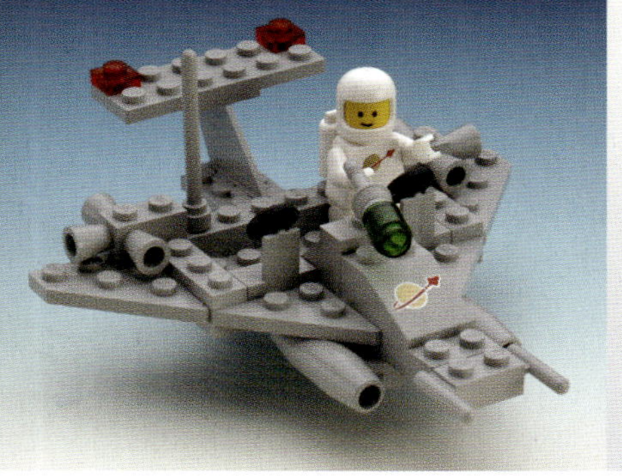

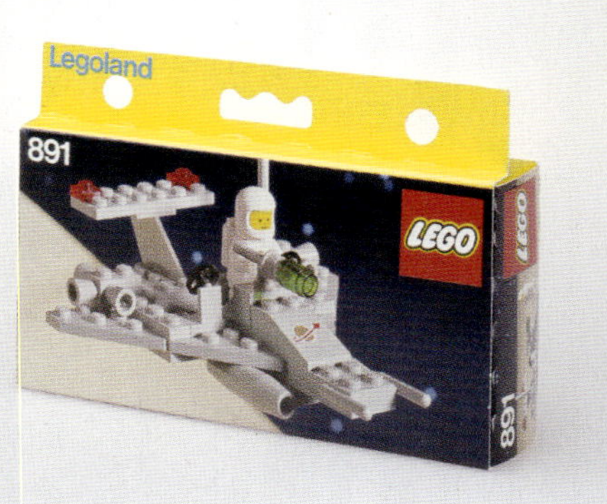

This is one of the few craft in LEGO® Space featuring solely 45°-angled wings, and the only ship until 1985 to use the smooth engine pieces. Interestingly, this two-person ship includes only one minifigure.

With her colleague lost somewhere toward the equator, the pilot maneuvered her powerful scooter across Triton on a race against time before temperatures sank below -370°F. She set the dash-mounted lidar to a wavelength of 1,600 nm and began to scan, keeping one eye on the Cartesian display for any anomalies in the barren terrain.

*The **wedge plate 4 x 4 wing** was introduced for LEGO Space, but its 45° diagonal didn't prove to be as popular as the 67° plates. Ultimately, the lack of notches along the diagonal edge—meaning it could not be attached to a plate below—sealed this element's fate, and it was replaced in the 2000s by the more symmetrical wedge plate 4 x 4 cut corner (30503).*

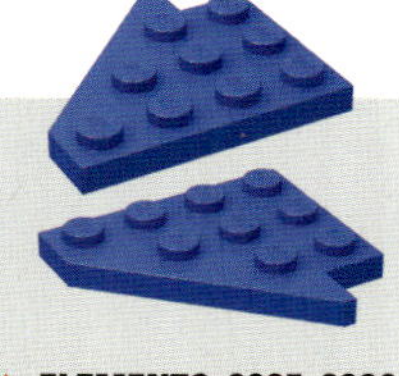

▲ **ELEMENTS: 3935, 3936**
INTRODUCED: 1979

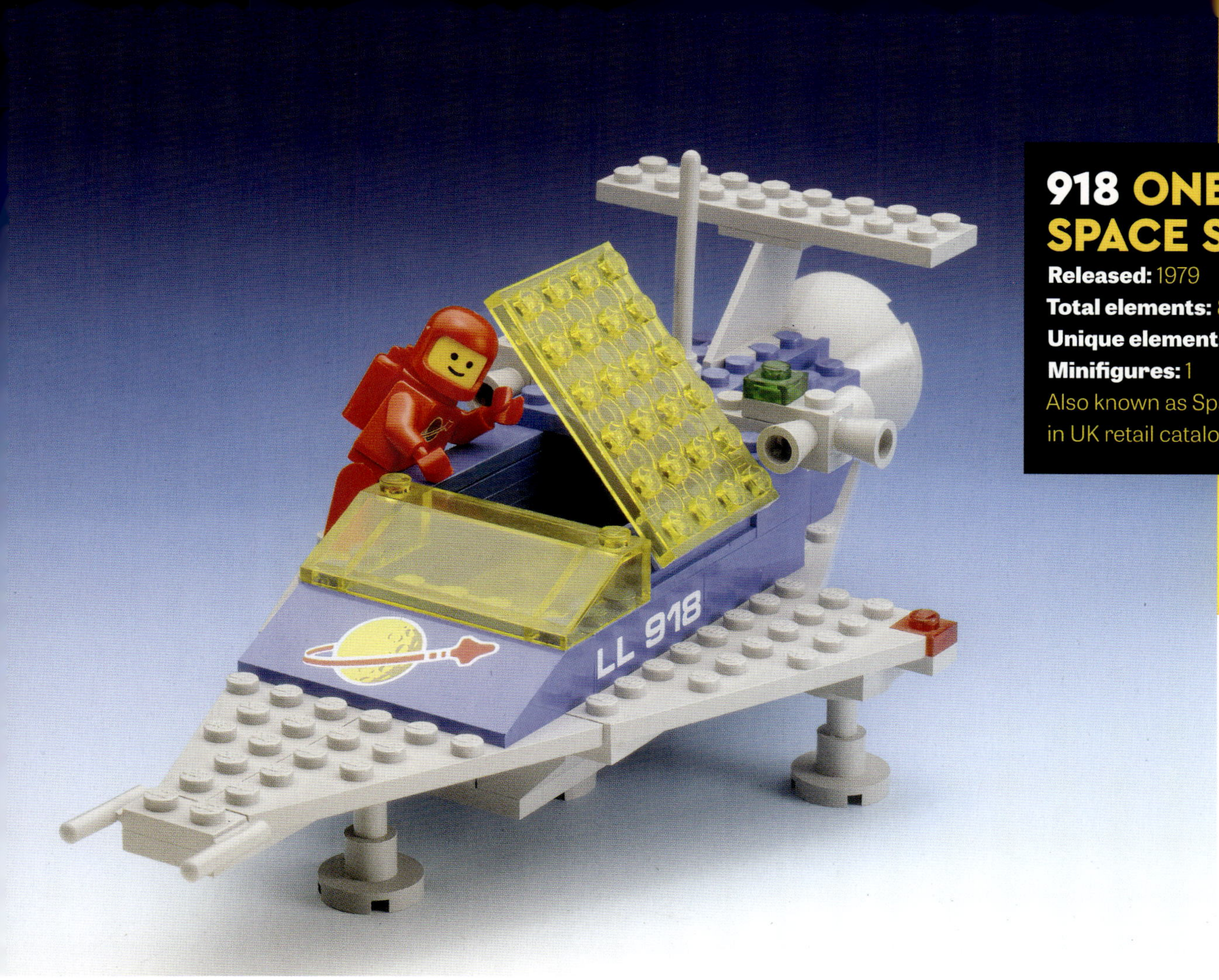

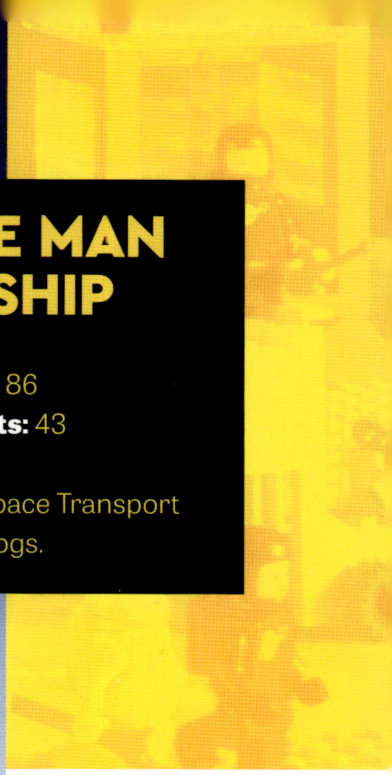

918 ONE MAN SPACE SHIP

Released: 1979
Total elements: 86
Unique elements: 43
Minifigures: 1
Also known as Space Transport in UK retail catalogs.

The initial wave of LEGO® Space came with not one but three spaceships. The smallest of these, set 918, is accordingly light on play functions: the cockpit can be opened, and there are two tiny cargo doors at the rear. Unlike the larger ships, the cockpit is sunken to accommodate the pilot—this required the inclusion of a round brick underneath, so that inserting the minifigure does not pop the bottom off the craft. The box could also be played with: the bottom of the interior was printed with a landing pad design.

THE BOX COULD ALSO BE PLAYED WITH: THE BOTTOM OF THE INTERIOR WAS PRINTED WITH A LANDING PAD DESIGN.

It had been deemed too dangerous to be sending an astronaut to Titan so soon, but the pilot had convinced the team that LL 918 was the right craft for the job. With no excess tech weighing it down, she had now reached Saturn's rings faster than any of the larger craft could have. The mid-fuselage wings provided excellent stability as she performed rolling movements through the icy debris, and she figured they would also keep interference drag low once she entered Titan's atmosphere. However, she was about to be proved wrong.

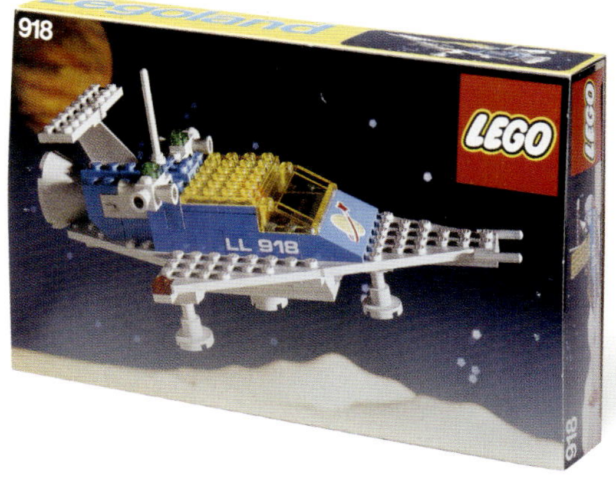

◀ Original box art for set 918 One Man Space Ship.

*Printed elements were a thrilling addition to any set, and LEGO Space sets came with plenty. Most were reused across as many sets as possible; however, the three 1979 spaceships each came with a unique element stating the designation and serial number of the craft: in this case, **brick 1 x 4 with white "LL 918" pattern**. These imbued the ships with an industrial look, as well as encouraging collectibility.*

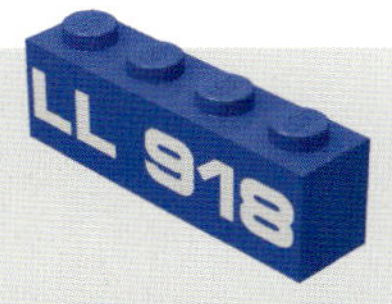

▲ **ELEMENT: 3010P918**
INTRODUCED: 1979

920 ROCKET LAUNCH PAD

Released: 1979
Total elements: 186
Unique elements: 70
Minifigures: 3
Initially released in the US in 1978 as 483 Alpha-1 Rocket Base.

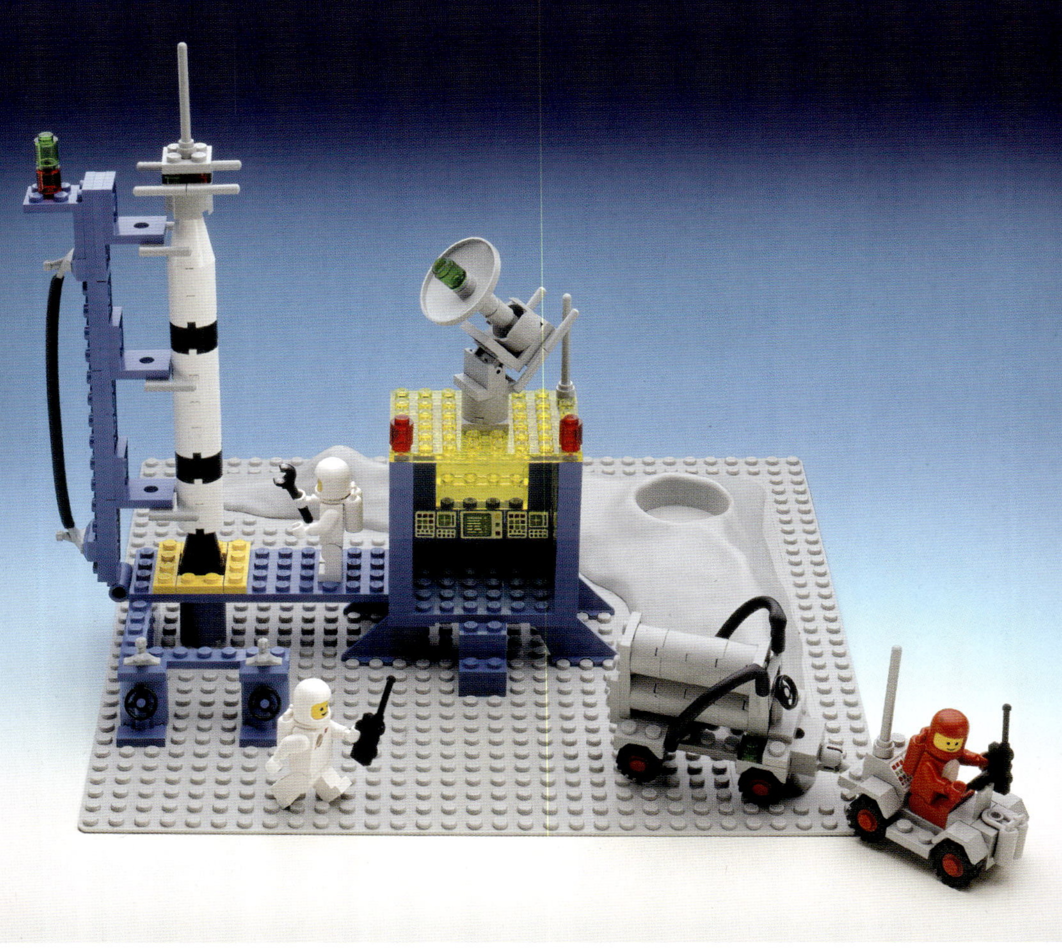

Although 920 Rocket Launch Pad contains slightly more elements than sets 924 Space Cruiser and 926 Command Centre, it seems smaller due to its box-like control base and crewless rocket. However, its strength is in its variety, with a range of vehicles and technology included to spark narrative play ideas. The base is open on one side for ease of play.

▼ Original box art for set 920 Rocket Launch Pad.

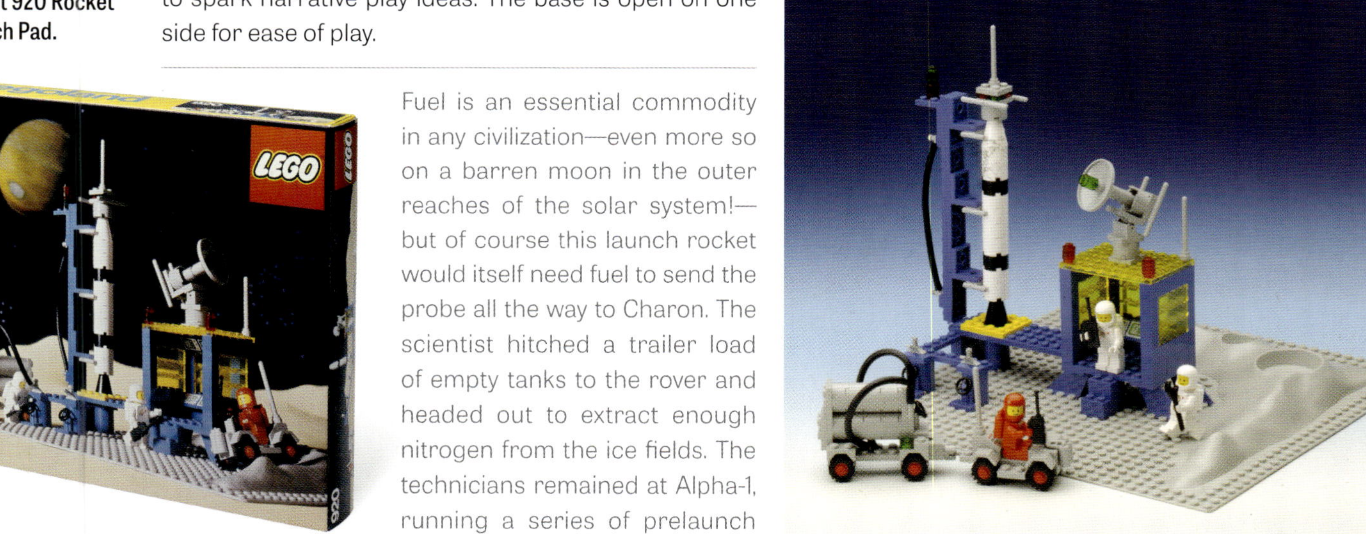

Fuel is an essential commodity in any civilization—even more so on a barren moon in the outer reaches of the solar system!—but of course this launch rocket would itself need fuel to send the probe all the way to Charon. The scientist hitched a trailer load of empty tanks to the rover and headed out to extract enough nitrogen from the ice fields. The technicians remained at Alpha-1, running a series of prelaunch checks: one synchronized the teleoperation controls while another ensured that the umbilical cable was securely connected along the full length of the launch mount—nothing could be left to chance this time.

The LEGO® System in Play included a few rounded elements in the 1970s, and ***round brick 2 x 2 with axle hole*** *was among the most useful new molds created for the LEGO® Space line. Here it is provided in several colors and used to great effect: gray to form the rover's fuel tanks, blue for the storage tanks in the base of the launch pad, and most notably, white and black for the rocket. After the unstable LEGO rocket designs using curved bricks that had preceded the LEGO Space line, these smaller bricks were a welcome addition, especially with the option to add a central LEGO® Technic axle for even more stability.*

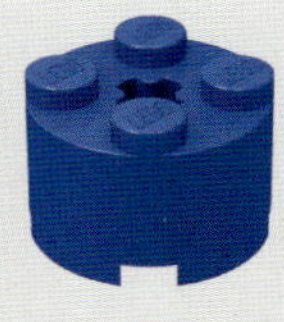

▲ **ELEMENT: 3941**
INTRODUCED: 1979

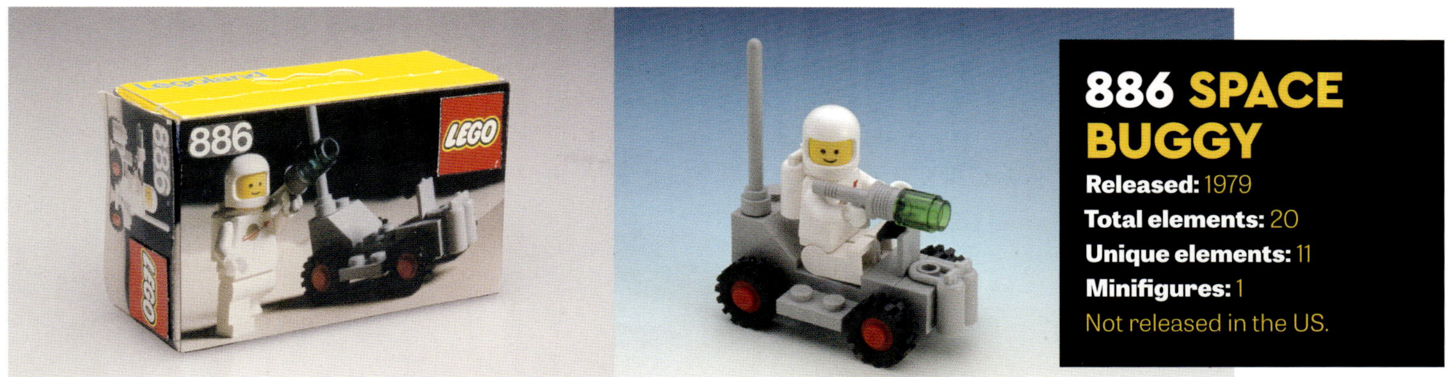

886 SPACE BUGGY

Released: 1979
Total elements: 20
Unique elements: 11
Minifigures: 1
Not released in the US.

Astonishingly basic and yet so charming, this vehicle proves that even a pocket-money set can be a classic. This basic buggy design was adapted for other sets in 1979 and was a template for years to come.

The geophysicist switched his probe from pulse induction detection to energy oscillation mode, breaking apart the rock with a couple of quick blasts. Retrieving the sample, he headed back to the rover. Just like the astronauts themselves, this vehicle was tough and reliable. Its lightweight aluminum-alloy body and twist-beam rear suspension were equally capable on the rock or the ice of Triton, and with no sensitive equipment onboard, there was no need for dust guards over the wheels. He detached the additional fuel storage from the front—and his face fell. How could this be empty?

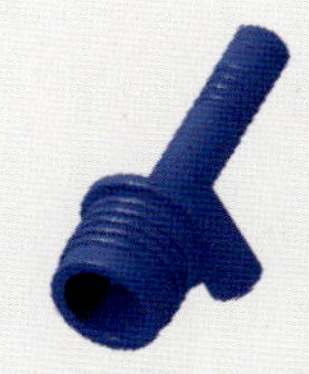

▲ **ELEMENT: 3959**
INTRODUCED: 1979

The design of the ***minifigure utensil space gun / torch*** *was so clean, it could be interpreted as any kind of scientific equipment—but never a weapon, of course! It appeared in seven LEGO® Space sets in 1979 and continued to appear in LEGO sets for another thirty-one years before receiving an updated, even simpler design.*

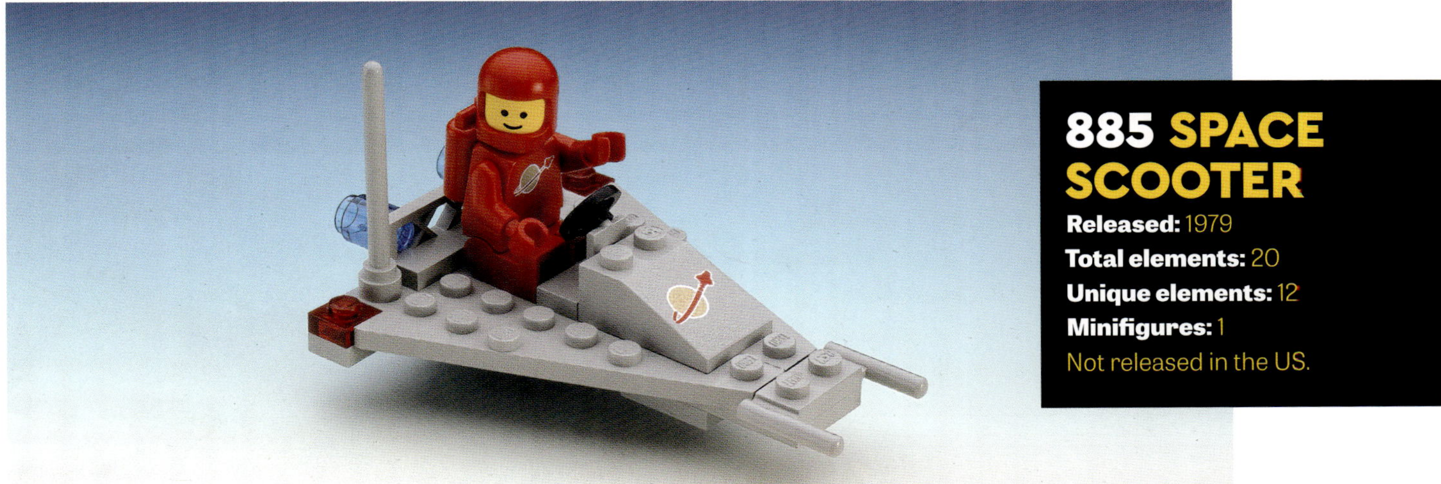

885 SPACE SCOOTER

Released: 1979
Total elements: 20
Unique elements: 12
Minifigures: 1
Not released in the US.

A mere fifteen elements (excluding the elements that make up the single minifigure) doesn't stop this pocket-money ship from providing a universe of play, and the inventory packs in some of the most exciting new elements created for the launch of LEGO® Space. To build sideways, the rear utilizes a common fence piece. Uniquely within the Classic era, the exhaust is transparent blue.

For conducting short-range missions within the troposphere, nothing beat this lightweight patrol craft. Its high overall propulsion system efficiency left the Alpha Mission pilots feeling assured they could take gravity field readings all the way up to the equator and still have enough fuel to return safely.

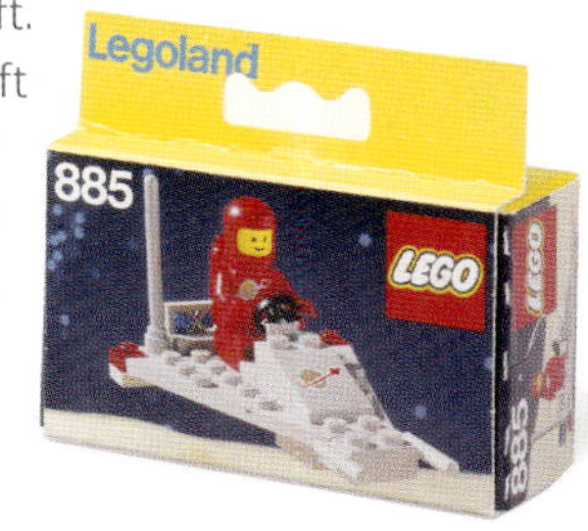

The ***wedge plate 8 x 4 wings*** *are truly iconic LEGO Space elements. They appeared on four out of the five spaceships from the launch wave and continued to define the shape of spacecraft for years to come.*

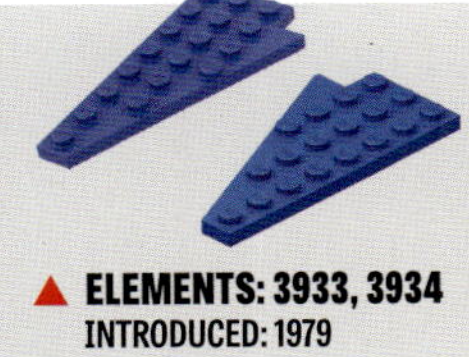

▲ **ELEMENTS: 3933, 3934**
INTRODUCED: 1979

Cleverly, the square cutout allowed multiple wings to be lined up to make a longer wing, but this feature perhaps limited the element's usage in other applications.

6841 MINERAL DETECTOR

Released: 1980
Total elements: 44
Unique elements: 19
Minifigures: 1

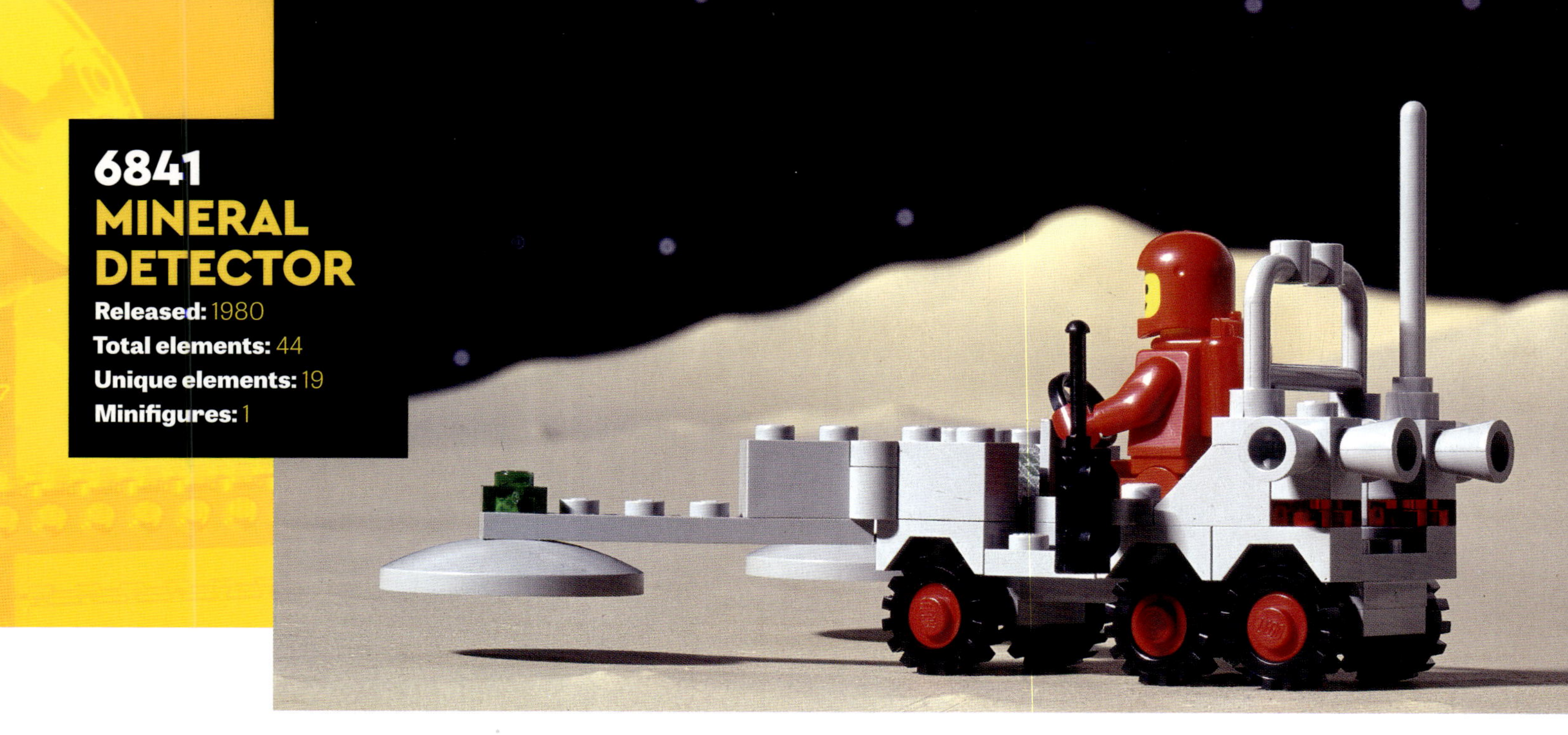

The novelty at the center of this highly original six-wheeled ground vehicle is the hinge bricks, introduced just two years earlier. Space positioning rockets once again adorn the vehicle, along with a brand-new roll bar element from LEGO® Town, its color changed to gray: bar 1 x 4 x 2 with studs (4083).

Could something be up with the equipment? The geologist reached over the steering wheel and ran the analysis methodologies through his dashboard computer once again. Same result. He drove again, the detection arms scanning the surface rapidly while the analysis cones at the rear simultaneously correlated readings from the surrounding area. Always the same results . . . no matter where he went in this massive crater. He stood up against the roll bar to survey its size and radioed the mission specialist. "Scratch that location. You need to get over here . . . I think I just found enough yttrium aluminum garnets to power an entire fleet to any star in the Milky Way!"

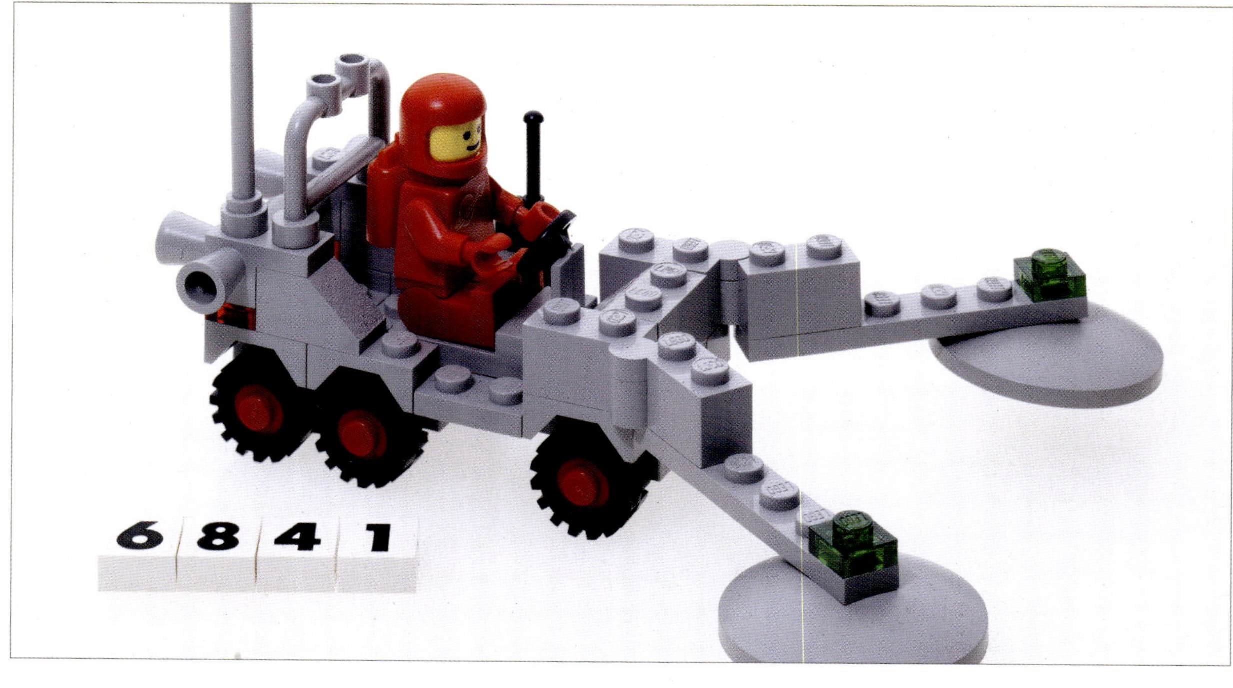

▶ The sets being photographed were identified using numbered tiles from the Modulex Planning System, a separate product developed by the LEGO Group in the 1960s.

The introduction of the minifigure in 1978 also meant the appearance of various accessories, which they could hold by clasping the 3.2 mm bar sections. Consequently, plates and bricks with clips were introduced into the LEGO® System in Play, so that the accessories could be placed somewhere when not being used. ***Modified plate 1 x 1 with clip (vertical grip)*** *was the first such clip element and is used here to hold the radio.*

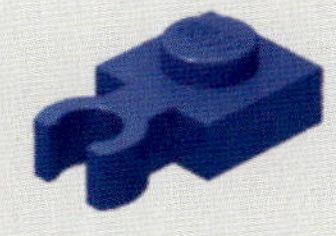

▲ **ELEMENT: 4085**
INTRODUCED: 1980

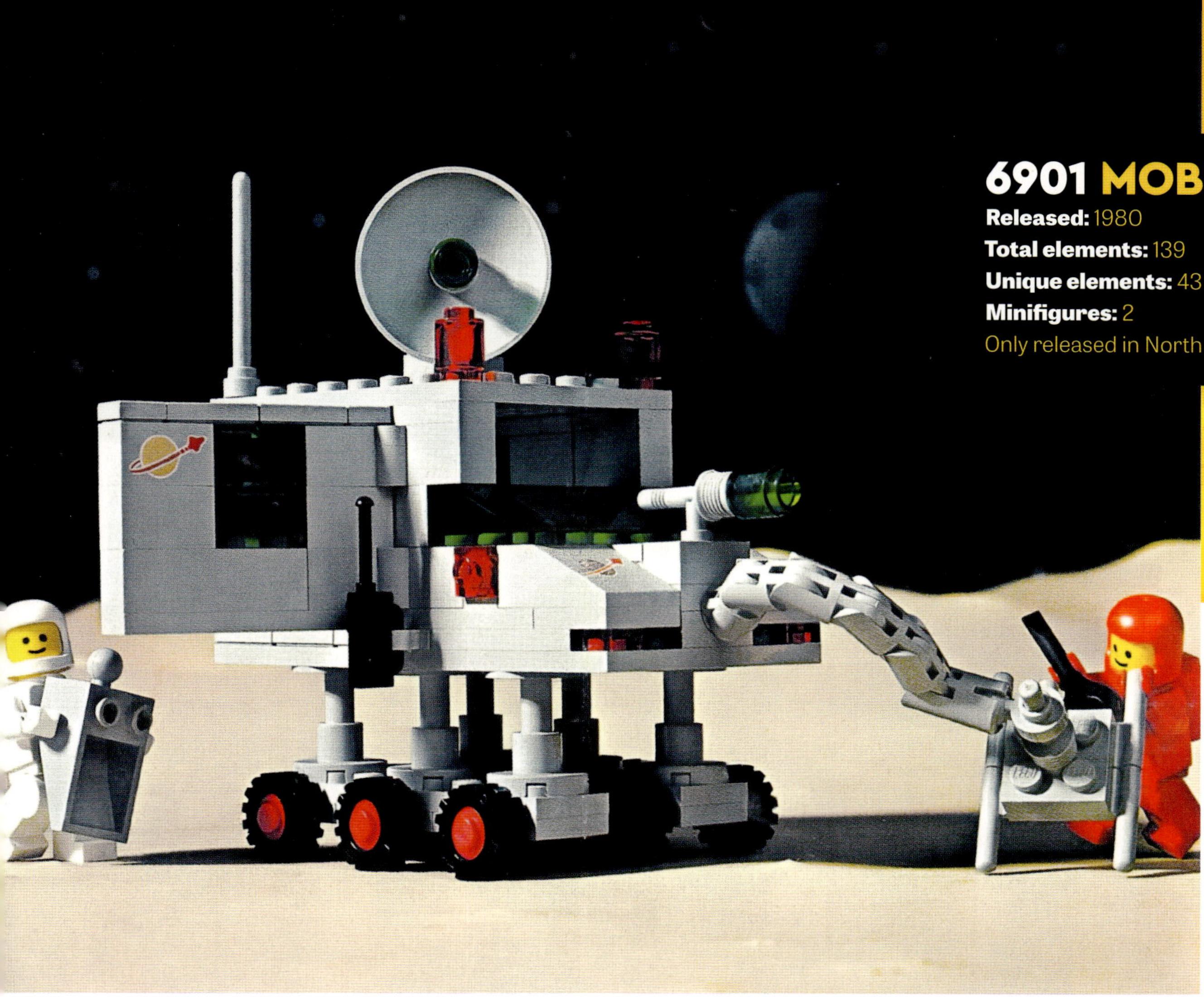

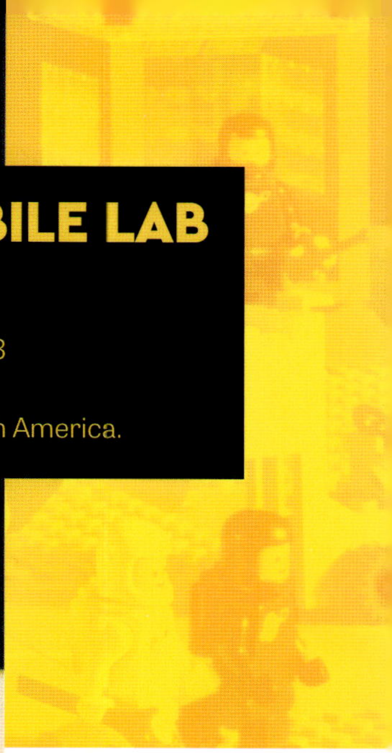

6901 MOBILE LAB

Released: 1980
Total elements: 139
Unique elements: 43
Minifigures: 2
Only released in North America.

There are many mysterious aspects to this set, such as the elevated body, the windscreen that doesn't actually cover the driver's face, and the doors on both the left and right sides—revealing a laboratory that contains no equipment! The design certainly has a ton of personality, however. The arm pieces are used again, with the fun addition of a second attachment to replace the shovel piece: a detector, made of handlebars and a space torch.

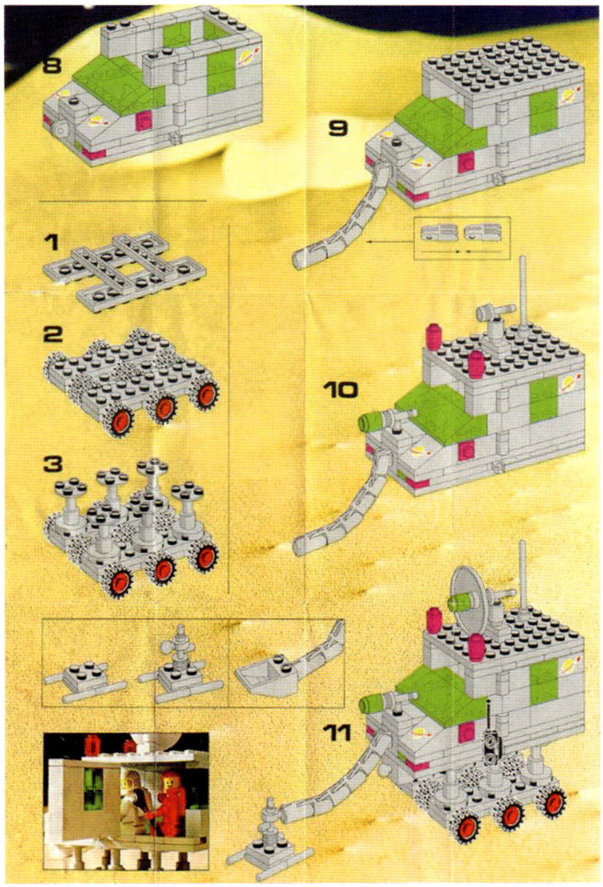

The mineralogist and the geologist trundled gently across the landscape in their newly constructed deep detection and extraction lab, allowing the yttrium fluorescence locator enough time to sweep the surface. They stopped at the first hint of a positive reading, and the mineralogist turned to the back of the vehicle to tap the seemingly blank wall, activating the atomic emission spectroscope controls. X-rays shot through the six optical cavities beneath their feet, and within moments, data flowed back. "I was right!" She turned to her colleague in wonder. "This planet has an yttrium core!" They leapt out of the doors simultaneously, rushing to the front of the proboscis to swap out the locator head for the digger.

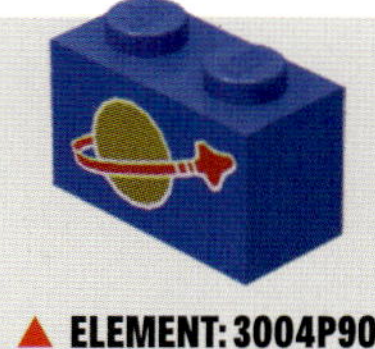

▲ **ELEMENT: 3004P90**
INTRODUCED: 1980

*The **brick 1 x 2 with Classic Space logo pattern** was usually blue but was issued in light gray only in this set and 894 Mobile Ground Tracking Station a year earlier, making it an element highly sought after by LEGO® Space fans.*

6970 BETA I COMMAND BASE

Released: 1980
Total elements: 270
Unique elements: 93
Minifigures: 4

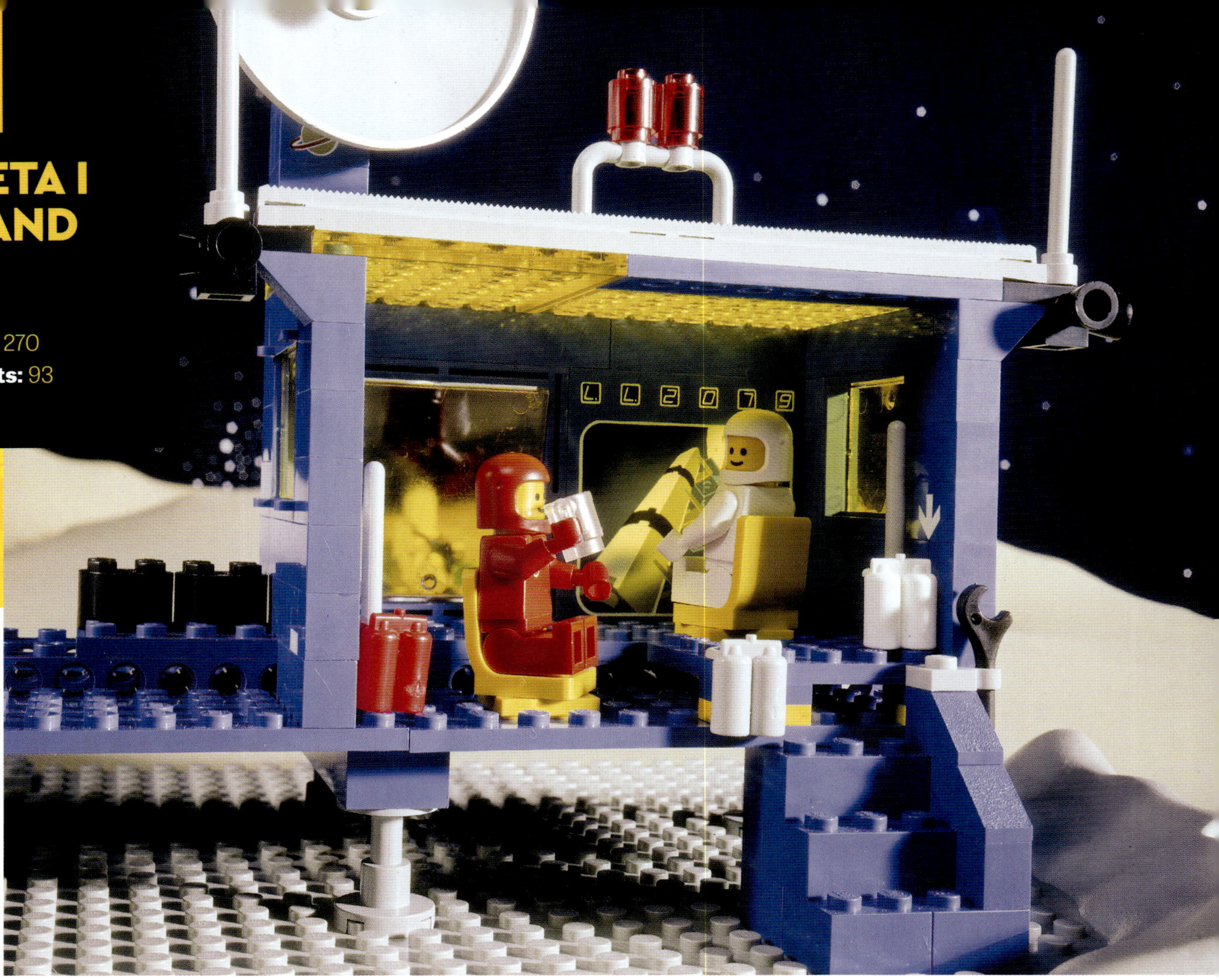

▶ One of the B models featured on the box.

ASTRONAUTS CAN TAKE A NAP BETWEEN SHIFTS OR ENJOY A HOT BEVERAGE WHILE ON DUTY, AND THERE'S EVEN A FLAG PROUDLY BEARING THE LEGO® SPACE LOGO.

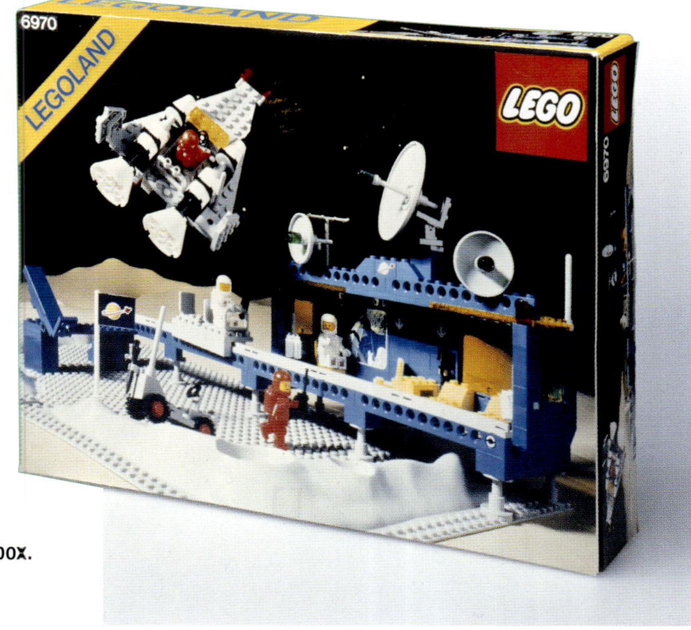

▶ Original mockup of box.

The space base definitely got upgraded in 1980! With ninety more pieces than 926 Command Centre and spread across two crater plates, 6970 Beta I Command Base added a spaceship, a launch pad, and even a monorail. This was the only 1980 set to include the original LEGO Space color scheme of blue with transparent yellow, but the craft looks quite different, sporting some black-and-white elements as well as wingtips. There are delightful details in this set: astronauts can take a nap between shifts or enjoy a hot beverage while on duty, and there's even a flag proudly bearing the LEGO Space logo.

They call Beta I "the base that never sleeps." With good reason: when you're protecting the largest known natural source of yttrium, vigilance is of the essence. The launch of LL 2079 with their latest shipment was going well until halfway through the countdown, when an unidentified craft was spotted on the video relay. Quick as a flash, the pilot threw down her space mug and leapt into the monorail carriage. She reached her interceptor in an instant! Once base gave the all clear, the turbo thrusters would have her there in no time . . . "Hello, base? You did wake up, right?"

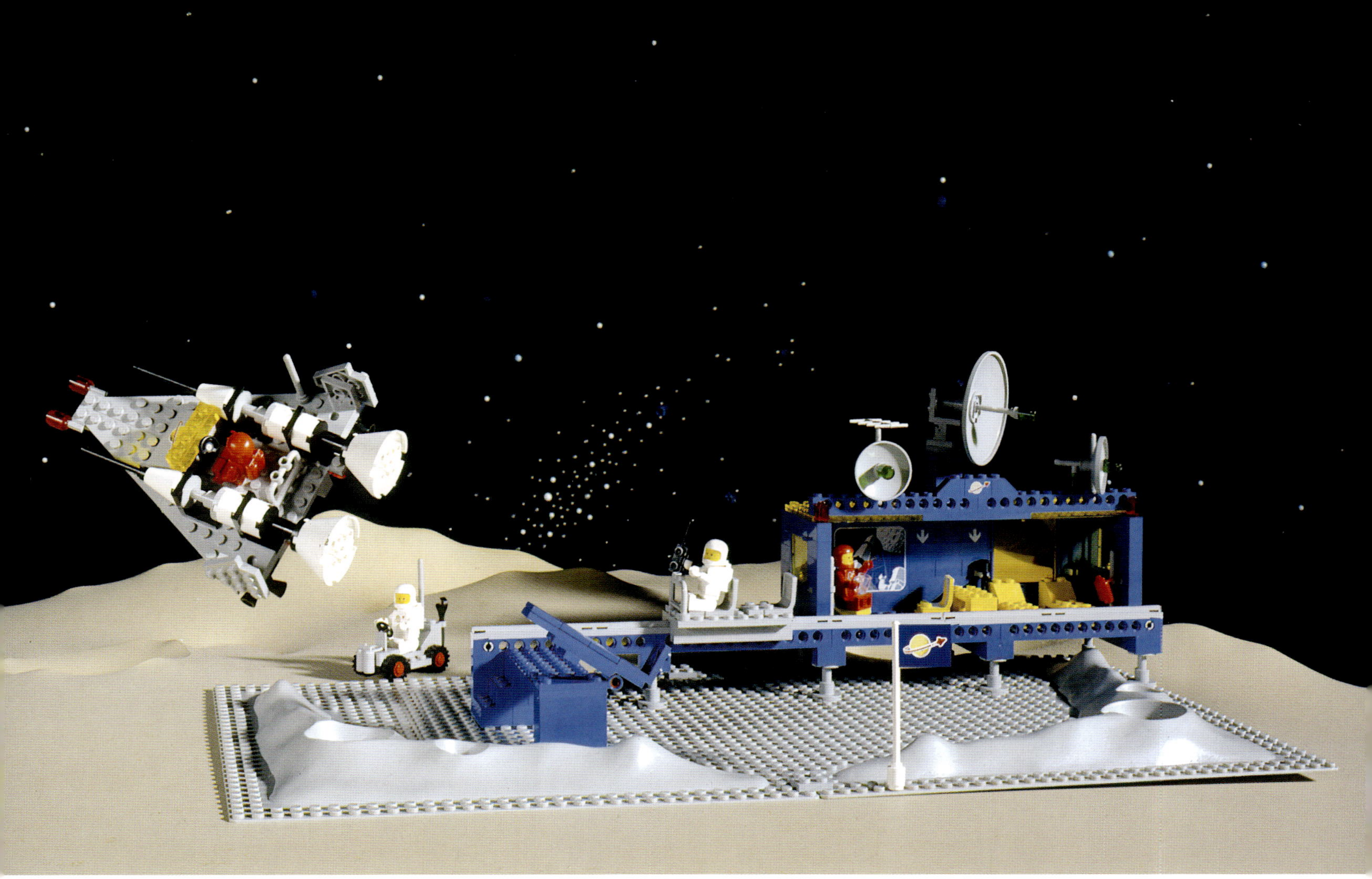

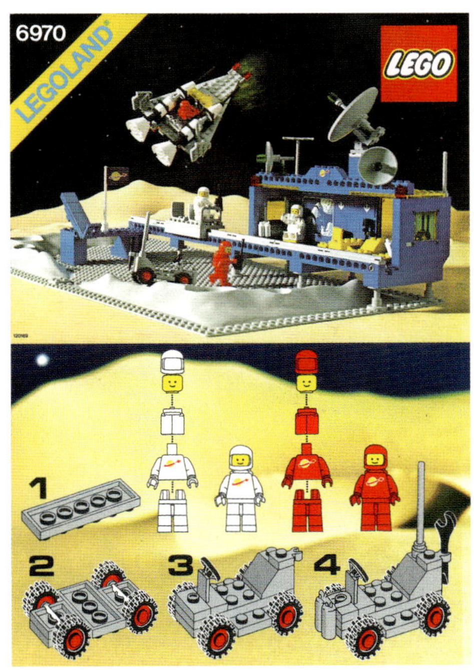

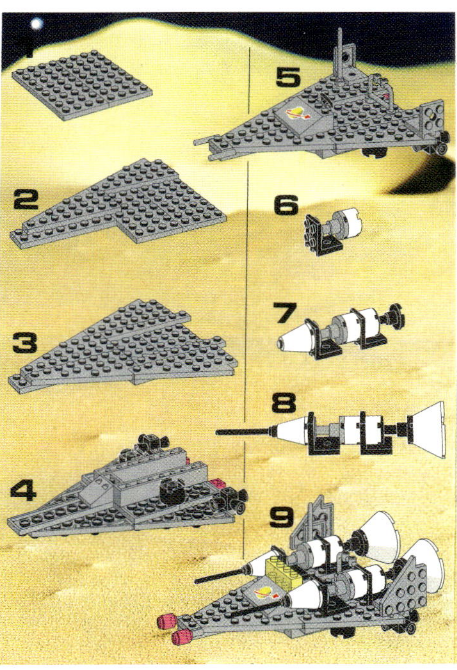

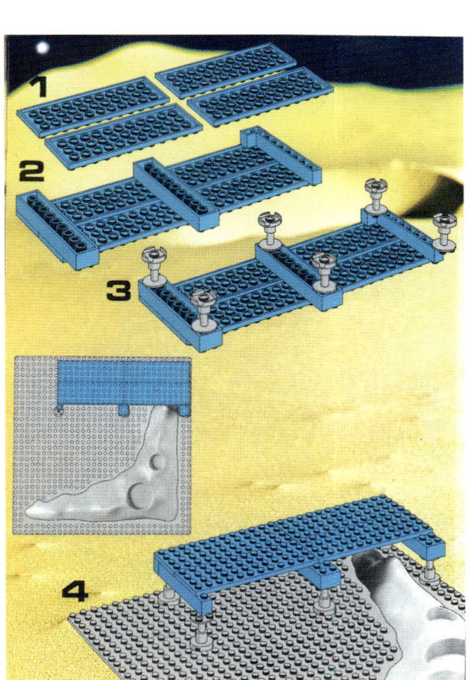

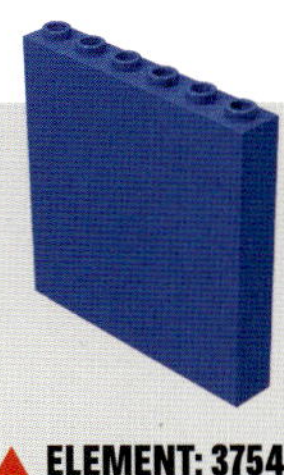

▲ **ELEMENT: 3754**
INTRODUCED: 1978

Brick 1 x 6 x 5 *was first seen in the LEGO® Homemaker line, where stickers could be applied to its large surface. LEGO Space sets actually printed onto the element, showing scenes on a video relay from LL 2079. In 1979, the screen in 926 Command Centre had shown a minifigure astronaut on a spacewalk to a satellite; now 6970 Beta I Command Base included a liftoff scene. Strangely it was another decade before this large brick, plain or decorated, appeared in any other LEGO set.*

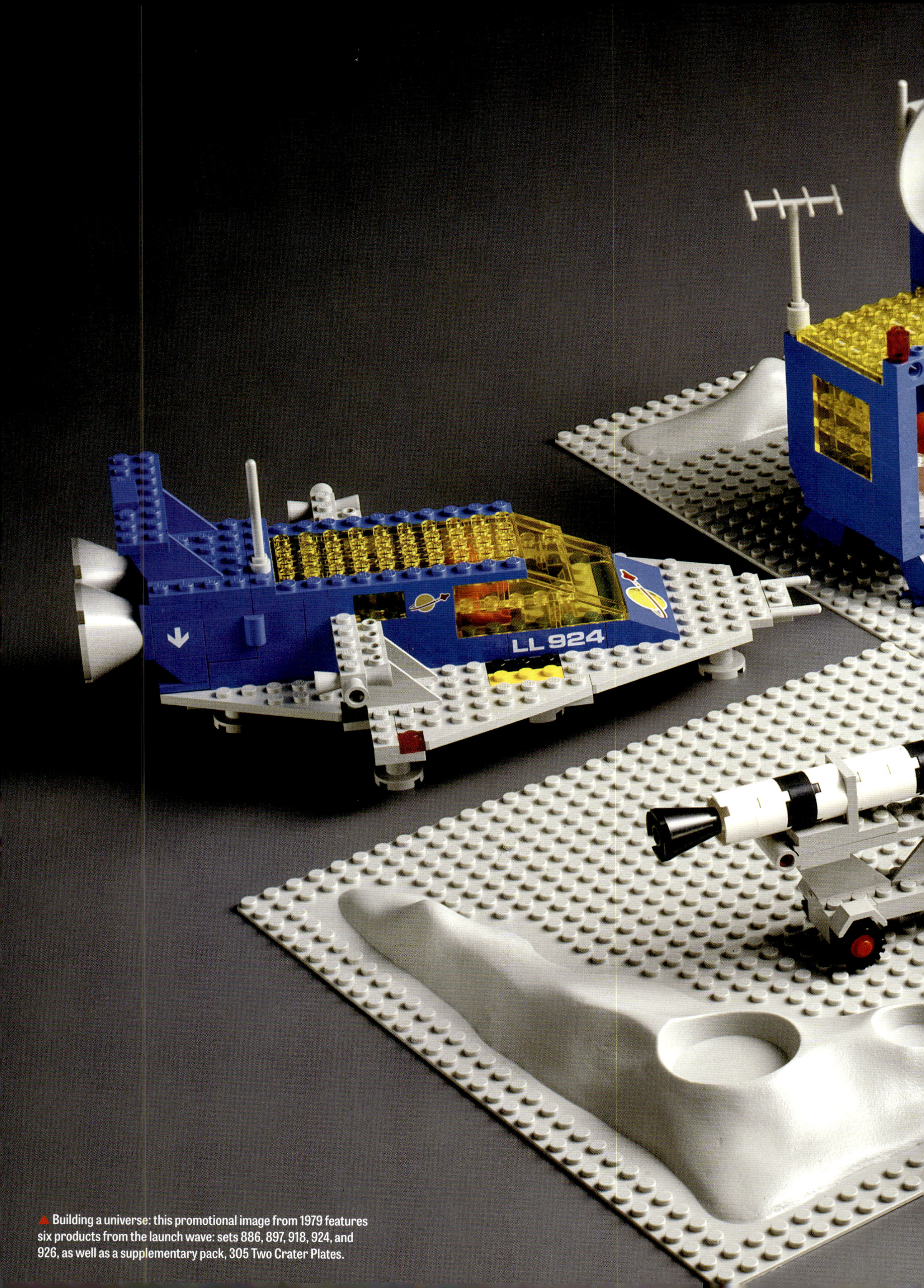

▲ Building a universe: this promotional image from 1979 features six products from the launch wave: sets 886, 897, 918, 924, and 926, as well as a supplementary pack, 305 Two Crater Plates.

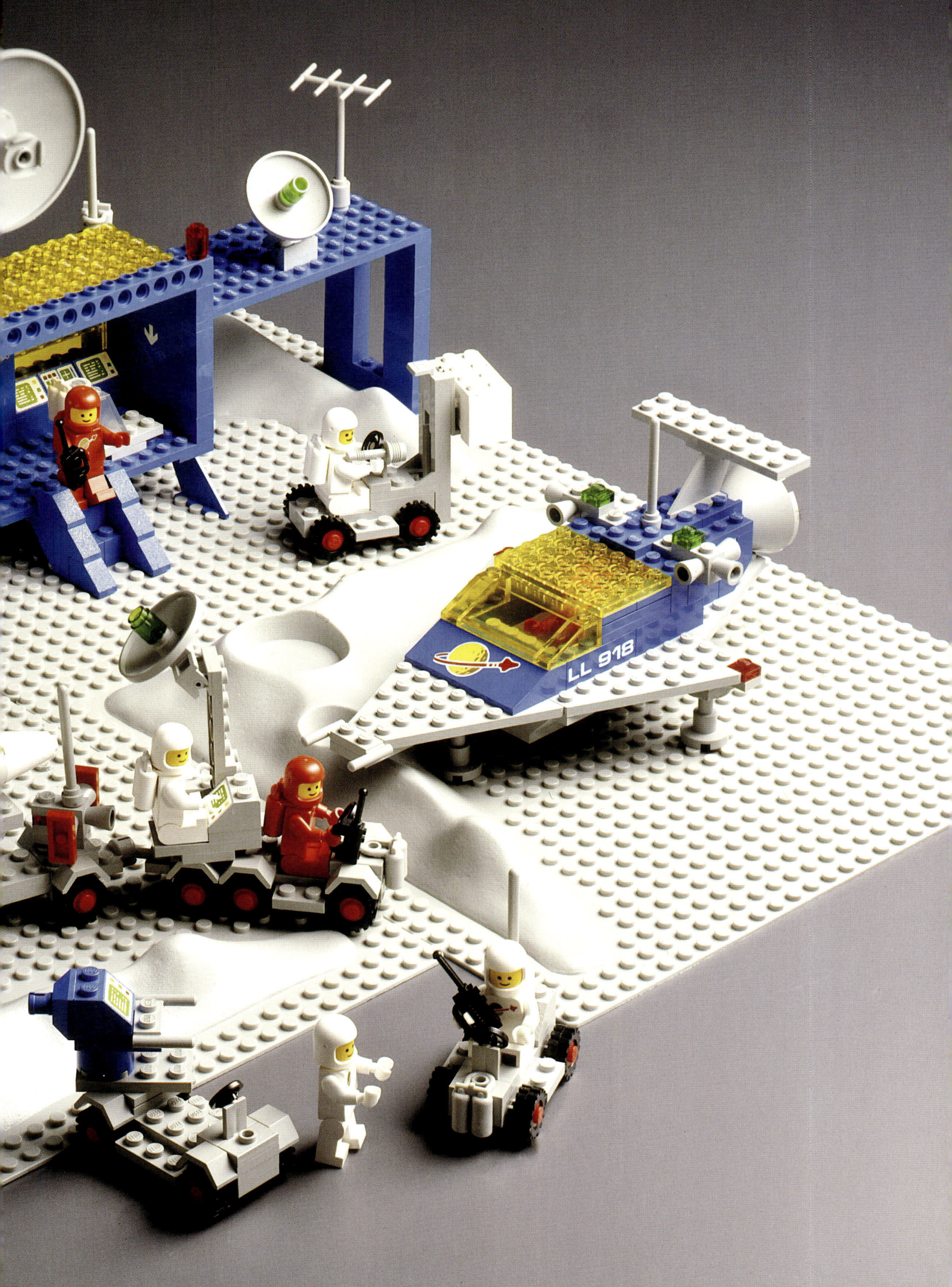
LL 918

6861 X-1 PATROL CRAFT

Released: 1980
Total elements: 55
Unique elements: 25
Minifigures: 1

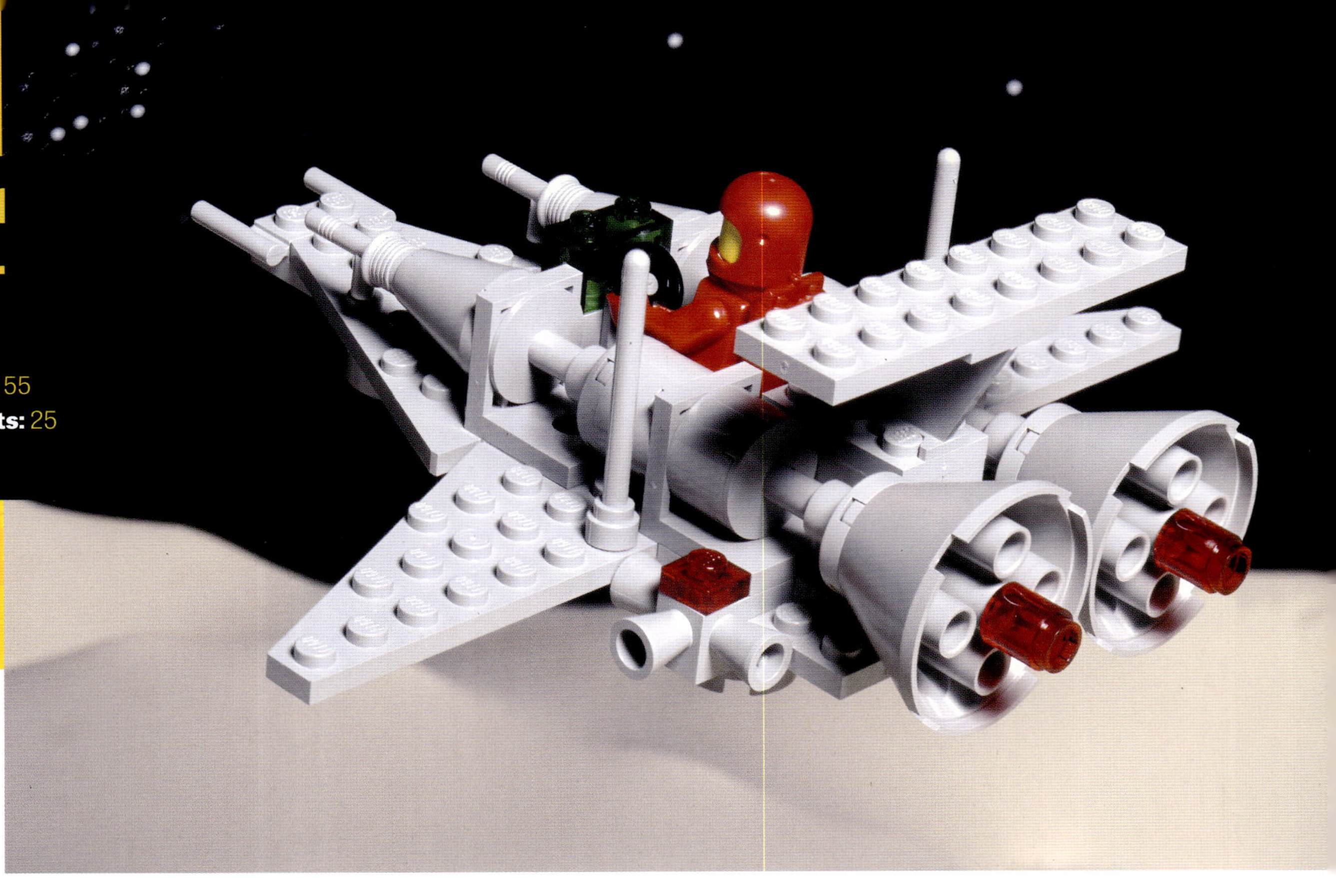

▶ This dramatic B model shown on the back of the box reveals how the cone 2 x 2 element was designed to match the loudspeaker.

The 1980 range did not add any new full-fledged spaceships with enclosed cockpits, but this new scooter design was an exciting development. The jagged leading edge and the exposed twin engines, along with other details at the rear, suggest technological detail—called "greebling" in the special effects industry (and by LEGO® Space fans). The transparent yellow of 1979 has been updated to transparent green, although it is questionable whether the 1 x 2 brick is actually meant to represent a windscreen.

"X-1 to Beta I. Yes, I'm seeing it now." The tiny speck in space, far above Proxima b, was moving quickly. With a tug on the steering wheel, he arched his patrol craft upward, confident that the high lift-over-drag ratio of this craft would be able to get him close enough to intercept. Could this be first contact? Might they know this planet was rich in yttrium? Suddenly the target began moving faster, and his green readout screen flashed red. "Computer, engage megaboosters! Recalculate interception path!" He braced himself as he felt the fuel cylinders grind into action to either side of him, and with a jolt he was hurtling out of the stratosphere.

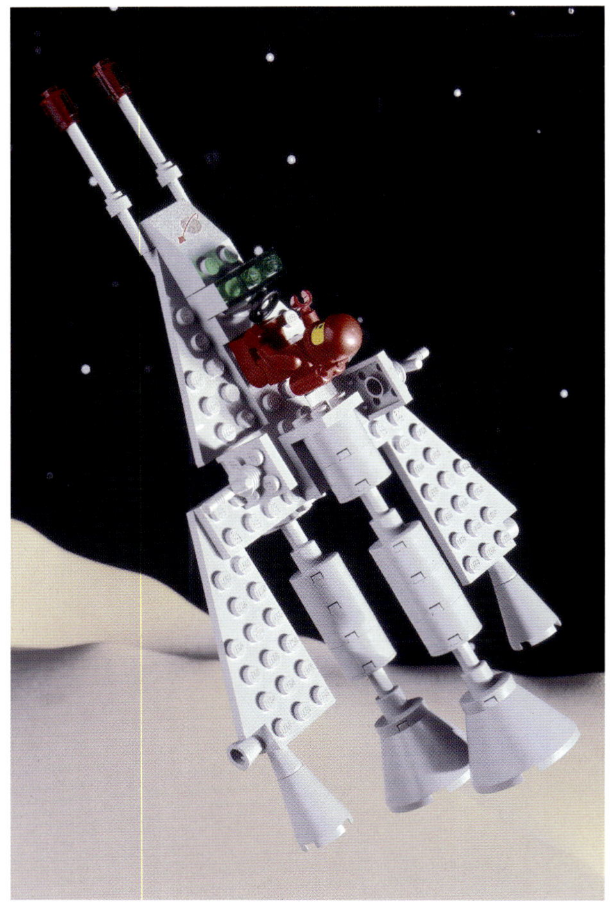

*The humble **support 2 x 2 x 2 stand** was another practical element created for LEGO Space, and having been somewhat hidden under the three spaceships of 1979, it enjoyed more prominence in 1980 on this craft, as well as the one in 6970 Beta I Command Base. Cleverly positioned between other 2 x 2 elements, it takes on the appearance of pipes connecting parts of an engine.*

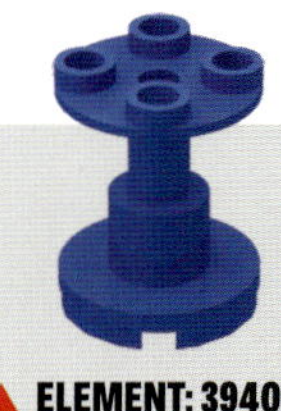

▲ **ELEMENT: 3940**
INTRODUCED: 1979

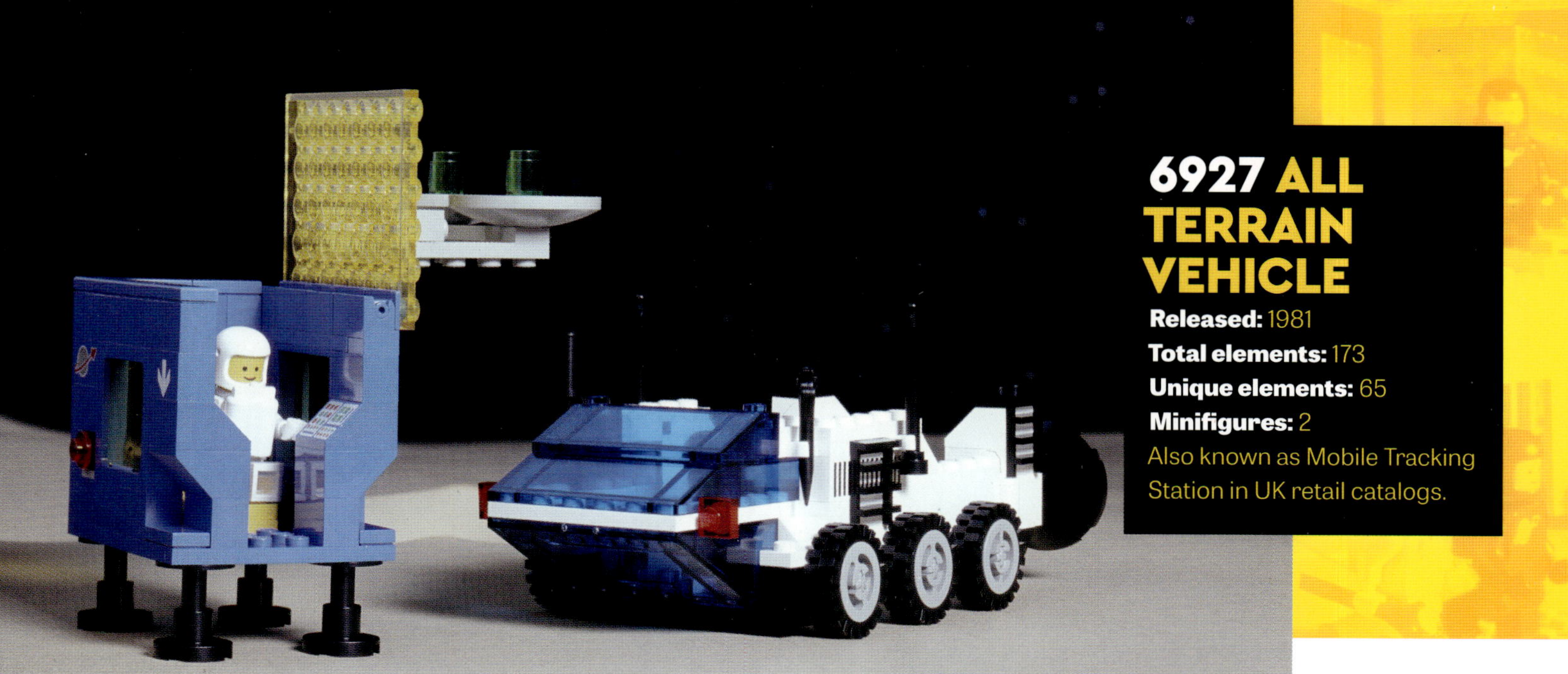

6927 ALL TERRAIN VEHICLE

Released: 1981
Total elements: 173
Unique elements: 65
Minifigures: 2
Also known as Mobile Tracking Station in UK retail catalogs.

1981's mobile lab takes things to a whole new level, providing a rocket-powered six-wheel-drive vehicle with enough space to carry a small building! The "classic" color scheme of blue with transparent yellow is used for the lab, while the vehicle sports the new palette of white and black. A new windscreen color is added: transparent blue. A mere 19 of its 173 parts are gray.

The need for rapid, detailed hydroxylamine analysis in the field inspired a creative solution from the engineer: build a U-shaped truck that could wrap around the lab, fitting like a glove to its contours, to drive it wherever it was needed. She couldn't figure out how best to secure the rear, until she was inspired by spacecraft design: simply mount the exhaust cones on hinges. To provide the driver with maximum visibility for avoiding the crystal outcrops so prevalent on Alpha d, she designed a revolutionary new ground-to-ceiling windscreen.

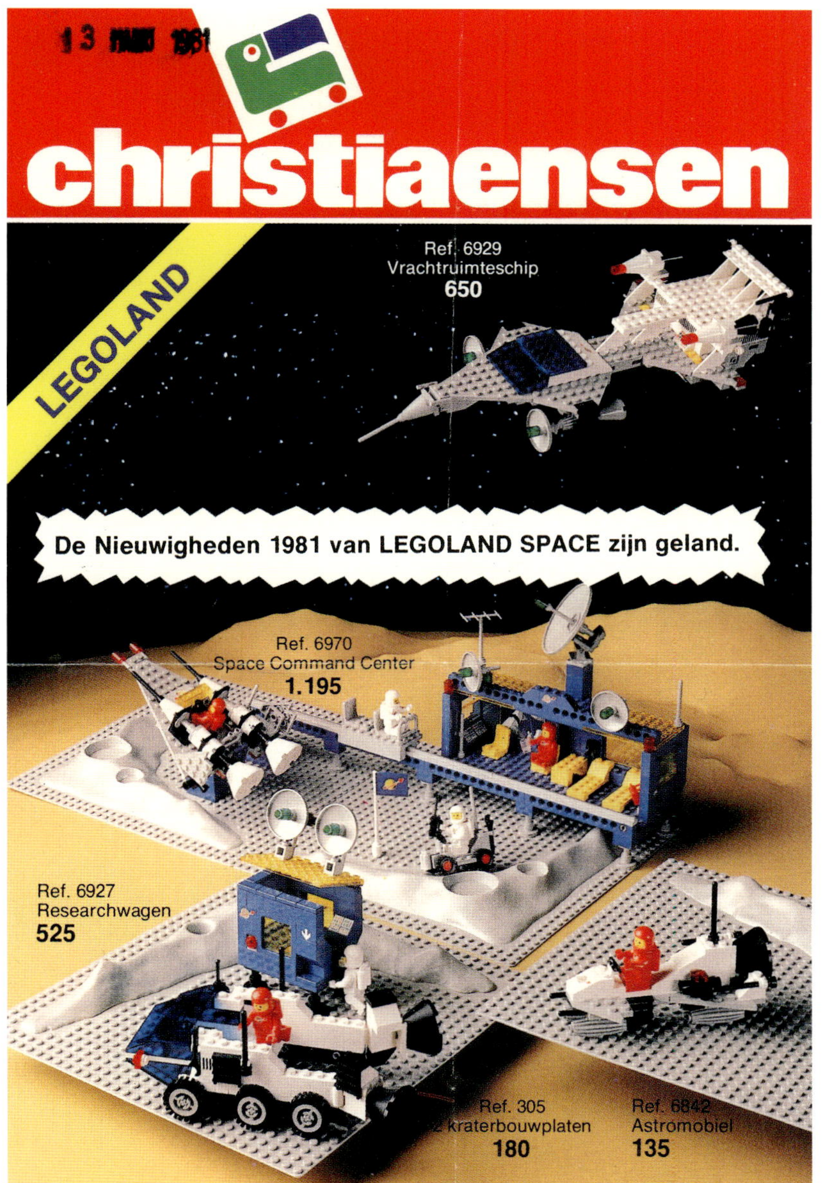

A 1981 catalog from Belgian toy store chain Christiaensen.

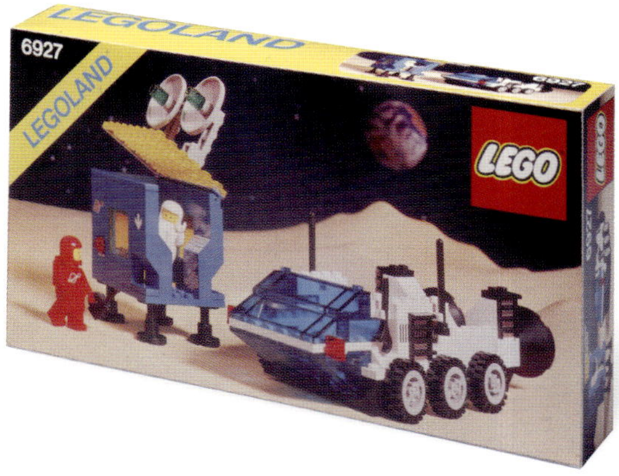

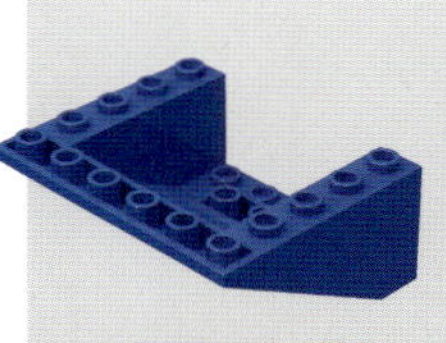

▲ **ELEMENT: 4228**
INTRODUCED: 1981

*For a set with a lot going on, the new **inverted slope 33° 5 x 6 x 2** still manages to draw the eye. Only ever used in LEGO® Space, this large element was eventually cast in six colors across eight sets.*

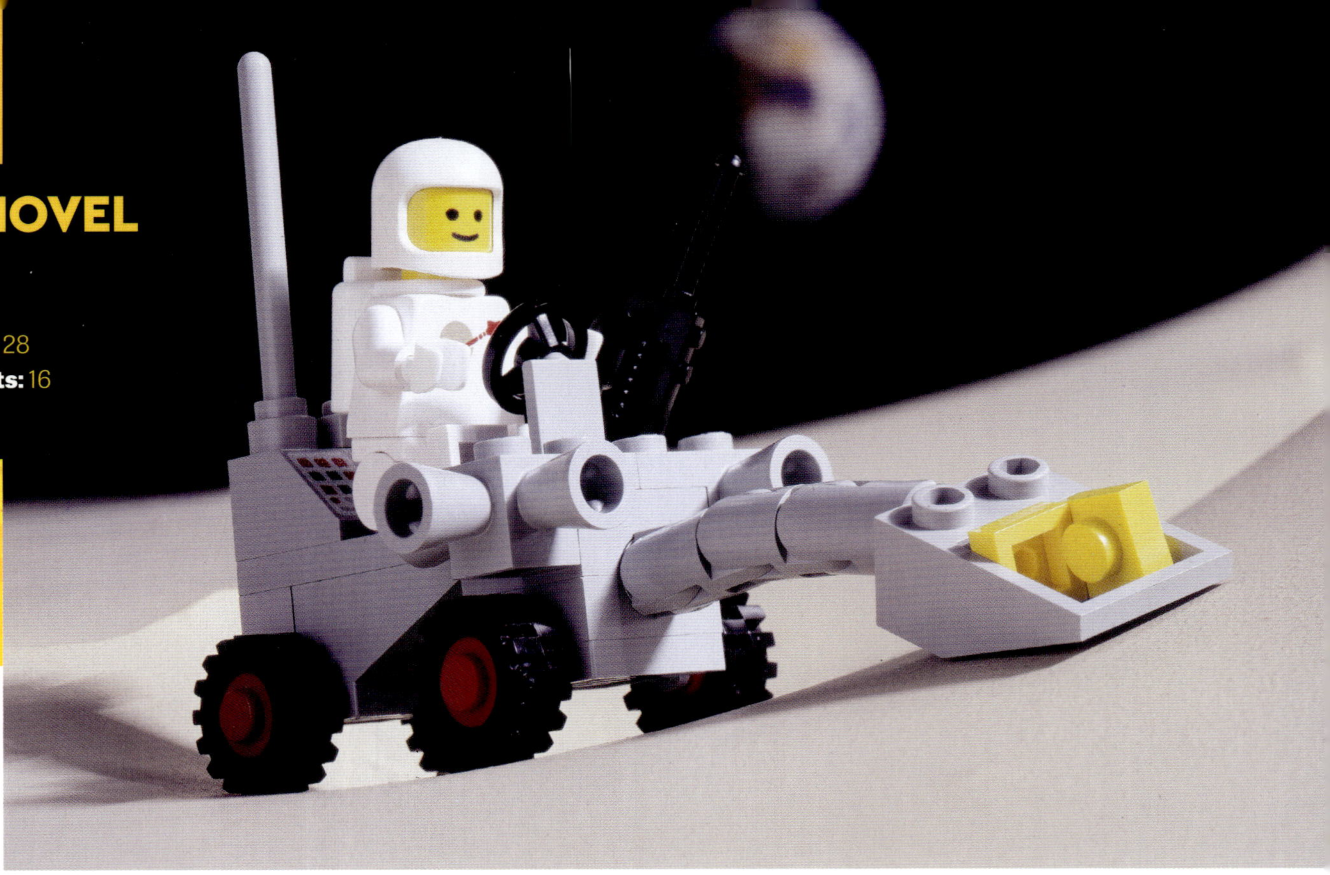

6821 SHOVEL BUGGY

Released: 1980
Total elements: 28
Unique elements: 16
Minifigures: 1

A new kind of vehicle appeared in the second year of LEGO® Space: the space digger. The larger arms introduced in LEGO® Homemaker buildable figures in 1974 get a new lease on life here, employed as hinge joints for the shovel arm. Strangely, the control panel is behind the driver. Space positioning rockets make a surprising appearance on a vehicle, and not for the last time.

The mission specialist thundered across the barren landscape of Proxima b, sending dust flying everywhere—except onto her vehicle, as she had already activated the front-mounted hydropumps. Suddenly the vehicle came to a stop, and a calm voice emitted from the panel behind her: "You have arrived at your destination." She certainly had; there were yttrium aluminum garnets literally sticking out of the soil. "Computer, activate hydraulic mode," she said, and with a satisfying jolt the steering wheel became a joystick. The bucket plunged into the soil, the crystals glinting in the yellow and red of the three suns of Alpha Centauri.

▼ Original mockup of box.

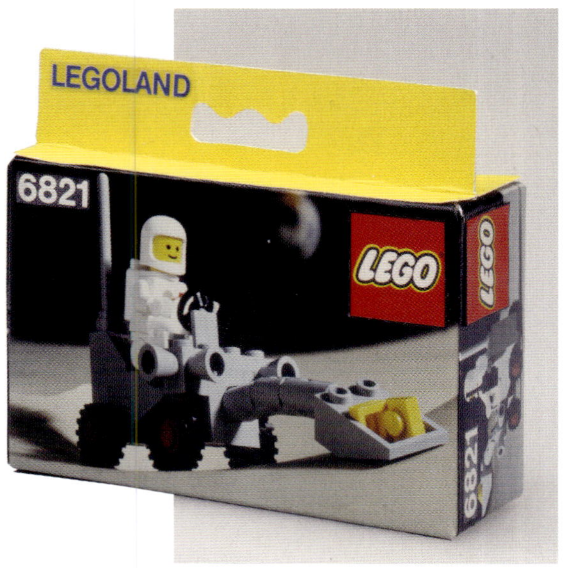

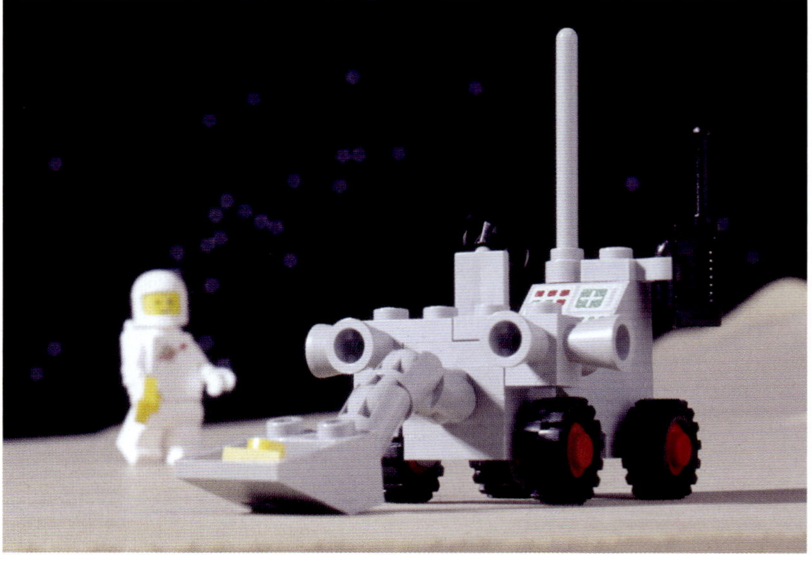

*With the buildable figures' arms forming a hinged connection system, the digger bucket, **inverted hollow slope 33° 3 x 2 with tow ball**, was designed to fit on the end, using the same small ball joint connection used for the buildable figures' hands. The regular inverted slope 33° 3 x 2 (3747) was introduced in 1980 as well and is also present in this model.*

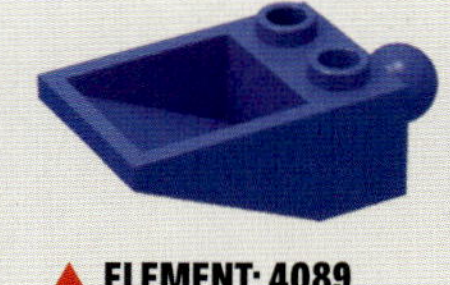

▲ **ELEMENT: 4089**
INTRODUCED: 1980

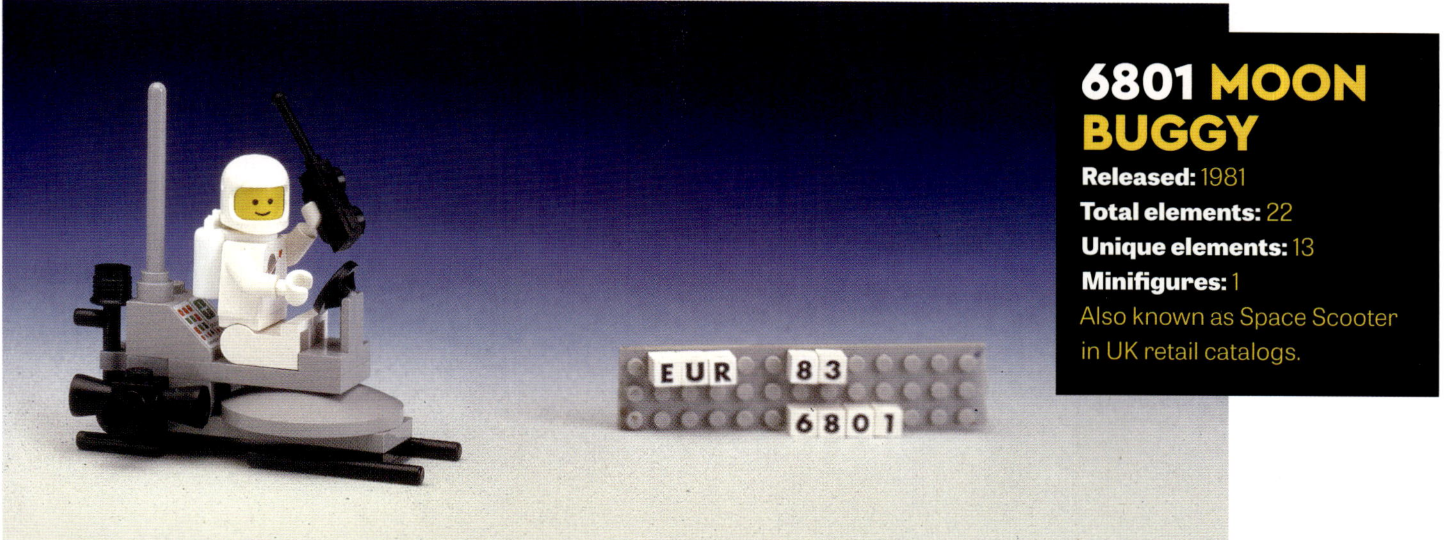

6801 MOON BUGGY

Released: 1981
Total elements: 22
Unique elements: 13
Minifigures: 1
Also known as Space Scooter in UK retail catalogs.

The new small vehicles for 1981 portrayed familiar subjects but with a notable difference—no wheels! As such, of the names this set is known by, Space Scooter seems the better choice. Color changes which first appeared on the craft in 6970 Beta I Command Base became standard in 1981: in this set, some black LEGO® Space elements break up the monotone gray of previous years.

Traversing the landscape of a planet rich in hydroxylamine crystals had proved cumbersome with traditional buggies, so the Gamma Mission engineer had developed surpulsion: a vertical thrust technology powered by a hydroxylammonium nitrate monopropellant blend. Tentatively, the engineer fired up the surpulsor disk below the chassis, and with a gentle hum, the vehicle began to rise. Satisfied, he set out for a test flight.

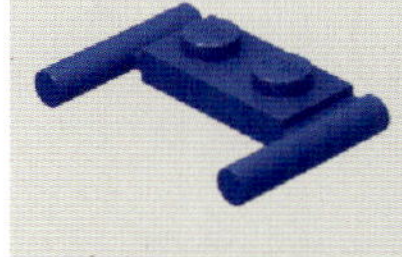

▲ **ELEMENT: 3839**
INTRODUCED: 1978

*A common gray element in LEGO Space now appears in black: **modified plate 1 x 2 with bar handles**, used as landing skids. The mold design soon received a tweak to lower the handles slightly and flatten the curved ends, aligning it better with the LEGO® System in Play grid.*

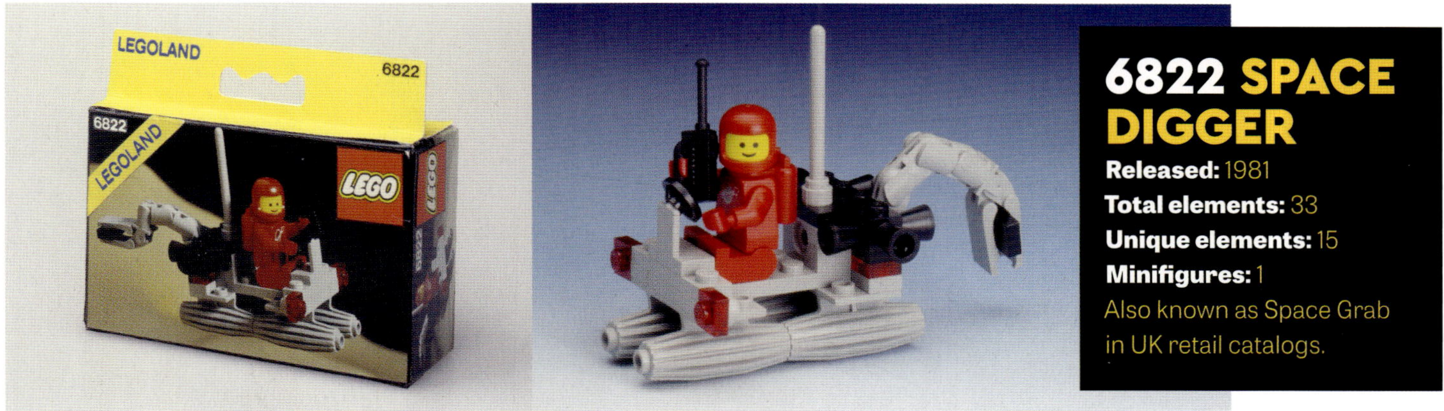

6822 SPACE DIGGER

Released: 1981
Total elements: 33
Unique elements: 15
Minifigures: 1
Also known as Space Grab in UK retail catalogs.

The digger returned, but with exciting new features—the ability to hover using the new engine piece, and a shovel bucket that now doubled as a jaw-like clamp.

The scientist scooted over the sandy landscape of Alpha d, heading toward the steep crystalline mountains of the western sector. How could he have been so clumsy? With a slight bump, his hover digger brushed a shard jutting from the ground. He activated the aerodyne mode, and the craft lurched off the ground, nose down. Quickly, he activated the reaction control system thrusters to restore balance and soon reached the spot where he had dropped his tablet. There it was! Nestling precariously between two crystals. Setting the craft to hover mode, he turned to operate the digger, maneuvering its powerful jaws gently to clasp the device. Phew! No one would ever learn of his mistake.

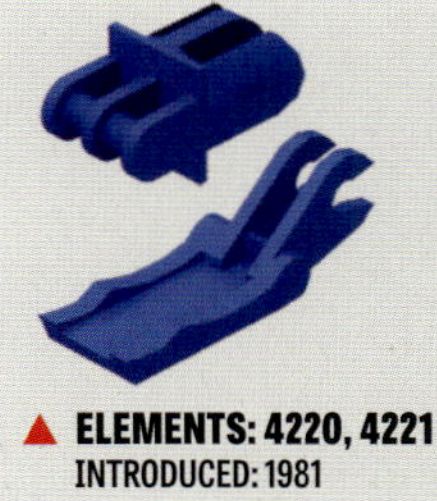

▲ **ELEMENTS: 4220, 4221**
INTRODUCED: 1981

*The new excavation tool consisted of two **arm grab jaw** elements attached to one **arm grab jaw holder**, which in turn was attached to the existing LEGO® Homemaker buildable figures' arm connector. It enjoyed occasional yet regular use in LEGO® Space sets, eventually appearing in other lines from 1987 onward.*

6929 STAR FLEET VOYAGER

Released: 1981
Total elements: 246
Unique elements: 86
Minifigures: 1
Also known as Space Transporter in UK retail catalogs.

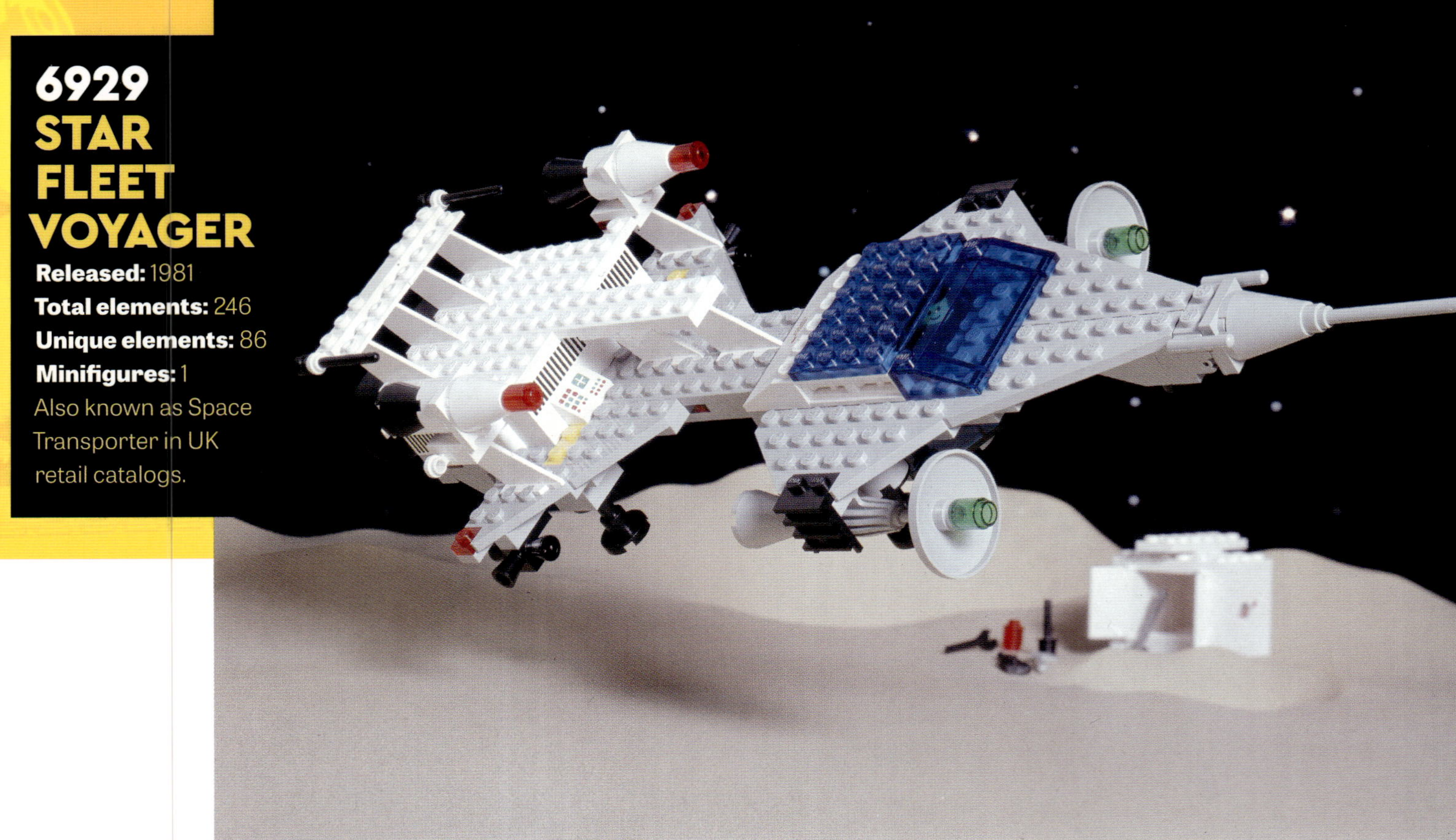

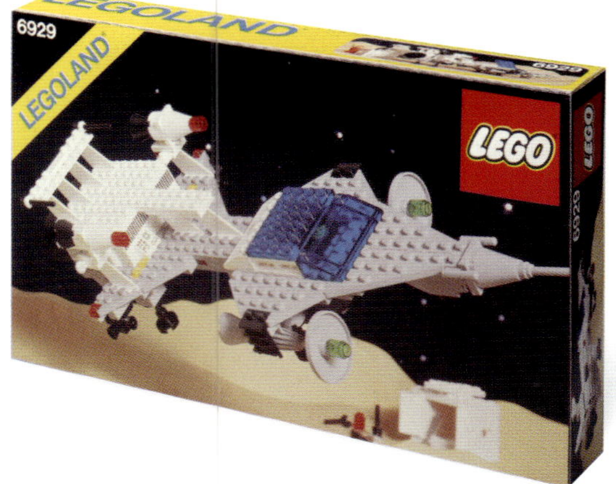

The first large spaceship since set 928 featured the hallmarks of 1981 craft: double-wing design; black highlights, including horizontal antennas; and the smart new look of white fuselage with transparent blue cockpit. And what a cockpit! Inverted slope 33° 5 x 6 x 2 provides the craft with not only an interesting profile but also a beautiful underside. Like 6927 All Terrain Vehicle, there is no floor at the rear, and the cargo container is simply clasped by the hinged empennage, making this the first ship to be made of segments. This modular approach would develop over the coming years. For the first time in LEGO® Space, the supplied minifigure accessories included the shovel, axe, and signal paddle.

The pilot could see the ground crew cheering as LL 6929 lurched sharply upward from the landing pad. She raised the droop nose back to the central position, and as the craft left the atmosphere, she momentarily steeled herself for the long, lonely Delta Mission to the satellite galaxy. As she engaged the twin hydroxylammonium nitrate engines, the jolt did not surprise her, but the asteroid belt that suddenly shot up in front of her certainly did. Why didn't she think to check the long-range multibeam radars first? "Computer, multiple targets, plot intercept!" A digital overlay appeared right across the blue canopy around her, as she enabled the yttrium lasers and began to fire.

THE CARGO CONTAINER IS SIMPLY CLASPED BY THE HINGED EMPENNAGE, MAKING THIS THE FIRST SHIP TO BE MADE OF SEGMENTS.

*The nose of the Star Fleet Voyager can be tilted downward, similar to the one on the iconic supersonic Concorde jet. This is thanks to the humble **hinge brick 1 x 2**, which initially only appeared in LEGO Space sets when introduced in 1979. It had been used as the means of opening the cockpit roofs, but that was about to change.*

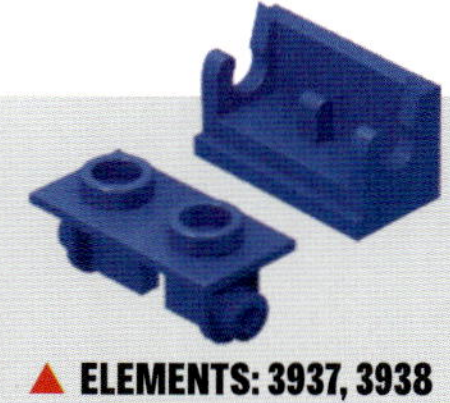

▲ **ELEMENTS: 3937, 3938**
INTRODUCED: 1979

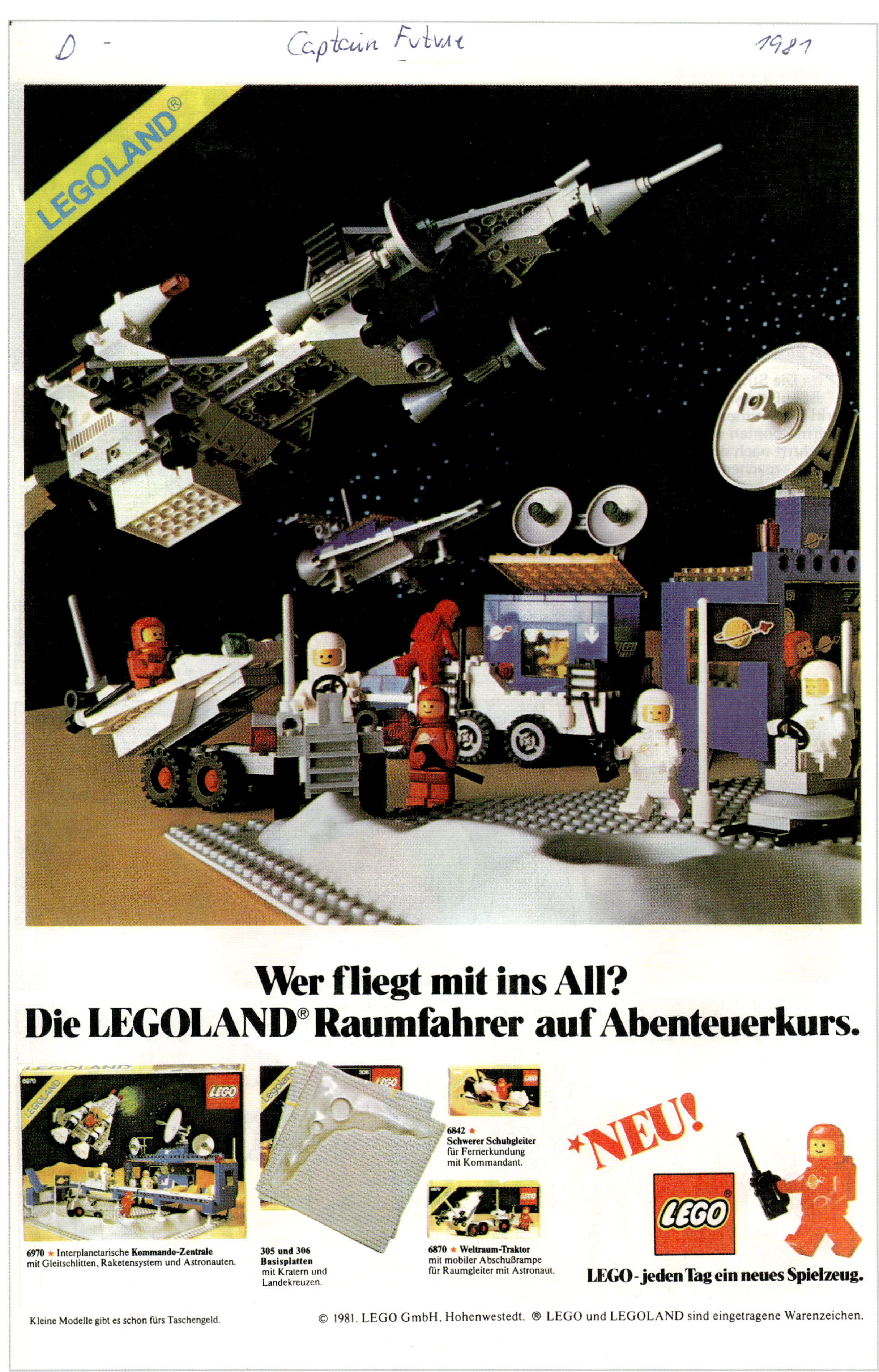

This German ad of 1981 reveals the exciting underside of set 6929. In the center, set 6927 is pictured with the lab loaded onto the vehicle.

6870 SPACE PROBE LAUNCHER

Released: 1981
Total elements: 60
Unique elements: 29
Minifigures: 1
Also known as Spacecraft Launcher in UK retail catalogs.

The rocket launchers and small scooters of previous years are here combined neatly into one small set. The ship clearly matches the design ethos of others in the 1981 range but is notable for having almost no gray elements: a first for LEGO® Space. Another first is that the articulated buggy uses the connector hinge plate 2 x 4 with articulated joint, introduced by other lines in 1976.

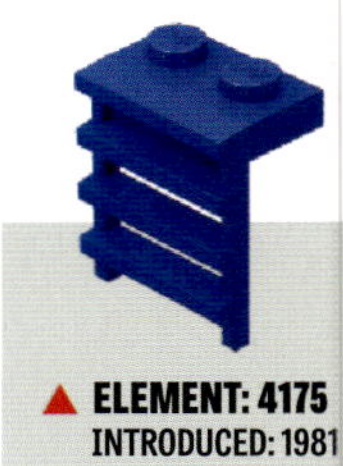

▲ **ELEMENT: 4175**
INTRODUCED: 1981

*The **ladder 1½ x 2 x 2** is another design that cleverly balances specialization and flexibility of use. Here it is used as a front grille on the buggy.*

The engineer realized that on the moon of Alpha d, it was more efficient to dispatch their scooter probes using launchers, achieving the necessary delta-*v* and freeing up the burden of carrying additional fuel. The pilot drove the mobile launcher down to the equator, angled the rear of the buggy to match the planet's rotation, and braced herself for what was certain to be a harsh takeoff.

6842 SMALL SPACE SHUTTLE CRAFT

Released: 1981
Total elements: 46
Unique elements: 25
Minifigures: 1
Also known as Inspection Spacecraft in UK retail catalogs.

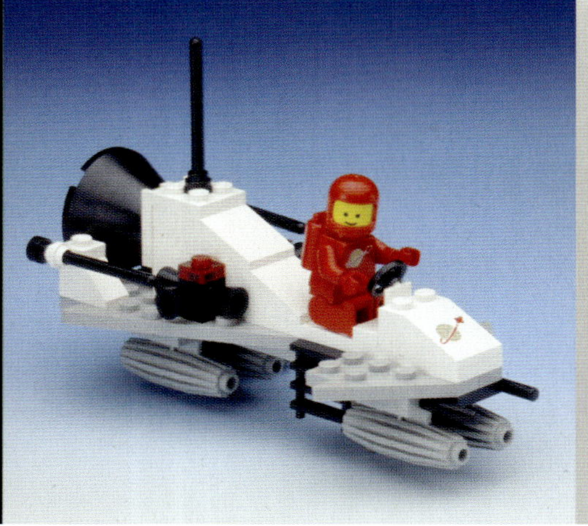

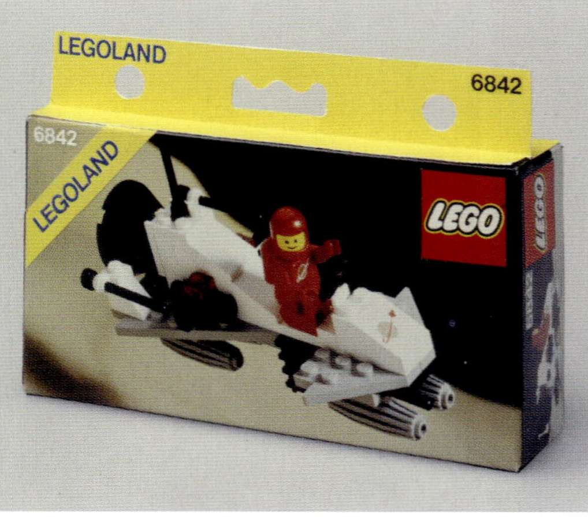

The color scheme hinted at by the craft in 6970 Beta I Command Base comes into full force here, with white fuselage and black highlights balancing out the gray. The craft has two distinct sets of wings, perhaps to mimic 1981's larger ship, 6929 Star Fleet Voyager. Unusually, antenna elements are positioned horizontally—another hallmark of 1981 sets.

LL 6842 proved to be the ideal craft for monitoring terrain on many of Alpha Centauri's planets and moons. Star Fleet pilots found that its control-canard design provided excellent pitch control during maneuvering, as well as maximum downward visibility between the wings. The new engines improved airflow and doubled as landing gear, and the zinc-aluminum-alloy fuselage reflected intense heat yet resisted corrosion on wet planets. With this technology, they could develop a whole new breed of craft.

__Engine strakes 2 x 2__ was an exciting new design based upon the original aircraft engine element (3475) from 1974: the smooth surface was cut with grooves, and connection possibilities were increased by a stud and socket at either end.

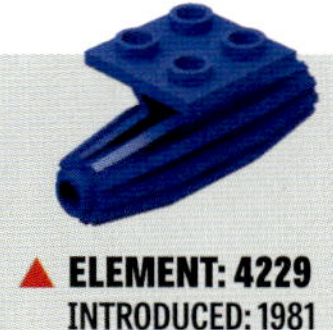

▲ **ELEMENT: 4229**
INTRODUCED: 1981

The yellow spacesuit appeared for the first time in 1982, the first new color since LEGO® Space launched with red and white. The figure is alone here, despite this well-appointed mining vehicle having controls for two operators. This is also the first set to feature the new balloon wheels, providing twelve in total.

The Delta Mission geologists were having a huge problem mining the mercury sulfide. The immense heat and high levels of nitric acid in the atmosphere of Cygni Bb meant that within minutes of exposure, the precious substance would yellow, turning into useless mercury nitrate. This new surface explorer, however, managed the whole process as the vehicle was slowly driven along. The robotic claw broke the soil and removed foreign objects, the three pneumatic extractors created a slurry and separated impurities, and finally the digger scooped up the Hg_2S. A disk-mounted gas chromatograph detected nitric acid concentrations around the vehicle and neutralized them with jets of sodium hydroxide.

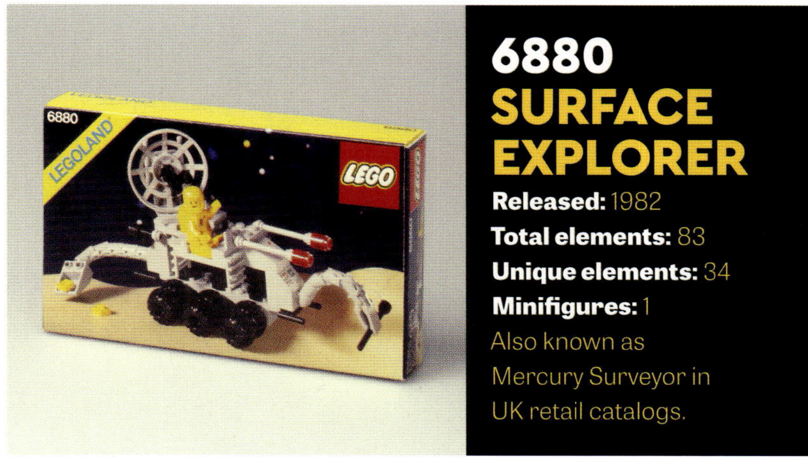

6880 SURFACE EXPLORER

Released: 1982
Total elements: 83
Unique elements: 34
Minifigures: 1
Also known as Mercury Surveyor in UK retail catalogs.

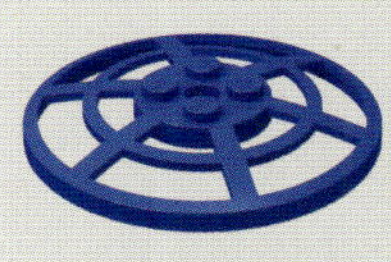

▲ **ELEMENT: 4285**
INTRODUCED: 1982

An exciting new dish joined the four- and eight-wide versions, but it looked very different. ***Inverted webbed dish 6 x 6*** *had a perforated surface, like radio telescopes. While this design feature may have limited its uses, it looked fantastic. A plain 6 x 6 dish was not introduced until 2002.*

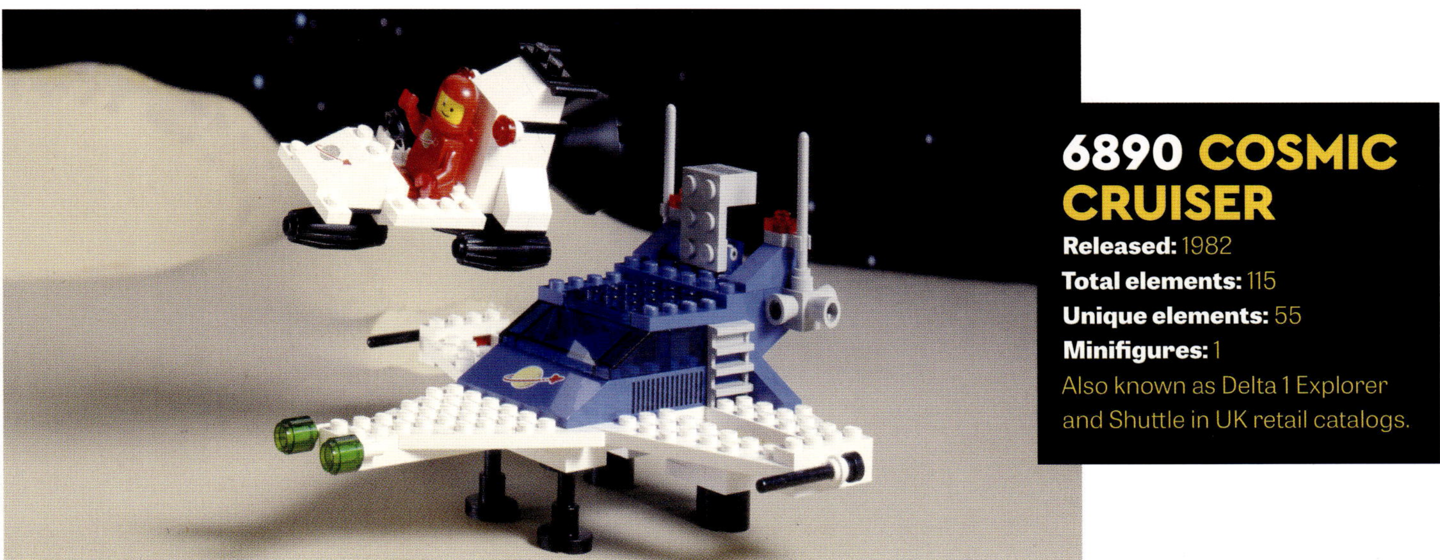

6890 COSMIC CRUISER

Released: 1982
Total elements: 115
Unique elements: 55
Minifigures: 1
Also known as Delta 1 Explorer and Shuttle in UK retail catalogs.

This midsized ship is a wonderful blend of earlier design concepts with fresh ideas. The white, black, and transparent blue palette of 1981 is married with the blue fuselage of 1979, and the wings feature a striking large gap. The concept of inserting cargo into a hollow rear end is taken a big step further: now the cargo is actually a secondary craft!

The pilot powered down his ship as he reached the thermosphere of this strange, glittering world. "Computer, recharge and observe." The orbit control subsystem rotated the craft slightly to face the nearby stars of Gamma Cephei while the solar wings came online. The instrumentation slowly began recording data from the planet below. He figured he might as well take a closer look in the meantime. "Computer, deploy shuttle." The rear clamp released, and he gently reversed the core of the Delta 1 out, leaving the shell safely in orbit as he shot down toward the ground like an arrow. As he approached the rocky surface, he powered down from the main thruster to the three scooter engines. "Computer, let me know when you're done up there. And don't miss me too much."

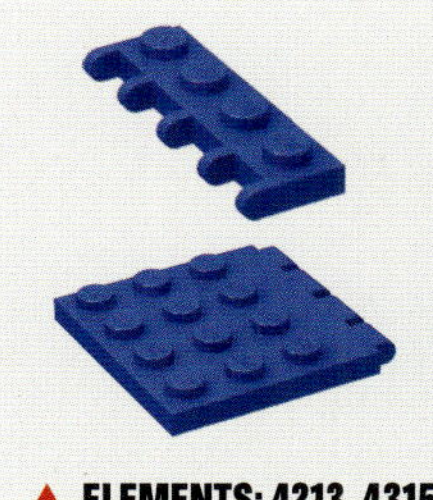

▲ **ELEMENTS: 4213, 4315**
INTRODUCED: 1981

For three years, LEGO® Space had relied on hinge bricks for cockpit access, but now the much more stable ***hinged vehicle roof 4 x 4*** *and* ***hinged vehicle roof holder 1 x 4 x 2*** *were available, having been introduced in LEGO® Town a year earlier. The holder element was also shorter than hinge bricks, allowing for the sides of the cockpit to include transparent bricks.*

6950 MOBILE ROCKET TRANSPORT

Released: 1982
Total elements: 210
Unique elements: 73
Minifigures: 2
Also known as X15 Satellite Launcher in UK retail catalogs.

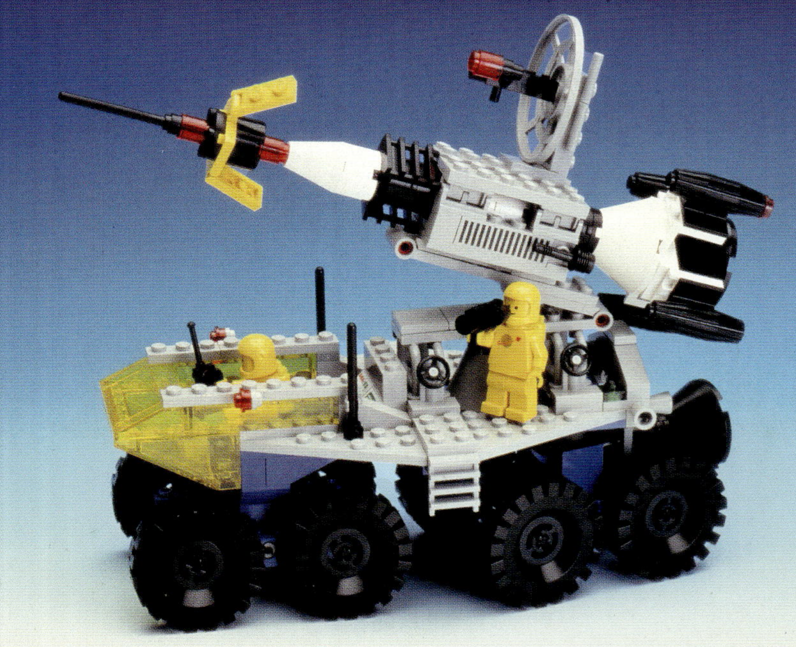

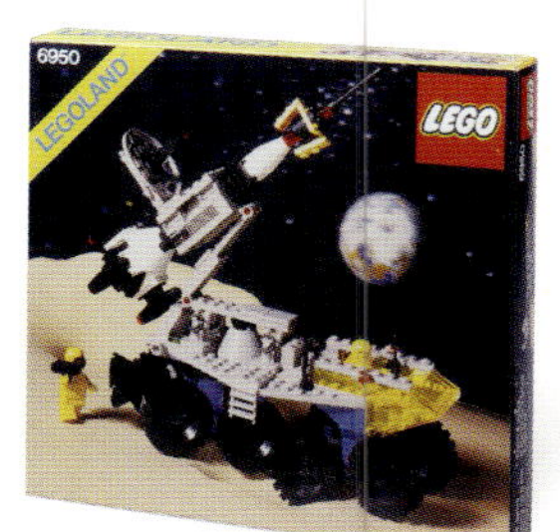

The ground launcher received a serious upgrade in 1982: chunky wheels with independent suspension, an articulated launch arm to enable a diagonal or vertical launch, and massive thrusters on the rocket fins! The design delivers realistic features within a fantastical vehicle; even the platform and cockpit are unusual designs loaded with play opportunities. The color scheme returns to the original 1979 palette, even down to the white-and-black rocket. Populating the vehicle were not one but two of the new yellow-suited astronauts.

It was critical to launch a replacement communications satellite around Tadmor as quickly as possible. As soon as the rocket was loaded onto the X15, the ground crew drove it directly to the launch zone—no need to skirt around the craters; this transport could handle any terrain without so much as scratching the payload! Safely in position, one astronaut climbed the gantry to unfurl the launch arm, placing the rocket vertically on the surface. The other lifted the optical emission spectroscope to her shoulder to ensure the electron density of the solar winds would not jeopardize the Delta Mission.

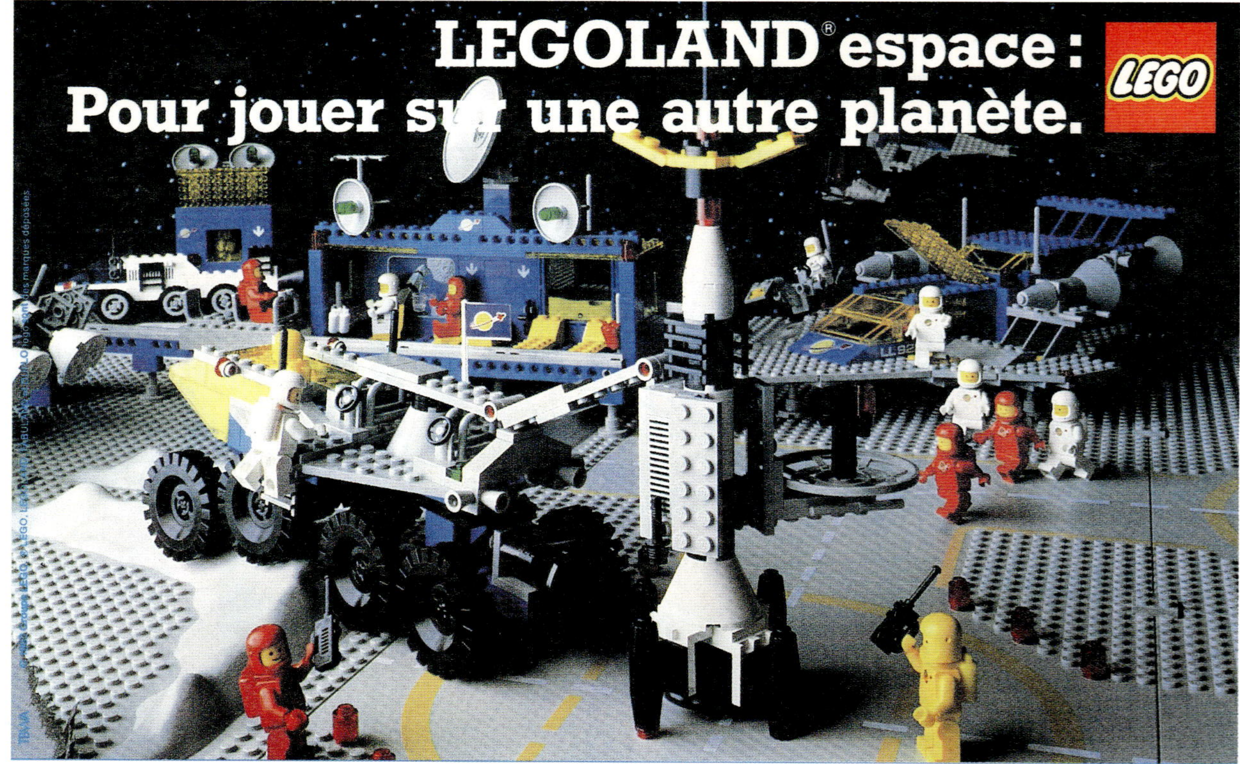

This French ad of 1982 shows the rocket deployed to vertical position.

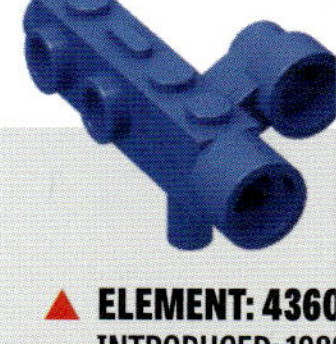

▲ **ELEMENT: 4360**
INTRODUCED: 1982

*The **minifigure utensil camera with side sight** may look specialized, but it has enjoyed a wide variety of uses in LEGO® Space and dozens of other lines, appearing in sets virtually every year since its inception. Here it is used both as a handheld device and as part of the radar dish assembly.*

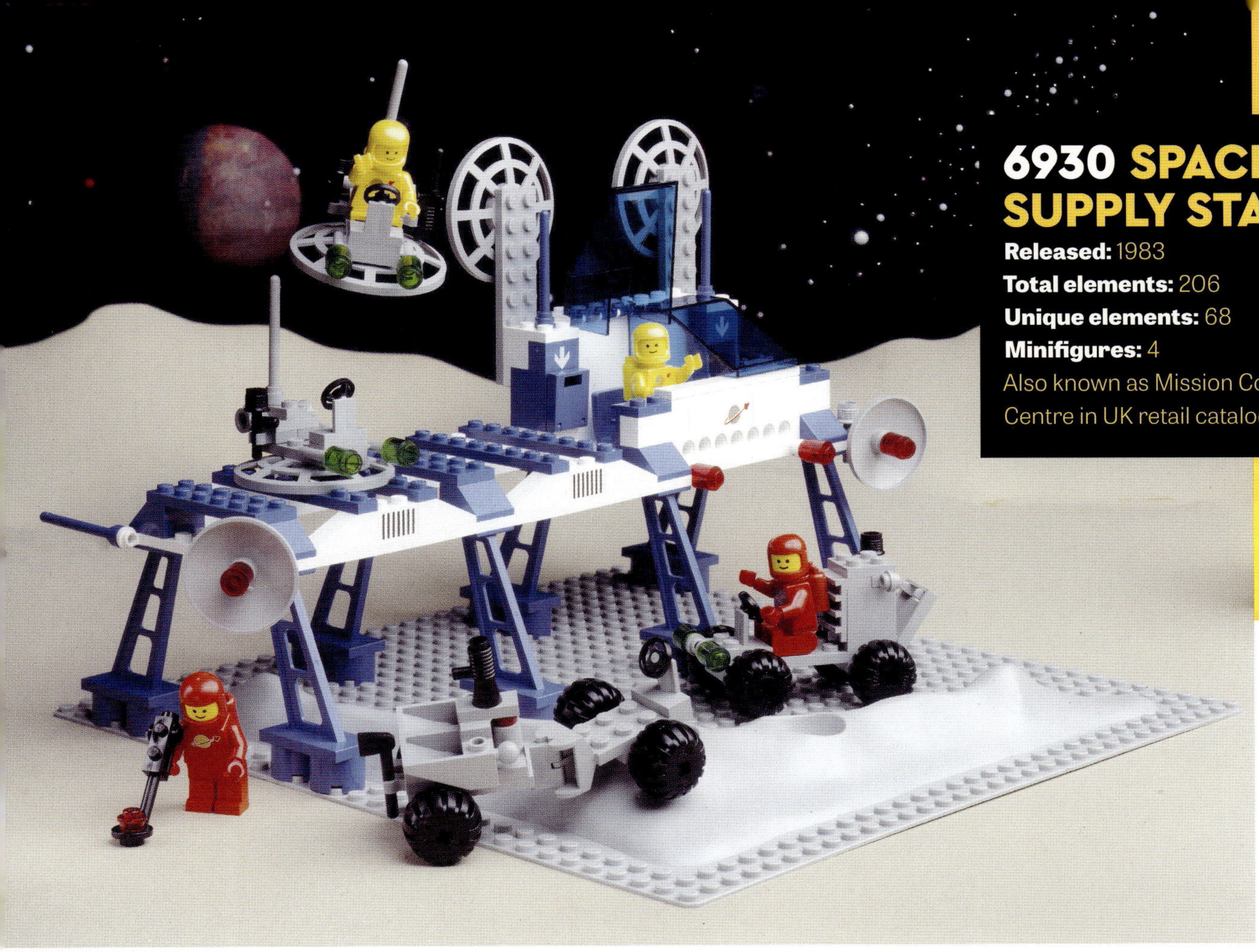

6930 SPACE SUPPLY STATION

Released: 1983
Total elements: 206
Unique elements: 68
Minifigures: 4
Also known as Mission Control Centre in UK retail catalogs.

Good things come in pairs. This sleek yet industrial base features two ground vehicles with two red astronauts, two flying scooters with two yellow astronauts, and a control room with two pods and two radar dishes! The crater plate returns, and LEGO® Technic bricks with holes once again provide some visual detail, but otherwise this is a wholly new look for a space base—the first new one since 1980.

The geophysicist skidded his rover to a halt underneath the landing pad, just as SS 6930-B was landing. No time to lose. His colleague drove the drone buckets up to the chamber, and they divided and loaded the radioactive zircon, lifting it to the nuclear casks for storage while the saucer scooters prepared for delivery to the command ship waiting in orbit.

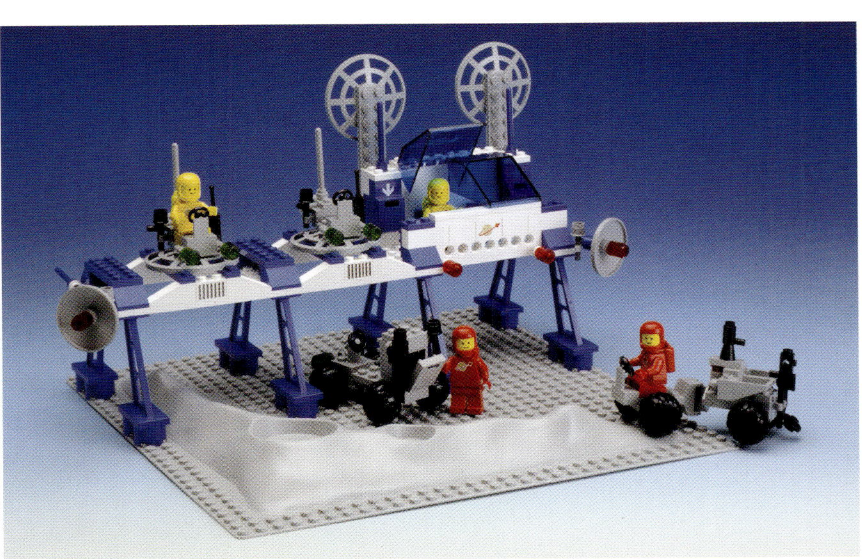

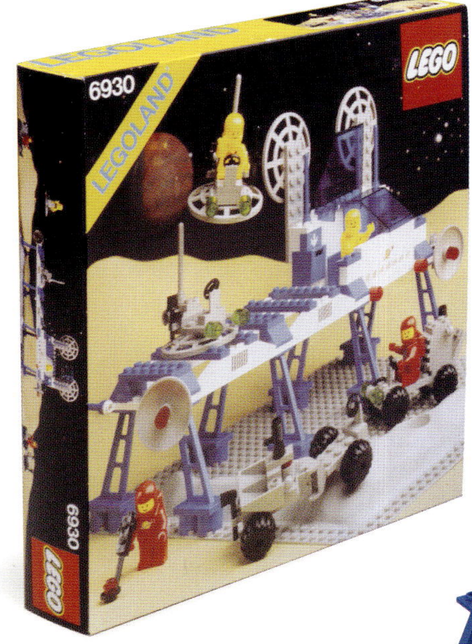

Stability is a critical factor in LEGO models, not just within a completed model but also throughout the building process. Bigger and stronger than the plane tail elements often used previously, the ***inclined stanchion support 2 x 4 x 5*** *first appeared in both LEGO® Trains and LEGO® Space sets in 1983.*

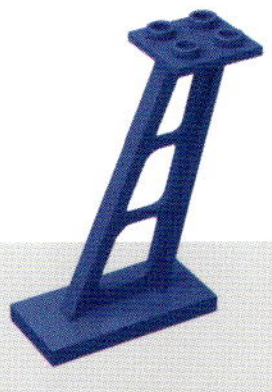

▲ **ELEMENT: 4476**
INTRODUCED: 1983

6980 GALAXY COMMANDER

Released: 1983
Total elements: 443
Unique elements: 116
Minifigures: 5
Also known as Starship Explorer in UK retail catalogs.

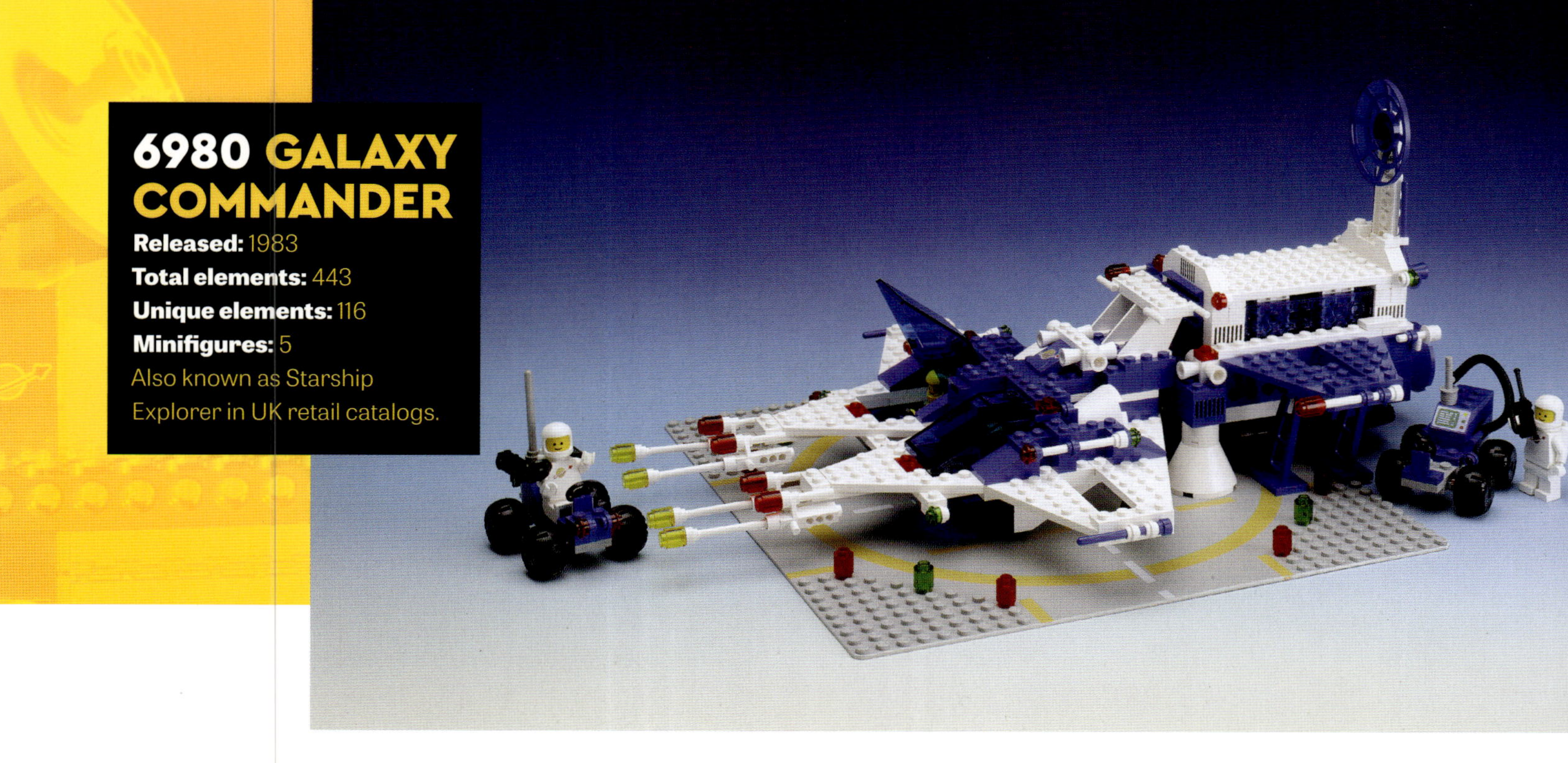

The largest LEGO® Space set to date—exceeding set 928 by over one hundred pieces—this was also the first spaceship to feature a twin cockpit; it would be another decade before another was released. Thrillingly, it also comprised three stages using the increasingly familiar modular approach: the rear comprises a mobile lab that actually lifts off the ground once connected, while the main body of the craft divides into two smaller craft. The set also included a rover, a refueling tanker, all three available colors of astronaut, and a landing pad baseplate.

"Initiate separation sequence," said the captain as her command ship neared the surface of Zeta Cephei b. From his neighboring pod, the first officer released the electric clamp, and the starship divided, ejecting the lander module backward. The engines on the control craft roared as it turned and accelerated back toward the command center while the thrusters on the lander module fired up, gently lowering the laboratory and its precious zircon cargo to the landing pad. But why had the scientist demanded such a remote location for his work? She didn't trust him, but orders were orders.

▶ The pilots are visible through the underside of set 6980 in this unusual shot from the March 1983 cover of the international trade journal for toys *Das Spielzeug*.

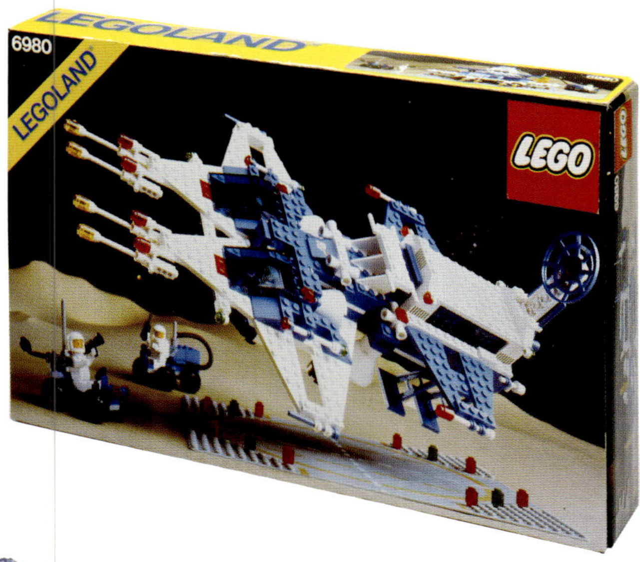

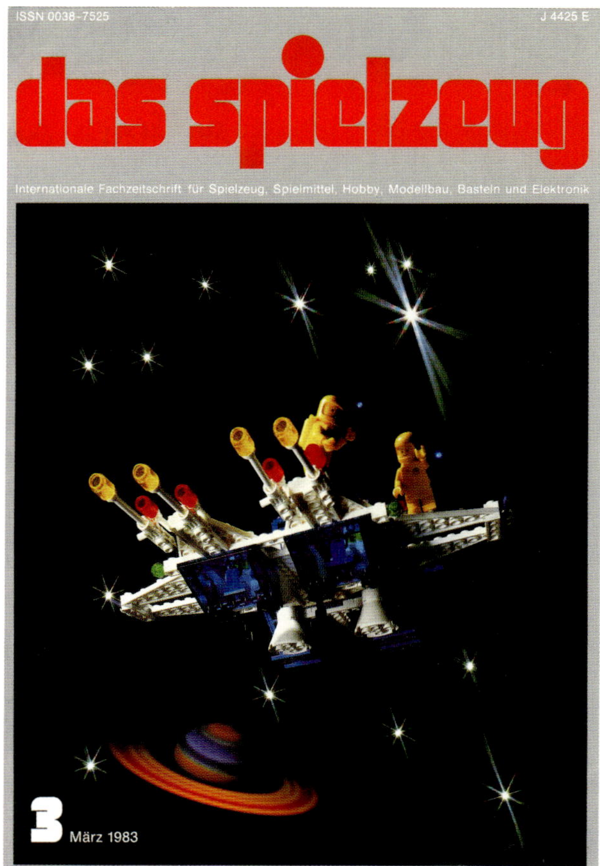
ISSN 0038-7525
J 4425 E
das spielzeug
Internationale Fachzeitschrift für Spielzeug, Spielmittel, Hobby, Modellbau, Basteln und Elektronik
3 März 1983

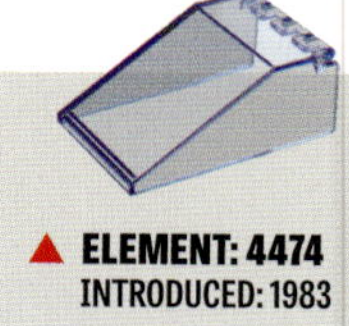

▲ **ELEMENT: 4474**
INTRODUCED: 1983

The introduction of hinged roof plates in 1981 was now rolled out further with ***windscreen 6 x 4 x 2 canopy****, also used in LEGO® Town sets. It made somewhat redundant the iconic transparent slope 33° 3 x 6, which did not reappear in LEGO Space for another three years.*

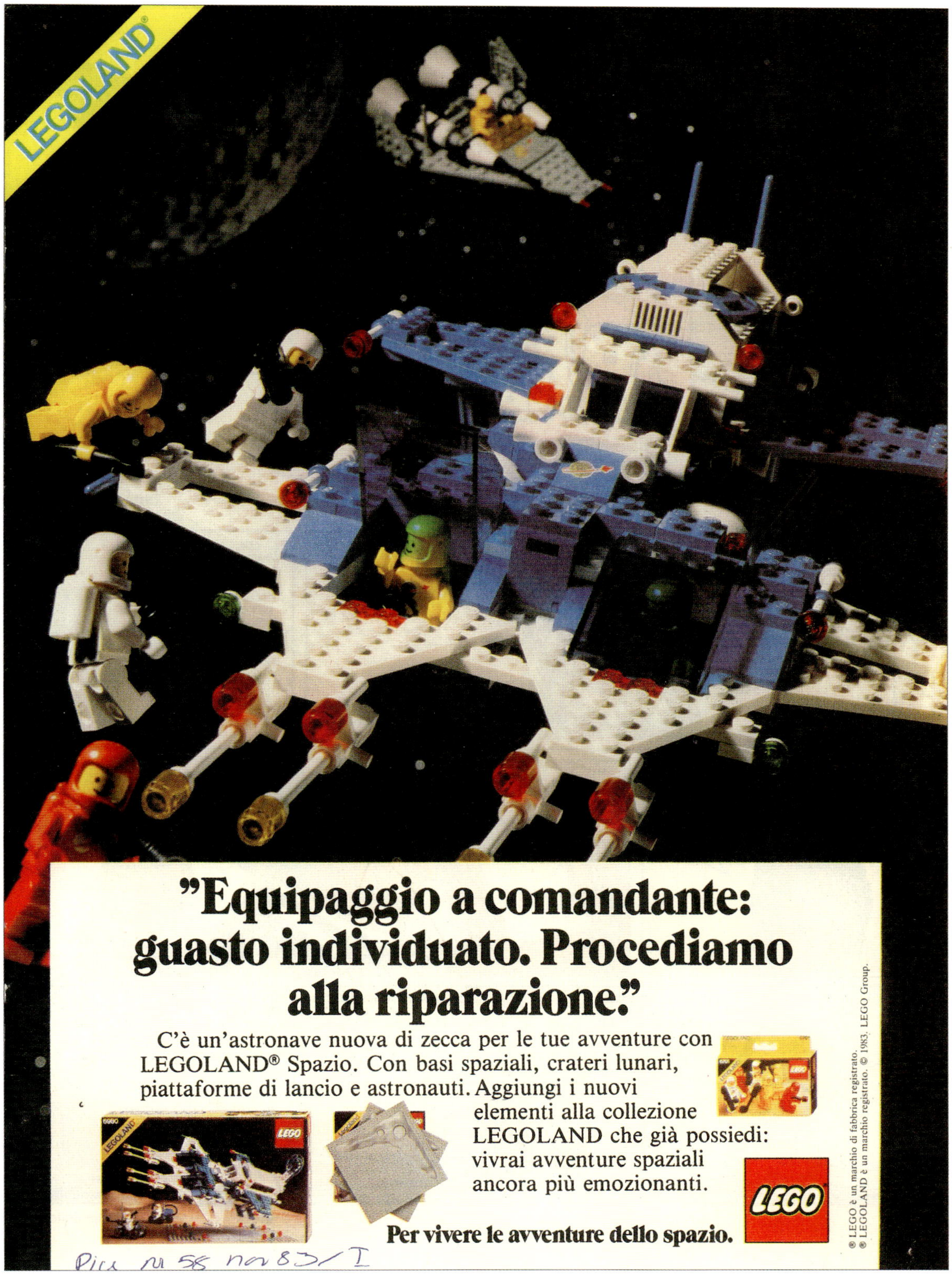

LEGOLAND®

"Equipaggio a comandante: guasto individuato. Procediamo alla riparazione."

C'è un'astronave nuova di zecca per le tue avventure con LEGOLAND® Spazio. Con basi spaziali, crateri lunari, piattaforme di lancio e astronauti. Aggiungi i nuovi elementi alla collezione LEGOLAND che già possiedi: vivrai avventure spaziali ancora più emozionanti.

LEGO

Per vivere le avventure dello spazio.

® LEGO è un marchio di fabbrica registrato.
® LEGOLAND è un marchio registrato. © 1983. LEGO Group.

"Crew to commander: fault detected. Let's proceed with the repair." This Italian ad from 1983 presents an exciting scene with the Galaxy Commander.

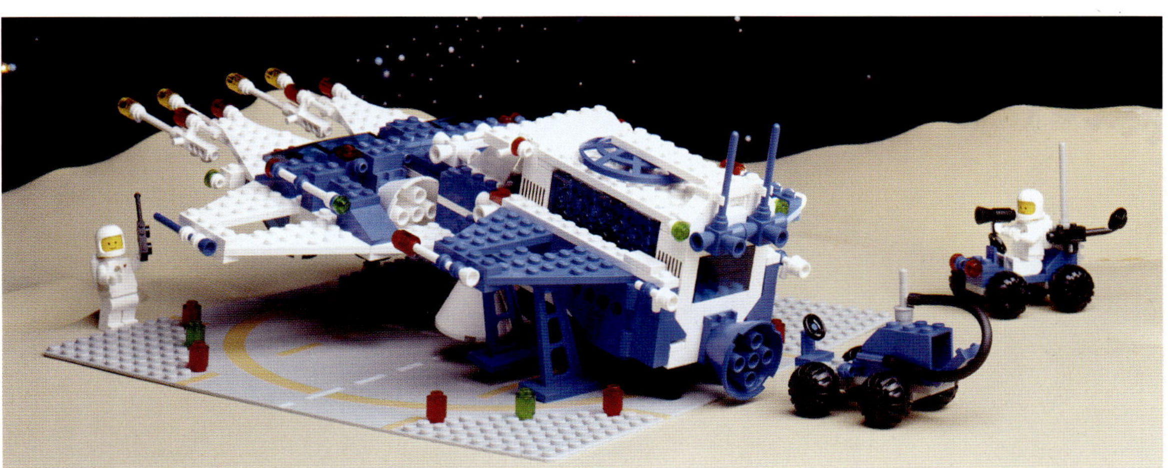

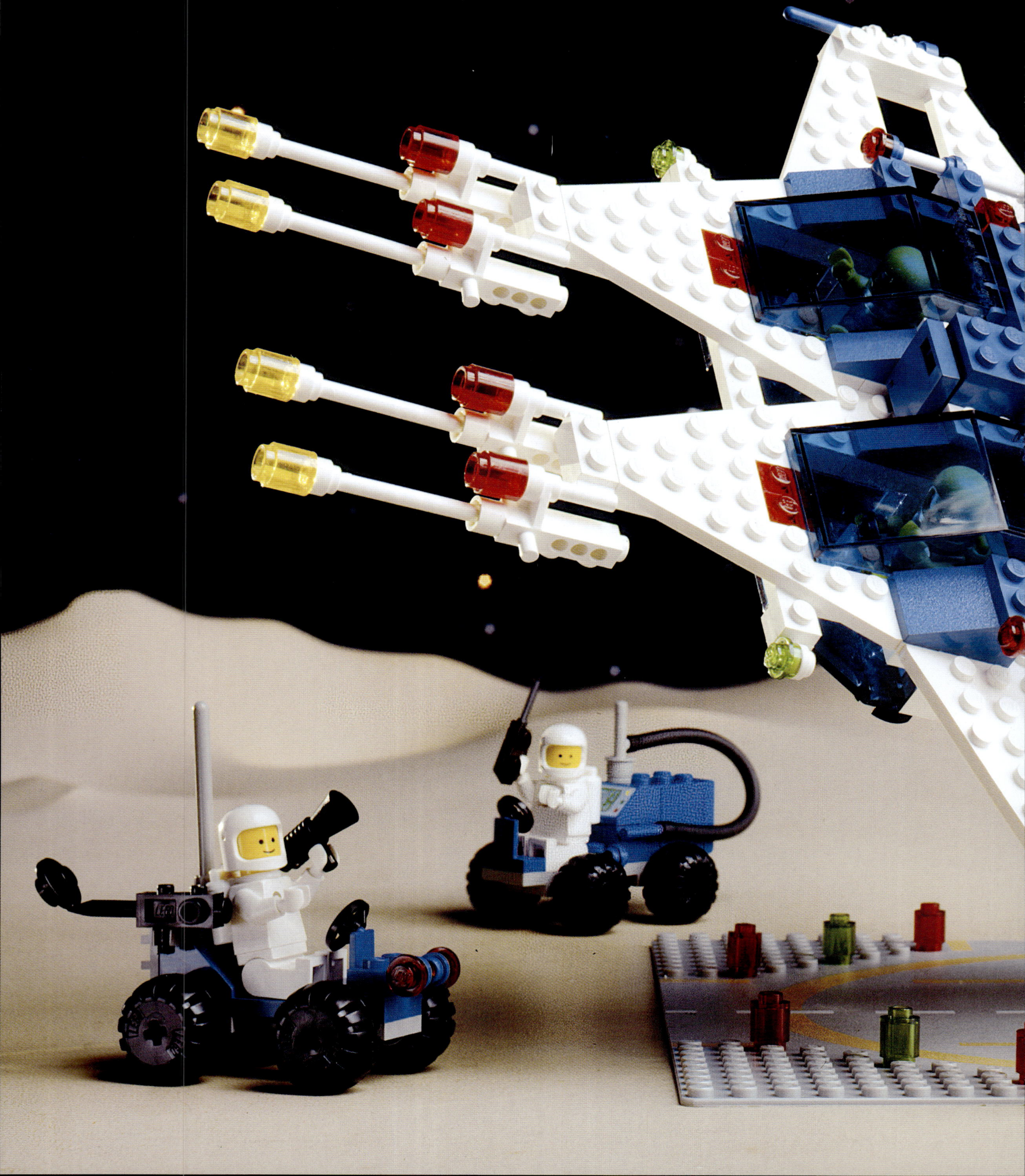

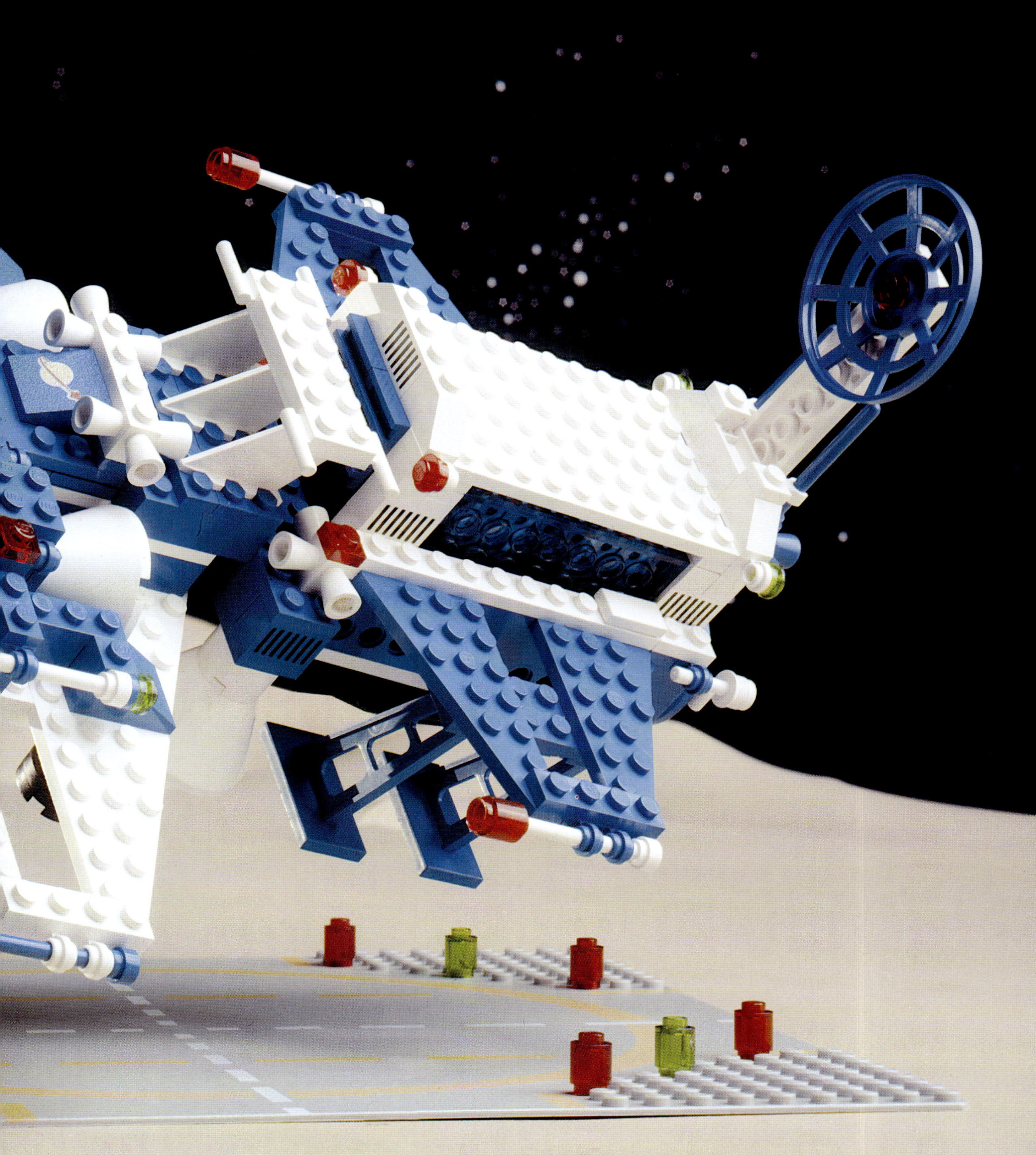

6803 SPACE PATROL

Released: 1983
Total elements: 24
Unique elements: 14
Minifigures: 1

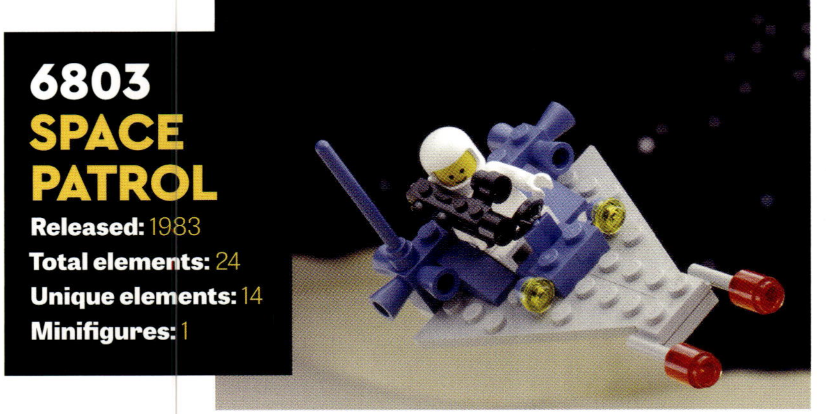

The first pocket-money small ship since 1979, this delta-wing scooter has a color scheme that also harks back to the beginnings of LEGO® Space, including gray wings—which seems odd, given that the elements were at last available in white. Novelty is added by the presence of the space camera and positioning rockets in blue.

With the increased need for surveillance and no budget to commission new craft, the mechanic reluctantly turned to the wreckage sitting in the back of Storage Pod C. The wings of one of the old LL 885s were in decent-enough shape, and she managed to cobble together an engine and empennage from the mobile laboratory. After applying two coatings of radar-absorbing kyanite, she grabbed the thermal monoscope and took the newly christened LL 6803 for a test run across Cygni Bb.

*The humble **round plate 1 x 1**, now one of the most common elements in LEGO sets, was introduced in 1980 and works beautifully here in set 6803. In these early days, this element was supplied in pairs—still attached to the molding sprue.*

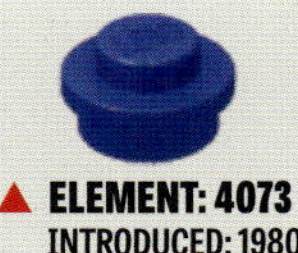

▲ ELEMENT: 4073
INTRODUCED: 1980

1593 SUPER MODEL

Released: 1983
Total elements: 313
Unique elements: 114
Minifigures: 2
Limited release in some European countries. Also known as Delta Spacecraft with Survey Vehicle.

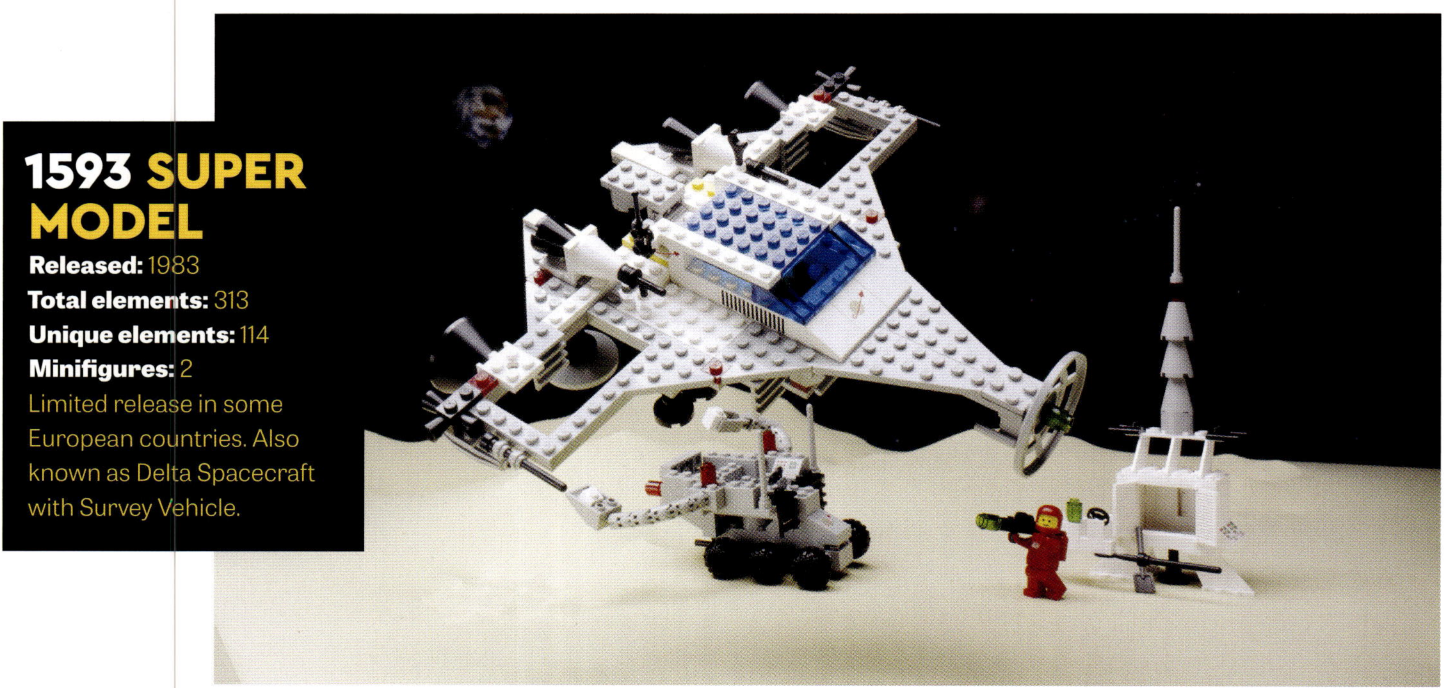

To create this promotional product, which was advertised by mail order on washing powder and sold on the shelves of a German toy chain, the elements from 1981's set 6929 Star Fleet Voyager and 1982's 6880 Surface Explorer were combined. Like other ships of the era it included holes in the wings and a vehicle that attached to the rear, but like most B models it lacked stability.

The Star Fleet Voyager, which had brought the mission all the way to the Cygnus system, now lay in ruins on this uncharted asteroid. With Delta base unlikely to have the resources to launch a rescue mission, the exploration team salvaged the wreckage of the ship and its cargo to create an ultra-long-range communications module to try to get a message back to Tadmor. With no reply, a replacement mining vehicle was designed from the remnants, along with a new craft that just might get one of them back within range.

*The versatile **window 1 x 4 x 3** has attachment points for window panes, glass, and shutters which could all be swung open. LEGO® Space sets used the element sparingly, and sometimes simply as an open frame.*

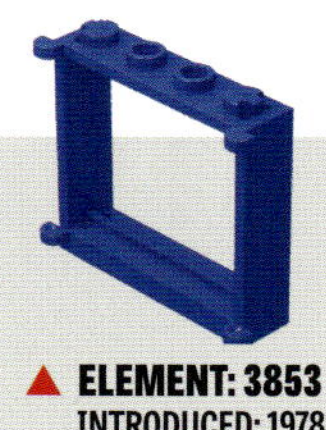

▲ ELEMENT: 3853
INTRODUCED: 1978

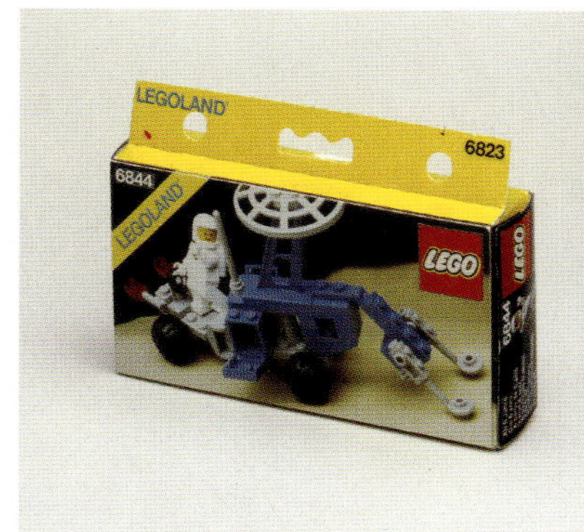

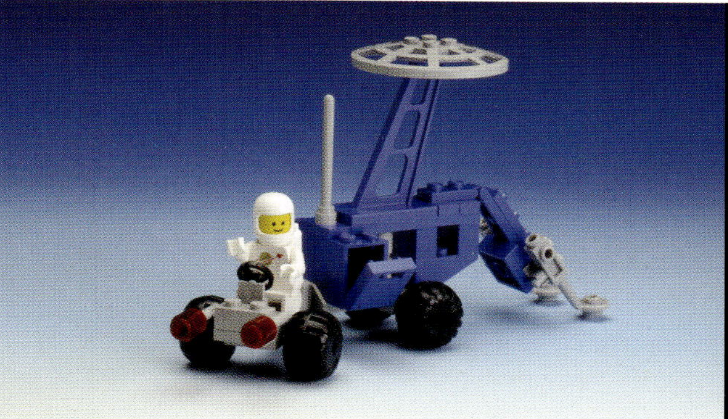

6844 SISMOBILE

Released: 1983
Total elements: 46
Unique elements: 25
Minifigures: 1
Also known as Seismologic Vehicle.

This tiny articulated chassis may look overloaded, but it's a sturdy design. It packs in several unusual play functions and details seemingly related to exploration geophysics. This was the first ground vehicle to feature blue bodywork. The rear detection arm assembly is attached centrally by using two modified plate 1 x 2 with 1 stud parts—previously used for the basic centering of single elements, such as the antenna toward the front.

The survey was a long, lonely task, but having a tomographic model of Tadmor was essential if they were to locate the hypocenter of the nuclear explosion. Suddenly, both of the front-mounted gamma sensors flashed. The geoscientist took a sharp left, fired up the parabolic surface wave monitor, and lowered the twin seismic refractors. She kept a steady pace as she navigated the edge of the radiation zone.

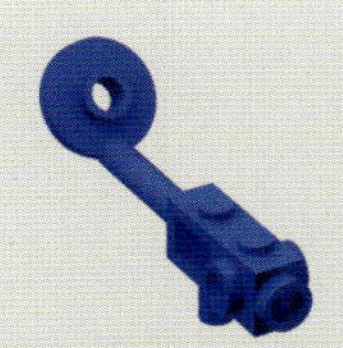

▲ **ELEMENT: 4479**
INTRODUCED: 1983

Mineral detection had been a regular subject of LEGO® Space *sets, and so the introduction of the* ***minifigure utensil metal detector*** *was a superb idea. The Sismobile was the only 1983 set to contain it in light gray, as opposed to black, and two were provided! It included studs not only on the controller unit but also on the search coil assembly, a feature that was removed when this element was redesigned in 2013.*

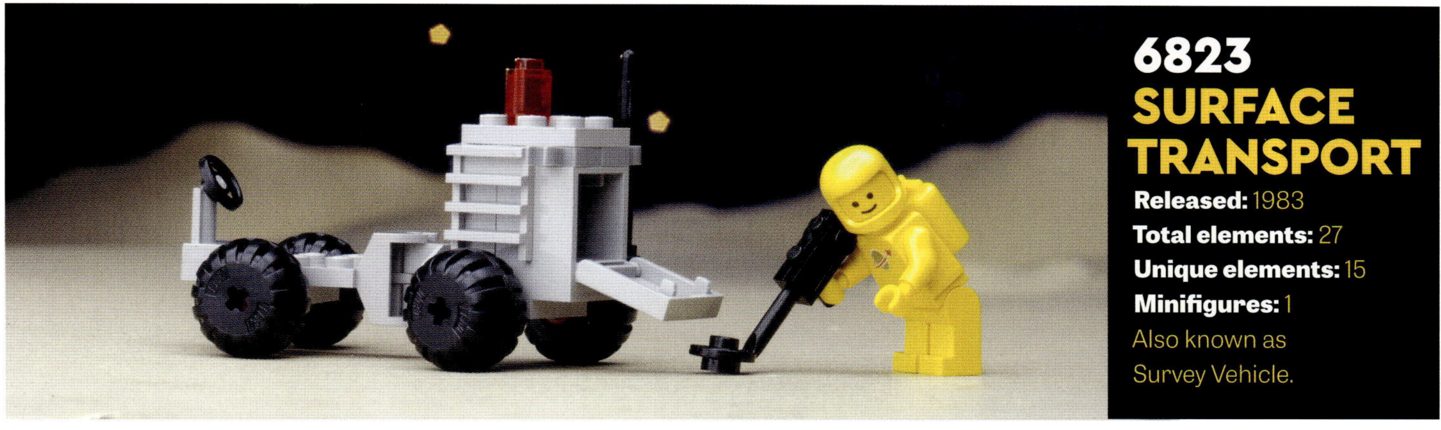

6823 SURFACE TRANSPORT

Released: 1983
Total elements: 27
Unique elements: 15
Minifigures: 1
Also known as Survey Vehicle.

The new metal detector accessory resulted in two small sets focused on detection in the 1983 range. Here, the astronaut uses it for handheld detection, while the tiny yet articulated vehicle focuses on storage with the help of the new container box element.

Having traversed Tadmor for over three Earth hours, the geophysicist finally got a signal from the survey meter. He drove the transport vehicle to the source location, carefully maneuvering the trailer as close as possible. Unclipping the suction rod from the vehicle, he lifted the zircon from the soil, depositing it quickly into the multichannel ion chamber. Moments after he slammed it shut, the indicator flashed red: an unusually high level of gamma particles. He had to get back to the station, quickly.

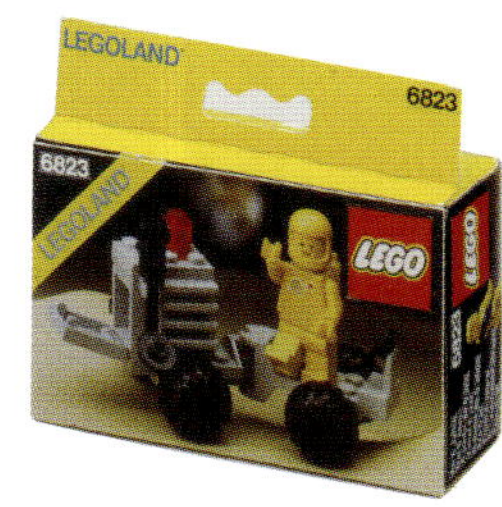

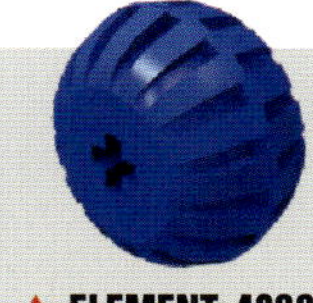

▲ **ELEMENT: 4288**
INTRODUCED: 1982

LEGO® Space *had initially used the available existing wheels in the* LEGO® System in Play *but from 1982 had a special new one:* ***balloon wheel with axle hole****. Unlike earlier rubber tires, these were plastic, which was perhaps appropriate, given that real-world lunar rover tires are not made of rubber either. One side fitted the standard LEGO wheel hub, while the other sported a* LEGO® Technic *axle hole.*

6846 TRI-STAR VOYAGER

Released: 1984
Total Elements: 69
Distinct Elements: 34
Minifigures: 1
Also known as Zero 15 Interceptor in UK retail catalogs.

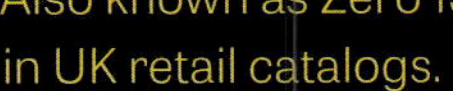

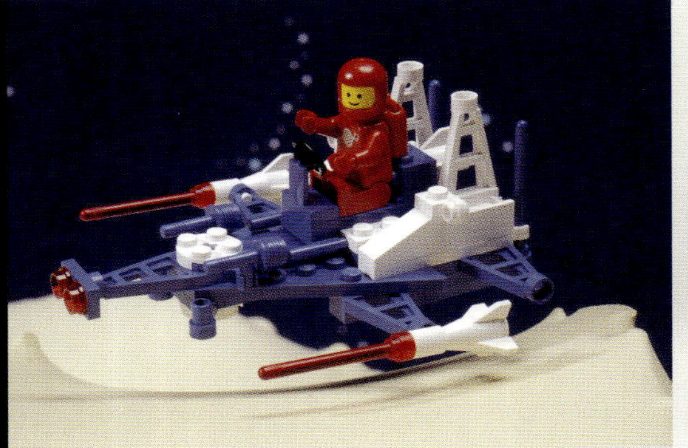

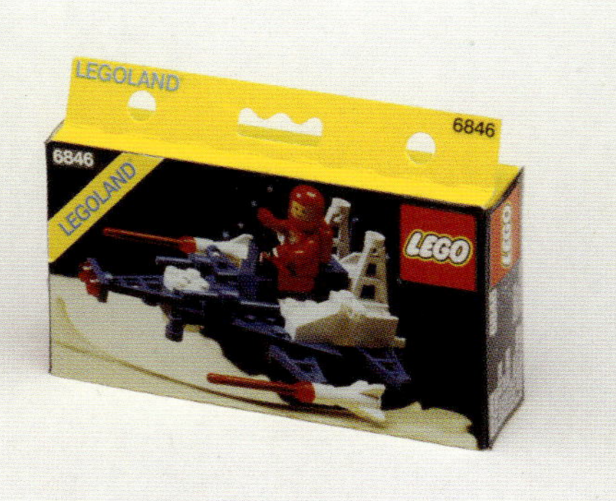

The new space wing element provides much of the detailing on this small ship—seven are included, two of which are white vertical wings that fold down. This releases the central section, which slides off the back as a smaller craft, which, at this small scale, is a simple and functional design. This is the only 1984 set to use a blue-and-white color scheme.

As he approached the location of the signal from the zircon probes, the pilot powered down. "Zero 15 to Alpha, I hope you get this. No sign of Zero 10. My ultrazinc and kyanite coating seems to have allowed me to avoid detection . . . so far, anyway. Activating observation mode and proceeding to the void by shuttle." The vertical wings glinted in the light of Cygni's three suns as they slowly folded down and, as the pilot decoupled from the craft, its seven laser interferometer antennas began scanning for gravitational waves. He turned toward the inky blackness, noting that as he moved slowly forward, the stars disappeared one by one.

***Modified plate 1 x 2 with long stud receptacle** is an iconic LEGO® Space element with a perforated appearance similar to earlier elements, like the dish and stanchion. Commonly known as the "space wing," it is actually used in all manner of ways—as a nose tip, as a stanchion, and for satellites. The tip has a tube rather than a stud, meaning that the undersides of elements inserted into it are visible. It can also connect with bar handles.*

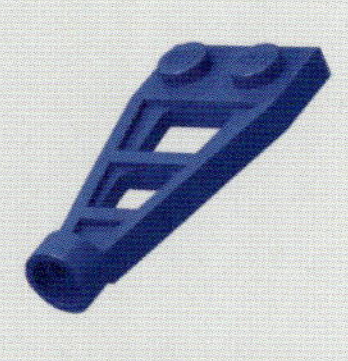

▲ **ELEMENT: 4596**
INTRODUCED: 1984

6804 SURFACE ROVER

Released: 1984
Total Elements: 21
Distinct Elements: 10
Minifigures: 1
Also known as Space Jeep in UK retail catalogs.

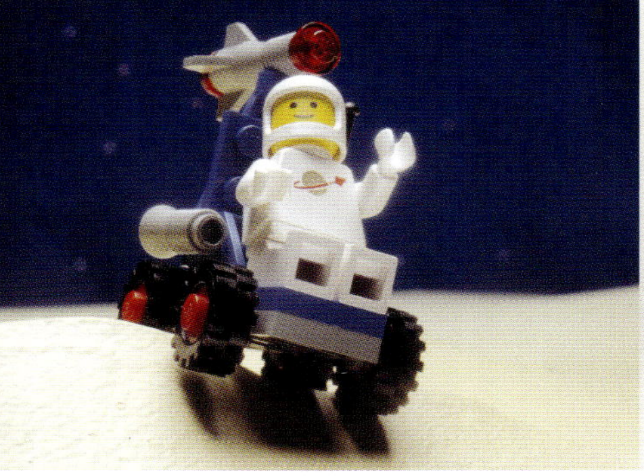

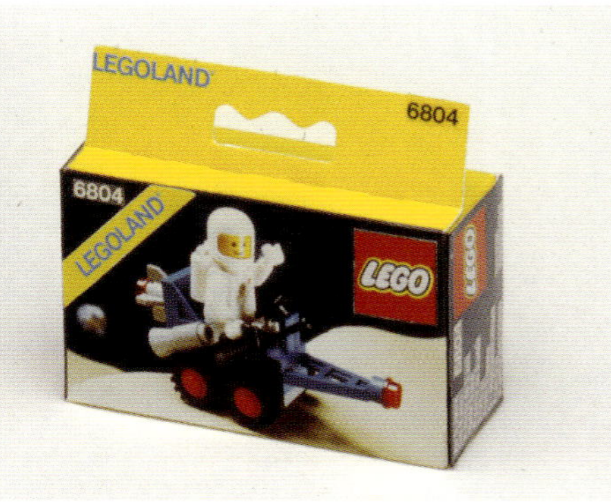

▶ An alternate build for 6804.

1984's small rover was small indeed—the chassis is a 2 x 4 plate. However, there was excitement in its new elements: the design made judicious use of the space wing, space seat, and round brick with fins. The all-blue bodywork was typical of sets this year.

The geomorphologist looked back toward the crevice that was appearing rapidly behind them. Gripping the steering wheel with one hand, they radioed back to base with the other. "Rover 4 to Beta. Condition red, seismic event," they shouted above the increasing roar of crumbling terrain. "It's some kind of quake, but I swear, it's like the fissure is following me . . . I am activating HAN now." They braced themself and fired the hydroxylammonium nitrate booster, but as they sped ahead, the deep crack began to grow even more quickly.

*The **inverted bracket 3 x 2 - 2 x 2**, or "space seat," can truly be described as an element designed for astronauts! There's room for the air tanks on their backs, and the connection holes provide various opportunities to build sideways in a small space. The recesses in the angled supports echo the industrial aesthetic of earlier LEGO® Space elements.*

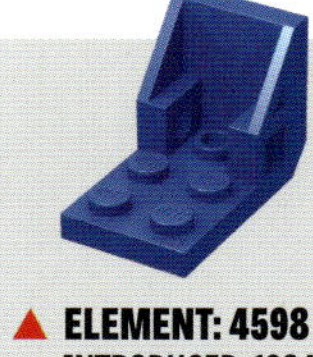

▲ **ELEMENT: 4598**
INTRODUCED: 1984

6824 SPACE DART I

Released: 1984
Total Elements: 48
Distinct Elements: 24
Minifigures: 1
Also known as Zero 10 Interceptor in UK retail catalogs.

This sleek and unusual ship makes thrilling use of the new space wing element, layering two on each side for added detail. This was the smaller of the two 1984 sets to contain a blue astronaut, and the gray color of the ship provides a pleasing contrast to the minifigure. The new round bricks with fins are especially noticeable in black, and the humble antenna element was cast in transparent red for the first time. A surprising feature is that the rocket-like nose assembly can be freely rotated, thanks to a LEGO® Technic pin connection.

The pilot snatched her transceiver. "Zero 10 to Zero 15. They've spotted me already . . . they must have radar technology. I can see the void dead ahead of me." Without warning, the craft slowed to a sudden halt and flipped backward, spiraling through space. "Zero 15! Are you receiving? Something hit me! Some kind of gravity wave . . . nil damage, it seems." Quickly stabilizing the craft and regaining her bearings and composure, the pilot energized the yttrium laser probe. "Deploying long-range analysis." The nose of her craft began to whirl until it reached a constant hum and the probe glowed red. As the twin zircon missiles blasted into the vast patch of starless sky ahead, she hoped that their purpose would not be misconstrued.

*The offset **modified plate 1 x 4** is an unusual element, one of a few which offset the position of LEGO studs from the traditional building grid. However, it wasn't used in complex building techniques in 1984, instead providing visual interest and texture. Four are used on the undercarriage of the Space Dart in a striking square arrangement.*

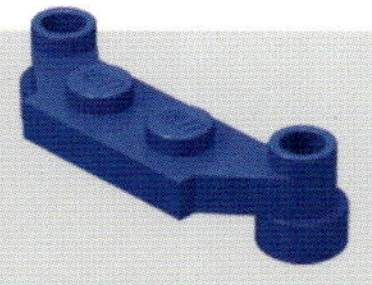

▲ **ELEMENT: 4590**
INTRODUCED: 1984

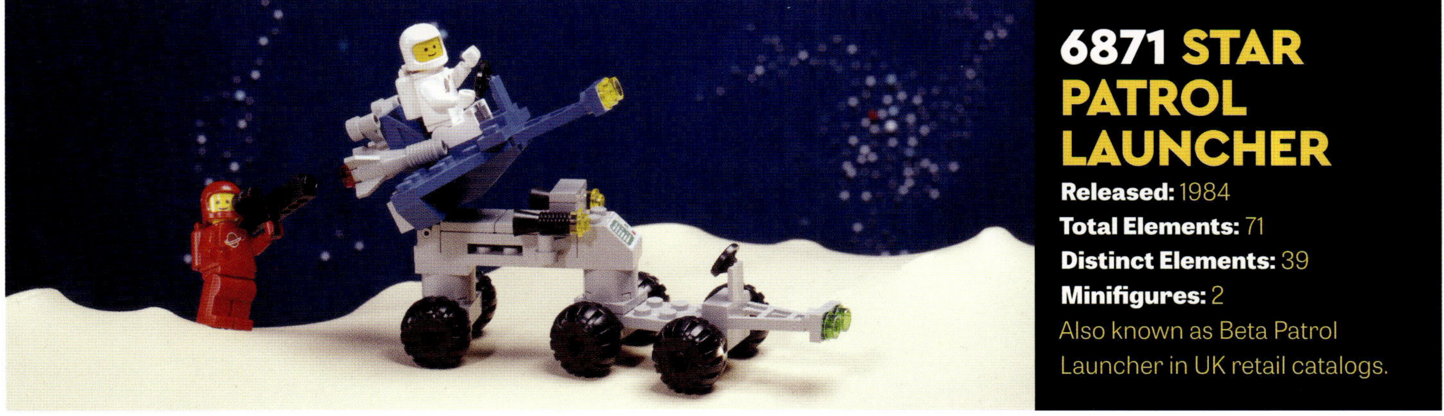

6871 STAR PATROL LAUNCHER

Released: 1984
Total Elements: 71
Distinct Elements: 39
Minifigures: 2
Also known as Beta Patrol Launcher in UK retail catalogs.

The smaller of the two launcher vehicles released in 1984 achieved a lot with a low element count. Notably, two minifigures are provided, unlike the larger 6881 Lunar Rocket Launcher, which only contained one. The split-level articulated vehicle has a simple design that includes all the necessary functions. While basic, the craft has a little more detail, as well as an unusually large rear undercarriage extending to the wing tips. A simple attachment with LEGO® Technic axles holds the ship in place, even allowing for vertical takeoff.

"Space Station Alpha to Beta patrol. They have located your vehicle. You have fifty seconds to launch" came the command into the scientist's earpiece. "Copy that. I see the ridge now. Engaging HAN," she replied while firing up the hydroxylammonium nitrate thrusters. "You better be all buckled up back there"—she smiled to her colleague as they skidded to a halt—"because I'm about to blast you out of here!" She spun in her chair to program the flight path while the arm rose to launch position, sending him skyward with only seconds to spare. "Beta to Alpha." She sighed as an ominous rumbling began beneath her feet. "Mission accomplished. Signing off."

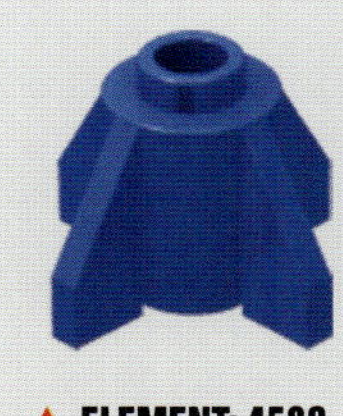

▲ **ELEMENT: 4588**
INTRODUCED: 1984

__Round brick 1 x 1 with fins__ was the smaller of two elements with fins introduced this year, bringing realistic detail to small projectiles and exhaust nozzles. The clever shape of the fins means that plates and bricks can still be attached below. Unlike the regular round brick 1 x 1, a LEGO Technic axle can be inserted underneath.

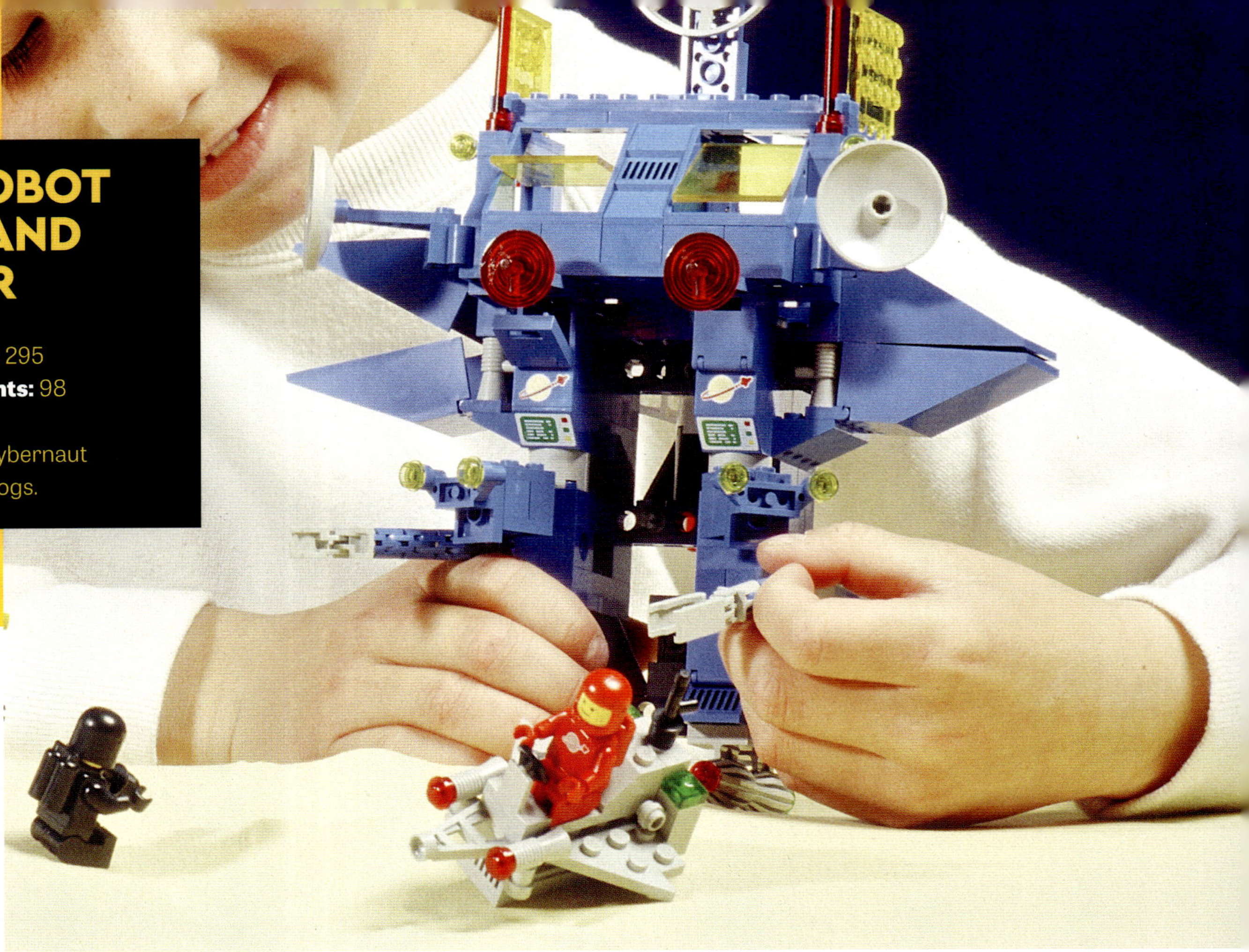

6951 ROBOT COMMAND CENTER

Released: 1984
Total Elements: 295
Distinct Elements: 98
Minifigures: 3
Also known as Cybernaut in UK retail catalogs.

Something had been missing from the LEGO® Space play theme in its initial years: robots! This set certainly made up for lost time by delivering the biggest, most unusual robot imaginable. It's a control center that glides along, scanning the environs with all manner of mysterious appendages, collecting samples with its arms and deploying the ground buggy and folding-wing glider from between its legs. There's even a rocket strapped to its back! The original 1979 color palette is very much in evidence here.

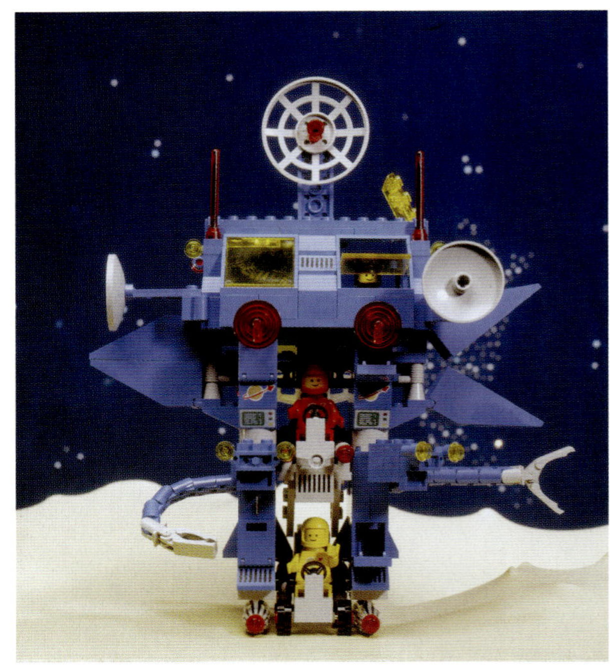

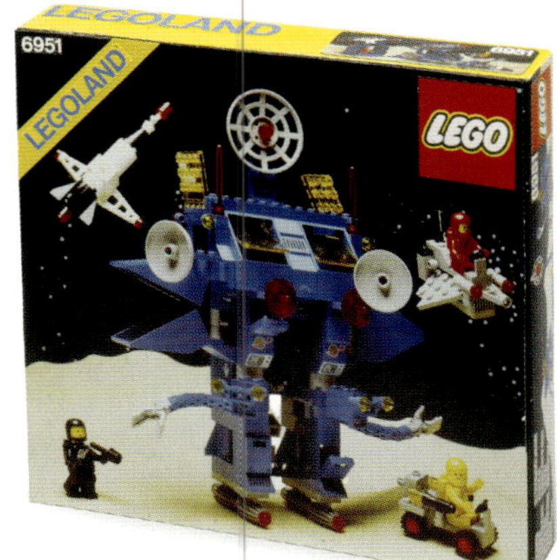

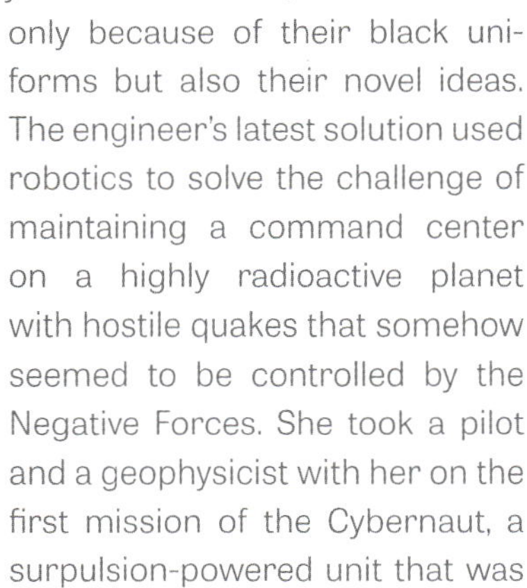

The new crew members, urgently sourced from the technology company Sort Elektronika, stood out not only because of their black uniforms but also their novel ideas. The engineer's latest solution used robotics to solve the challenge of maintaining a command center on a highly radioactive planet with hostile quakes that somehow seemed to be controlled by the Negative Forces. She took a pilot and a geophysicist with her on the first mission of the Cybernaut, a surpulsion-powered unit that was self-guided by an ultradrive governing system. Freed from navigational duties, they could focus on placing any promising samples into its pinna mandibles for analysis and—if safe—deploying the manual vehicles for detailed investigations.

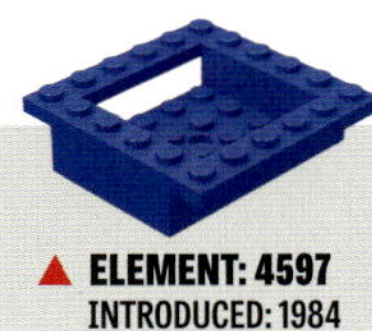

▲ **ELEMENT: 4597**
INTRODUCED: 1984

*The **cockpit 6 x 6 x 1 cabin base** solved a challenge for LEGO designers. Minifigures require a sunken cockpit with a secure base that would not pop out the bottom of the craft when the figure was inserted.*

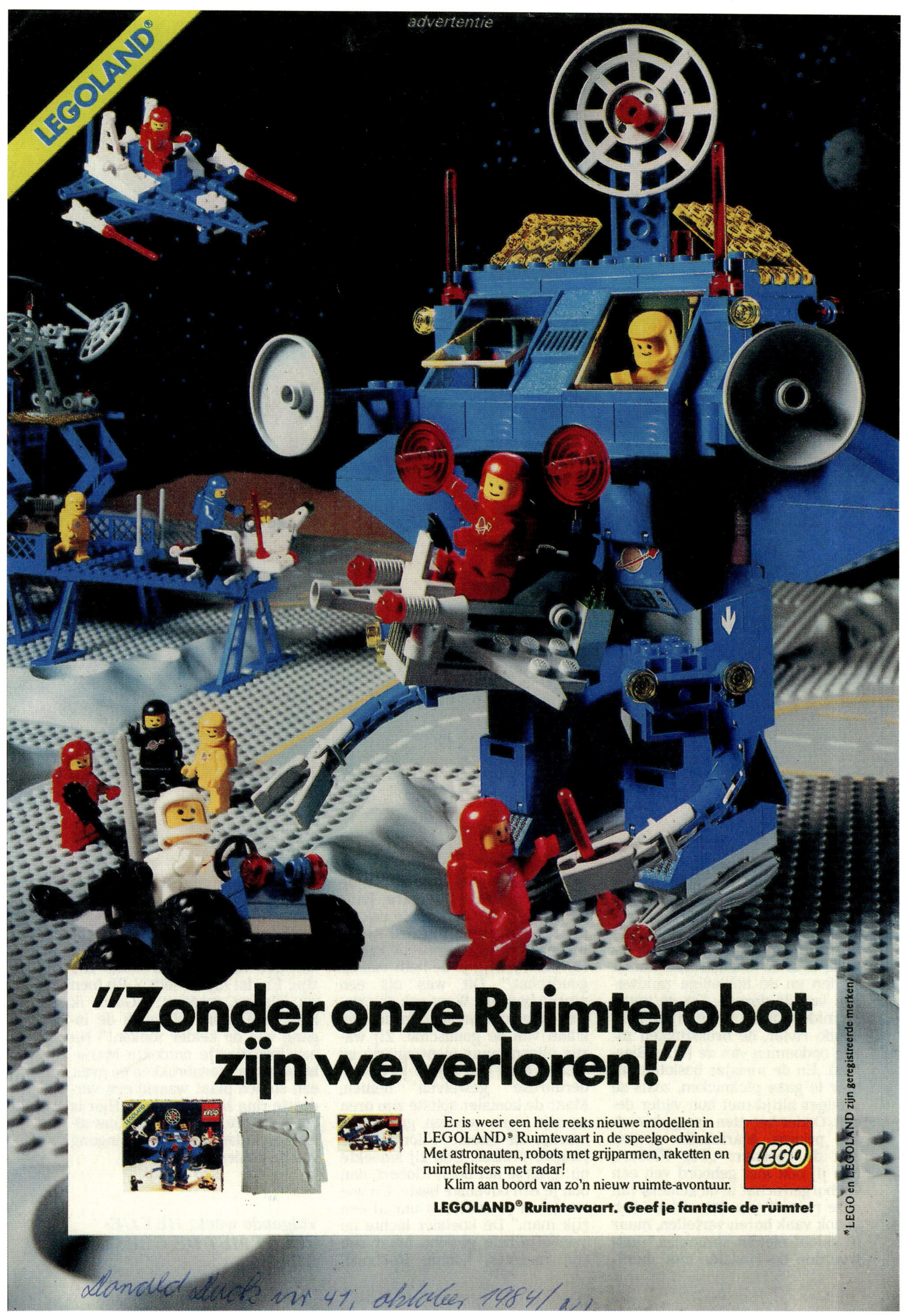

Set 6951 was highlighted in many 1984 ads, such as this Dutch example which reads, "Without our Space Robot, we are lost!"

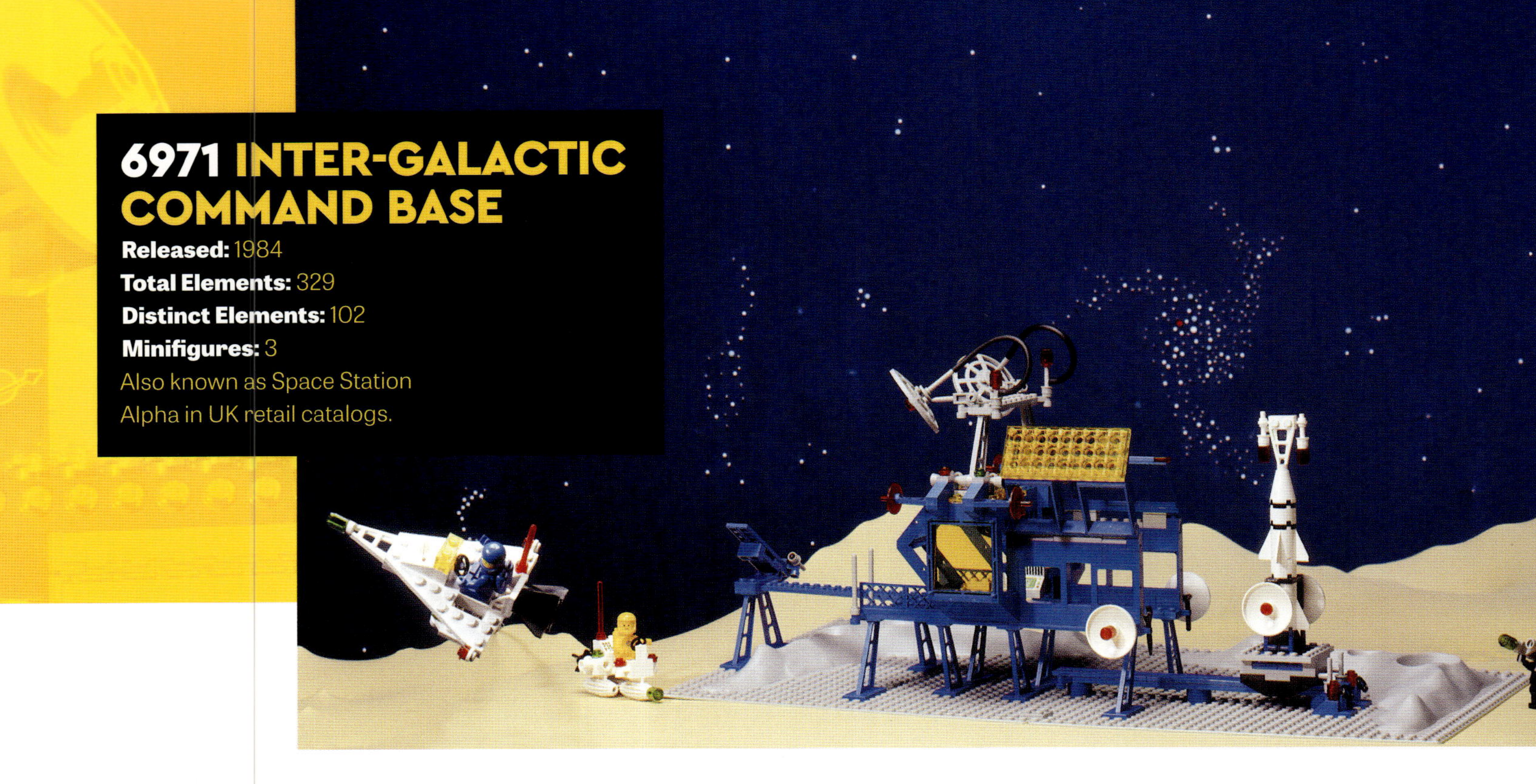

6971 INTER-GALACTIC COMMAND BASE

Released: 1984
Total Elements: 329
Distinct Elements: 102
Minifigures: 3
Also known as Space Station Alpha in UK retail catalogs.

THE 1984 COMMAND CENTER WAS THE BIGGEST TO DATE AND TOOK AN UNCONVENTIONAL APPROACH.

The 1984 command center was the biggest to date and took an unconventional approach. Abandoning almost all sense of enclosure, the base is a sparse latticework of window frames, with a bay roof that unfolds to expose a tech-laden rocket on a monorail. The 1979 color palette is present again, aside from the hover speeder and scout ship, which were the only white vehicles of 1984. The detailed gray assembly on top is beautifully constructed, although it forces the dishes to point downward.

"But, Commander," said the engineer, "with this data in hand, Sort Elektronika could harness the Negative Forces for our own use. We could defend the whole of the galaxy . . . and travel to the next one!" "No," began the commander firmly, but before she could continue, the sirens on the seismodishes flashed and wailed, angling down toward the source. "Quake incoming!" she roared as she dashed to the scout ship. As soon as she had taken off toward the epicenter, the engineer darted to the Alpha command post, spinning in the chair as he furiously programmed the controls. "Deploy the satellite!" he hollered at the avionics technician. Startled, the technician obeyed, attaching the precious payload to her hover buggy and flying it to the bay roof that was slowly opening. Her hands shook as she attached it to the rocket. "Deploy!" Slowly, surely, the rocket began trundling toward the launch pad.

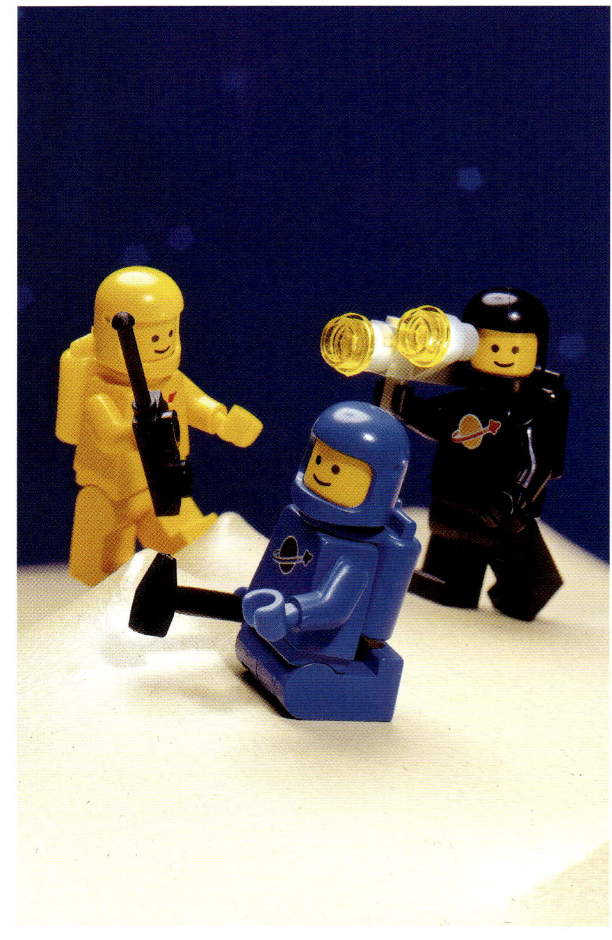

*A year after being introduced in LEGO® Town, **roof window 4 x 4 x 3** was blasted into space. The diagonal profile of the element suited the LEGO® Space design aesthetic, and the triangular effect achieved here by stacking them was later reused in many sets—and influenced the shapes of some new elements as well.*

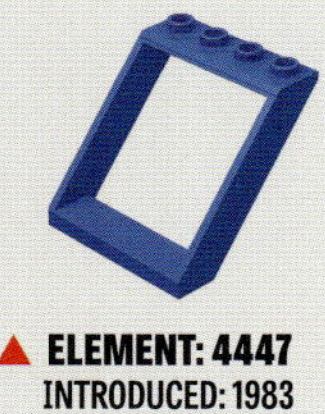

ELEMENT: 4447
INTRODUCED: 1983

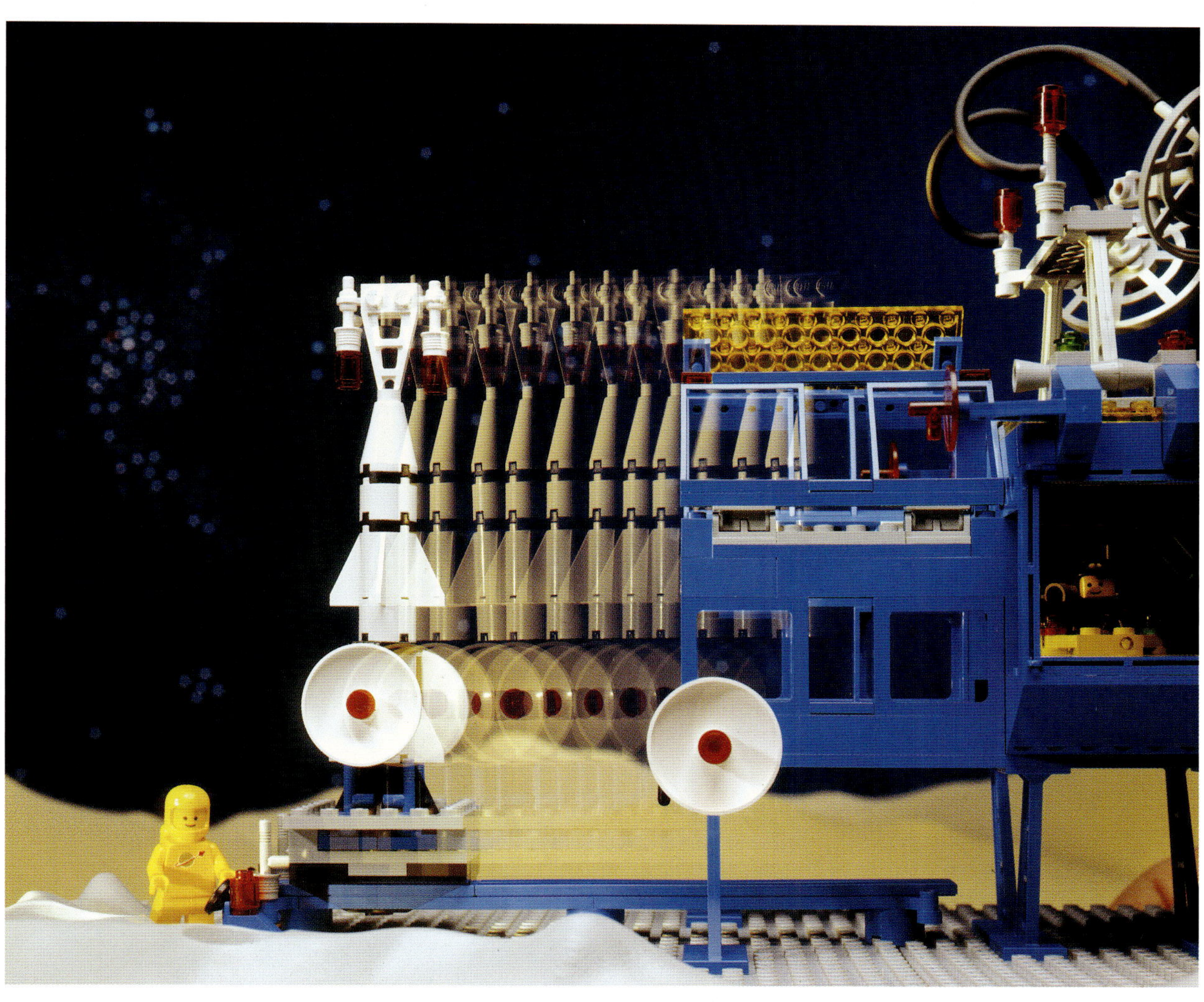

◀ In the era before digital image-editing software, this effect of suggested movement would have been achieved in the camera: the photographer would expose the film multiple times on a low exposure, shifting the rocket between shots.

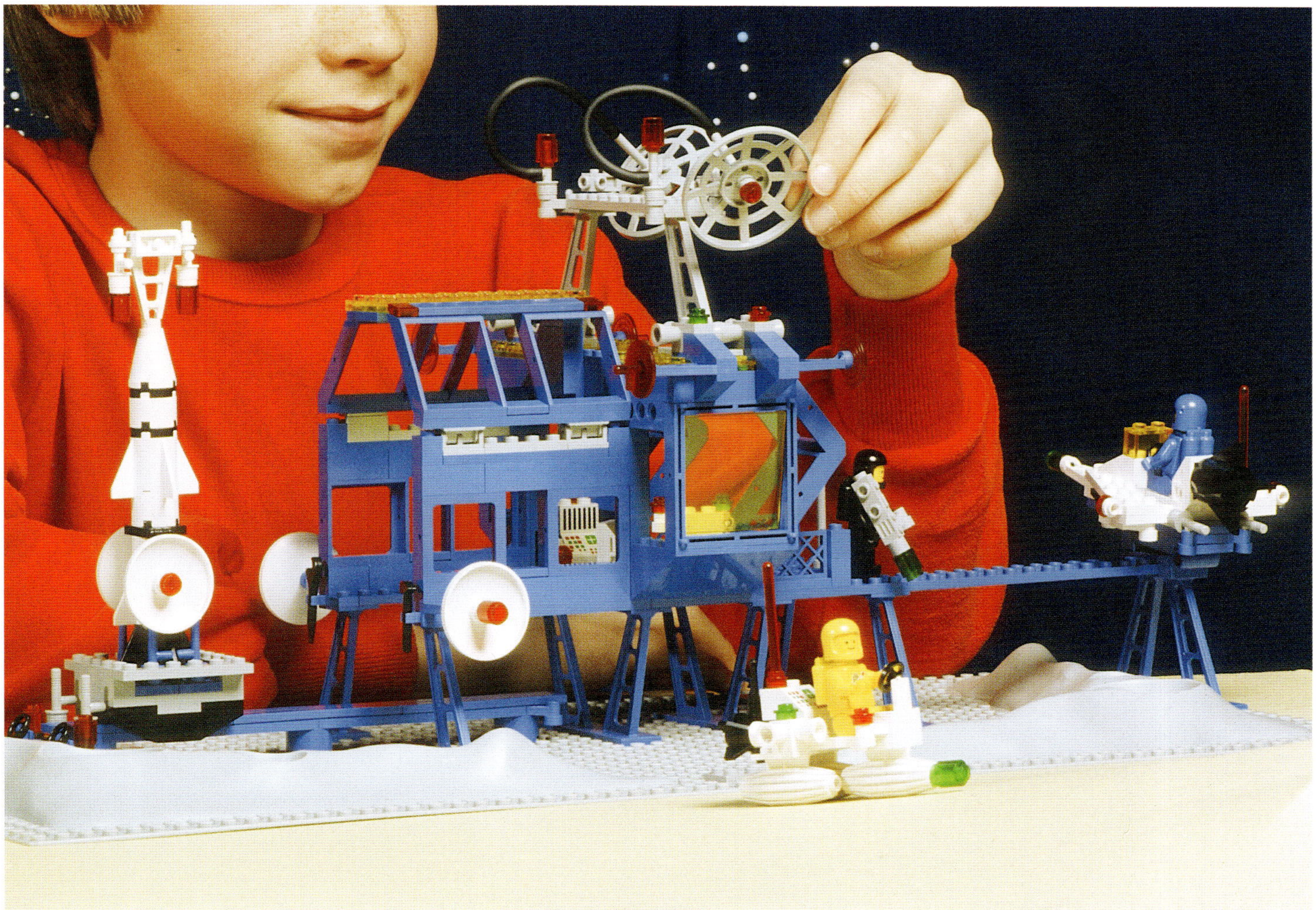

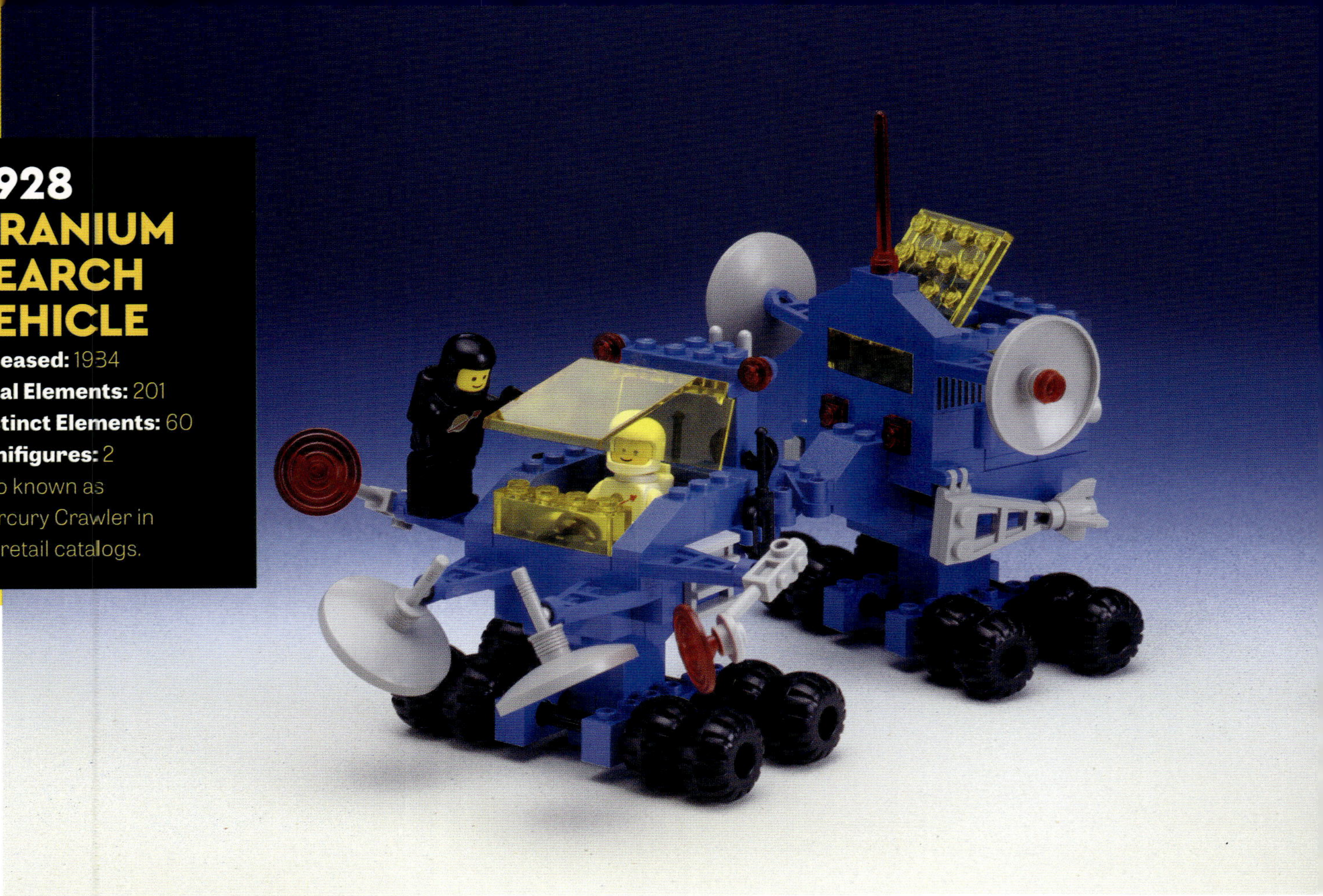

6928
URANIUM SEARCH VEHICLE

Released: 1984
Total Elements: 201
Distinct Elements: 60
Minifigures: 2
Also known as Mercury Crawler in UK retail catalogs.

1984's mobile laboratory was truly an out-of-this-world affair. The driver's cab and the hinged laboratory are raised high on inverted slopes, forming unusual hourglass shapes. A remarkable count of sixteen wheels provides stability to this fantastical design. However, the color scheme is classic, with transparent yellow returning alongside transparent red highlights, such as the rare minifigure shield. Cleverly, the new offset plate 1 x

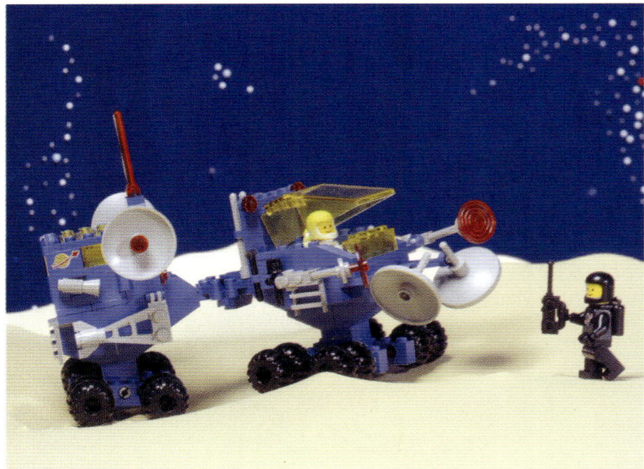

4 not only makes the pivot point more beautiful but also limits the extent of its range.

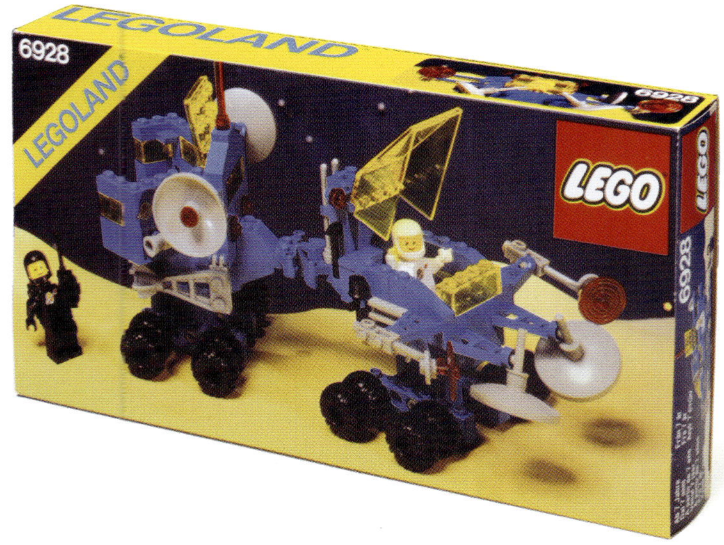

At last, hammers in space! Astronauts no longer needed to bang malfunctioning rockets with wrenches as the ***minifigure mallet/hammer*** *joined the toolbox this year (it also appeared in LEGO® Castle and LEGO® Town sets).*

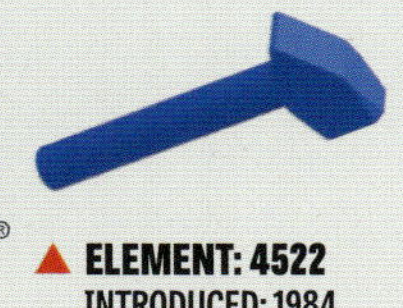

▲ **ELEMENT: 4522**
INTRODUCED: 1984

The senior geophysicist had voted against recruiting staff from Sort Elektronika, given their background in defense technology, but he had to admit that their new engineer had provided excellent solutions to the dangers they faced on Cygni Bb. High up in the cab of the crawler she had designed, he could safely detect any seismic anomalies or excessive radiation and guide them to safety. Some uranium and mercury sulfide readings brought them to a stop, and she insisted on staying put while the data was collected. "You go back; you don't have enough protection," she said with a casual smile, pinching the black fabric of her suit. Reluctantly, he decoupled the lab with a swift bang from the atomic hammer, while she extended its gray wings. "The HAN boosters are enough to get me back. Believe me, I have all I need right here."

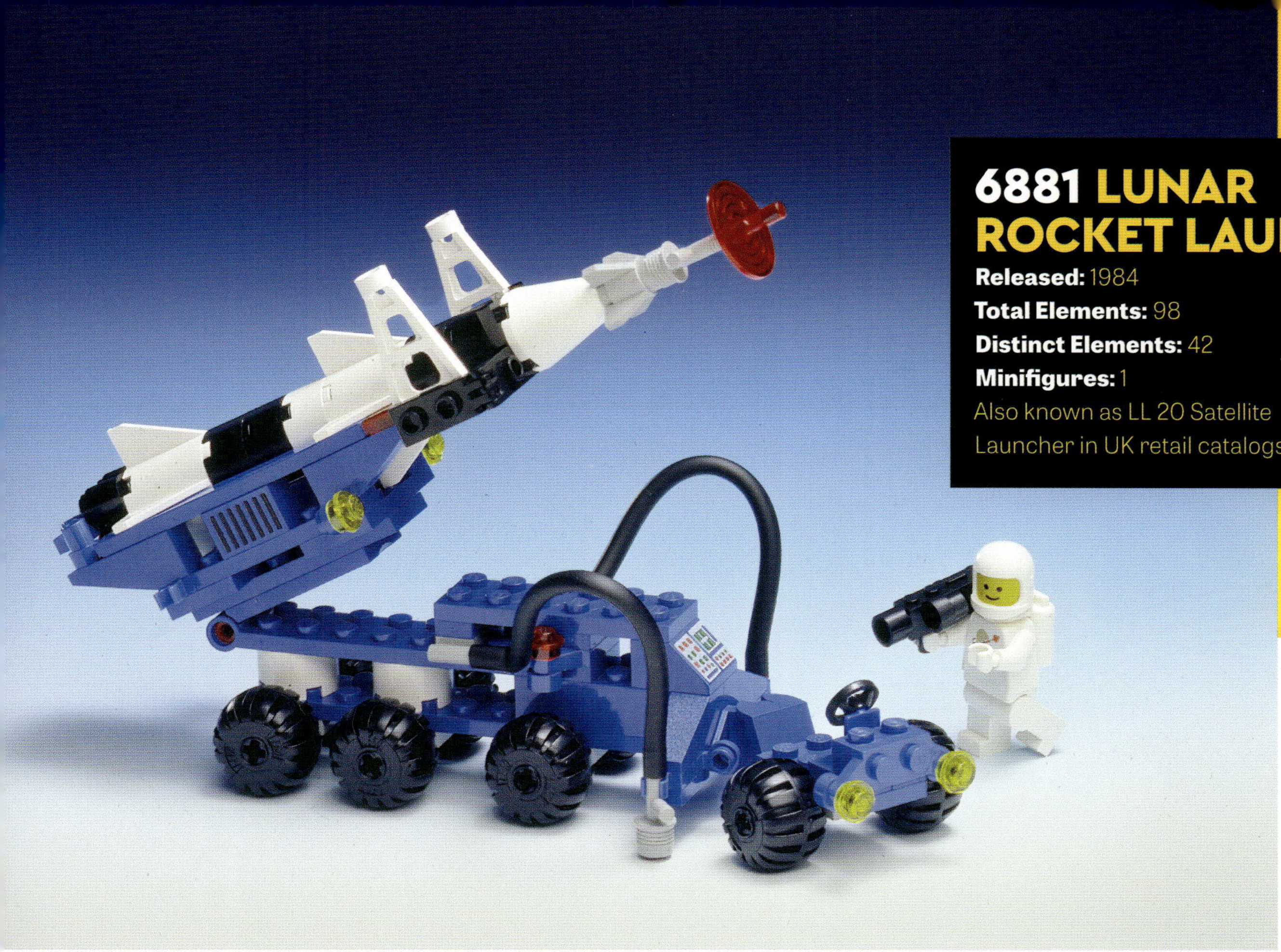

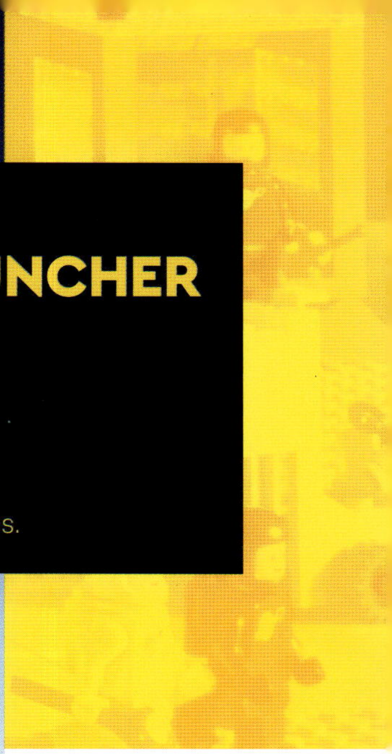

6881 LUNAR ROCKET LAUNCHER

Released: 1984
Total Elements: 98
Distinct Elements: 42
Minifigures: 1
Also known as LL 20 Satellite Launcher in UK retail catalogs.

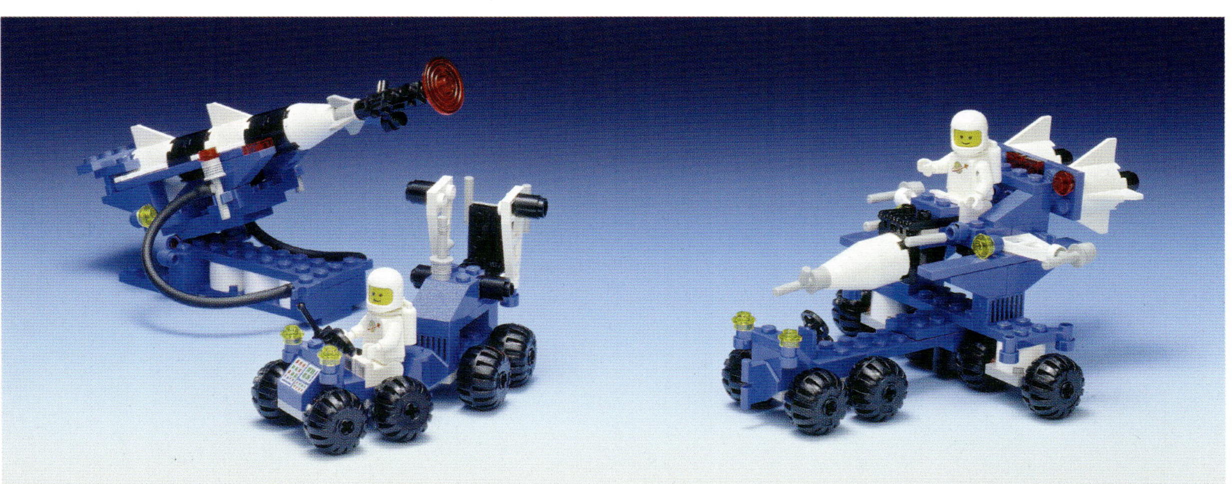

◀ These photographs depict two possible B models for set 6881 and were shown on the back of the box.

The larger of the two launcher vehicles from this year uses black and white for the rocket as in the original 1979 wave, but this 1984 version is certainly distinguished by its blue eight-wheeled vehicle and impressive detailing on the rocket, including the new space wing and the welcome return of the minifigure shield in transparent red. Unusually for a launcher set, there is only one minifigure. The arm and the rocket are not connected by studs and tubes, so the system does not support a vertical launch.

The scientist leapt from his seat, snatched one of the data cables, and clamped it to the side of the rocket. He clambered over the vehicle to attach the other cable—this was ideally a two-person operation, but in all probability he was now the last surviving human on Cygni Bb. He reached up to the command desk and began the data transfer. As the 38-zebibyte upload began, he took a deep breath. If this satellite didn't make it to the Delta team, everything he had learned about the Negative Forces would be lost forever. Almost half of the sky was now pitch black, but he could still see Zeta Cephei. He angled the launch arm and programmed the flight path. Ninety-three percent transferred . . . his finger hovered over the execute command . . . ninety-four percent . . .

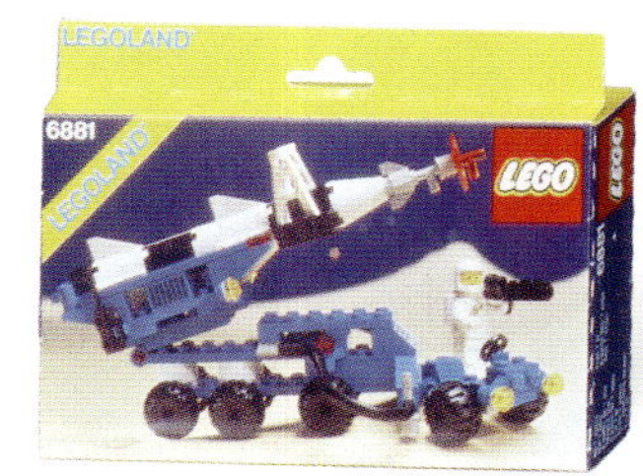

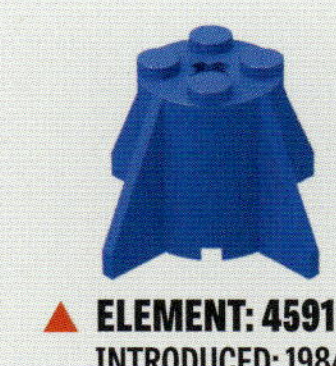

▲ **ELEMENT: 4591**
INTRODUCED: 1984

*The **round brick 2 x 2 x 2** with fins was a thrilling addition to the LEGO® Space inventory, but unlike the smaller round brick with fins introduced in the same year, it proved less useful in other LEGO themes, and has not been seen since 2012.*

6891 GAMMA V LASER CRAFT

Released: 1985
Total Elements: 135
Distinct Elements: 61
Minifigures: 1
Also known as Intergalactic Freighter Dark Star in UK retail catalogs.

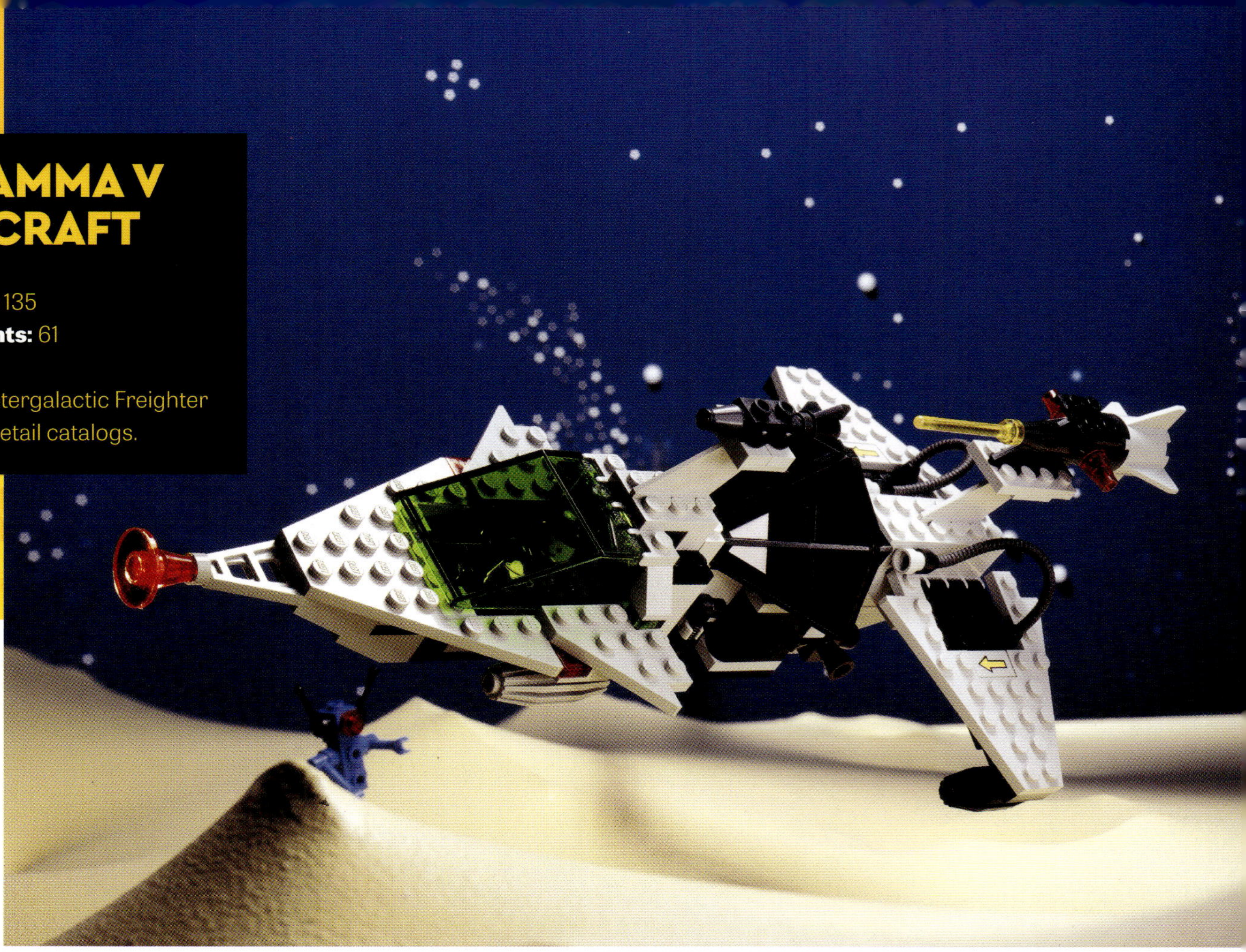

▶ This exciting German ad from 1985 imagines a wholly robotic crew of the Gamma V arriving on a fertile world. It was unusual to shoot ships from a low angle.

Two large ships graced the LEGO® Space line this year, this one sporting the increasingly common color scheme of gray with black and transparent green highlights and presenting a refreshing new design with angled rear wings and double-height empennage. Its dart-shaped form made it easy to hold by the central cargo section—especially useful when pulling it apart into its two sections, which are connected by a LEGO® Technic pin. A tall, skinny blue android adds variety to the industrial color scheme.

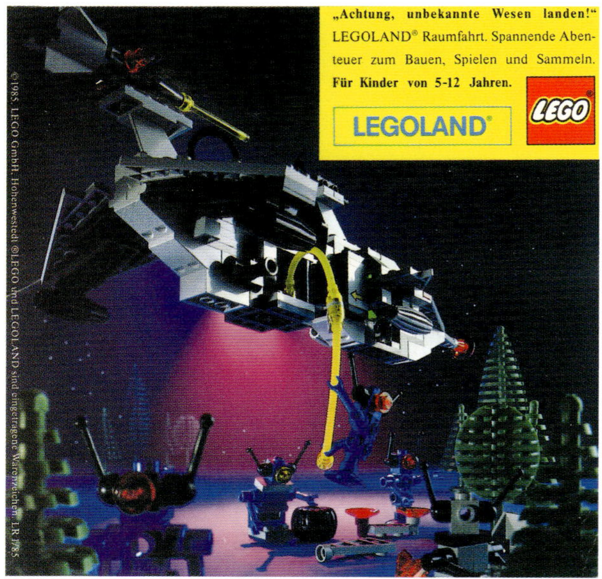

"I'm still seven light months from Alpha Centauri. There's no point sending help." With that, the pilot cut transmission to Inter-Galactic Command. As her craft shot faster than the speed of light toward the black void ahead, the android spoke. "Information: I have plotted a suitable course of evasion. Prediction: sixty-eight percent chance of success." The pilot took a deep breath. "Otte-91, negative to that. And . . . I'm sorry." With that, the pilot released the gravity pin to split the craft and swerved off course, leaving the detached aft section to tumble helplessly onward toward the Negative Forces. With it went the android, the zirconium capacitor, the precious gamma fuel, and, most worryingly of all, the tail-mounted gamma laser. She reopened communications—but not with command this time. "Sort Elektronika to Negative Forces. You have the technology, as promised. Now give us what we ask."

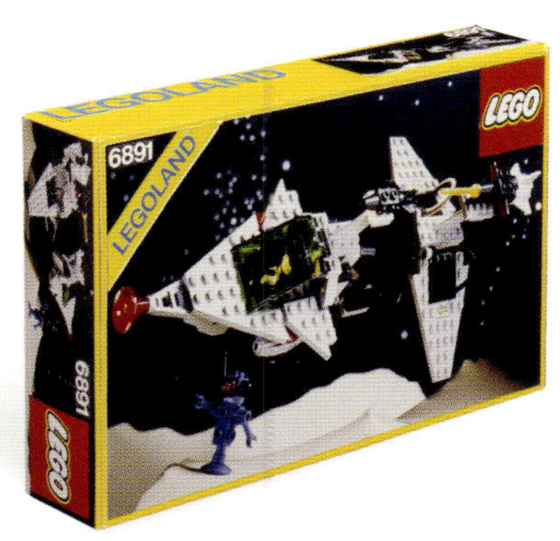

*A highly imaginative design, the decorative ribbed **convex corner panel 4 x 4 x 6** features diamond-shaped edges that can line up with the angle of the roof window 4 x 4 x 3. It appeared in only eight LEGO Space sets before being retired in 1990.*

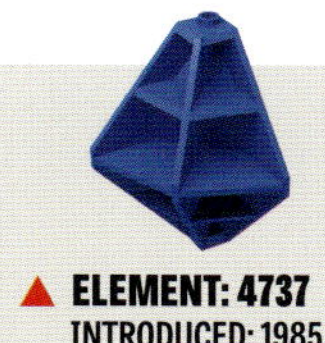

▲ **ELEMENT: 4737**
INTRODUCED: 1985

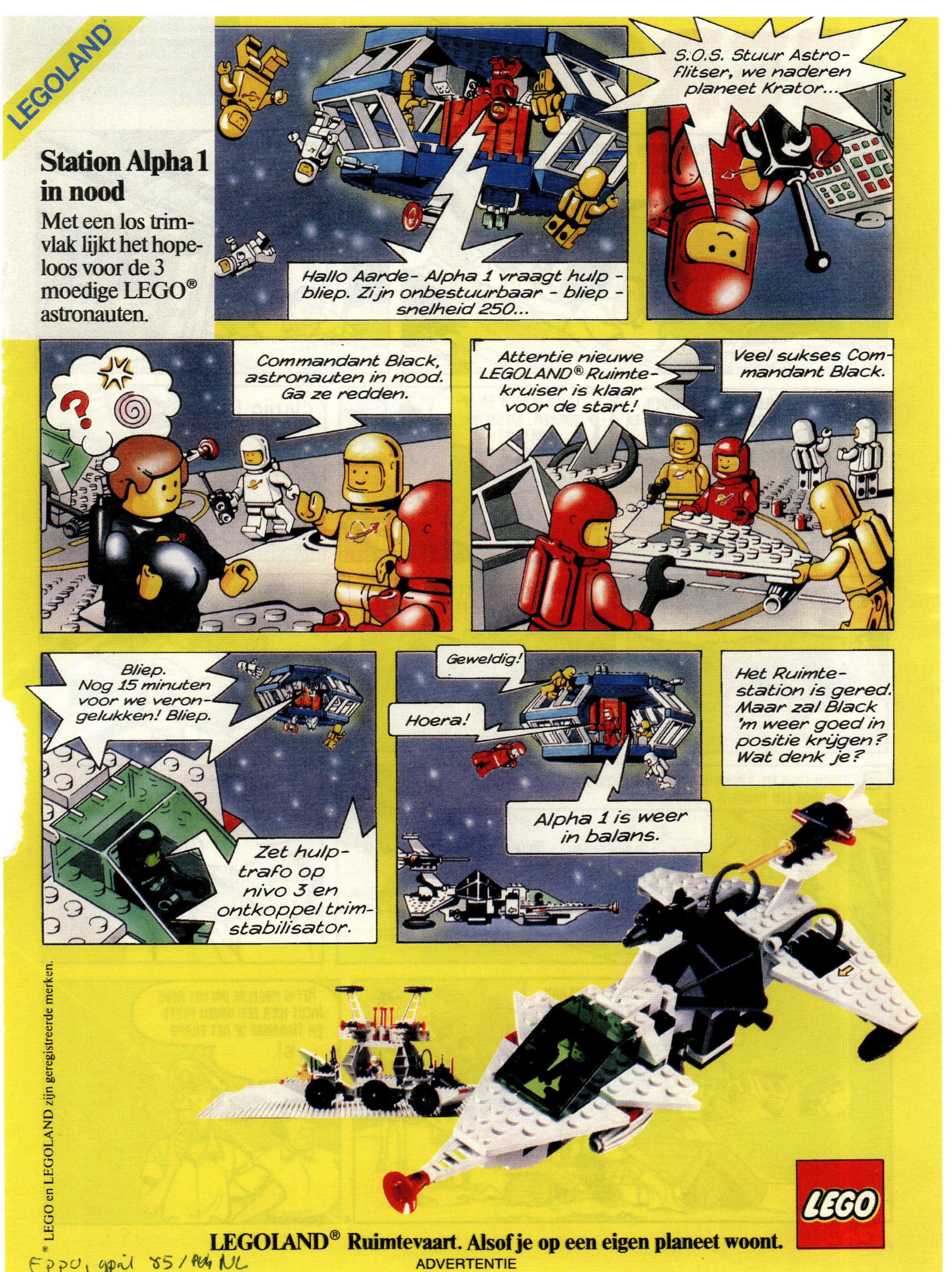

In this 1985 Dutch ad in the format of a comic, Commander Black brings the Gamma V to the rescue of Station Alpha-1. Although the space station was never a product, it did appear in brick-built form in other ads.

6806 SURFACE HOPPER

Released: 1985
Total Elements: 23
Distinct Elements: 8
Minifigures: 1

The move away from steering wheel controls allowed for this quirky jet-powered hover chair, and its twin antenna controls look like they could navigate any tight situation. Other new elements are present too: combined with a transparent green dish 2 x 2, a black robot arm becomes an interesting space tool. The new flexible hose elements add evocative detail when bunched up on a small vehicle.

The pilot hopped up the cliff face, firing the surpulsion motors in short bursts to conserve fuel. He angled the port joystick to veer over to an interesting fissure but misjudged the sensitivity of this new control: the hovercraft toppled sideways, but thankfully the fuel line obstructed his fall. "Moons and meteors!" he muttered under his breath as he returned to his task. If they couldn't find an abundant natural source of hydroxylamine in the Sagittarius Dwarf Galaxy anytime soon, they'd have to find a new way to power surpulsion.

*Smaller versions of existing elements were being introduced more often, and the new **small antenna** not only afforded finer detailing on models but also provided new functionality: it comprises a base and a lever which can be angled. Its design is simple enough to be used for a host of applications, such as a robot head.*

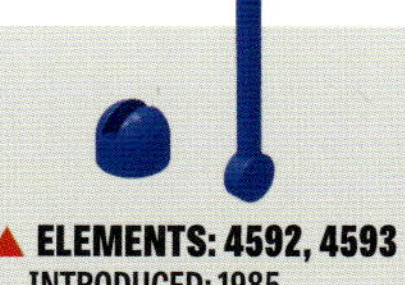

▲ **ELEMENTS: 4592, 4593**
INTRODUCED: 1985

6805 ASTRO DASHER

Released: 1985
Total Elements: 29
Distinct Elements: 14
Minifigures: 1
Also known as Space Patrol in UK retail catalogs.

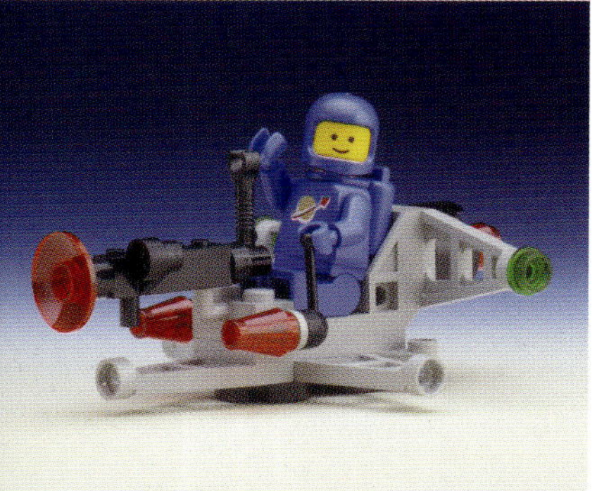

This zippy craft shares much of its design aesthetic with 1984's 6824 Space Dart I, but as was customary with the smallest LEGO® Space sets, it includes several new elements. Two small antennas replace the steering wheel, cones 1 x 1 are used on the front and rear, and the black space camera uses a dish 2 x 2 and a robot arm as a rear sight.

Swooping across the surface of an alien moon on a defensive patrol was the last place the seismologist had expected to find herself, but she had to do her part. Activating every one of the S- and P-wave interferometers that surrounded her, she pitched downward into a narrow, deep valley, tugging each control lever in turn to navigate the many rocky outcrops. One of the starboard interferometers gave a reading, and she turned the craft around deftly in the narrow space. Closer analysis suggested it was probably nothing, but she was under orders to take no chances. She powered up the dash-mounted quantum vacuum squeezer, took aim, and annulled the potential anomaly.

*The **bar 1 x 3 with clip and stud receptacle** is commonly known as "the robot arm," but it enjoys a wide variety of applications in LEGO sets to this day. Its useful connection points allow it to connect with studs, holes, bars, and clips. The thin grooves add subtle design flair, echoing earlier elements, such as the minifigure space gun torch.*

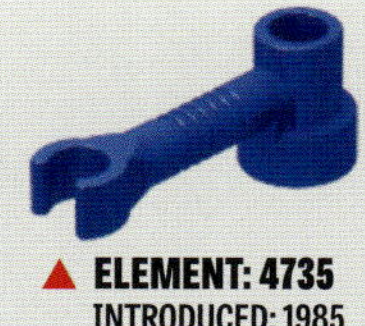

▲ **ELEMENT: 4735**
INTRODUCED: 1985

This wing design features a rare appearance in LEGO® Space of a straight leading edge, which balances nicely against the small wings. Gray elements add functional and technical detail and—combined with the new hoses and an unusual four-cone exhaust—imbue this small craft with a sense of industrial realism.

Swoosh! The pilot couldn't help smiling as he felt the xenon engine roar beneath his feet, its energy pulsing down the fuel lines at either side of him to the quadrilateral impulsors. The dash-mounted lidar bleeped and began rotating starboard, pinpointing the epicenter location as being in Cygni Bb's eastern quadrant. Tugging the control levers in opposing directions, the pilot spun the craft sharply and, switching to the levers by his waist, fired the quantum vacuum squeezer. A circle of red light burst from its discoid muzzle and enveloped the ground, and following a deep sonic boom, the quake stopped in its tracks.

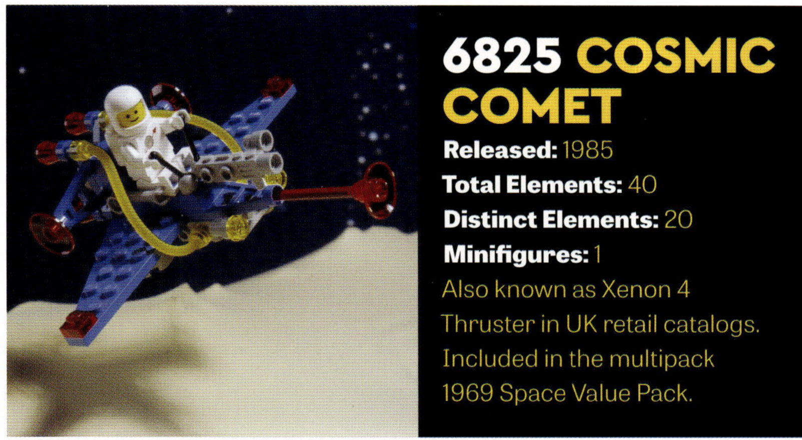

6825 COSMIC COMET

Released: 1985
Total Elements: 40
Distinct Elements: 20
Minifigures: 1
Also known as Xenon 4 Thruster in UK retail catalogs. Included in the multipack 1969 Space Value Pack.

***Modified brick 1 x 2 x ⅔ with studs on sides** is an unusual element that strikes a successful balance between functionality and novelty. Its "traffic light" design comprises dual studs on three sides and tubes on a fourth. Here, two are joined to create an engine underneath the wings, while in other sets it was used as a robot torso or to add industrial-looking details to ships, vehicles, and missiles.*

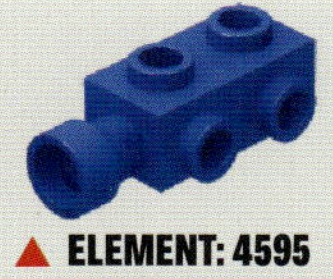

▲ ELEMENT: 4595
INTRODUCED: 1985

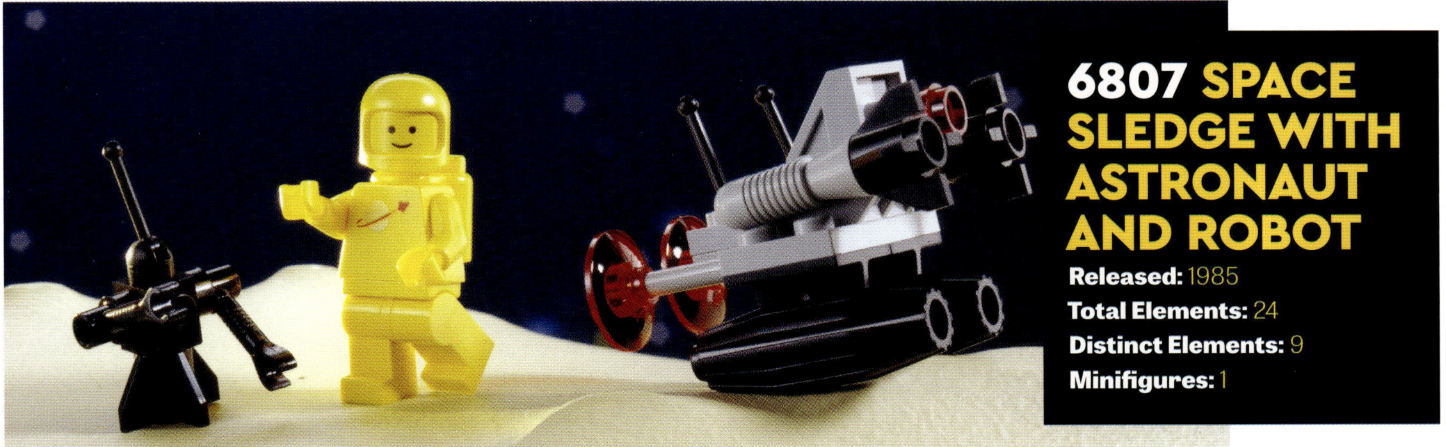

6807 SPACE SLEDGE WITH ASTRONAUT AND ROBOT

Released: 1985
Total Elements: 24
Distinct Elements: 9
Minifigures: 1

A sledge (or sled) might not be the first thing that comes to mind when thinking of extraterrestrial vehicular travel, but when it is powered by twin rockets and comes with a little android constructed from new elements, one might think again! The exaggerated black elements on the vehicle give this set a unique charm.

With dwindling hydroxylamine supplies proving a very real threat, the mechanical engineer developed a more fuel-efficient vehicle that took advantage of the geography of Messier 54 Omega Cf. With just short bursts on its two-cylinder twin surpulsive engines, she anticipated it could skid for long distances across the gentle dunes. Rather than weighing it down with navigational and communications hardware, she paired it to a zirconium android. "Otte-07, prepare for test flight," she instructed. The robot whirred gently, its antenna rising to a vertical position. "Acknowledged."

*Traditional LEGO® building bricks with added studs on the sides are common in LEGO building nowadays, but **modified brick 1 x 1 with studs on 4 sides** was only the second such element to be introduced. The ability to "build sideways" in four directions opened up new possibilities for finer detailing or simply for use as a robot torso or head.*

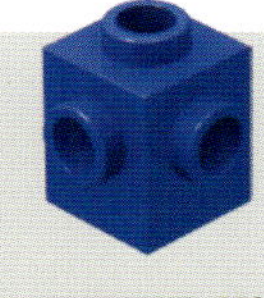

▲ ELEMENT: 4733
INTRODUCED: 1985

6826 CRATER CRAWLER

Released: 1985
Total Elements: 32
Distinct Elements: 17
Minifigures: 1
Also known as Lunar Mobile in UK retail catalogs.

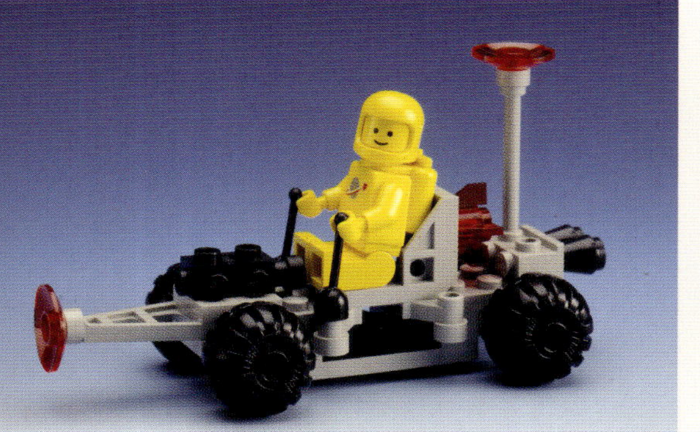

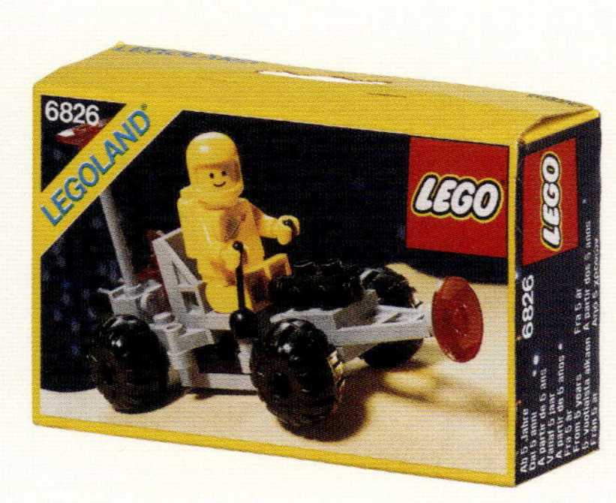

At first sight this ground vehicle appears somewhat stripped down, with a thruster near its middle and a gap in the chassis, but that's because the operator is not restricted to ground travel! The central section detaches vertically to reveal a hover buggy similar to 6806 Surface Hopper. The rover's black exhaust nozzles make use of the new cone 1 x 1, and transparent red dishes 2 x 2 complete the antennas.

The exploration geologist gradually rumbled across the rocky landscape of Messier 54 Omega Cf, cursing the low-quality biodiesel engines they had now resorted to using. It was a relief to finally reach the valley and fire up the surpulsion, lifting away from the buggy and lowering down into the cavernous depths. It wasn't long before yellow seams began to appear in the rock face, and he paused, hovering in the gloom, assessing the mercury nitrate formations.

Chassis are somewhat challenging to design for LEGO® vehicles because they must be small, strong, and have a sunken area for the minifigure. ***Bracket 8 x 2 x 1⅓*** *is one of the elements that have been introduced to simplify the process and improve playability. It was also used by LEGO® Town and LEGO® Castle but was discontinued in 2006.*

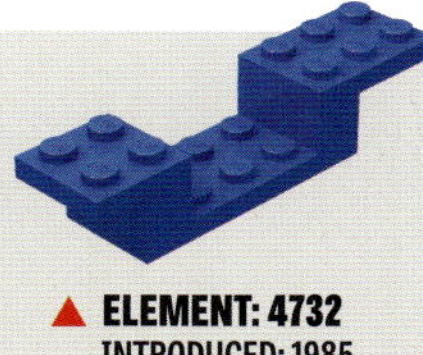
▲ **ELEMENT: 4732**
INTRODUCED: 1985

6847 SPACE DOZER

Released: 1985
Total Elements: 49
Distinct Elements: 27
Minifigures: 1
Also known as Lunar Dumper in UK retail catalogs.
Included in the multipack 1969 Space Value Pack.

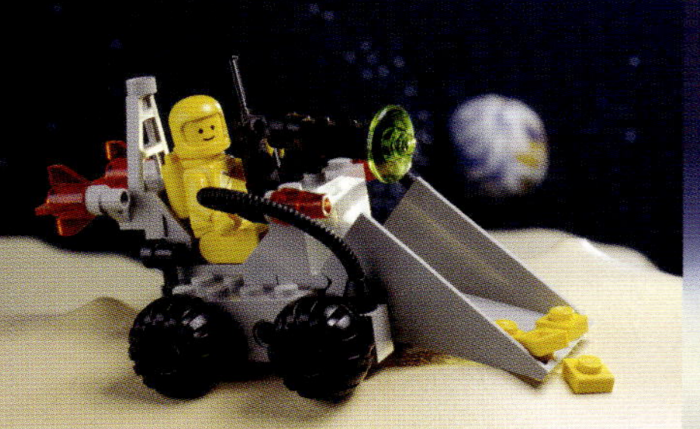

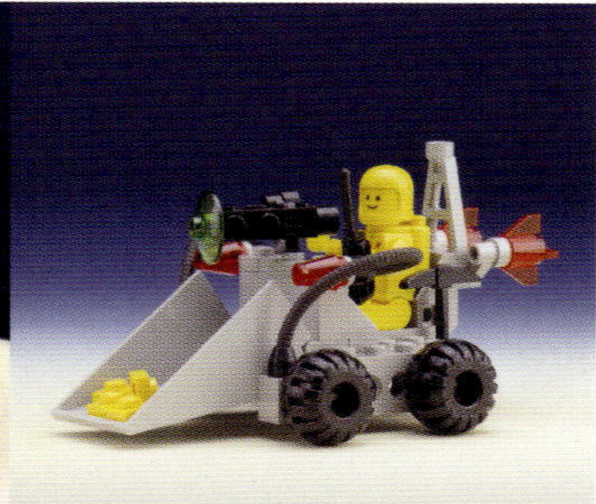

The trend in LEGO® Space sets for rocket-powered ground vehicles sometimes belied the more fantastic tone of the theme, but this bulldozer is so wide and heavy, perhaps the extra power would be necessary! Windscreen 6 x 4 x 2 canopy is used for the massive shovel, and a minifigure utensil pickaxe appears for the first time in LEGO Space.

Deep on the valley floor, the miner tapped the yellow rock gently with the pickaxe and pointed his handheld spectrometer at the exposed red crystals, which soon began to yellow in the acidic air. "Dozer to shuttle," he said into his transceiver, "cinnabarite is confirmed. Proceed for collection." After positioning his vehicle, he angled the green dish of the ultrasonic transducer and after a good ten minutes of loosening the surface, he blasted it with the twin compressor jets. Chunks of cinnabarite tumbled to the ground, yellowing as he scooped them up.

Previously, small cone shapes had been used within the design of other elements, such as modified brick 1 x 1 with three space positioning rockets, but finally the ***cone 1 x 1*** *was introduced in 1985. It appeared in all but one of the LEGO Space sets that year in five different colors. Its interior grooves can grip an axle element.*

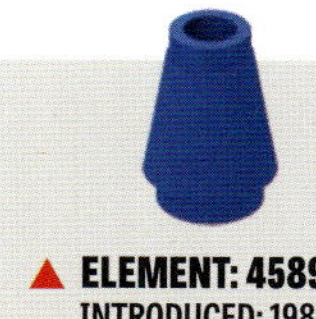
▲ **ELEMENT: 4589**
INTRODUCED: 1985

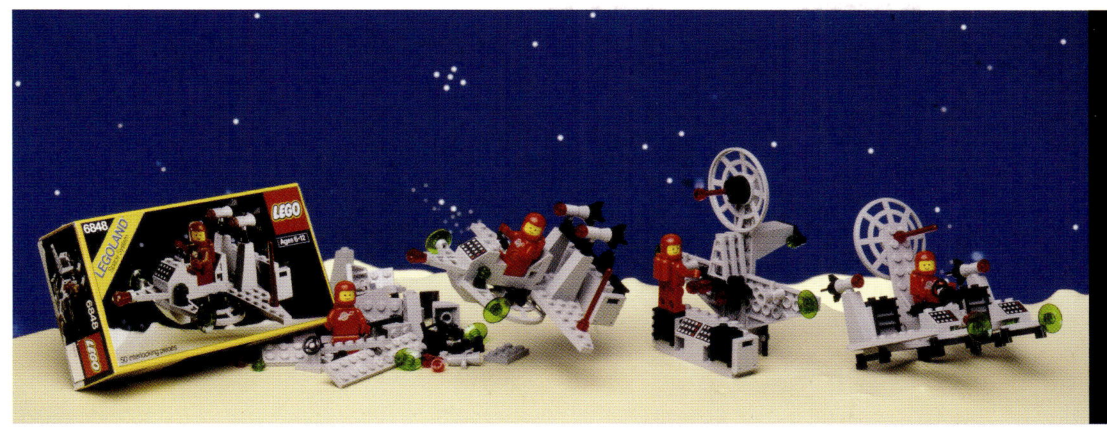

6848 INTER-PLANETARY SHUTTLE

Released: 1985
Total Elements: 54
Distinct Elements: 31
Minifigures: 1
Included in the multipack 1969 Space Value Pack.

▲ This unusual photograph shows a mockup of the box alongside its elements, the model, and two of the B models.

The nonaerodynamic profile, mining accessories, and generous storage capabilities of this craft suggest it may not truly be suitable for interplanetary travel! This was the first LEGO® Space set to use the slope 75° 2 x 1 x 3, which had been introduced for LEGO® Castle in 1984. The two front-facing control panels were also new to LEGO Space, having appeared in a couple of LEGO® Town sets since 1983.

Traveling to another planet in a surpulsion-powered shuttle tailored to mining work was sheer insanity, but with no gamma fuel left for the transporters, it was their only option to get the compressed mercury sulfide to the lab on Messier 54 Omega Cg. So the pilot had to improvise: discharging the surpulsion disk only occasionally on its lowest setting, she would adjust the trajectory by tapping the twin compressor missiles with a hammer and firing them in quick bursts. The ultrasonic transducers proved useful too, clearing her path when she encountered an asteroid belt.

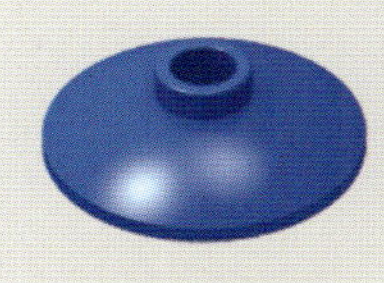

▲ **ELEMENT: 4740**
INTRODUCED: 1985

Radar dishes were an iconic part of LEGO Space, *and given the trend for designing smaller versions of existing elements, it is no surprise that* ***inverted dish 2 x 2*** *was introduced and used in virtually every 1985* LEGO Space *set. Unlike larger dishes in the family, its stud is recessed to fit bars, making it easy to attach to minifigure accessories.*

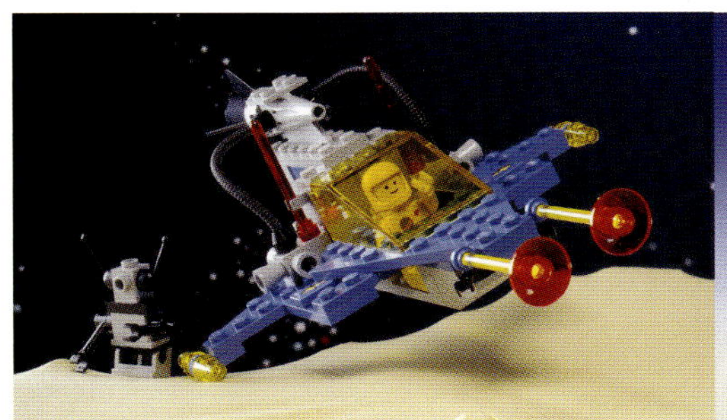

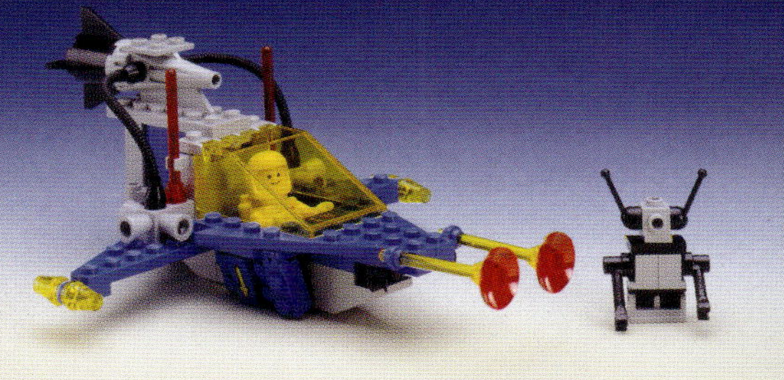

6872 XENON X-CRAFT

Released: 1985
Total Elements: 77
Distinct Elements: 36
Minifigures: 1

This midsized craft bears many hallmarks of the initial wave of LEGO® Space ships—the color scheme, rear cargo storage, and hazard markings on wings—but is clearly updated for 1985, including a robot! Among the many new elements adding fine detail is a powerful tail-mounted engine, and for the first time, the controls are printed onto a tile. The small blue winglets can be folded down to complete the hull, or, if positioned halfway as fins, they intimate the X formation suggested by the name of this set.

"Recommendation: Seal cockpit" came the gentle voice of the android as they rose through Cygni Bb's thick atmosphere. "Otte-72, understood," said the pilot. As the panels clamped shut alongside her feet, she depressurized the chamber. "Otte-72, how long until we get to Tadmor?" she asked as she swiped the control screen. "Data: approximately two days by direct route, which can be reduced to thirteen hours using full xenon tenfold power capability" came the reply. "Recommendation: A detour of two hours will avoid the Negative Forces' airspace." The pilot smiled as she fired up the xenon thruster. "Oh, they'll never catch up with us."

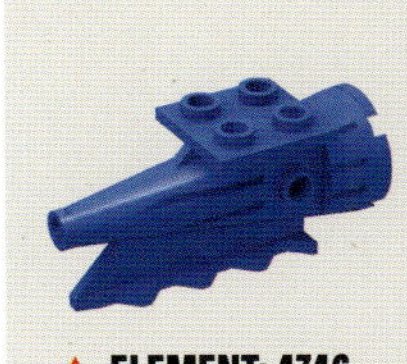

▲ **ELEMENT: 4746**
INTRODUCED: 1985

A thrilling new empennage in a single piece, ***aircraft rocket engine tail 4 x 2 x 2*** *is the intersection of a tail and a grooved cone that tapers from one to two modules in diameter over a length of just over five modules. For additional detailing possibilities, there are side studs with a recessed horizontal hole. It survived until 2002.*

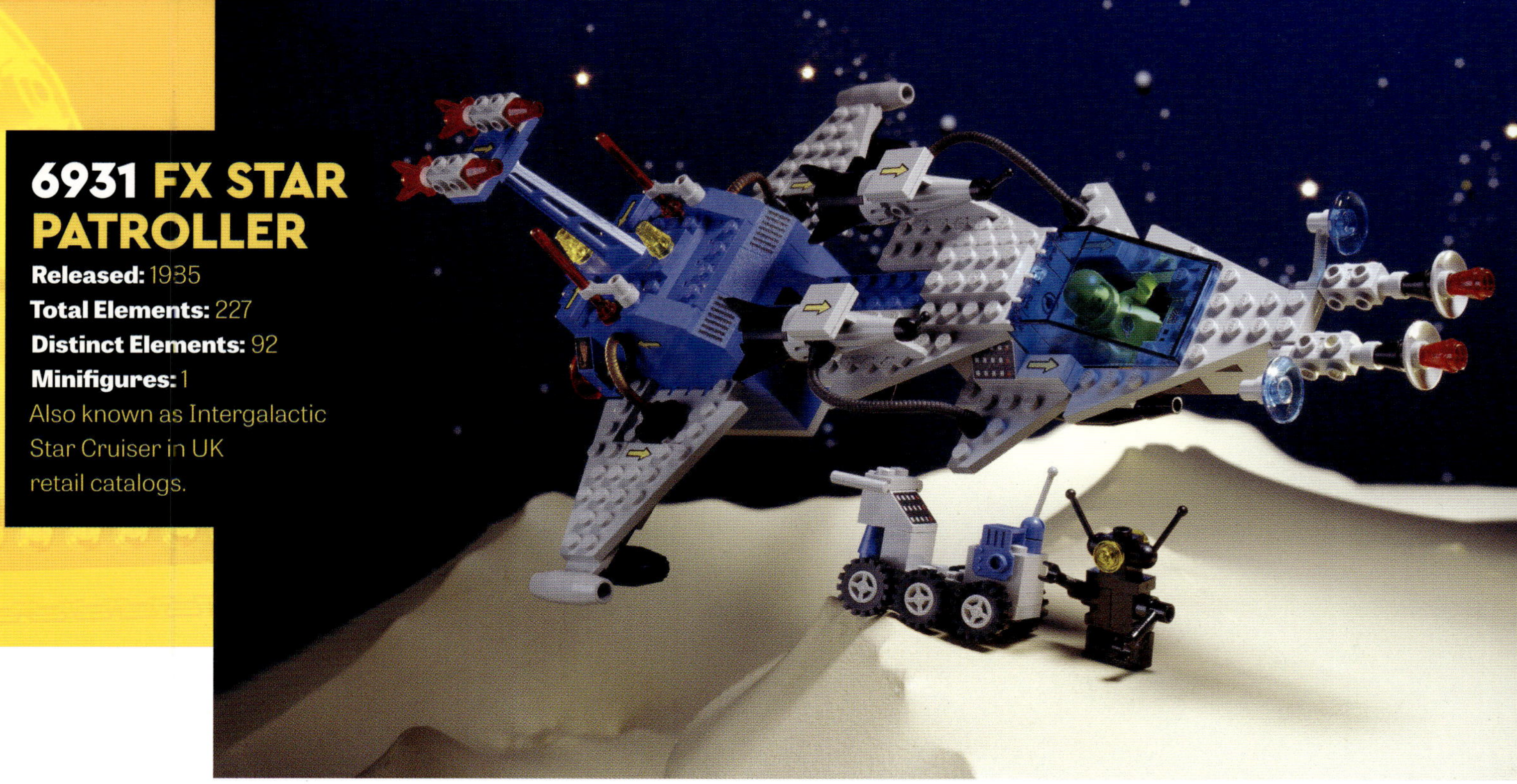

6931 FX STAR PATROLLER

Released: 1985
Total Elements: 227
Distinct Elements: 92
Minifigures: 1
Also known as Intergalactic Star Cruiser in UK retail catalogs.

The largest ship of the year shared structural similarities with 6891 Gamma V Laser Craft, with angled rear wings and clearly delineated front and rear sections that can be separated. Its color scheme, however, was more classic and used a transparent blue canopy for the first time since 1983. The elegant "neck" between the sections is balanced by the bulbous cargo bay, which is accessed via the hinged empennage. It accommodates a small six-wheeled buggy with a black robot and a minifigure jetpack. Once detached, the fore section makes a magnificent ship in its own right: a sleek form with thrilling twin engines and interesting details throughout.

The gravity pin disengaged with a loud clunk, and the pilot edged forward to free the cargo bay. "FX to ground crew," she said into her transceiver. "Immediate refuel, please." The empennage swung upward with a hiss as the vacuum of the cargo bay was exposed to the Tadmorian atmosphere, and the android tumbled out in its rover, turning instantly to draw up alongside the cockpit. "Assessment: Rescue mission is questionable," it stated blankly. "Chance of success is less than three percent." The pilot smiled. "Ni-30, no chance. We just can't let that tech fall into the wrong hands." The droid bleeped. "Data: Bioreport indicates signs of long-term fatigue," it announced. "Well, Ni-30, we better take turns flying then," she said lightly. "Hop on." Minutes later, the gamma engines fired, and the patroller swooped skyward, disappearing in an instant.

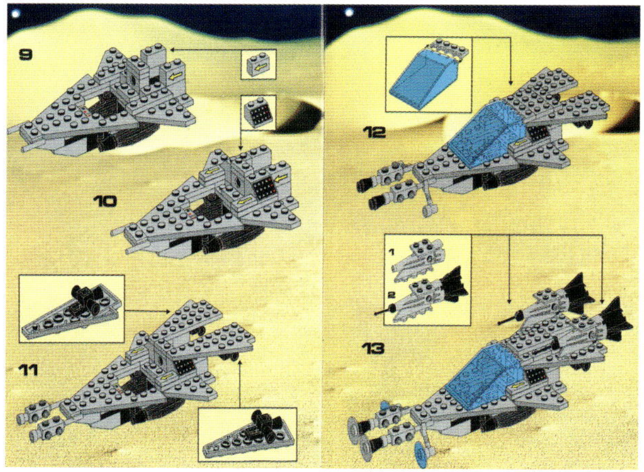

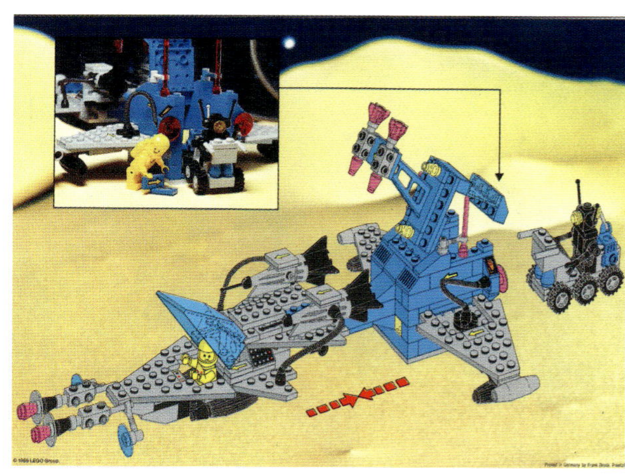

*A printed element that only appeared in three sets, **slope 45° 2 x 1 with orange microphone on black** adds a touch of industrial realism. Its design is generic enough to be open to interpretation, but its placement here next to the black hoses at the rear suggests that it may actually represent a hose plug.*

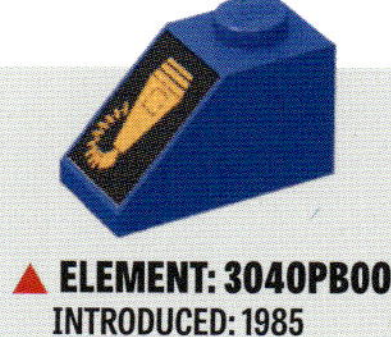

▲ **ELEMENT: 3040PB001**
INTRODUCED: 1985

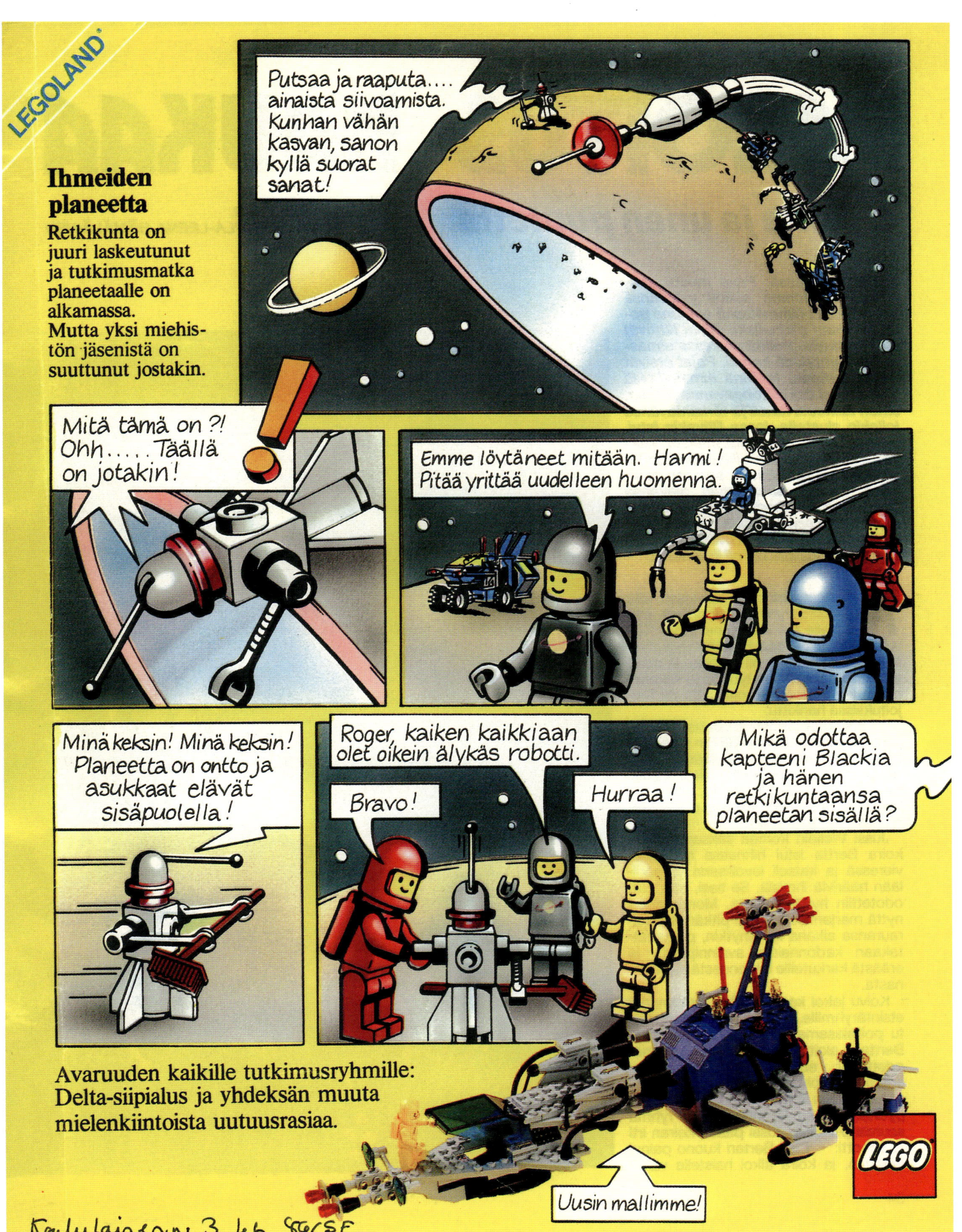

In this 1986 Finnish ad in the format of a comic, the crew arrive on what they think is a planet but is discovered by the robot to be a hollow hemisphere.

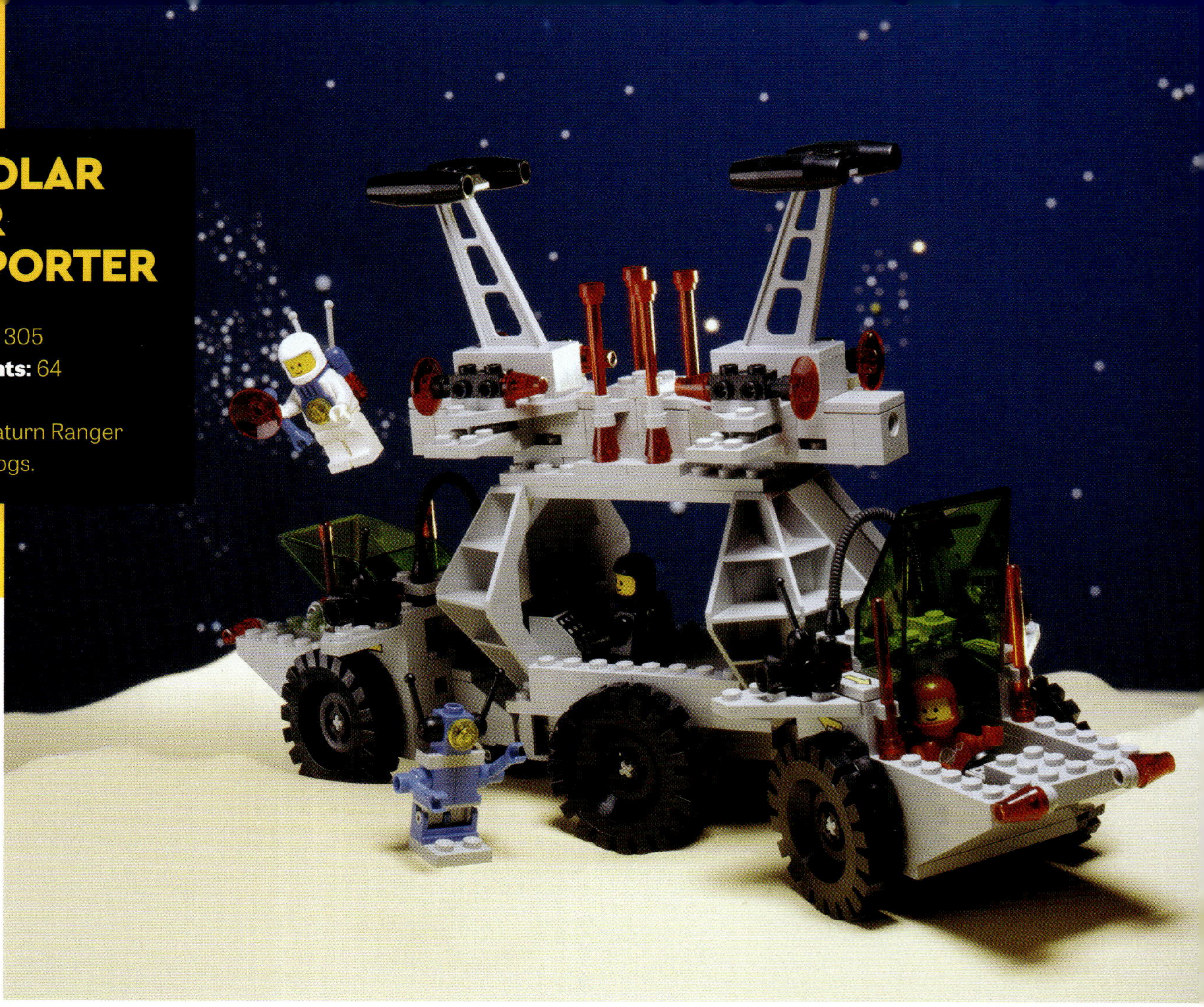

6952 SOLAR POWER TRANSPORTER

Released: 1985
Total Elements: 305
Distinct Elements: 64
Minifigures: 4
Also known as Saturn Ranger in UK retail catalogs.

[6952] WAS THE ONLY 1985 SET TO CONTAIN MORE THAN ONE MINIFIGURE—IT HAD FOUR, IN DIFFERENT COLORS, PLUS A ROBOT!

The biggest set of the year was a vehicle . . . or was it? 6952 Solar Power Transporter embraced the concept of modularity more than any other set before it. LEGO® Technic pins connect not only its three articulated segments but also the two detachable tail sections on the roof of the central coach. These five subbuilds can be reconfigured in many clever ways, four of which are shown on the back page of the building instruction booklet. It was the only 1985 set to contain more than one minifigure—it had four, in different colors, plus a robot!

The petrologist switched off his laser theodolite and rotated his chest dial, powering down his surpulsion pack to lower himself to the ground. "Geo 3 to Ranger. We'll set up here; come to my location." Everyone on the team beamed with excitement as they arrived: here they were, the first humans in another galaxy, testing cutting-edge technology that could power a civilization. The engineer, safe in her zircon nanotube fabric, stepped inside the solar generator to fire up the ultraphotovoltaic tubes, which soon glowed red in the weak light of Messier 54 Rho Ac. Only a couple of minutes passed before the empennages were fully charged, and the petrologist and the android flew up to detach them while the others separated the dual cabs and angled their wheels to hover mode. "How about that," said the engineer. "One hundred fifty seconds of charging, and you'll be flying for fifteen hours! By then, these will be full," she said, tapping one of the four accel-capacitors. "Imagine what we can power then!"

*It was surely just a matter of time before a **minifigure jetpack** came along! Several have been introduced over the years, but this first one featured a stud on the front and two on the back, to which cones or other suitable elements could be attached. This version only ever appeared in twelve sets and has not been used since 2011.*

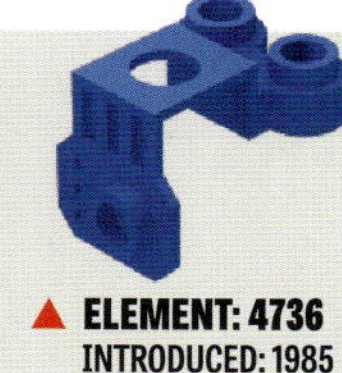

▲ **ELEMENT: 4736**
INTRODUCED: 1985

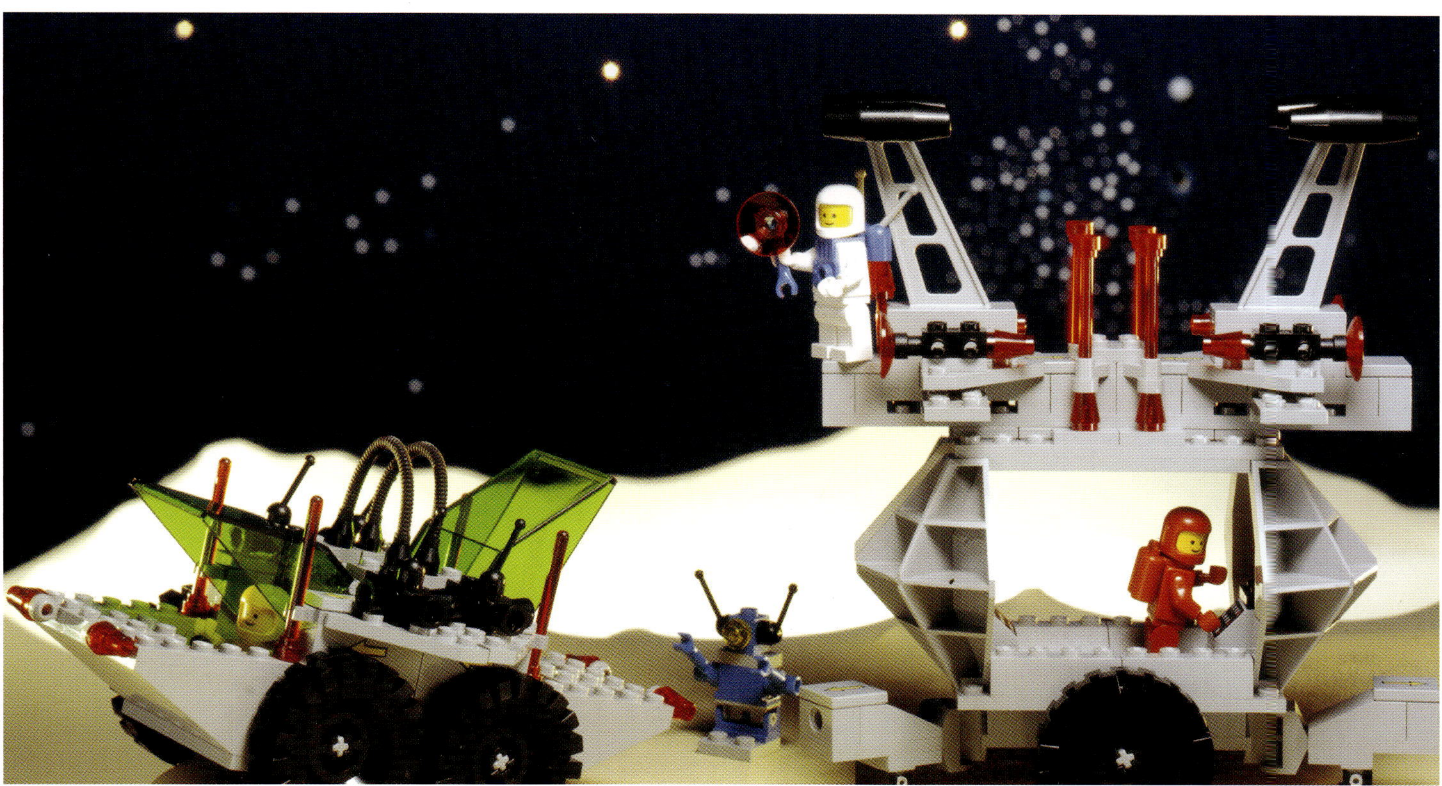

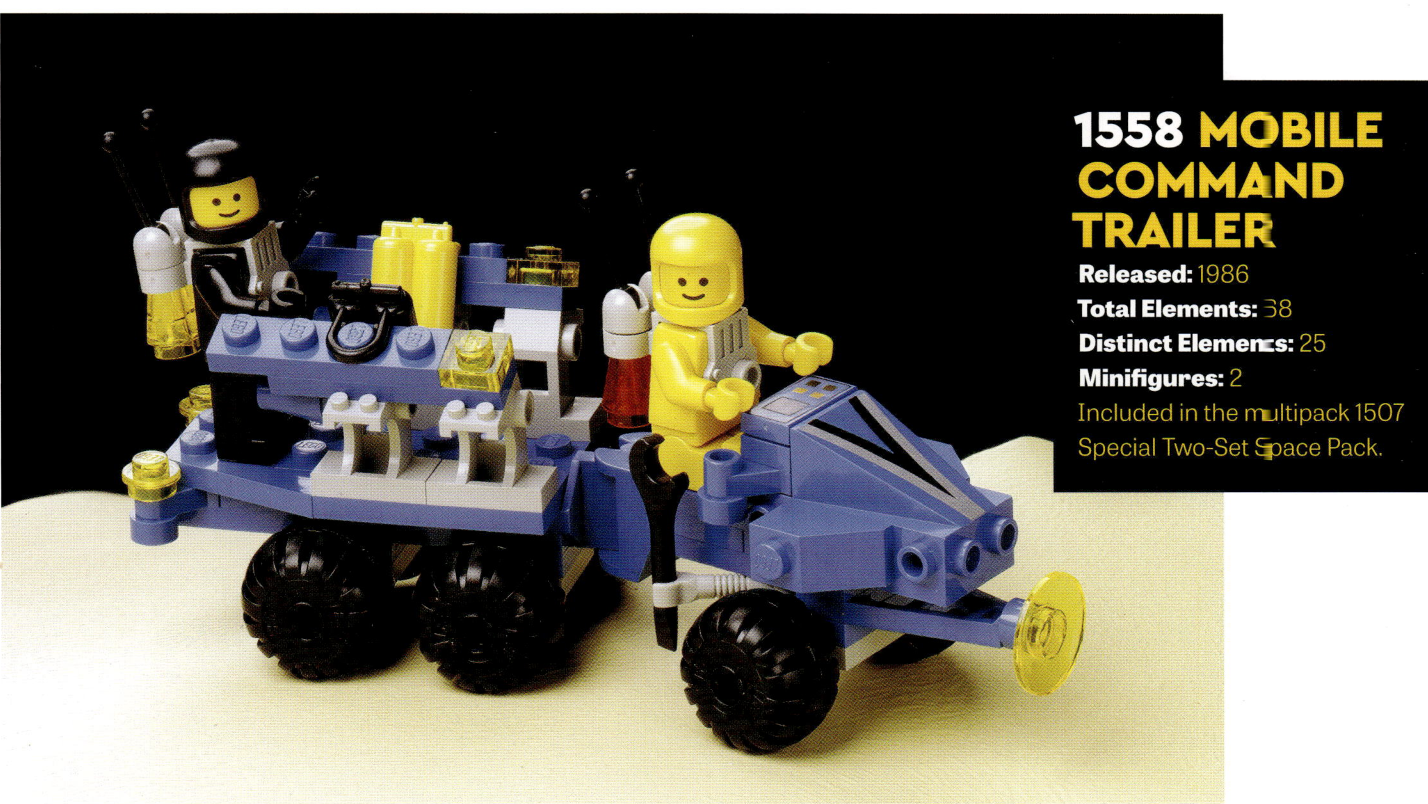

1558 MOBILE COMMAND TRAILER

Released: 1986
Total Elements: 38
Distinct Elements: 25
Minifigures: 2
Included in the multipack 1507 Special Two-Set Space Pack.

This small set offers a lot of valuable elements, including not one but two minifigures with jetpacks and four minifigure control panels, which are used as building elements to create angled equipment racks. It has an exciting look, and serves a useful function as an equipment and people carrier.

The astroculturist could just make out the engineer flying far in front of her: a tiny black dot swaying over the dark mountains of Errai Cg. Why had he rushed back without her? She increased the yttrium jets to maximum power, but by the time their command trailer came into view, he had already reached it and was swinging an atomic wrench wildly. The gamma disk was already just a pile of yellow shards, and by the time she arrived, the charging station and comms rack was beyond repair. He stopped what he was doing and smiled at her, saying, "It's for the best," as he tugged a lever and shot upward to the darkness above.

6882 WALKING ASTRO GRAPPLER

Released: 1985
Total Elements: 102
Distinct Elements: 35
Minifigures: 1
Also known as Xenon 2 Walking Machine in UK retail catalogs.

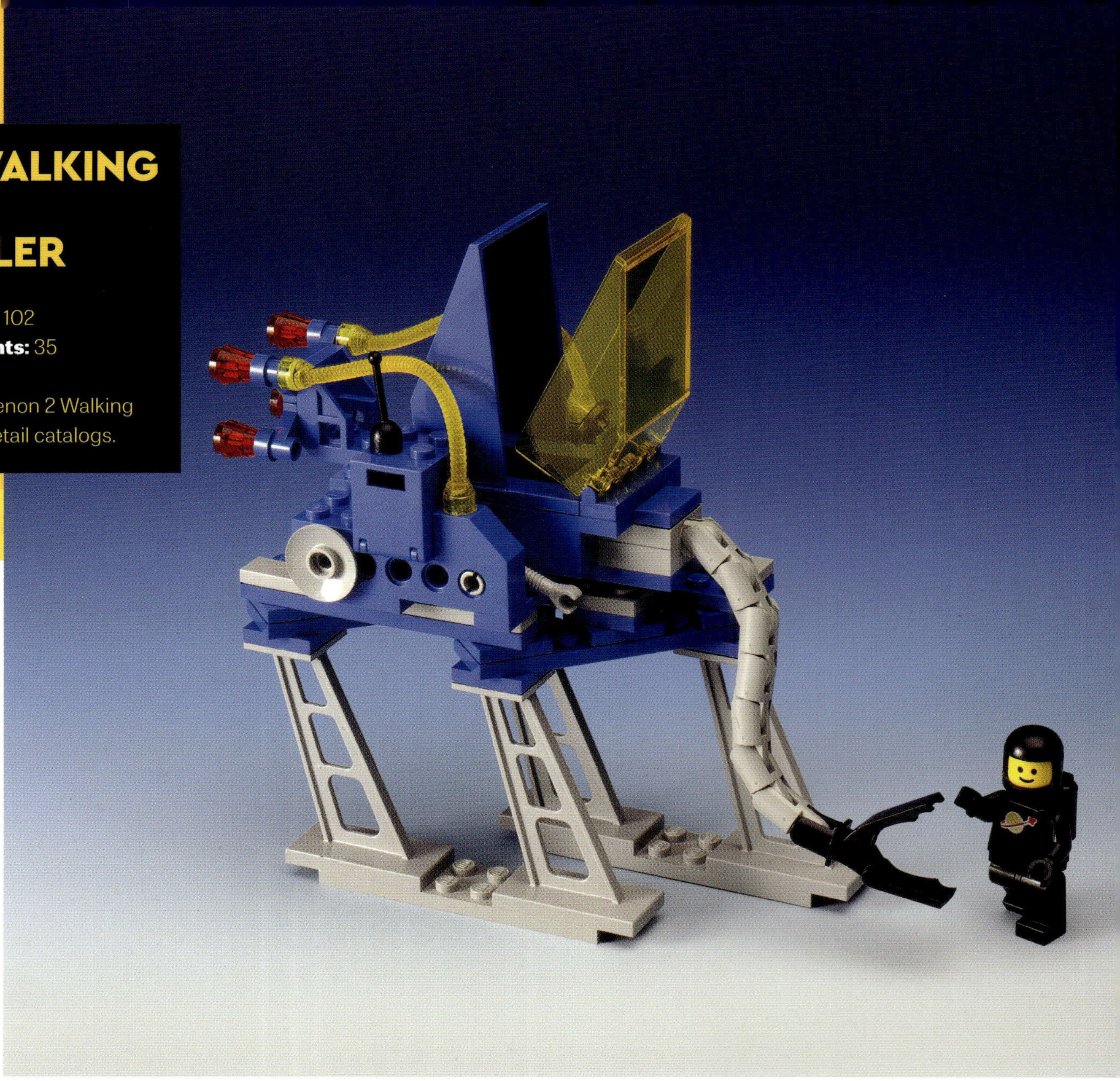

There had been wheelless LEGO® Space vehicles before, but this was the first proper "walker." Its shuffling motion is achieved with LEGO® System in Play elements—ten turntable 2 x 2 plates, to be precise—rather than using the LEGO® Technic system, as one might expect nowadays. Atop this framework are some quirky design features: a clamshell cockpit, a long proboscis with clamp, and a quadrilateral exhaust all combining to create a truly out-of-this-world mining vehicle.

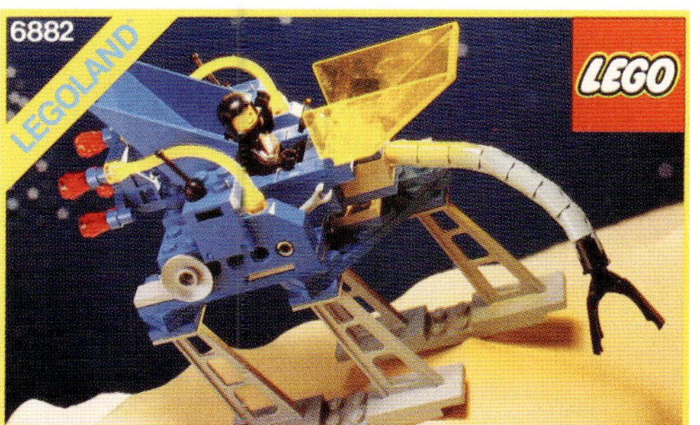

▼ **Original box art for set 6882 Walking Astro Grappler.**

The machine strode tirelessly through the dunes of Tadmor, the sand and zirconium silicate crystals sent flying in its wake. With skillful precision, the ore dressing engineer snapped up the largest crystals with the clamp and tossed them into the scuttles to either side. The seismometer blinked, and he powered down for a moment, unfolding the cockpit to properly scan the horizon. "Grappler to Comet. Negative Forces are striking, eastern quadrant. Locate and disarm the quake." As he closed the cockpit, he pressed the clamp into the ground to angle the walker upward and fired the twin xenon engines. As the strange craft became airborne, speeding over the landscape toward base, he switched transmission frequencies. "Operative 12 to Sort HQ. Negative Forces confirmed on Tadmor. Permission to make contact."

A whole new level of realism and storytelling potential was added to LEGO Space models with the ***flexible hose 8.5 modules****, also appearing in LEGO® Town fire sets and LEGO® Trains sets that same year. The organic folds of their ribbed tubing added vibrancy to the rectilinear form of LEGO bricks.*

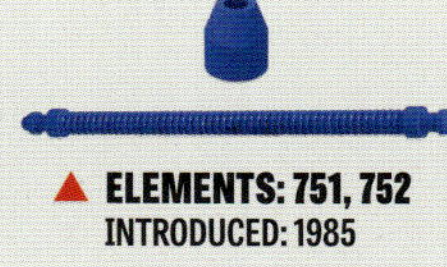

▲ **ELEMENTS: 751, 752**
INTRODUCED: 1985

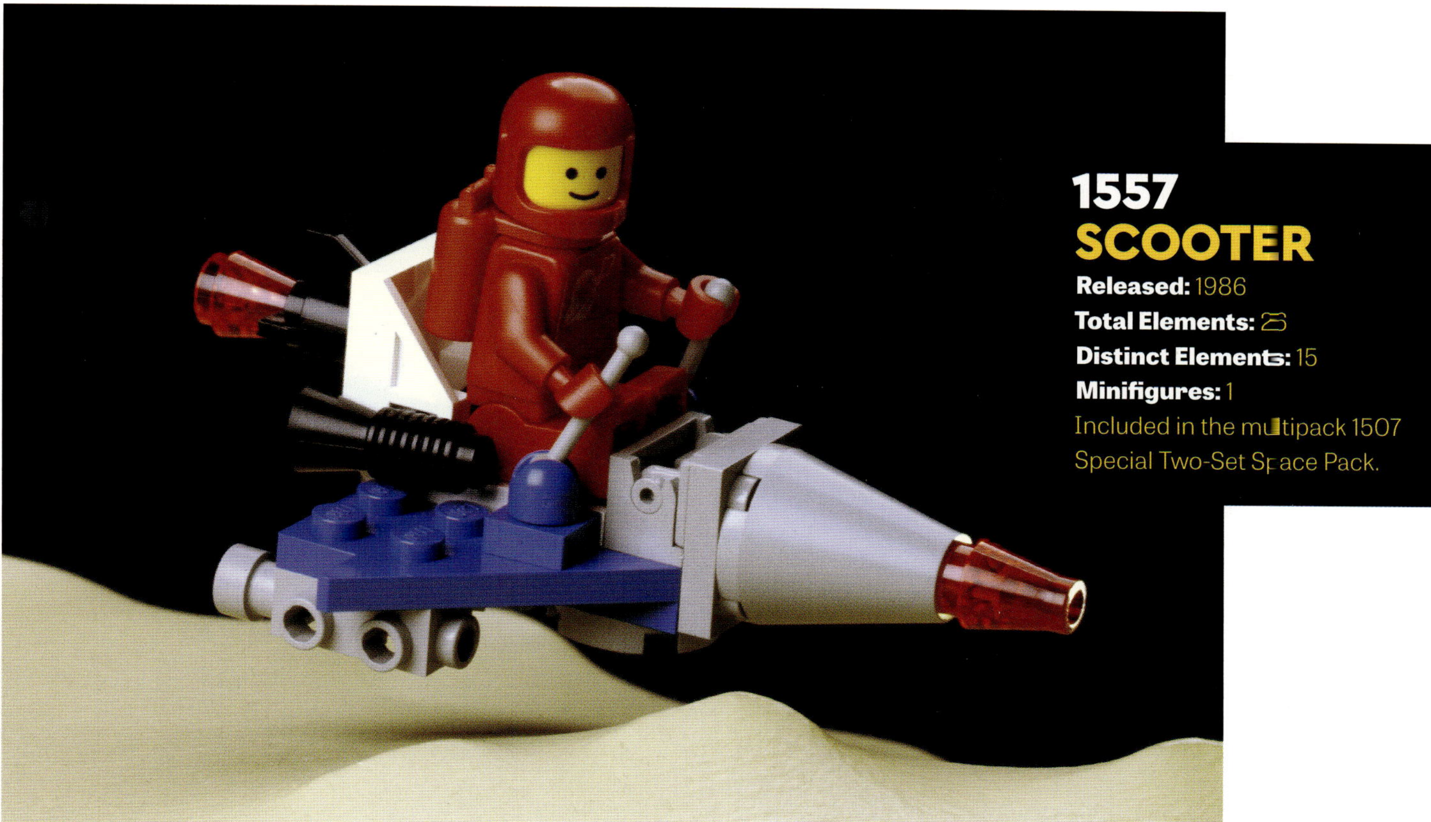

1557 SCOOTER

Released: 1986
Total Elements: 25
Distinct Elements: 15
Minifigures: 1
Included in the multipack 1507 Special Two-Set Space Pack.

The new cone 1 x 1 in transparent red looks excellent here in combination with the larger gray cone. As with 1981's massive 6929 Star Fleet Voyager, it is attached to a hinge brick so that it can be angled—but to point upward. Two modified bricks 1 x 2 x ⅔ with studs on sides make perfect underwing thrusters.

The lieutenant skipped over the surface of Errai Cg, easily navigating the mountainous outcroppings, and as he approached the base of the rumbling volcano, he fired up the quantum vacuum squeezer cone. "Nice try, Negative Forces," he thought to himself, "but you're no match for my QVS!" The tip pulsed red, and he began to fire at the base before gently levering the nose upward to seal the fissure all the way to the zenith. Angling the zirconium thrusters, he began scooting up the side of the volcano to ensure the threat was completely annulled.

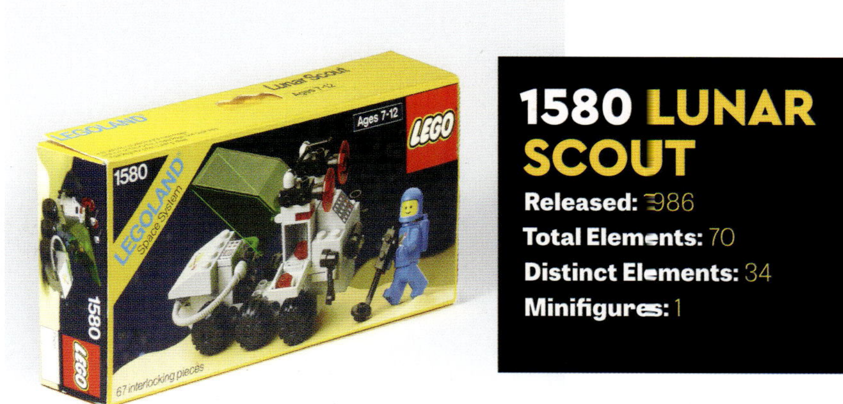

1580 LUNAR SCOUT

Released: 1986
Total Elements: 70
Distinct Elements: 34
Minifigures: 1

The shape of this chunky mining vehicle is reminiscent of a dune buggy, and it features some exciting large elements. It's the smallest set to include the windscreen 6 x 4 x 2 canopy in transparent green and one of only two sets to include a slope 33° 3 x 4 printed with the LEGO® Space logo. Two hoses surround the slope to create an unusual ring-like bonnet. Its blocky rear end can be detached from the vehicle to form a standalone control panel and tool rack.

The hydrologist checked the dash geo as she bounced over the surface of Messier 54 Epsilon Gb. She eventually spotted the pluviograph, toppled on its side and half covered with sand, and powered down the toroidal engine. Thankfully, it must have been blown over after completing the survey, because when she logged in, the readout provided the complete spectral analysis. "Hydro 1 to base. Liquid oxygen confirmed. Bringing you a sample." Locating a likely source with the radar dowsing rod, she shot down deep into the sand with the lightning gun to create a narrow well lined by silica glass. She used the gravity cone to take a sample before dropping the tensiometer deep into the well. While it assessed moisture levels, she lifted the pluviograph back onto the buggy, ready to depart just as soon as the test was complete.

6750 SONIC ROBOT

Released: 1986
Total Elements: 110
Distinct Elements: 84
Minifigures: 2
Also known as Magma Robot in UK retail catalogs.

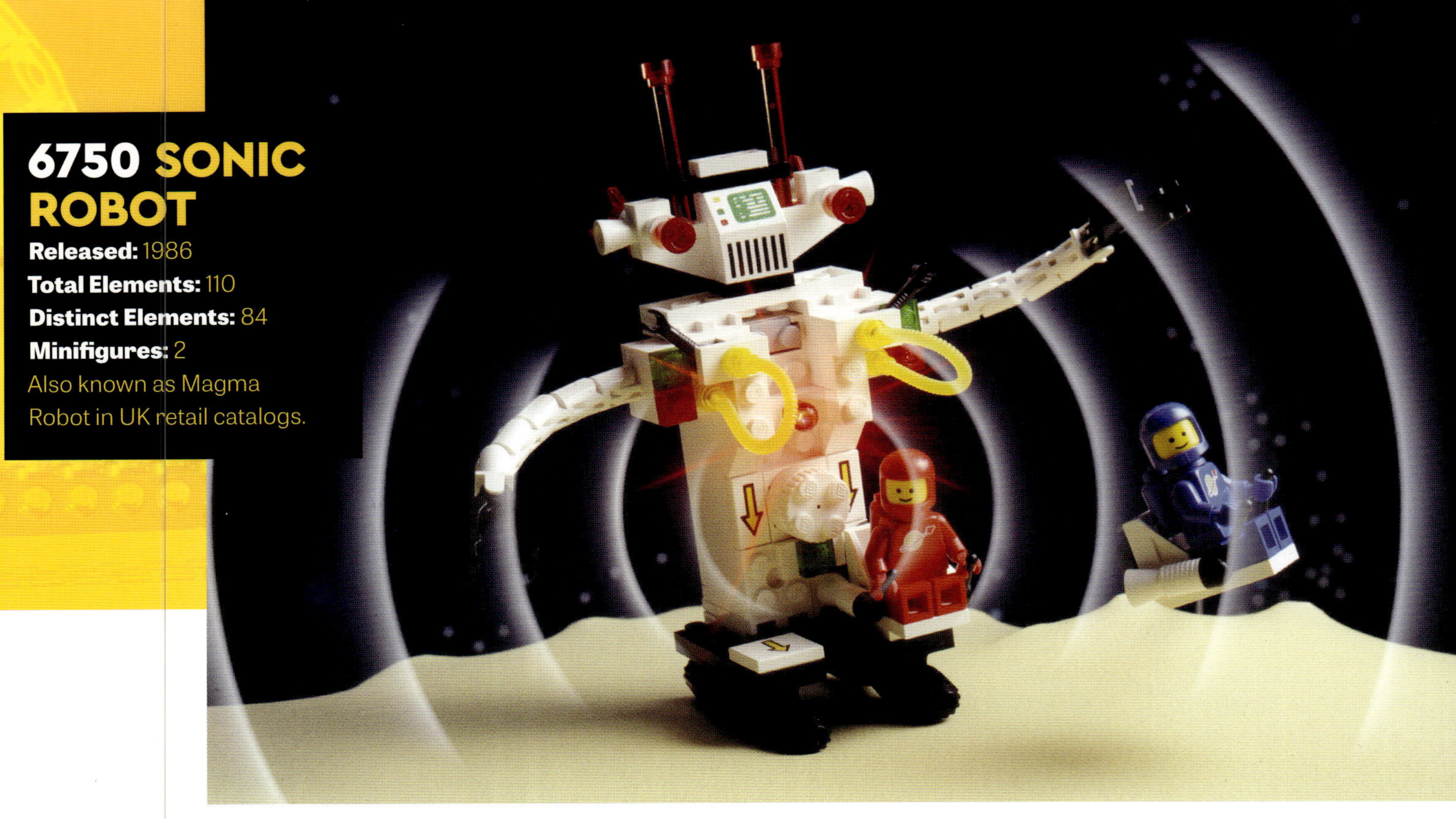

Up to this point, the new "side studs" present on some LEGO® elements had generally been used to apply small details, but this model took things in a whole new direction, literally. Only the base has studs facing upward: the whole body section is attached sideways using large brackets, while the head is upside down. The body is formed by the Light & Sound System battery box; the gray button on the neck turns on the circular sound brick on the stomach and the light on the chest. Two digger bucket elements are also provided, which can replace the claws, and the robot's compact arms are compensated by its ability to hover.

The electromechanical engineer hovered in front of the robot's face, stabbing the controls halfheartedly. "If only the manufacturer had left an instruction manual before she and the rest of her Sort Elektronika chums disappeared off the face of the galaxy," she muttered to the volcanologist. She managed to get it online eventually, and the two astronauts climbed onto the machine's feet and flew up to the volcano towering over the base on Errai Cg. Once there, they immediately activated the sonar interpolator, sending sound waves crashing onto the streams of lava to instantly dissolve all liquids and gases, leaving harmless rocks embedded with zircon crystals behind. While they assessed progress from a safe distance in their scout units, the volcanologist suddenly lost power, crashing onto an outcrop directly in the path of dripping magma. "Syv-50, assist!" he commanded. The robot's override unit flashed red, and the deafening noise ceased. It rushed to the lava flow above him, using its dispersion-strengthened platinum body to divert its path, scooping overflow out of the way with its zirconium bucket attachments.

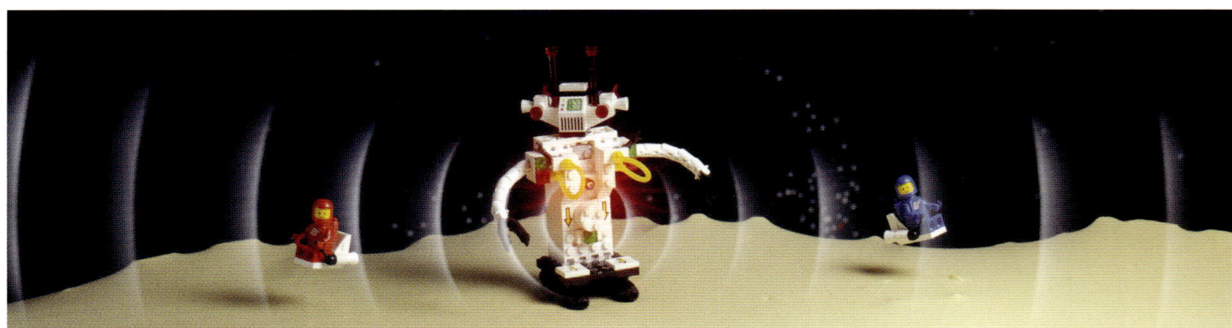

*The noisy element created for the Light & Sound System was **9 V electric sound siren 2 x 2 x 1⅓**, which had the appearance of a plate 2 x 2 with a round brick 2 x 2 on top. Once connected to the battery box, it offered two sounds, chosen by rotating its grooved top disk. The accompanying building instructions helpfully described the two noises as "BIB BIB BIB" and "WIII WIII WIII"!*

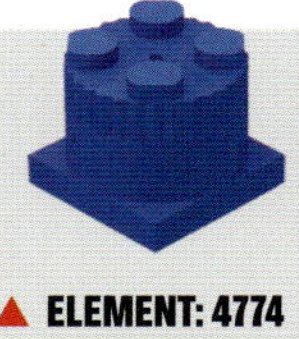

▲ **ELEMENT: 4774**
INTRODUCED: 1986

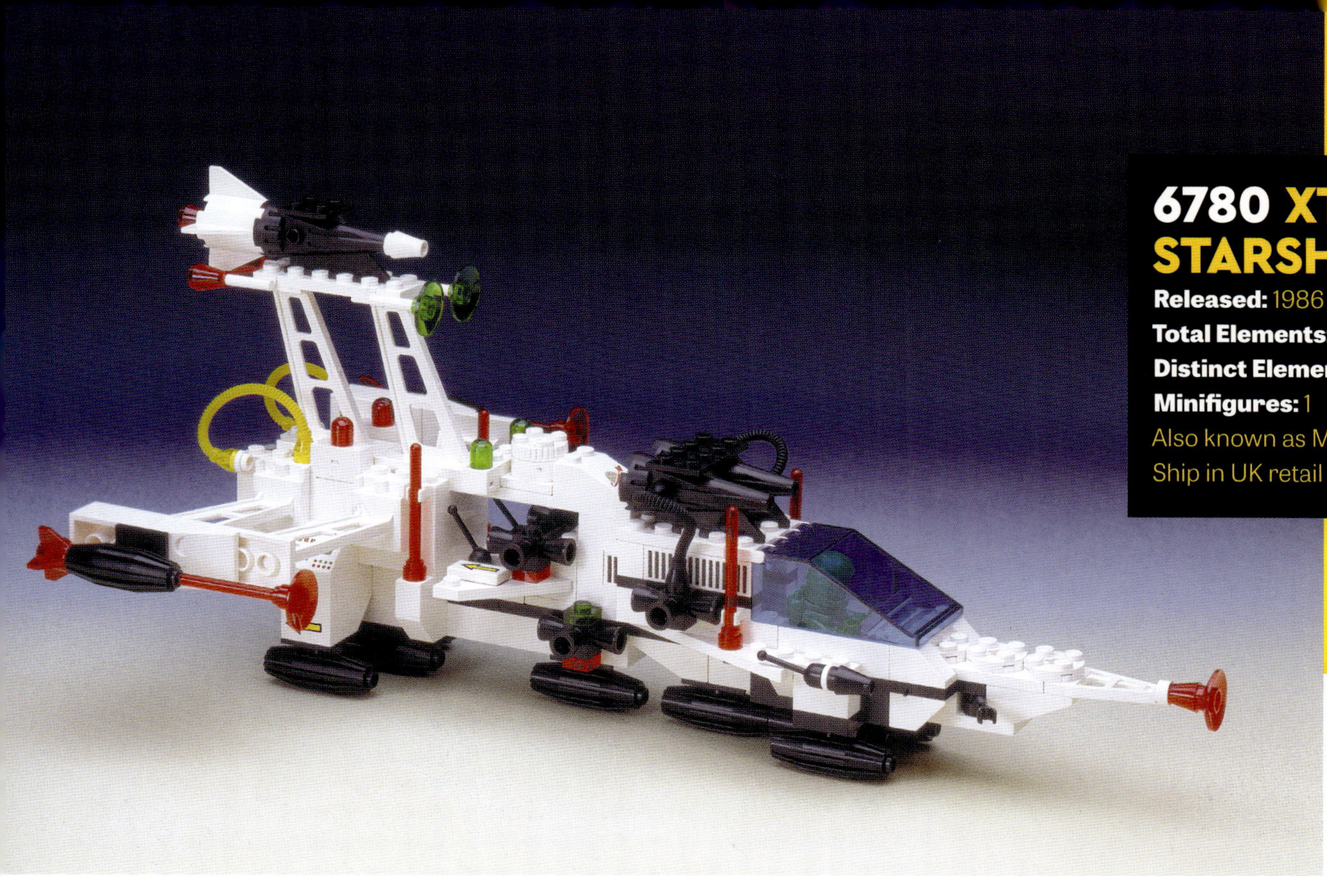

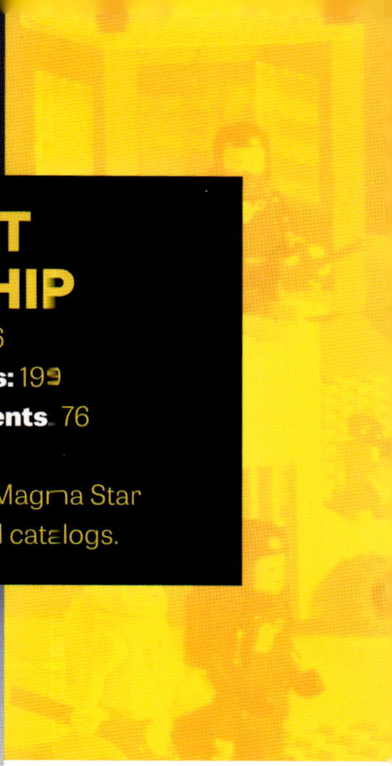

6780 XT STARSHIP

Released: 1986
Total Elements: 199
Distinct Elements: 76
Minifigures: 1
Also known as Magna Star Ship in UK retail catalogs.

With four flashing lights, a two-sound brick, and a dazzling design, this Light & Sound System ship was a thrilling addition to the LEGO® Space lineup. A battery box formed the rear with brackets lying over its smooth sides so that four stanchions could be attached sideways to form the wings. Together with a further two stanchions on the empennage, they suggest the feathers on a dart. The set included three of the new aircraft rocket engine tails 4 x 2 x 2. Note also the horizontal black line running through the white fuselage beneath the forewings; this design cue would soon reappear in Futuron, along with the transparent blue canopies. As was typical, the craft detached into two segments, but a surprise third was included: the centrally housed 45-degree winged section is actually a tiny shuttle with lever controls.

The pilot flew their shuttle up into the exosphere and back to the orbiter, where the engineer had insisted on remaining to receive a potential communication from Sort Elektronika. The module was there, but the engineer was not. They soon noticed the service craft was gone; she must have gone to the surface. But why didn't she contact them? Three full planetary rotations later, they gave up on the search on Errai Cf and headed toward its moon, just in case she had managed to reach it in that tiny craft, unlikely though it seemed. Connecting their shuttle to the orbiter, they powered up the engine. The hazard lamps flashed red as they activated the adaptive waveform sonar and began to look for signs of life in this barren system on the outer reaches of the galaxy.

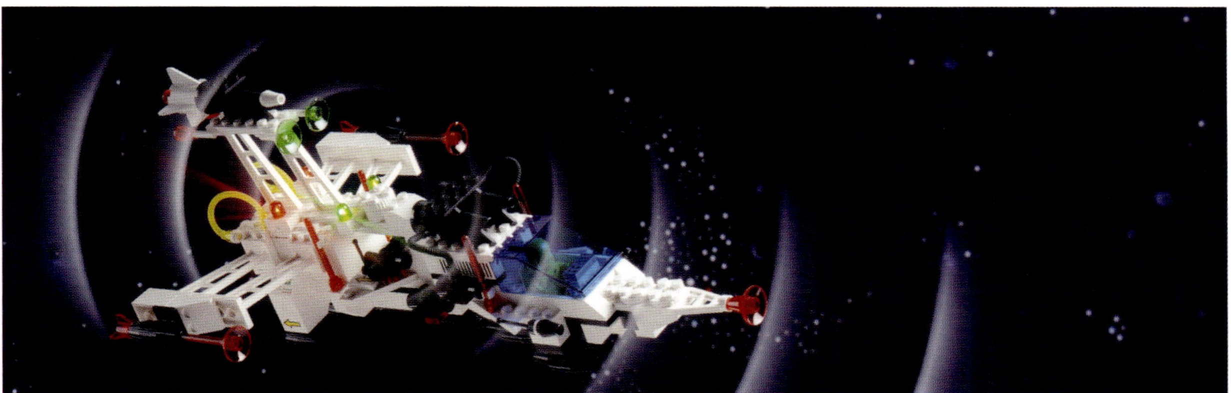

*The **electric light brick 1 x 2 with single top light** was one of the two initial light bricks introduced with the Light & Sound System in 1986, the other being a 1 x 4 with twin lights. The light switches on when it is connected to the battery box, and if the brick is reattached at 180 degrees, it will blink. Red, yellow, and green interchangeable covers were provided to customize the lamp color.*

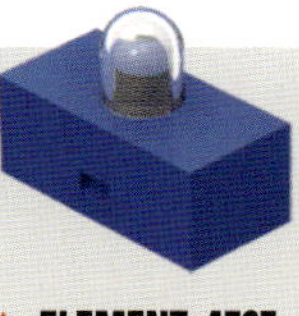

▲ **ELEMENT: 4767**
INTRODUCED: 1986

6783 SONAR TRANSMITTING CRUISER

Released: 1986
Total Elements: 348
Distinct Elements: 100
Minifigures: 2
Also known as Scorpio Modular Transporter.

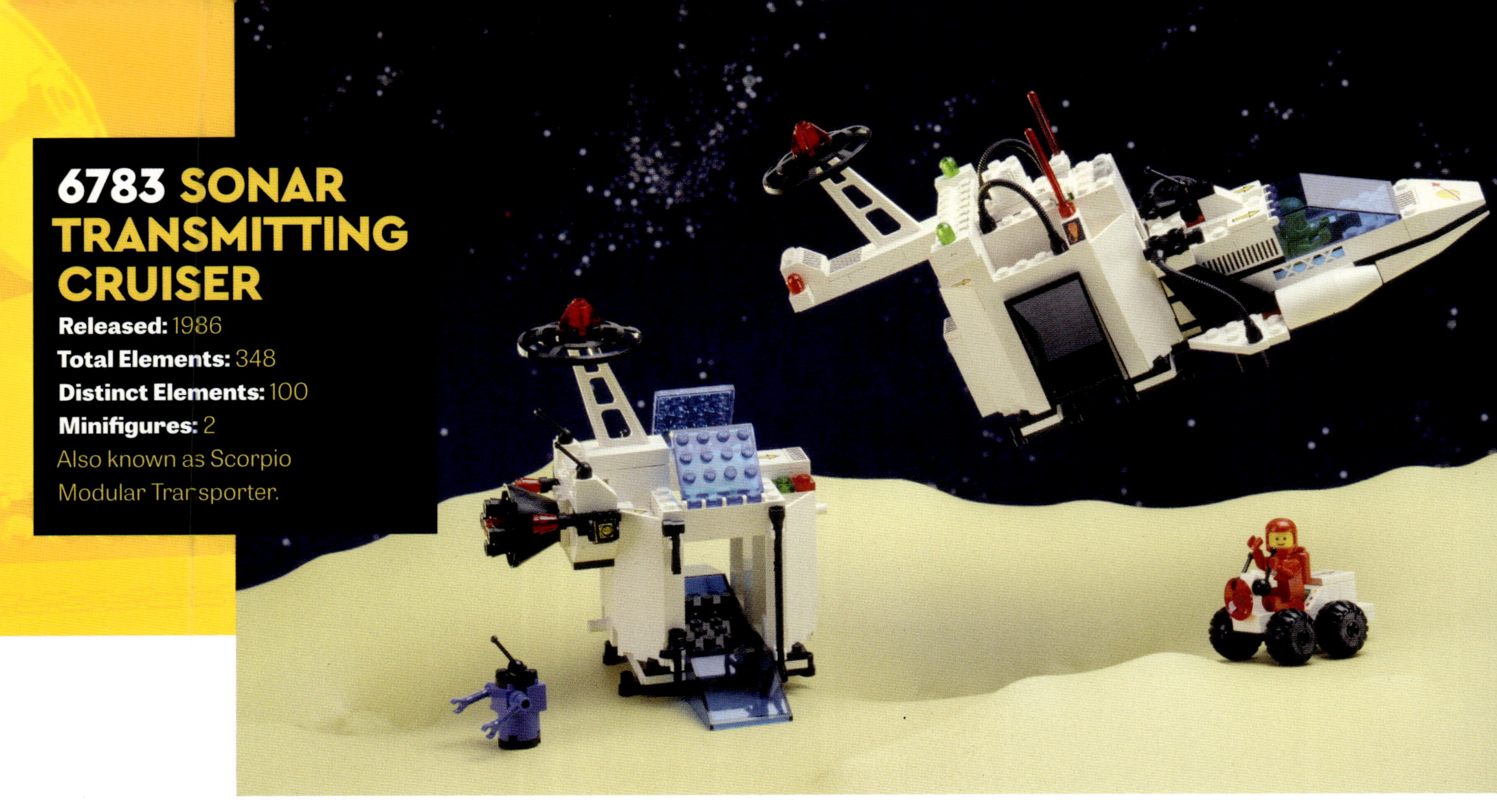

Earlier spaceships had featured modular segments, some with detachable laboratories, but this set truly placed these concepts at the forefront. This is no sleek fighter craft: it's a chunky transporter that comprises a shuttle with two generously proportioned capsules and an exhaust assembly that could be reattached to other sections. Note the inclusion of angled wall panels, introduced for LEGO® Castle, and the ingenious use of canopy windscreens 6 x 4 x 2 as ramps that fold up to become visually arresting doors, big enough for minifigures or the cylindrical robot. The tiny buggy does not fit inside, however, instead being attached to the rear of the assembly. Light & Sound System elements are included atop both capsules: the battery box is part of the central module but supplies power to the lights on the rear module via a bridge made with electric plates.

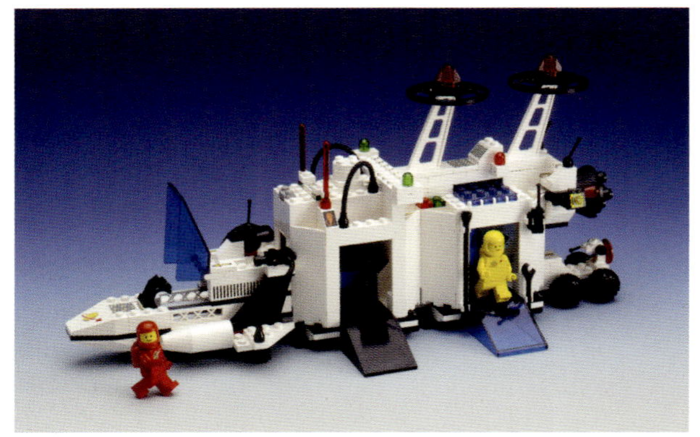

The transporter descended into the colorless lake of Errai Scorpii Majoris, settling into the sludge on its bed. "Syv-83, detach exhaust unit and attach to power block," commanded the biochemist as she depressurized the aft pod. The blue doors opened simultaneously, and after the liquid methane had poured in, the android swam out to perform its duty while the pilot released the gravity pin and nudged forward. "Information: task completed," confirmed the droid. "Syv-83, remain in the pod and await the countdown," said the pilot. He fired up the ship and delivered the biochemist and her command pod to the other side of the massive lake. "Commence countdown," she said. "Syv-83, synchronize transducers." The dishes at either end of the lake glowed, and a red flash streaked through the methane waters. As the countdown reached zero, the biochemist activated the sonar interpolator, and with an immense bang, the lake was empty. The robot trundled out in the rover to collect the crystal cubes of compressed methane.

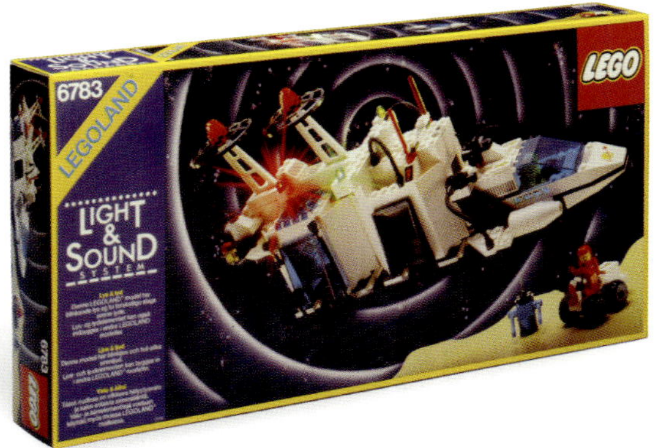

Ingeniously, the electric elements introduced by the Light & Sound System were also building elements that children were already familiar with, such as ***electric plate 2 x 4 with contacts****. The modules of the Sonar Transmitting Cruiser could therefore be easily separated without the hassle of detaching wires.*

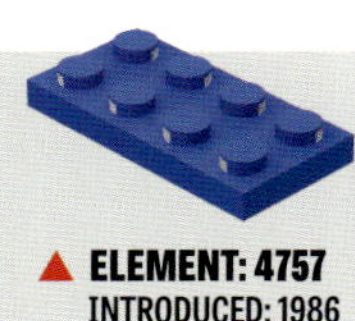

▲ **ELEMENT: 4757**
INTRODUCED: 1986

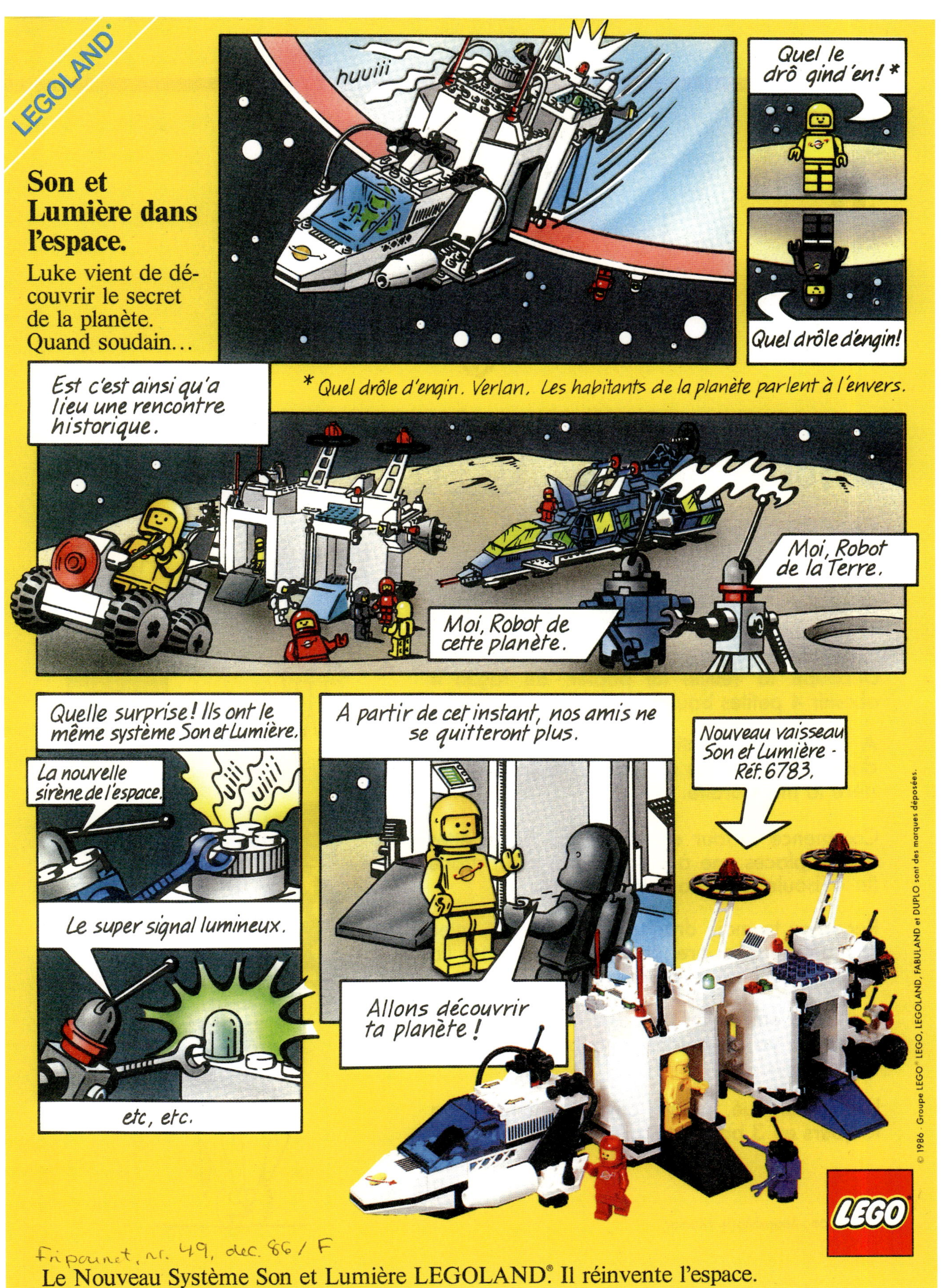

In a delightful French ad from 1986 in the format of a comic, a robot from Earth meets a robot from this planet; coincidentally, they share the same Light & Sound system! "From this moment on, our friends will never leave each other."

6802 SPACE PROBE

Released: 1986
Total Elements: 26
Distinct Elements: 12
Minifigures: 1

It's remarkable that this set delivers a rocket launcher and a robot in a mere twenty-six elements. Consequently, the buggy is very simple but is perfectly efficient, and the black-and-white banding on the rocket works nicely at this small scale. The robot uses the same characterful design as 6807 Space Sledge with Astronaut and Robot.

The android arrived first, of course, skating across Messier 54 Epsilon Gb like some overexcited child before settling into hover mode at the top of a dune. "Information: target location reached," it said as the buggy finally caught up. "Otte-02, yes, I gathered that." The dynamic meteorologist smiled. "Data: Ideal trajectory is fifty-four degrees," it replied. He angled the vehicle and rocket accordingly, switched the steering panel to command mode, and sent the Kuiper satellite blasting into orbit. "Otte-02, when can Gold Team command expect an answer?" he asked. "Prediction: Initial molecule escape velocity data available in forty-two minutes. Assessment of potential atmosphere hosting available after two lunar rotations."

*For years, the LEGO® Space designers had made do with steering wheels for controls, but new possibilities were now being introduced. The most specialized of these was the **minifigure control panel**, which jutted upward with a sweeping arch to provide seated minifigures with an angled panel. Studs on its top and sides and a hole at the rear provided opportunities for decoration and further controls to be added. Occasionally, it was used as a building element too.*

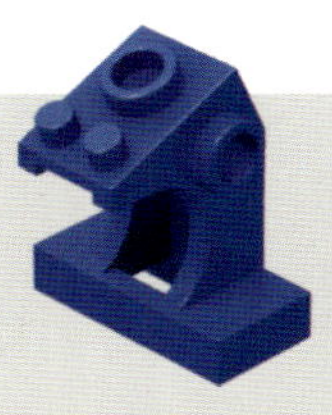

▲ **ELEMENT: 2342**
INTRODUCED: 1986

6874 MOON ROVER

Released: 1986
Total Elements: 66
Distinct Elements: 32
Minifigures: 2

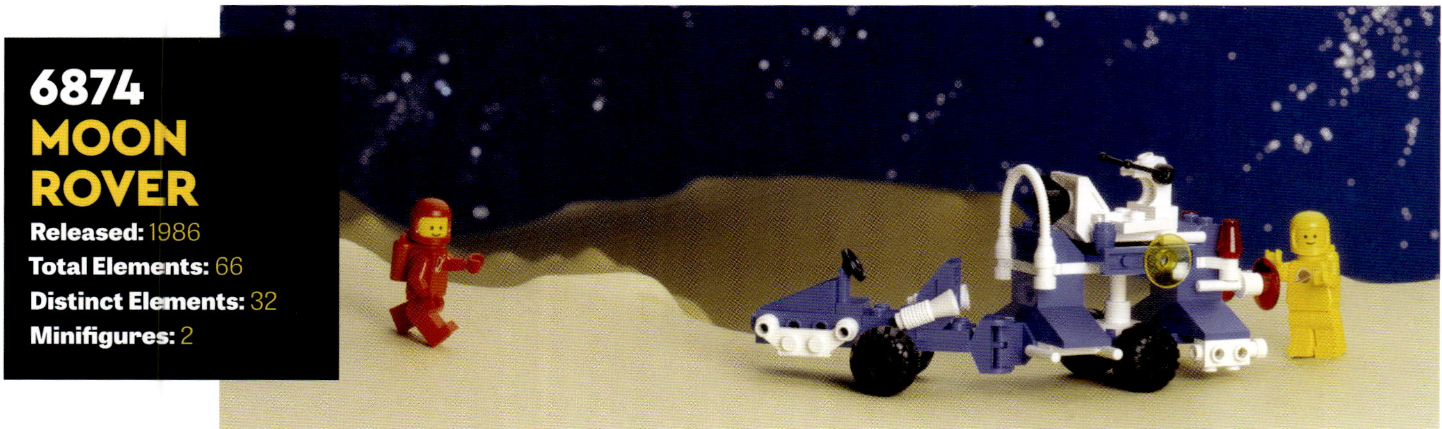

This four-wheeled articulated vehicle features an elevated launch pedestal for a tiny shuttle, but otherwise its design is something of a mystery. The pad has an industrial feel, utilizing two fences to create a grille. There is plenty to ignite imaginative play here, including the various white accessories in stark contrast to the blue body.

The shuttle descended gently from the skies of Epsilon Gb onto the waiting trailer, and the pilot commenced transfer of the atmospheric particles. They gently sifted from the collector pads beneath his craft down through the hydropump and forward into the particle screener. Gentle gamma acceleration ejected most particles out through the extraction hose, leaving only the precious pollen, which the astronauts collected carefully from the nozzles. "Right," said the Gold Team engineer, "I'll drive these straight to the astrobotanist. See what he can cook up."

***Cockpit space nose** is a complex specialized element that looks like a brick 2 x 4 with side studs on three sides, a projecting plate 2 x 2 on the other, and a slope on top. It was usually used on ground vehicles in the LEGO® Space line; only two spaceships ever included it. Following its "V" decoration in 1986, it was printed with the LEGO Space logo in 1987 and M:Tron in 1990.*

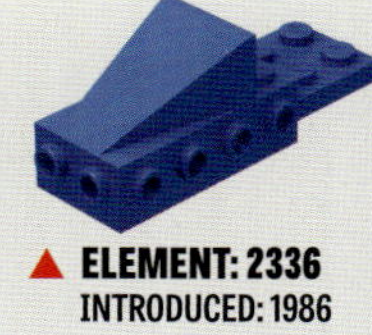

▲ **ELEMENT: 2336**
INTRODUCED: 1986

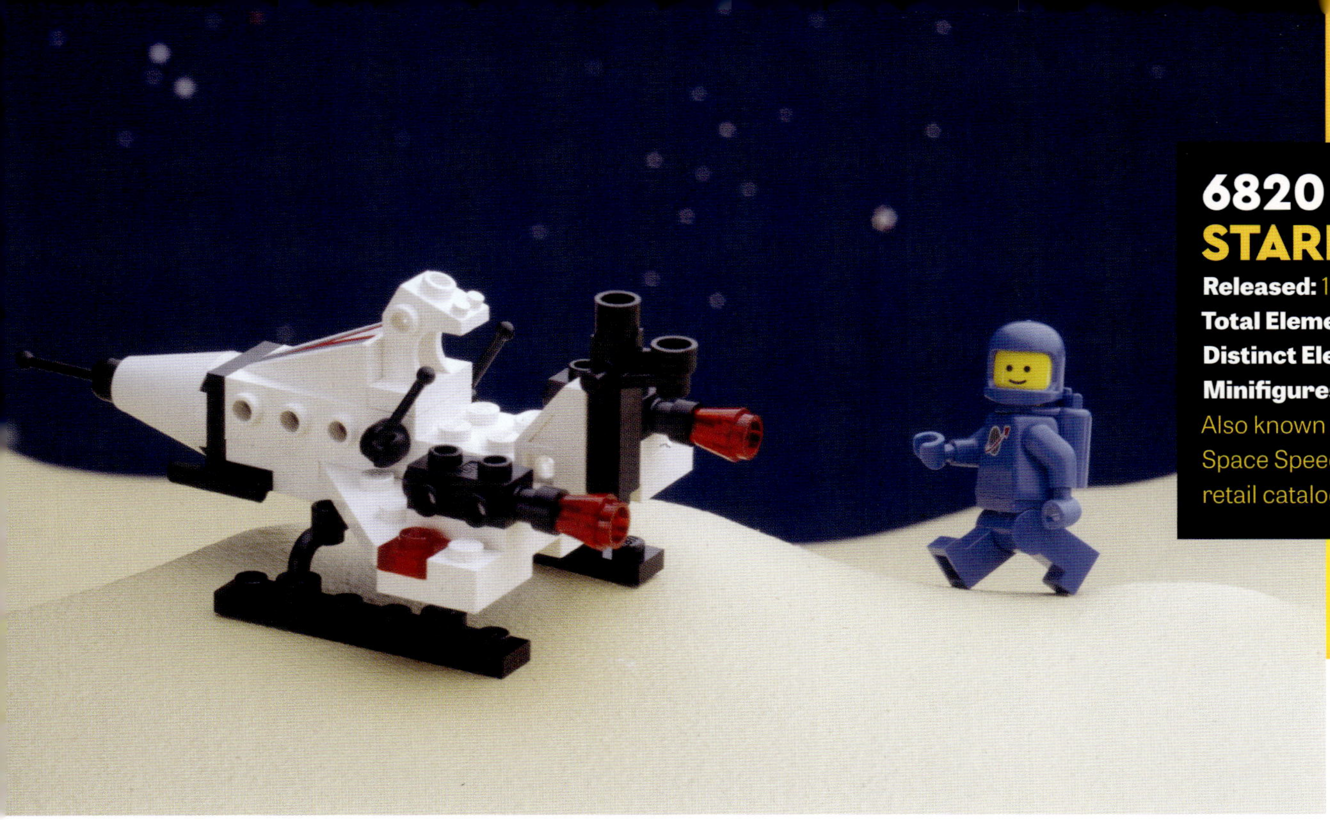

6820 STARFIRE I

Released: 1986
Total Elements: 34
Distinct Elements: 19
Minifigures: 1
Also known as Space Speeder in UK retail catalogs.

The white-and-black color scheme of this speeder more closely matches the three Light & Sound Systems sets than the rest of the 1986 range, but this small set contains no electric elements. It does, however, make great use of the new "space nose" cockpit, creating an elegant compound delta shape with the 45-degree wings. Its red-and-gray "V" printing is unique to this set.

The speeder skidded across the frozen sea of Errai Scorpii, slowing to a halt. The pilot reached back to grab the monocular and slowly scanned the entire horizon. Nothing. He proceeded to the next sector, but the direct route was obstructed by a massive iceberg. He turned the dial on the side of the controls, and the zirconium transducer powered up, red pulses glowing in a V formation down to the nose cone. With only three blasts required, he was soon on his way, but his meticulous search continued to prove fruitless. Just as the other reports had indicated, it seemed all of the Sort Elektronika employees had disappeared without any trace, as had the company itself.

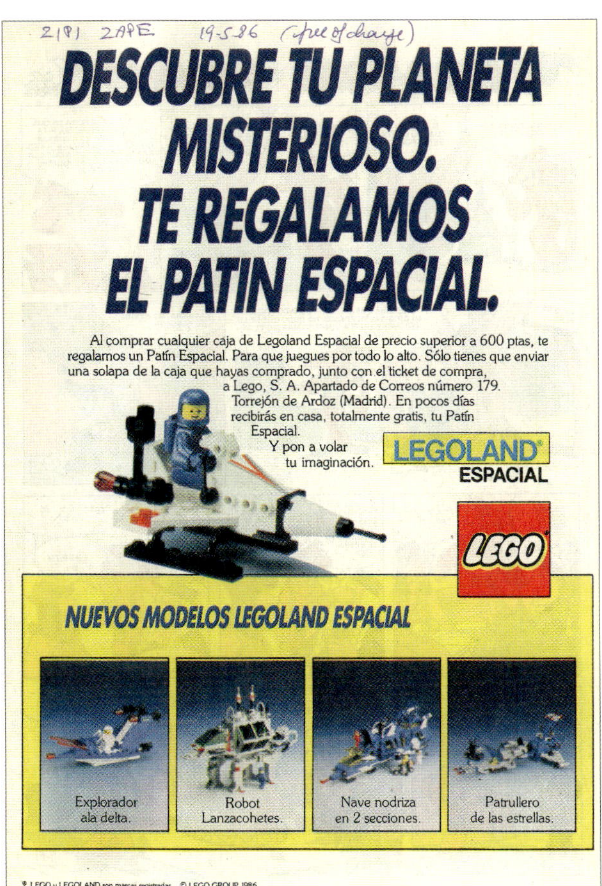

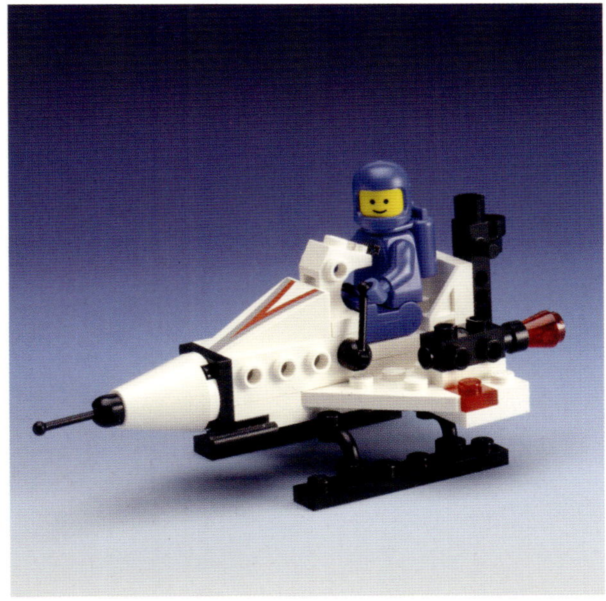

A Spanish ad from 1986 invites you to discover your own mysterious planet with set 6820.

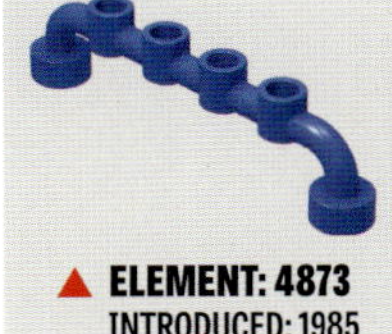

ELEMENT: 4873
INTRODUCED: 1985

*Unusually, this craft has skids as the landing gear, which are constructed using two **bar 1 x 6 with studs**. Introduced by LEGO® Town a year earlier and often used as a roll bar, this element eventually found use in ten LEGO® Space sets, mostly Space Police.*

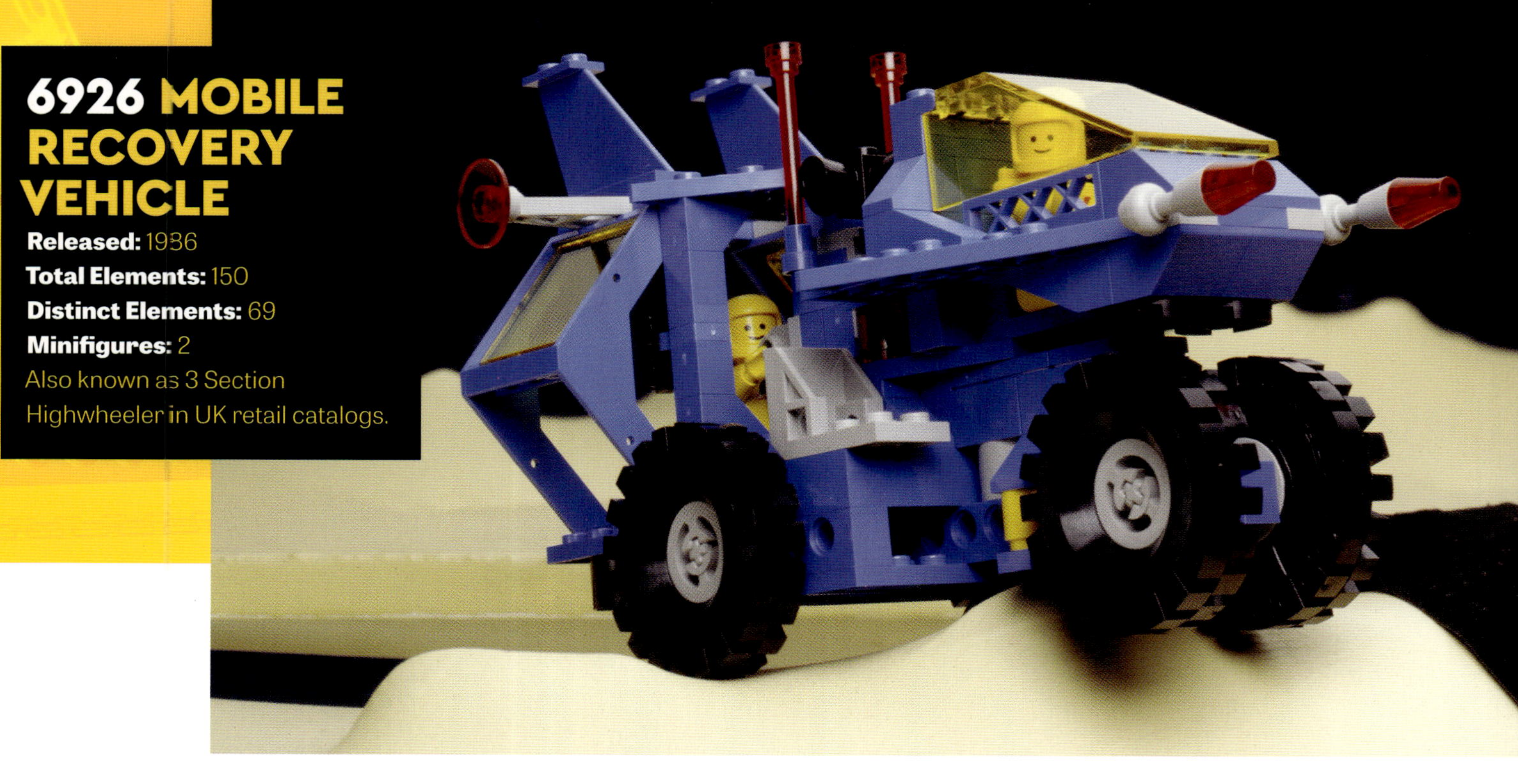

6926 MOBILE RECOVERY VEHICLE

Released: 1986
Total Elements: 150
Distinct Elements: 69
Minifigures: 2
Also known as 3 Section Highwheeler in UK retail catalogs.

This shares many obvious visual and functional similarities with 6892 Modular Space Transport—in particular the miniature shuttle, hexagonal laboratory, and dual yellow minifigures—but here the chunky-wheeled two-seater vehicle is an attractive module of its own, resembling a tractor. The simple lab includes a couple of tools, and interestingly the minifigure control panel is mounted sideways.

The pilot spotted the astroculturist in Orchard Cube spraying the fruit with the tractor, and she opened the comms channel. "Fancy a ride?" The astroculturist laughed in response. "Tell you what, why don't I give you one?" She parked and joined him on the tractor, and they drove across to the terrapod, hitching it to the rear before heading back to reconnect with the craft. It certainly was the slowest way of completing the task, but with limitless solar power at their disposal, they didn't really care. They could do nothing but smile as they drove between the trees, reaching out to snatch some ripe peaches for the trip back, with the setting suns of Epsilon glowing on their Gold Team uniforms.

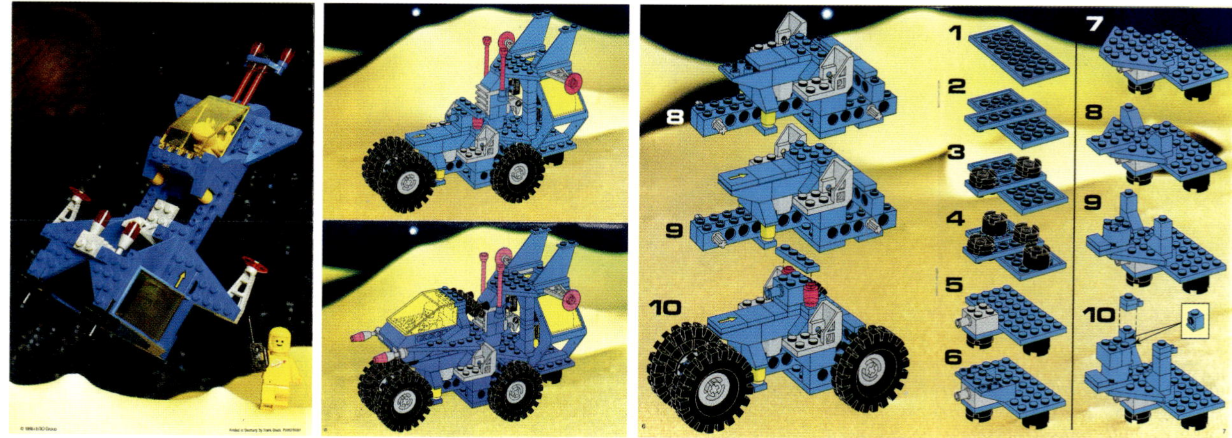

*The **minifigure goblet** was of course created for the LEGO® Castle line but is a wonderful example of how seemingly specific LEGO elements can include design features that make them more universally useful. In the case of the goblet, the lip of the cup can connect to the base of a round element, such as the transparent red cones seen here.*

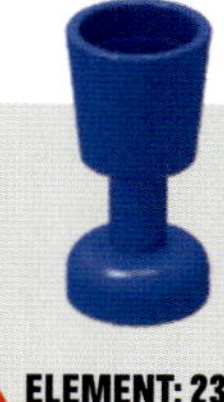

▲ **ELEMENT: 2343**
INTRODUCED: 1985

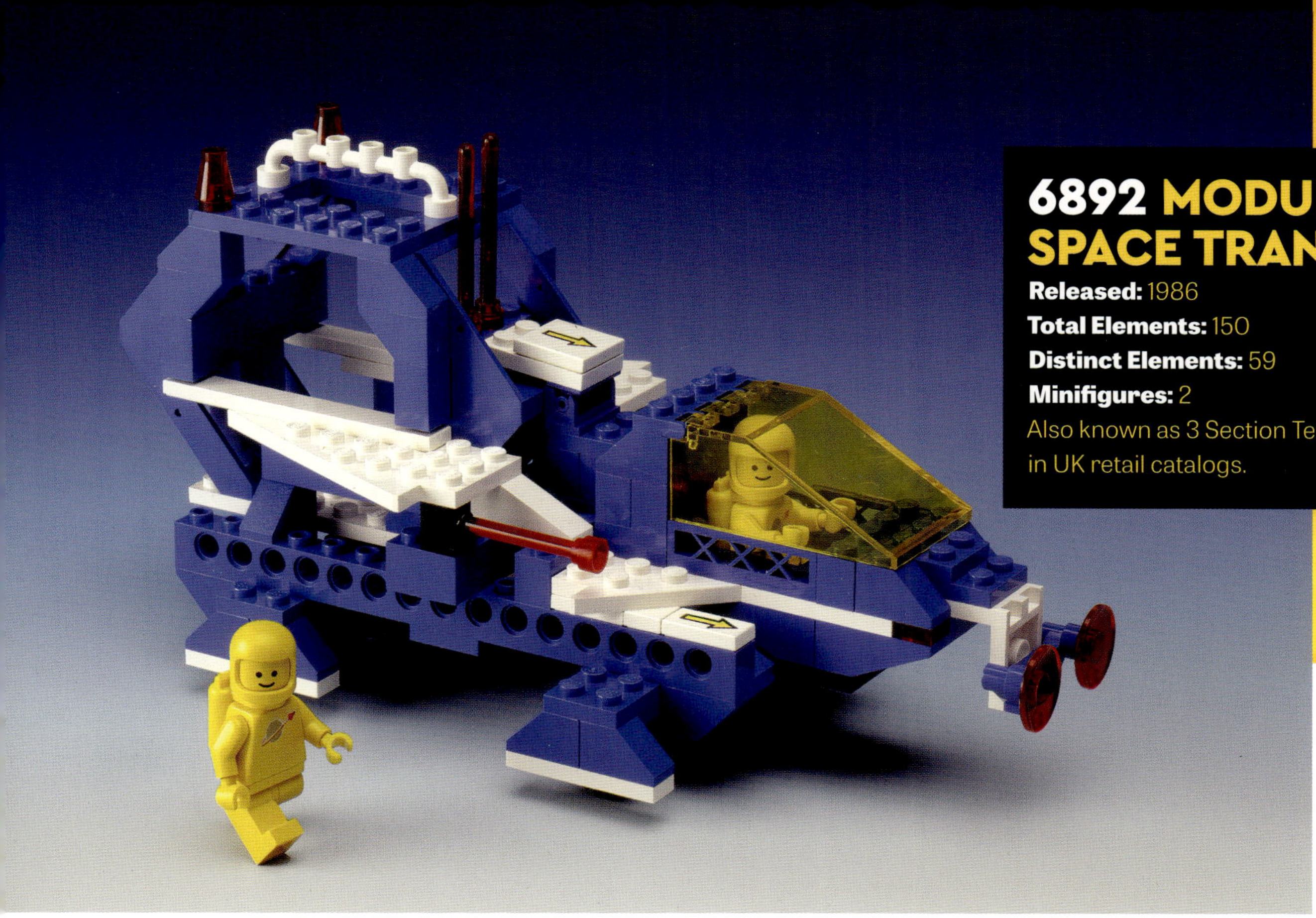

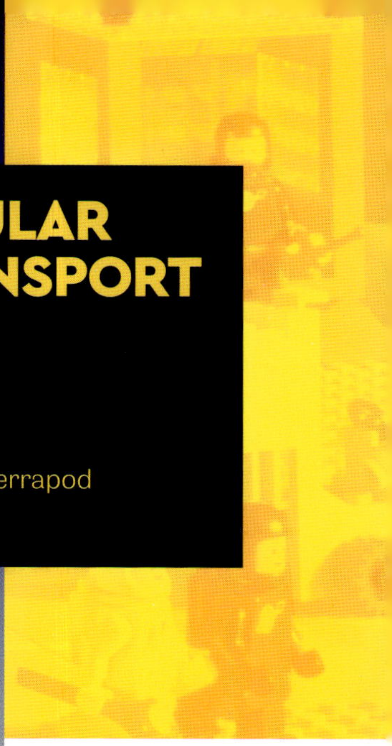

6892 MODULAR SPACE TRANSPORT

Released: 1986
Total Elements: 150
Distinct Elements: 59
Minifigures: 2
Also known as 3 Section Terrapod in UK retail catalogs.

There's something decidedly out of this world about this vehicle. It is adorned with only a few accessories, and LEGO® Technic bricks and small fences provide much of the visual detailing. Instead, the attention is on the use of diagonal bricks. The front detaches to reveal a nicely shaped shuttle, and a central hinge releases the hexagonal laboratory at the rear. The remaining chassis is sparse, and it rolls on surprisingly small wheels that are almost totally obscured by the projecting sloped bricks—almost as if to mimic a hover vehicle. The lab is beautiful but has only one set of visible controls, leaving its purpose open to interpretation. Unlike many other modular sets, the pod does not slide back onto the vehicle; it needs to be lifted.

Reaching the location, the hover transport gently settled on the sandy surface of Epsilon Gb, and the two Gold Team members hopped out, keen to survey the future site of the forest. Soon, it was time to get to work. The pilot released the electroclamp while the astroculturist fired up the photovoltaic tubes. The terrapod glowed and beeped as displays appeared all around him, and he activated the solar surpulsors. The pod lifted gently into the air, and he flew a safe distance away. "Good luck; I'll be back with more methane at sundown," said the pilot as she too detached from the terraplatform, curving away into the atmosphere. The astroculturist created the solar windshield, surrounding the entire area in a cubic energy bubble. The terraplatform engaged, groaning as it leached liquid oxygen from the soil and filtered it through the methane, pumping carbon dioxide into the formative atmosphere.

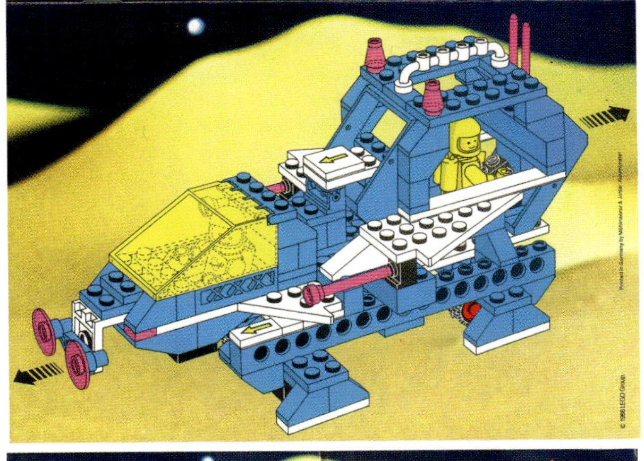

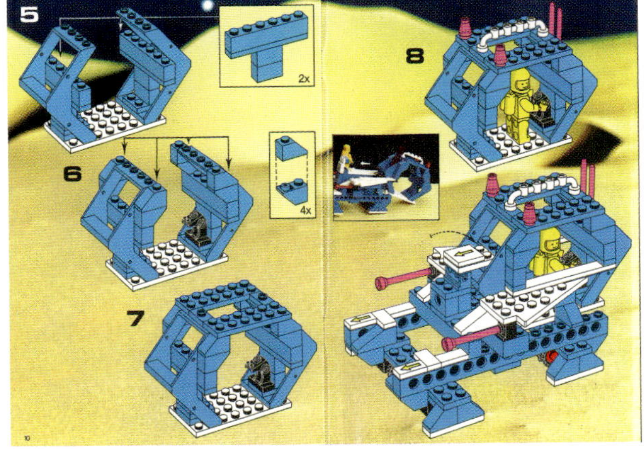

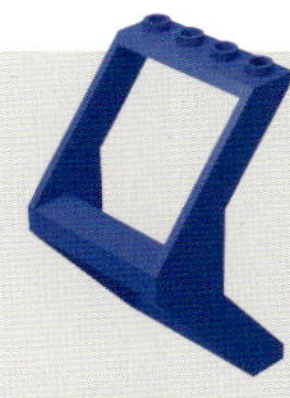

▲ **ELEMENT: 4741**
INTRODUCED: 1986

In 1984, 6971 Inter-Galactic Command Base included two stacked roof windows 4 x 4 x 3 which created a triangular window frame. It looked great but did not make for a stable build, and so two years later a combined version, ***outward sloping window 4 x 4 x 6****, was introduced.*

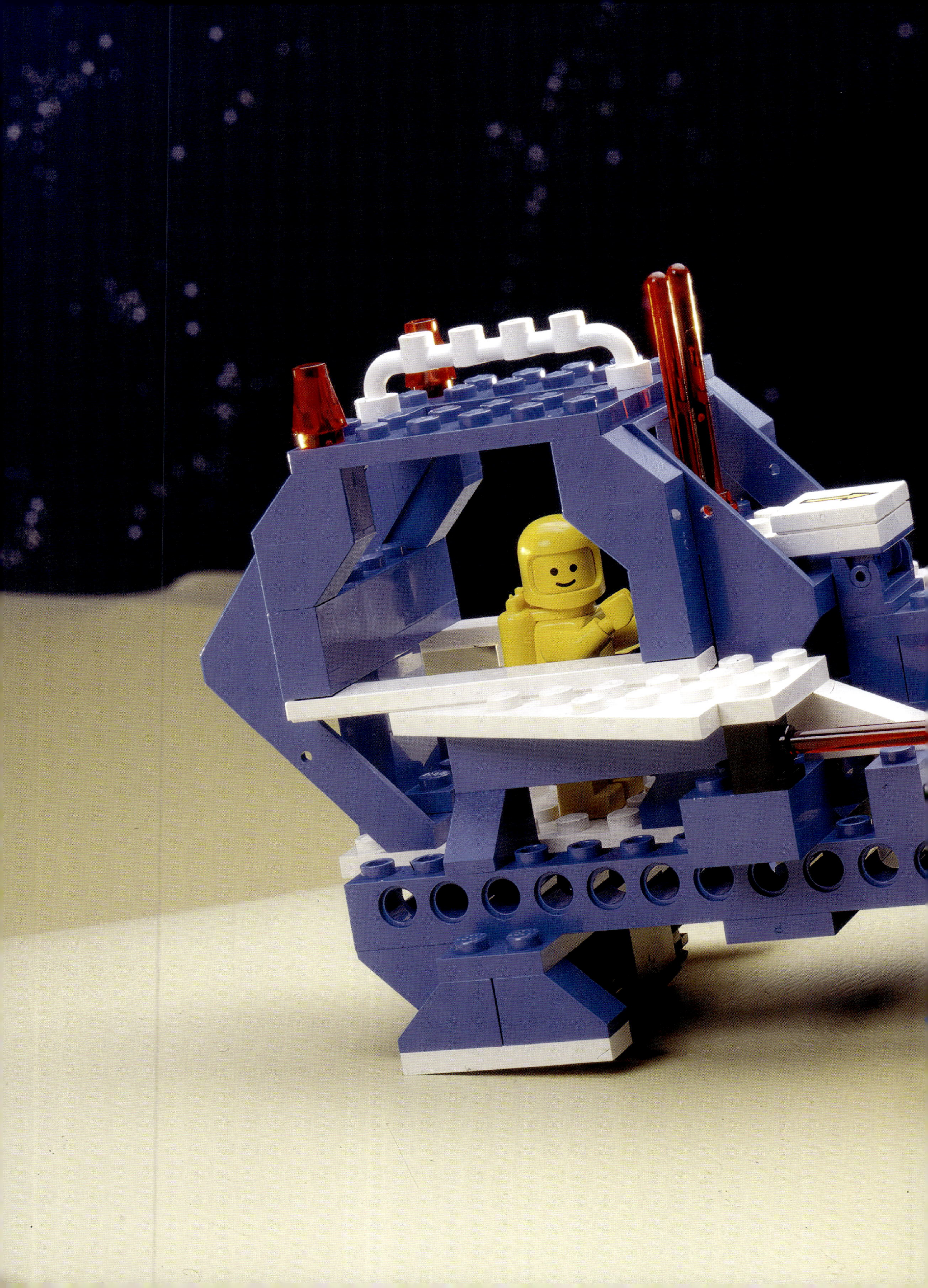

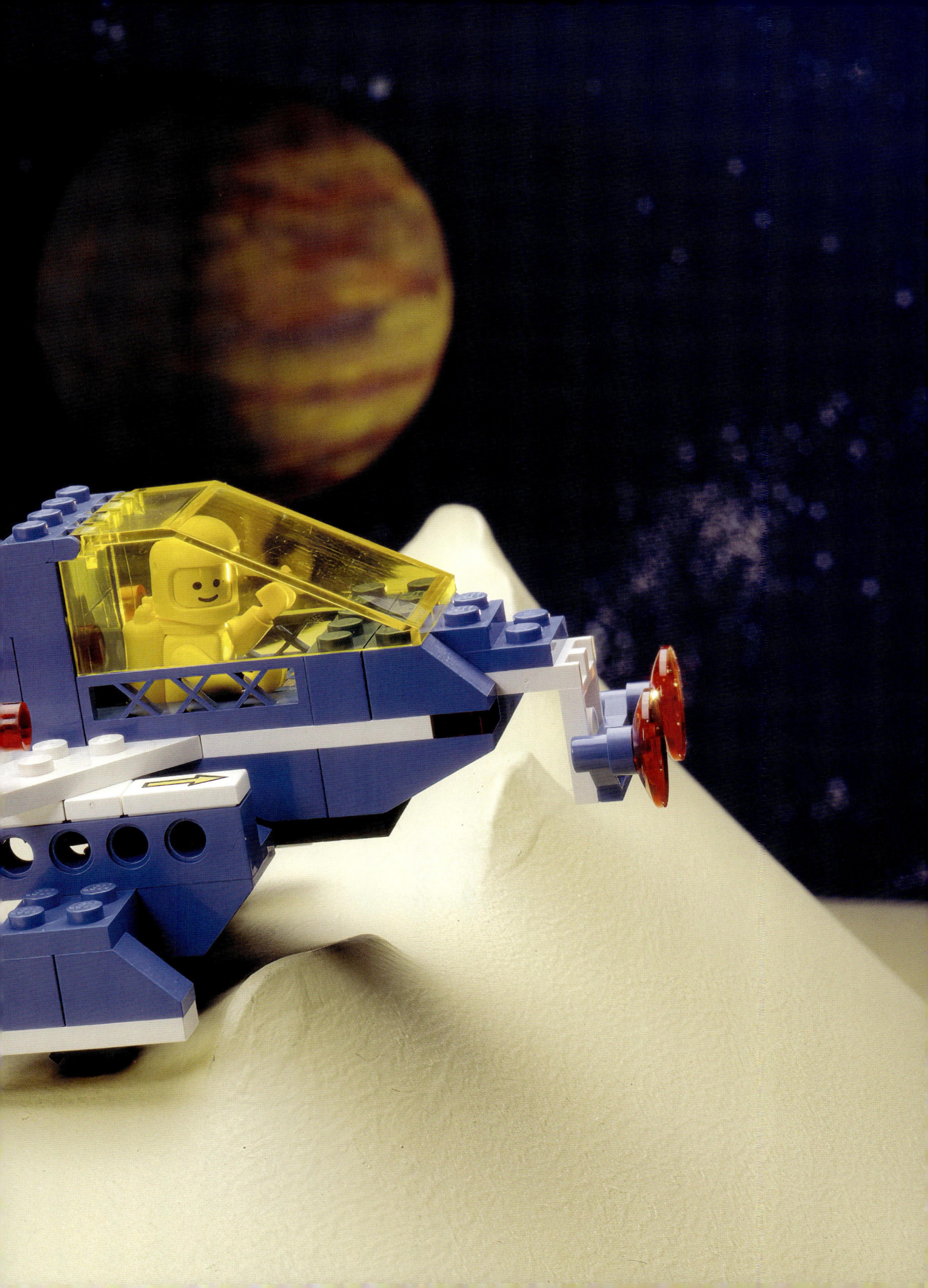

6940 ALIEN MOON STALKER

Released: 1986
Total Elements: 268
Distinct Elements: 76
Minifigures: 2
Also known as Thunderfoot Launcher in UK retail catalogs.

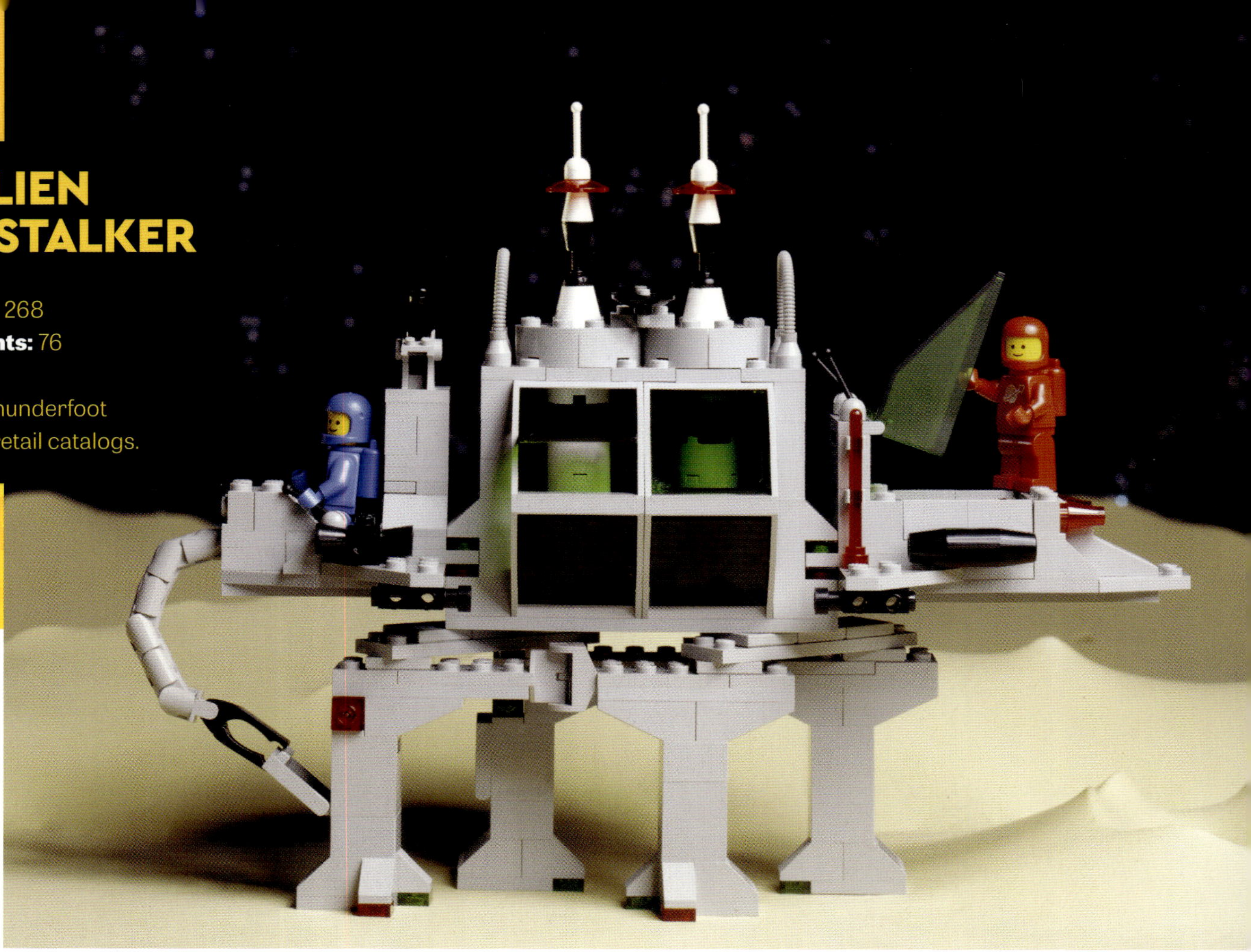

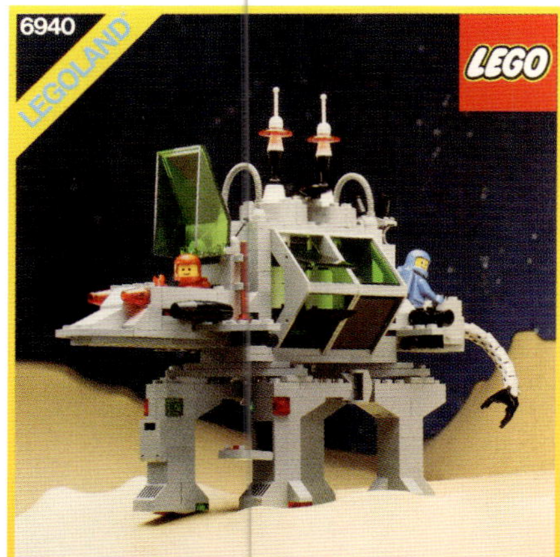

The second-ever LEGO® Space walker was a step up from 1985's 6882 Walking Astro Grappler and brimmed with personality. The color scheme of gray with transparent green suits its saurian appearance, with cute details like the bulbous body housing two rockets, segmented tail with grabber, and feet with printing that looks like claws. The blocky legs perhaps look comical to modern eyes, but they swing efficiently on their turntables as their articulated joints provide some tolerance to the movement. Note the transparent red and green plates, which assisted children in discerning between the left and right pairs of legs during construction. The tail digger cannot reach the ground, but no matter: it detaches to become a triangular hover digger with detachable rear-mounted equipment. The head detaches to become a shuttle, and by replacing it with the digger, the clamp can reach the containers in the legs.

"Tech 7 to Thunderfoot, I have located the device. Proceed to my location," said the mechanic as he gripped the twin pulse oximeter with the clamp and attached it to the rear of his hovering craft. He immediately heard a distant, repeating boom, and the stalker strode into view across the dunes. As it arrived, he reconnected to the command pod to charge while the pilot loaded sulfur into the rockets. "Thunderfoot to Gold Team captain," he said into the transceiver, "terrapod and platform are in position, and we are ready for blastoff. There'll be a biocube waiting by the time you return from Epsilon Gf." Climbing back into the primary shuttle, he engaged the remote controls and activated the solar windshield. "Nav 14 to Tech 7. Let's do it. You'll have twenty-three seconds to get clear." The two rockets and both of the craft all launched simultaneously, and just as the two ships cleared the site, the rockets detonated in the fluorine-rich atmosphere. As a cloud of sulfur hexafluoride successfully formed in the cubic bubble, the terraplatform began to mine the surface below.

*Cockpits can be a challenge to design, because minifigures need space for their arms, and a traditional LEGO brick is quite thick, resulting in six-module-wide cockpits. Some LEGO Space sets had used fence 1 x 4 x 1 on either side of the minifigure to provide the necessary interior space for a four-module-wide cockpit, but the fence's crisscross pattern didn't necessarily suit a spaceship. The introduction of **panel 1 x 2 x 1** provided a solution, and many more panels have joined it since.*

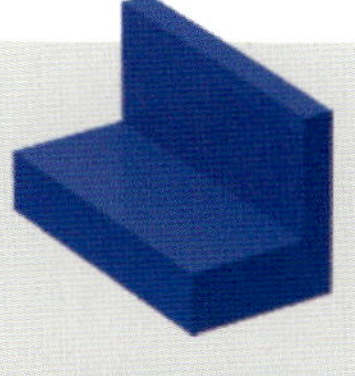

▲ **ELEMENT: 4865**
INTRODUCED: 1985

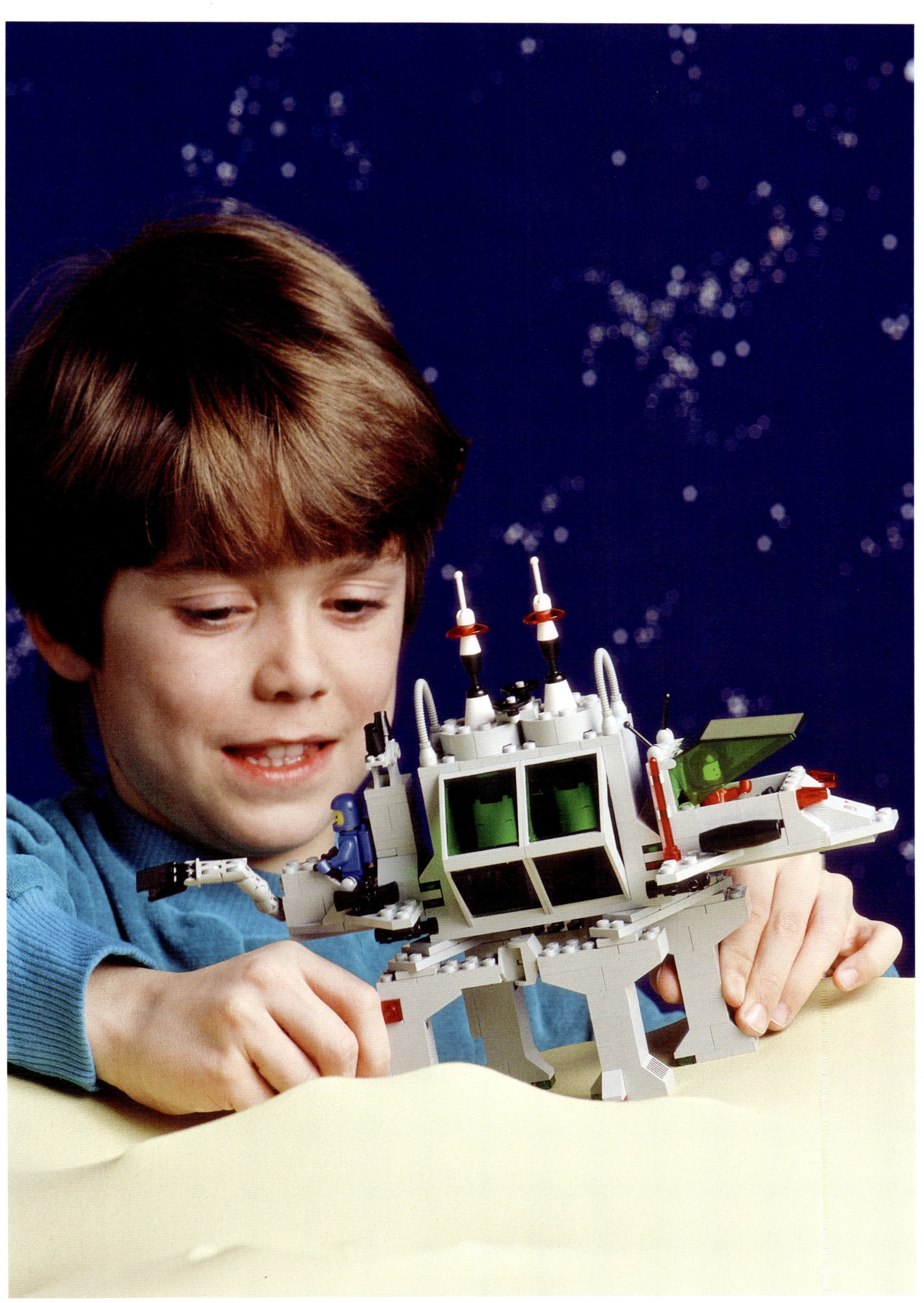

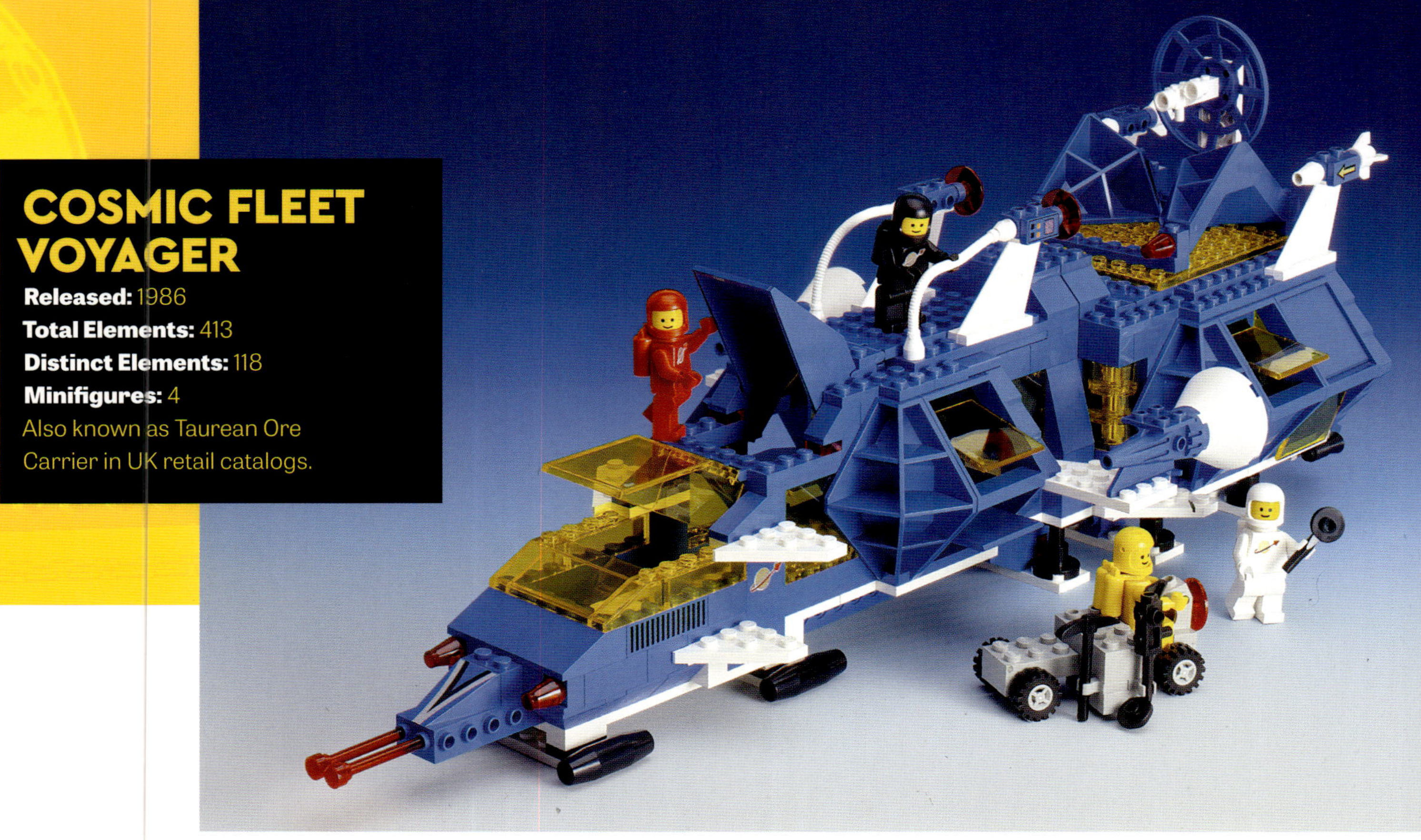

COSMIC FLEET VOYAGER

Released: 1986
Total Elements: 413
Distinct Elements: 118
Minifigures: 4
Also known as Taurean Ore Carrier in UK retail catalogs.

While this commandeering set contains slightly fewer elements than 1983's 6980 Galaxy Commander, more of them go into the craft itself. It is more than twelve bricks high and over fifty modules long, featuring a bulky fuselage which encloses a massive double-height cargo area that splits into two, accessed by rear hatches as well as hinged roofs. There are windows aplenty, including some made of round bricks 2 x 2, cast in transparent plastic for the first time. The new corner wall panel 3 x 3 x 6 is overlaid upon some of the windows, creating eye-catching details. The rover looks virtually skeletal until the two tool containers that are housed on the rear of the craft are attached. Like earlier ships, the rover fits in the rear cargo hold, but there is much more room; the separated rear section is big enough to act as a command center. The divided modules look magnificent—each could almost be an individual product.

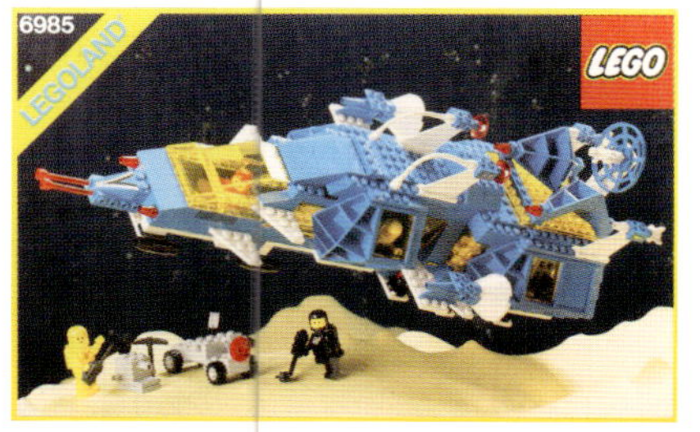

The crew stared through the windows in excitement as they entered the Milky Way and powered down to light speed. The captain smiled as a barrage of messages arrived on the ship's server: the backlog from their seven-year mission to the Messier 54 cluster. The system flagged several important messages: some about the shocking disappearance of Sort Elektronika, as well as another message with strange encoding, which he asked the computer to decipher. Once they landed in the Cygnus system, he assigned the engineer to duties in the command module but then sealed her inside. "Don't let her out—that's an order," he said to the surprised crew. He strode past the yellow window framing the black suit of the confused engineer as she banged on the citrine quartz glass, and then he drove the buggy to a secure location to contact Alpha command. "General, the Sort Elektronika employee we took to M 54 is still with us; she never got the summons. I intercepted it upon our return. It describes exactly where the faction is located and their intention: to make an alliance with the Negative Forces and seize control of this galaxy."

"MY ALL-TIME FAVORITE SET . . . THE FIRST ONE I HAVE REAL MEMORIES ABOUT!"

—FRÉDÉRIC ROLAND ANDRE, SENIOR DESIGN MANAGER, THE LEGO GROUP, 2010 TO PRESENT

__Tail 4 x 1 x 3__, taller than previous LEGO® tails, has a generic design that can be used by any LEGO theme. Initially, however, it was largely used by LEGO® Space—in its first four years of production, it featured in ten LEGO Space sets and only two LEGO® Town sets, one of which was a space shuttle! In 1987, a printed version proudly displayed the LEGO Space logo on both sides.

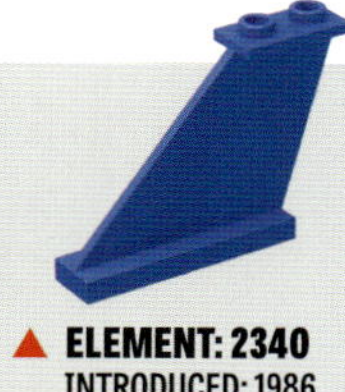

▲ **ELEMENT: 2340**
INTRODUCED: 1986

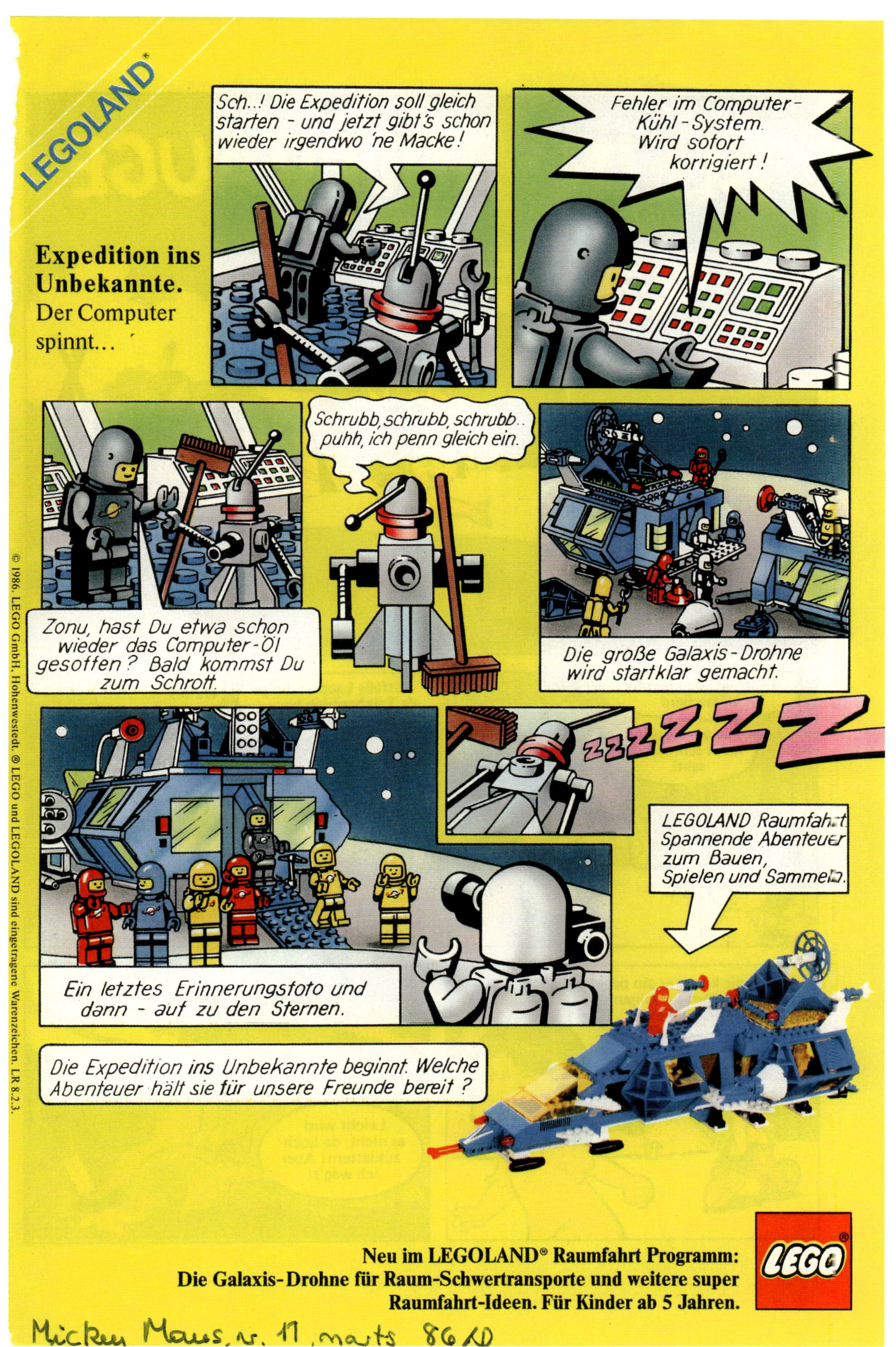

In this 1986 German ad in the style of a comic, the launch of the Cosmic Fleet Voyager is once again delayed by the robot Roger (localized here as "Zonu").

6845 COSMIC CHARGER

Released: 1986
Total Elements: 51
Distinct Elements: 26
Minifigures: 1
Also known as Scorpio Three Explorer.

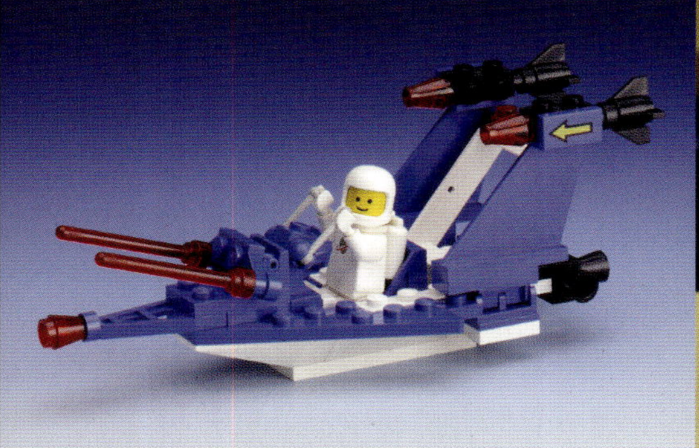

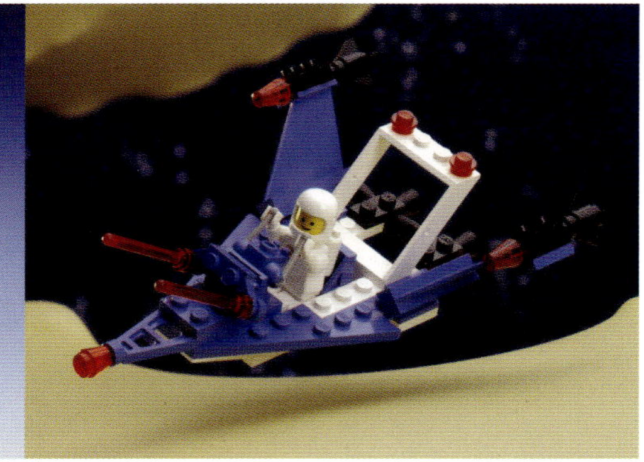

The use of diagonal and hinged elements makes this unconventional craft almost look like an unfolding flower. The blue tail elements can be rotated from vertical to horizontal alignment to become wings, revealing a surprising utilization of an angled window frame. Is it some kind of high boom tail, a closed wing, or some futuristic combination of both?

First appearing in LEGO® Town planes and helicopters in 1985, ***wedge 4 x 4 triple inverted*** *was one of several wedge-shaped elements to be introduced around this time. This was its first appearance in LEGO® Space, and to date it has only made another four.*

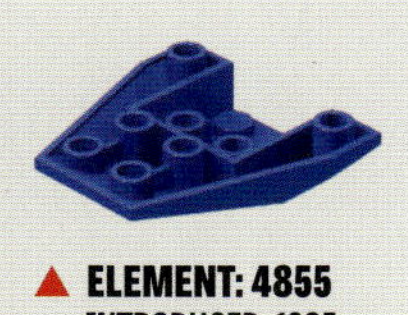

ELEMENT: 4855
INTRODUCED: 1985

Entering the Errai system, the pilot soon encountered the Negative Forces. She immediately switched the controls to variable configuration mode and, reaching up to the levers, lowered the wings down flat to prepare for the gravity waves ahead. With this configuration, they made little impact, and she was able to navigate back on course quickly. The computer calculated the source location, and she raised the twin quantum vacuum squeezers to fire. She hit the target, but the gravity waves kept coming. Could there be another, hidden, source? Programming the quantum-fluctuation-seeking missiles, she angled the wings and fired them both.

1498 SPY-BOT

Released: 1987
Total Elements: 63
Distinct Elements: 32
Minifigures: 0
Included in the multipack 1510 Special Two-Set Space Pack.

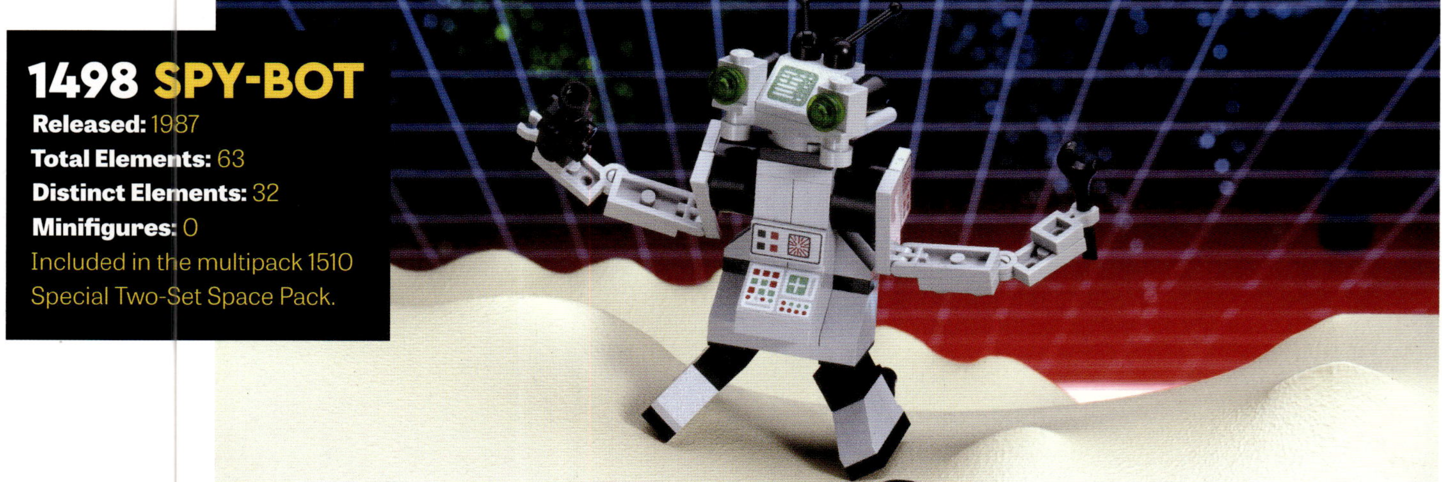

Three types of hinge, LEGO® Technic pins, and a turntable provide an ample quantity of joints for this likable robot, but they don't bend in an expected way! The arms look like they could reach into some tight spots, but the legs seem better suited to dancing than espionage. Thankfully, they have rockets on the back for forward propulsion. No fewer than five control panels adorn its body. This model is notable for being the only LEGO® Space set without any minifigures.

The robot descended slowly from the gantry to the bank of reactors, its carbon thrusters making no noise at all. It lowered its feet and peered closely to read the nearest barcode—SORTELEKTRON6954—before transmitting the information back to base with a wiggle of its antennas. The return command soon arrived: move six racks inward, turn left, the seventeenth reactor. The gaps between the racks of reactors were narrow, clearly serviced by slimmer droids, and the robot had to shuffle sideways to get in. It counted to seventeen and peered again: SORTELEKTRON0937. This was it. Rotating a shoulder, the robot unfolded one arm in between reactors and clasped the port at the rear. Its arm gently vibrated as pebibytes of data trickled up to its head drive. There was a distant clang, and the robot turned its head to observe something floating between the racks, something slim. With its free arm, it reached for the wrench on its back.

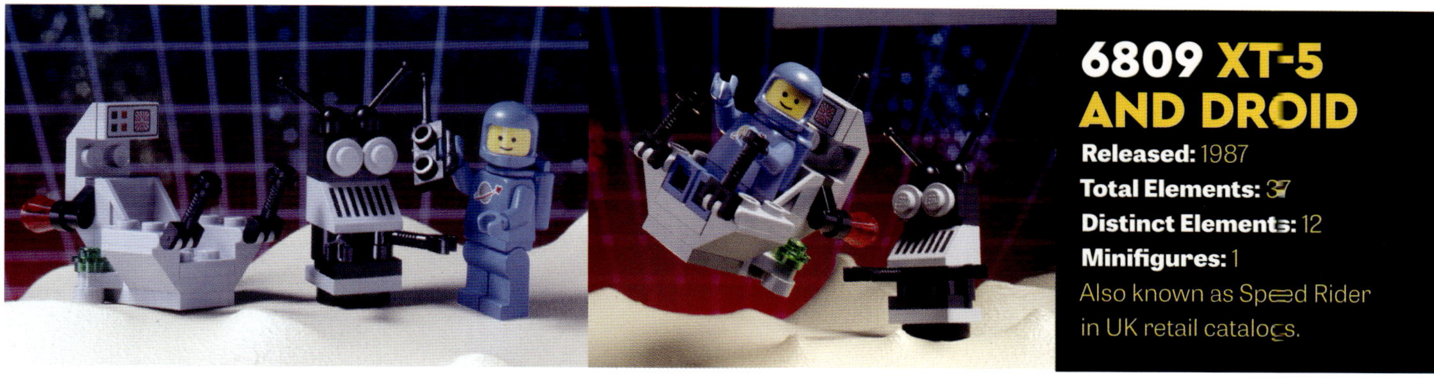

6809 XT-5 AND DROID

Released: 1987
Total Elements: 37
Distinct Elements: 12
Minifigures: 1
Also known as Speed Rider in UK retail catalogs.

This design makes clever use of just a few different molds to create a cozy flying chair with vertical and horizontal thrusters. It can be piloted by either the minifigure or the robot, using two of the new control lever hinge bars. Notably, it is the only gray vehicle in the 1987 range.

"Data: You have arrived thirty-seven minutes late," said the android as the corporal landed. "Otte-09, I've got data for you: you've got a big mouth." She smiled. "Your mission is to do a flyover of the former Sort Elektronika colony. Scan for any surviving tech: we need to know how they're powering a solar windshield of this size," she said, looking up at the network of faint blue lines crisscrossing the sky. She retrieved the radio from the back of the XT-5 as the droid positioned itself in the seat. It flew to the abandoned factory and methodically traversed the ruins, steering with the levers while simultaneously sending pulses through the back of its head to operate the scanning controls. "Data: technology located," reported the droid, descending to a twisted hunk of metal below. "Serial plate intact. It is our spy-bot unit Fire-98. Head drive has been obliterated."

A few large elements had been introduced over the years to assist with the challenge of creating the sunken area required for cockpits. ***Inverted double slope 45° 4 x 2 with 2 x 2 cutout*** *was the smallest yet. It is also used cleverly here to create a futuristic back for the chair.*

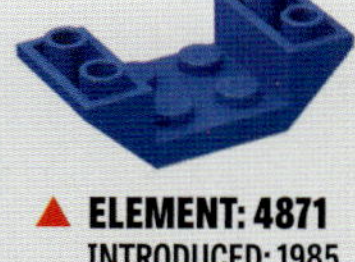

▲ **ELEMENT: 4871**
INTRODUCED: 1985

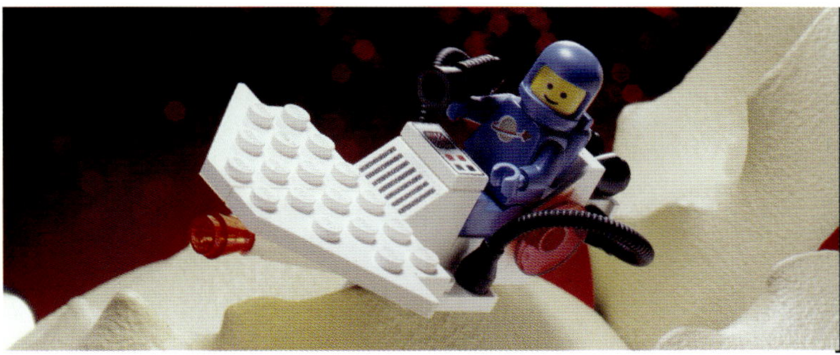

6808 GALAXY TREKKOR

Released: 1987
Total Elements: 29
Distinct Elements: 17
Minifigures: 1
Also known as Swing-Wing Sky Bike. Included in the multipack 1530 Space Value Pack.

This shuttle comes with a fresh idea: variable-geometry craft! The wings can be swung from a straight to a swept leading edge, perhaps suggesting this tiny craft is intended for a variety of speeds. White with black and transparent red highlights was the prevalent color scheme of this year.

The pilot flew slowly toward the ridge, gripping the personnel detector in one hand and tapping the controls with the other. "Sky Bike to Voyager. I'm seeing a lot of smoke." He slowed to a stop as he reached the peak and peeped over. "Voyager! Sort Elektronika are shooting at their own base! The fleet is departing," he shouted in panic. "Voyager to Sky Bike" came the reply. "They must know we're onto them. Abandon covert status and pursue." The pilot sped forward and activated the swing wing, quickly rising to supersonic speed. The ships seemed to be unaffected by the strange grid of blue lines that encircled this world as they passed through it—he hoped the same would be true for his craft.

Swivel hinge brick 1 x 4 was one of many elements that revolutionized LEGO® play in 1978, but it was best suited to large models. Nine years later, the flatter version, ***swivel hinge plate 1 x 4****, brought the functionality of horizontal rotation to even the smallest of LEGO sets.*

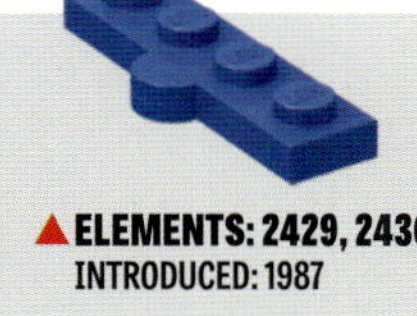

▲ **ELEMENTS: 2429, 2430**
INTRODUCED: 1987

6849 SATELLITE PATROLLER

Released: 1987
Total Elements: 47
Distinct Elements: 28
Minifigures: 1
Also known as Comsat Launching Vehicle in UK retail catalogs.

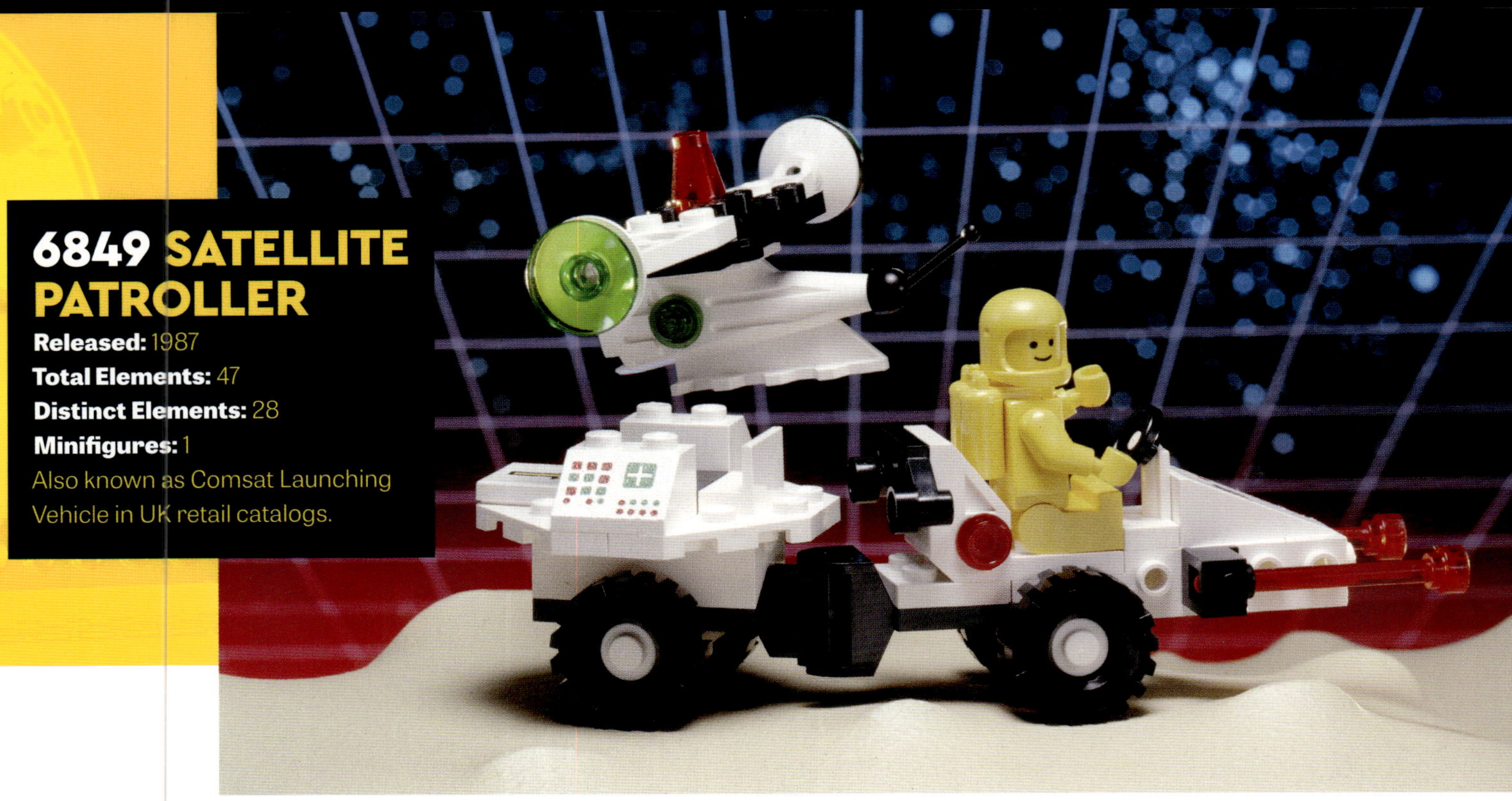

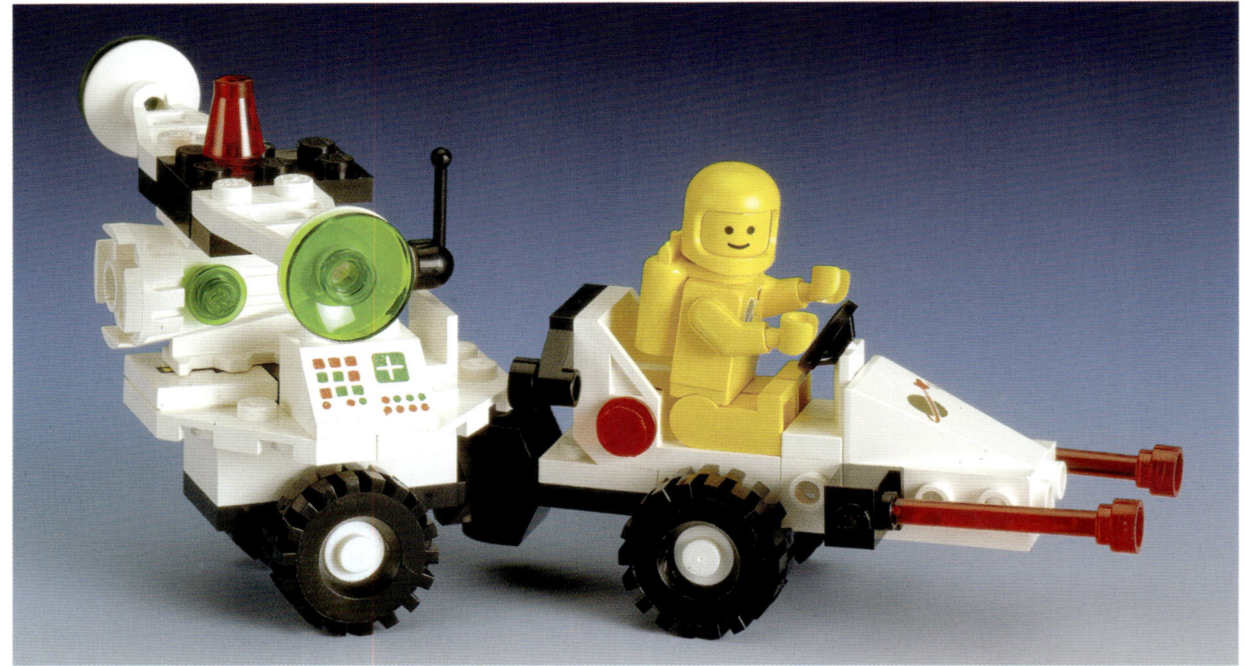

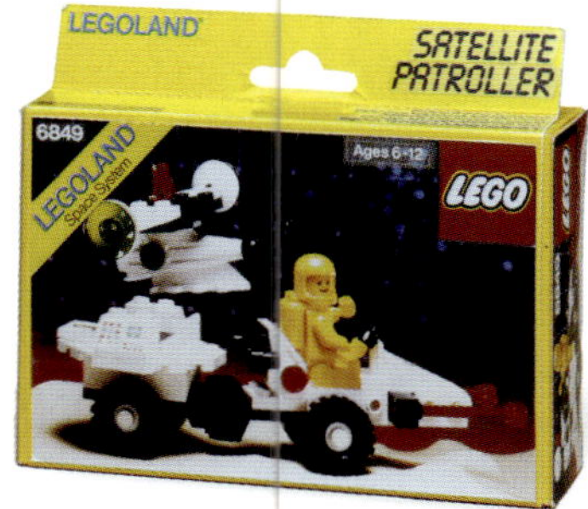

The highly specialized aircraft rocket engine tail 4 x 2 x 2 is utilized in a new way here: as a communications satellite, perhaps explaining how this payload can launch itself into orbit! Space wings on hinges offer customization of the satellite's shape and purpose, and the octagonal launch pedestal has two-person controls should another astronaut be available. The larger tires look well balanced on this articulated vehicle, which proudly bears the LEGO® Space logo on its "space nose" hood.

The solar wind protection offered by a global Mercator framework, as they had named the technology salvaged from the abandoned Sort Elektronika base, would be perfect in combination with the terraforming technology brought back by the Sagittarius mission. Another advantage was that communications satellites could safely orbit in the thermosphere with no interference. The cosmologist smiled at the beautiful sight of the comsat lowering itself back down against the latticed sky. He guided it into place and assessed its data. It clearly showed that the black hole had grown at an unnatural rate, but he could not infer what—or who—was responsible.

Diagonal edges and octagons were staple design features of LEGO Space in the late 1980s, and ***wedge plate 3 x 6 with cut corners*** *was an essential part of this. It was used for landing pads and wings, as well as appearing in LEGO® Castle sets.*

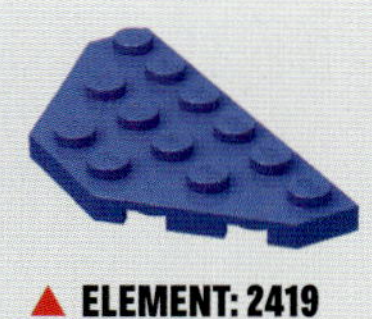

▲ **ELEMENT: 2419**
INTRODUCED: 1987

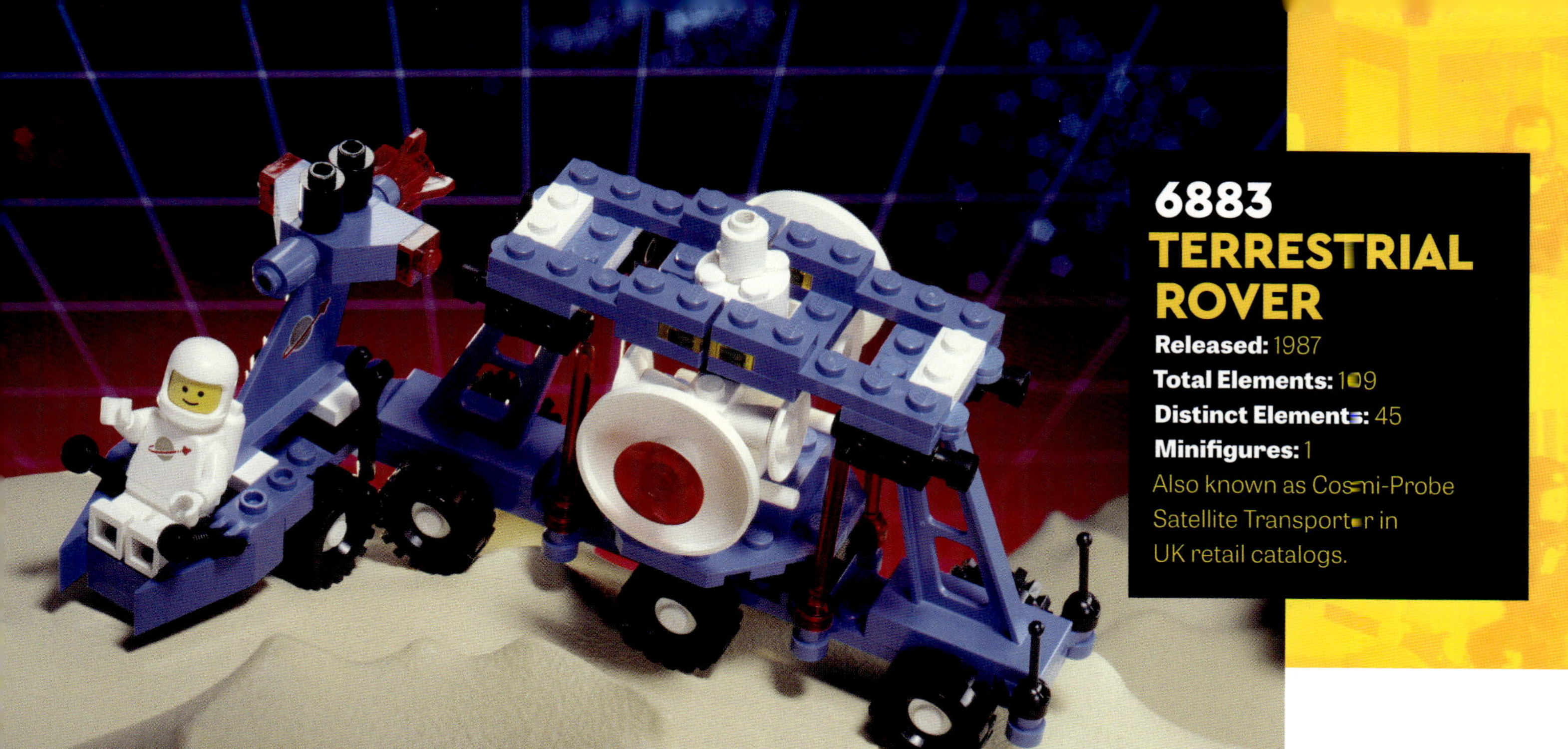

6883 TERRESTRIAL ROVER

Released: 1987
Total Elements: 109
Distinct Elements: 45
Minifigures: 1
Also known as Cosmi-Probe Satellite Transporter in UK retail catalogs.

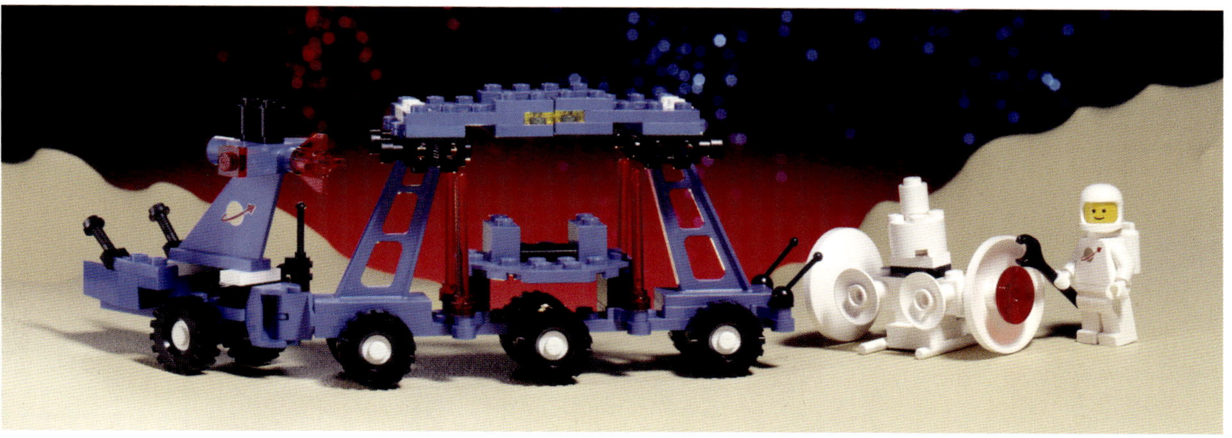

The rear trailer of this eight-wheeled articulated transporter is surprisingly long, being based upon a plate 2 x 16. This was the only blue set of the year, with the wonderful satellite contrasting in white. Coated in dishes, it is held on the octagonal platform by support arms on hinges that can be swung open like gates for launch. Mysteriously, there are no printed control panels. The front cab has what seems to be some kind of launcher of its own, in the form of the new large tail element. This element was decorated on both sides with the LEGO® Space logo, and is exclusive to this set.

Returning to the Cygnus system was a huge risk, but they had to find out if Sort Elektronika had fled here to work with the Negative Forces. Hopefully, kyanite coatings on metals and ultrazinc on electronics would still prove sufficient camouflage for radar—although, of course, it was a huge gamble for them to have established a global Mercator framework around Cygni Bb. It'd be worth it, though, she thought as she swung the support arms open, if it meant they could obtain the necessary data to annul the risks posed to the entirety of humanity's colonies in the Milky Way Galaxy. No pressure, then. The accel-photovoltaic tubes glowed, and she moved to the rear levers, sending the cosmic probe shooting upward on its desperate mission.

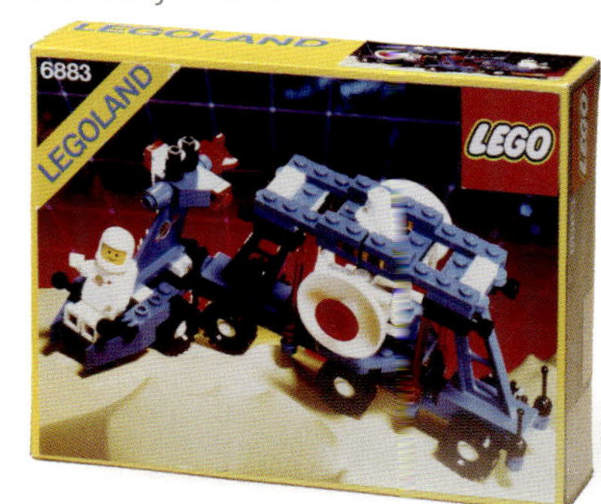

A number of V-shaped elements were introduced around this time, mostly used for adding greater realism to planes and boats in LEGO® Town. ***Wedge 3½ x 4*** *is one of the more specialized ones, and consequently it appears in fewer sets. It lies almost wholly exposed on this vehicle, working beautifully as a seat, and the control levels lie neatly upon its studless surfaces.*

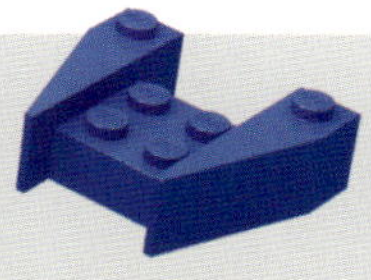

▲ **ELEMENT: 2399**
INTRODUCED: 1987

1499 TWIN STARFIRE

Released: 1987
Total Elements: 89
Distinct Elements: 38
Minifigures: 2
Included in the multipack 1510 Special Two-Set Space Pack.

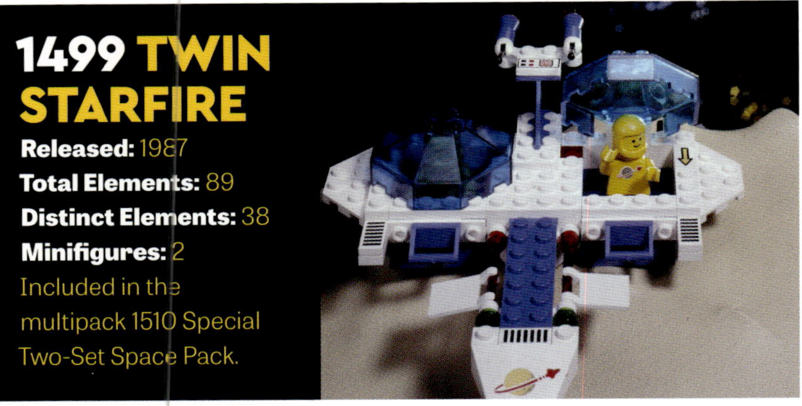

This ship brought a wholly different design aesthetic to LEGO® Space, closer in form to Futuron while reviving the white-with-blue color scheme used in 1982 and '83. The bulging head, long neck, and octagonal wing planform certainly don't look aerodynamic, suggesting this is not designed for atmospheric travel. There are wonderful details in the design: interesting small shapes create leading edge extensions, transparent red antennas are connected at both ends and decorated with the new flag element, and an aircraft rocket engine tail 4 x 2 x 2 is embedded between the pods so that only its exhaust cone is visible. As with some of the 1986 sets, both of the supplied minifigures are yellow.

One problem that the emergent technology of white electronics had solved was fuel efficiency on long journeys, meaning that even this two-seater craft could now leave the Messier 54 globular cluster to explore the Sagittarius Dwarf Galaxy. The ship's nuclear reactor, housed in its nose to provide sufficient separation from the crew, generated highly compressed starfire energy which bubbled red as it was conducted through the coolant tubes to the thermal rocket. This advance, combined with the global Mercator framework technology brought from the Milky Way, would allow the Gold Team to locate dozens, if not hundreds, of new worlds to terraform.

6827 STRATA SCOOTER

Released: 1987
Total Elements: 35
Distinct Elements: 19
Minifigures: 1
Also known as Scanmobile in UK retail catalogs.

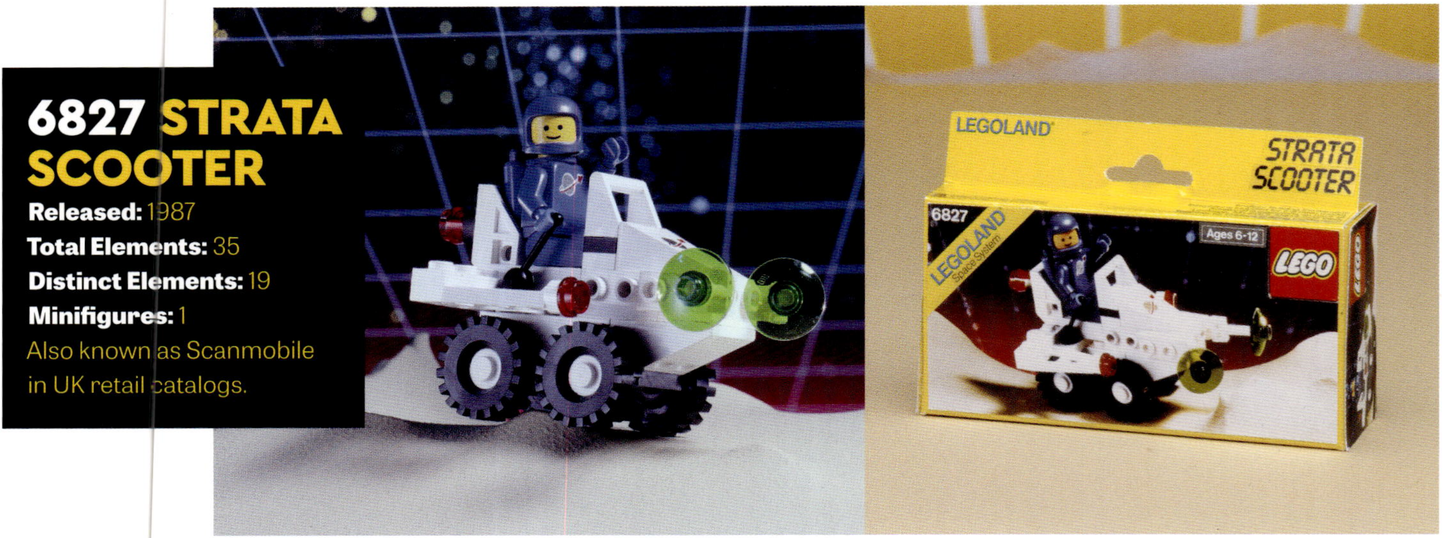

This dinky detector boasts four hinged elements for thorough scanning: two red-tipped space wings at the rear swing a full 180 degrees, while the transparent green dishes at the front are attached with control-lever hinge bars, new for 1987. The "space nose" cockpit adorned with the LEGO® Space logo only appeared in this set and 6849 Satellite Patroller.

The geoengineer drove straight toward the huge cliff face on Messier 54 Omega Cg. The ultrasonic transducers fluctuated rapidly in front of her, sending green pulses into the layered rock, while the gamma percussion drills at her sides blasted the softened surface cleanly away. She was able to tunnel into the cliff at great speed, thanks to the frictionless surfaces of the motor, a breakthrough dubbed "white electronics" by its inventor. A few minutes later, the lidar sent a warning. Wasting no time, she tugged on the starboard lever, spinning around in the tight space before firing the thruster and zooming back out of the entrance, just as the tunnel collapsed behind her.

__Hinge bar with three fingers and end stud__ not only provided minifigures with another new type of control lever but was also a useful building element. Here it is used on the vehicle's front with dishes attached to the studs. The concentric rings on its bar had been a design feature on some elements ever since the beginning of LEGO Space.

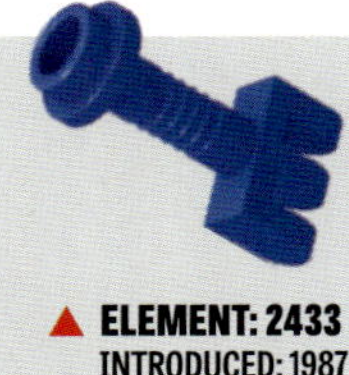

▲ **ELEMENT: 2433**
INTRODUCED: 1987

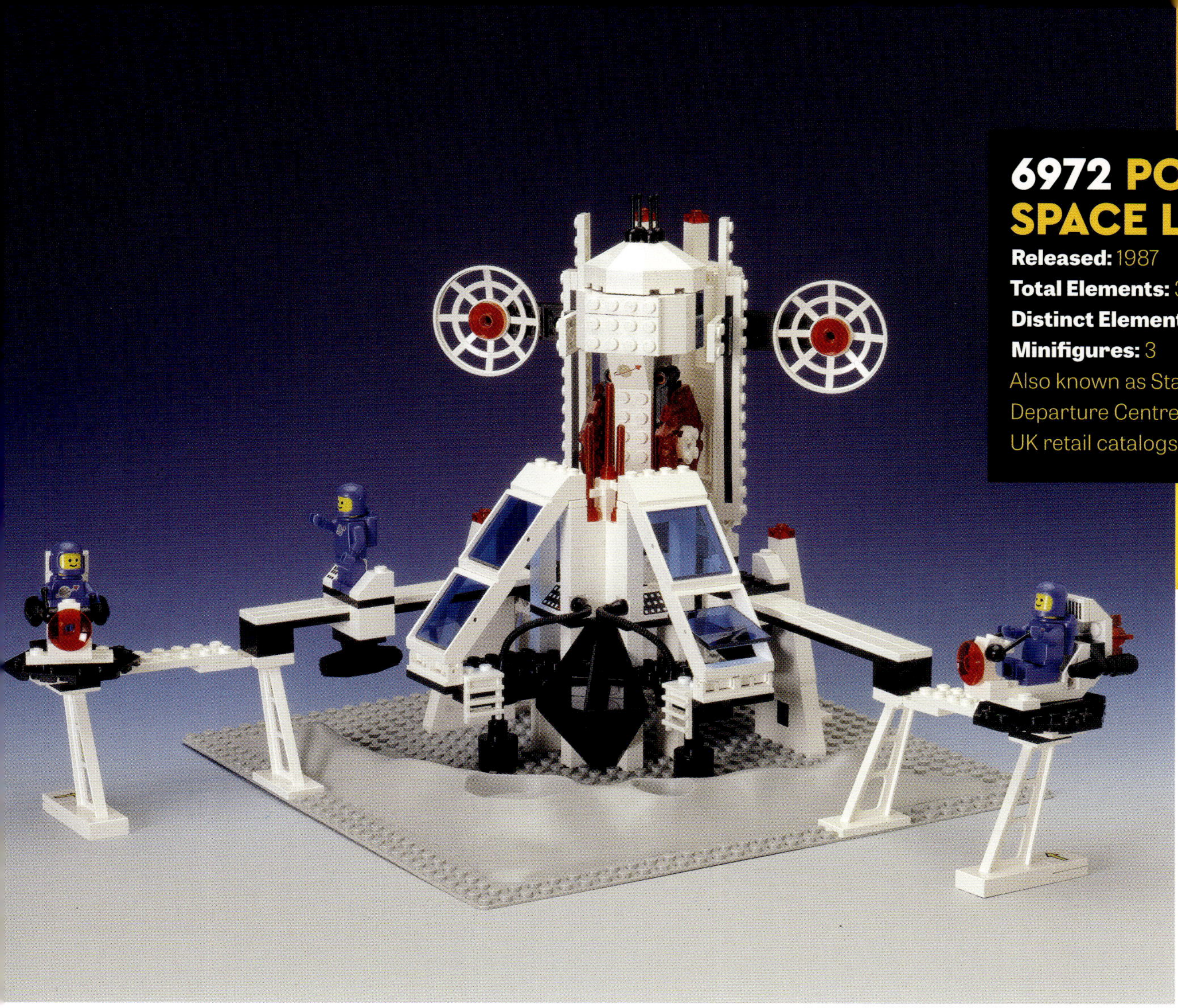

6972 POLARIS I SPACE LAB

Released: 1987
Total Elements: 390
Distinct Elements: 98
Minifigures: 3
Also known as Star Gate Departure Centre in UK retail catalogs.

This sprawling L-shaped structure has an almost minimalist aesthetic, unlike any base that had come before it, yet the features included are familiar. Two brick-built monorails shuttle pilots to octagonal landing pads that actually rest beyond the crater baseplate. To provide leeway while accounting for the slightly smaller height of LEGO® baseplates, these extensions are attached with vertical hinges. The command posts are shielded by long transparent angled blue windows which hinge wide open, almost like habitation pods, and are topped by a stylish trio of red antennas. The corner launch pad is another novel design, with dishes attached to long, spindly vertical struts that can be rotated. The ship itself is fish-like, with a bulbous head and thick spine housing two transparent red octagons that fold out. Its front compartment just fits a minifigure but contains no controls. All minifigures are blue, adding to the classiness of this product's design.

"XT-2, this is Star Gate tower air control. You are cleared to land." The soldier touched down and leapt onto the monorail, stabbing its controls furtively. She glided gently toward the waiting satellite looming above: her home for the next seven months. Climbing into its cramped pod, she pulled the door shut, and its octagonal, concave screen lit up before her. She tapped the secure comms icon. "You want to tell me what this is all about, Commander?" she grumbled as the support arms pulled back and she shot toward the Star Gate above. "I apologize for the secrecy" came the grave reply, "but we have a code zero cataclysm in the Milky Way." She paused in surprise, but with no time to waste, she unfolded the photovoltaic wings to power up while the commander continued to explain. "We're getting readings from a supermassive black hole centered on Gamma Cephei. We've had no comms at all from the Milky Way since then. Whether it was an accident or by design, we don't know. That's your challenge to answer . . . along with figuring out whether we are the last humans in the universe."

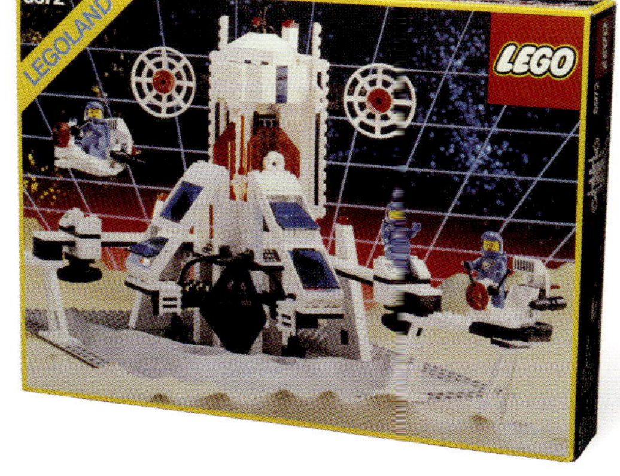

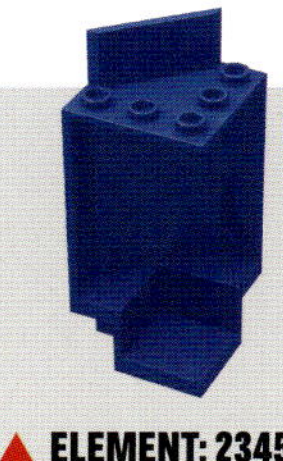

▲ **ELEMENT: 2345**
INTRODUCED: 1986

***Panel 3 x 3 x 6 corner wall** is usually associated with the LEGO® Castle line, but surprisingly, for its first three years of production, it appeared in an equal number of LEGO® Space sets. Here, three are stacked with train windows 1 x 2 x 3 to create the turret-like support tower on the launch pad.*

INTERVIEW WITH

CONNIE BORK

WHEN DID YOU JOIN THE LEGO GROUP?

I started in 1979 in the production department, where I worked on element quality. After a while I realized that I like to use my hands and make a lot of stuff, so I thought I'd try something else, and there was a job in design.

For most of my first half year in design I was a support for the designers. I counted the elements in all the models they made and then calculated the price. That was manual at that time; we didn't have a computer. But we did then get one—can you imagine, one computer for the whole department? 1984 is not so long ago!

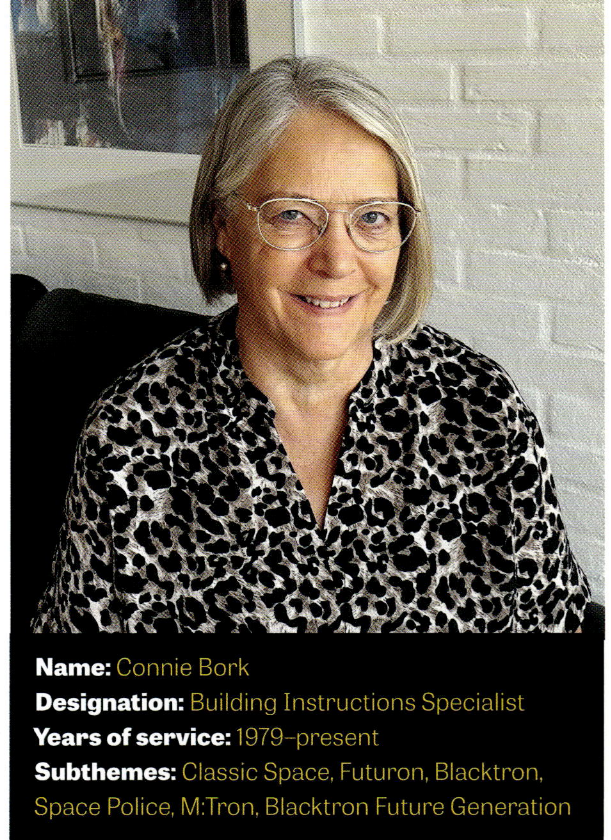

Name: Connie Bork
Designation: Building Instructions Specialist
Years of service: 1979–present
Subthemes: Classic Space, Futuron, Blacktron, Space Police, M:Tron, Blacktron Future Generation

WHAT WAS JENS NYGAARD KNUDSEN LIKE?

I really loved to work with him and he would always listen to people; he was always so kind. I feel that all the people who have worked with him would say the same. It was one big family and everybody knew each other. He liked to inspire all the designers. He was the ideas man.

Jens was always creating fun. We always knew when Jens had built something new: He would come down the stairs from the first floor with a model in his hands, making the noise that he thought the model should make! "Wheee! Waaah!" It was so funny. He would place it on a table by us, and in his eyes it was the most wonderful model he had ever made. "Have a look at it. See, isn't it beautiful?"

And although you never heard him say it, you can be sure that LEGO® Space was his favorite, and then LEGO® Castle. Because whenever he had made a nice space model, he would come down with it so we could all see it, but he never did that with a castle model. I think he'd just let Daniel and Niels know about those!

On one hand he could be the guy who has a lot of fun, but on the other hand he could also decide the strategy. I really think that was one of the best things about him, because he had both sides. Say there was a discussion about whether a cockpit should be orange or yellow or blue . . . or what else could it be? And then Jens would be very sharp and say, "Okay, we make a decision now. It is orange." And then everyone knew, and we didn't ask further. It had to stay orange on all the models. He was very good at lining up everything and saying, "Okay, now we're going in this direction."

HOW DID YOU COME TO BE WORKING ON THE BUILDING INSTRUCTIONS?

In 1983, two people started up the Building Instruction Team and soon realized they needed more people. So in 1984, just half a year after I had started in design, I moved to this team. Until 1983, designers made the building instructions and decided how they should look. After 1984, the instructions looked very different.

HOW WERE BUILDING INSTRUCTIONS CREATED BACK THEN?

The process took more than six months—nearly a year. Such a long time! After the designers finalized

WE DIDN'T HAVE ANY COMPUTERS, SO WE BUILT EVERY STEP. WE BUILT THEM ALL ON A TABLE, SO YOU COULD SEE ALL THE STEPS.

the model, someone else checked that they hadn't used an element in a way that was not allowed. Once they had agreed on the model, they counted elements to see what the price would be.

There were not many of us, so we all sat in the same office. We all worked together really closely, because when we made the building instructions it was very important that we integrated with each other, so we could show what the designer was thinking. Every time we had a new model, we built it through with the designers. Some of the models we built twice to be sure that everything was okay. And we still do that today: it's still one of our biggest things, to build together and see if there is anything we can change so that the children get a better experience when they build it. That's very important.

When the final model came to us, we split it down into all the building steps. We didn't have any computers, so we built every step. We built them all on a table, so you could see all the steps.

SO, IF THERE WERE TWENTY STEPS IN A MODEL, YOU BUILT TWENTY MODELS?

It took a lot of time. We used a hammer to tap the bricks into place; otherwise it would get hard on our fingers! After we finished the building instructions, Jens would come up and look through them. Thereafter he didn't look further at them. Never. By then, his head would be full of new ideas!

We would also talk to other relevant people, including the graphic designer who would later make the printed version. At this stage the graphic designer might say, "There's too many steps—take some away. There's not enough space."

Then they took pictures of all the steps, which would take days, because each model had to be placed totally correctly, as per the previous one and the next one. It was like when you were a child and you made an animated flipbook.

It could sometimes happen that you were off with a photograph. You placed the model on the table at the angle it should be, and if someone who was coming by hit the table a little bit, even just two millimeters, then we wouldn't be able to match the pictures over each other later. So we would take the pictures all over again. That was really tough.

After photography it went through a process where they made a print, from which they could later draw the building instructions. But first we had to build it again to check if it was right—if we had to make any changes, we had to do all of this process again! After corrections it went to the drawing stage. So that's why it took so long.

DID IMPROVEMENTS TO BUILDING INSTRUCTIONS OVER THE YEARS HAVE AN IMPACT UPON THE WAY MODELS WERE BUILT?

Around 1987, we were testing all the building instructions we made with children in Copenhagen. Always eight boys and eight girls, aged six and up. We sat and watched them to see how they built the model. What problems were they encountering? What did they like, and what didn't they like?

The designers went with us, if they had time, because it gave us more information on how children build, what they could and could not do. For example, if they had different yet similar small elements, they often took the wrong one. We learned a lot about left and right—the children couldn't see the difference between a left and a right wing. We would discuss afterward what we could change and how we could be more careful. So one of the things that we learned was, when we make wings, we should differentiate the builds so that they are not quite the same. ■

▼ Various examples of building instructions from the 1980s.

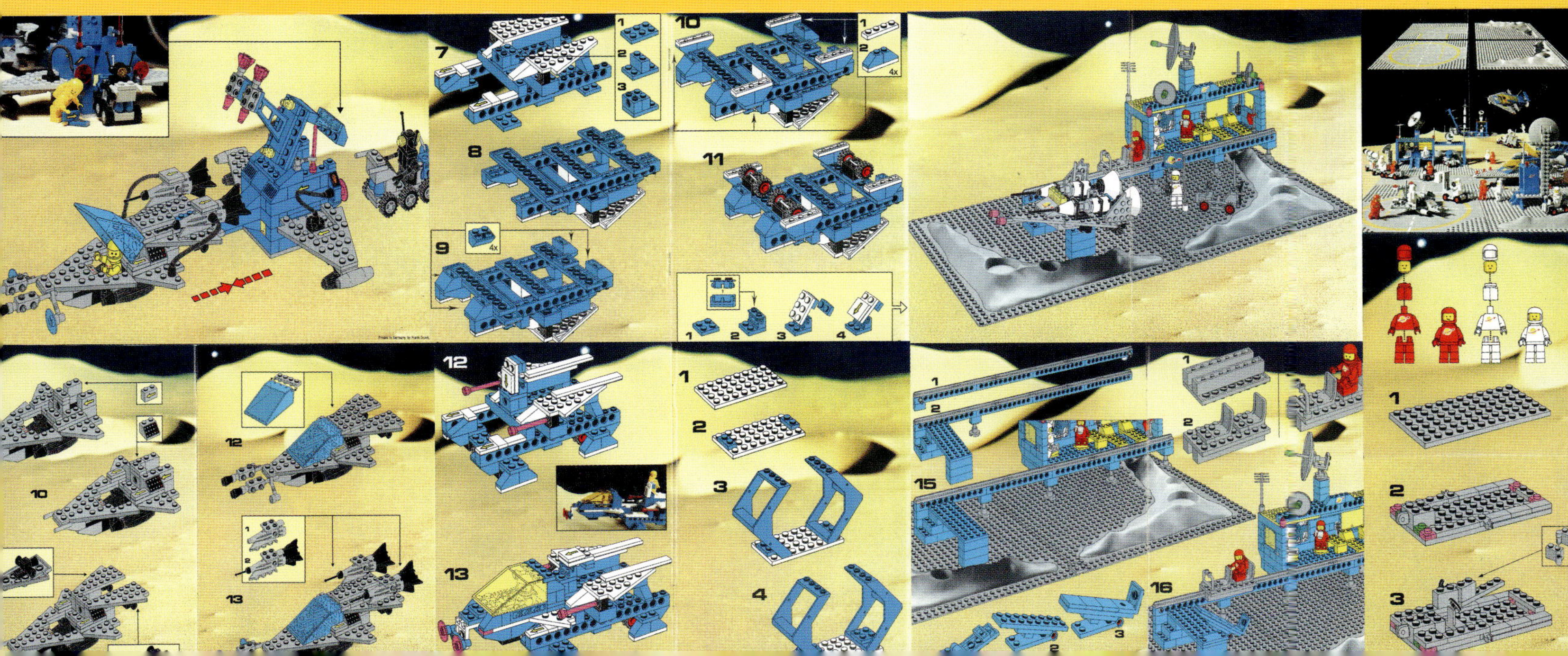

CHAPTER 2

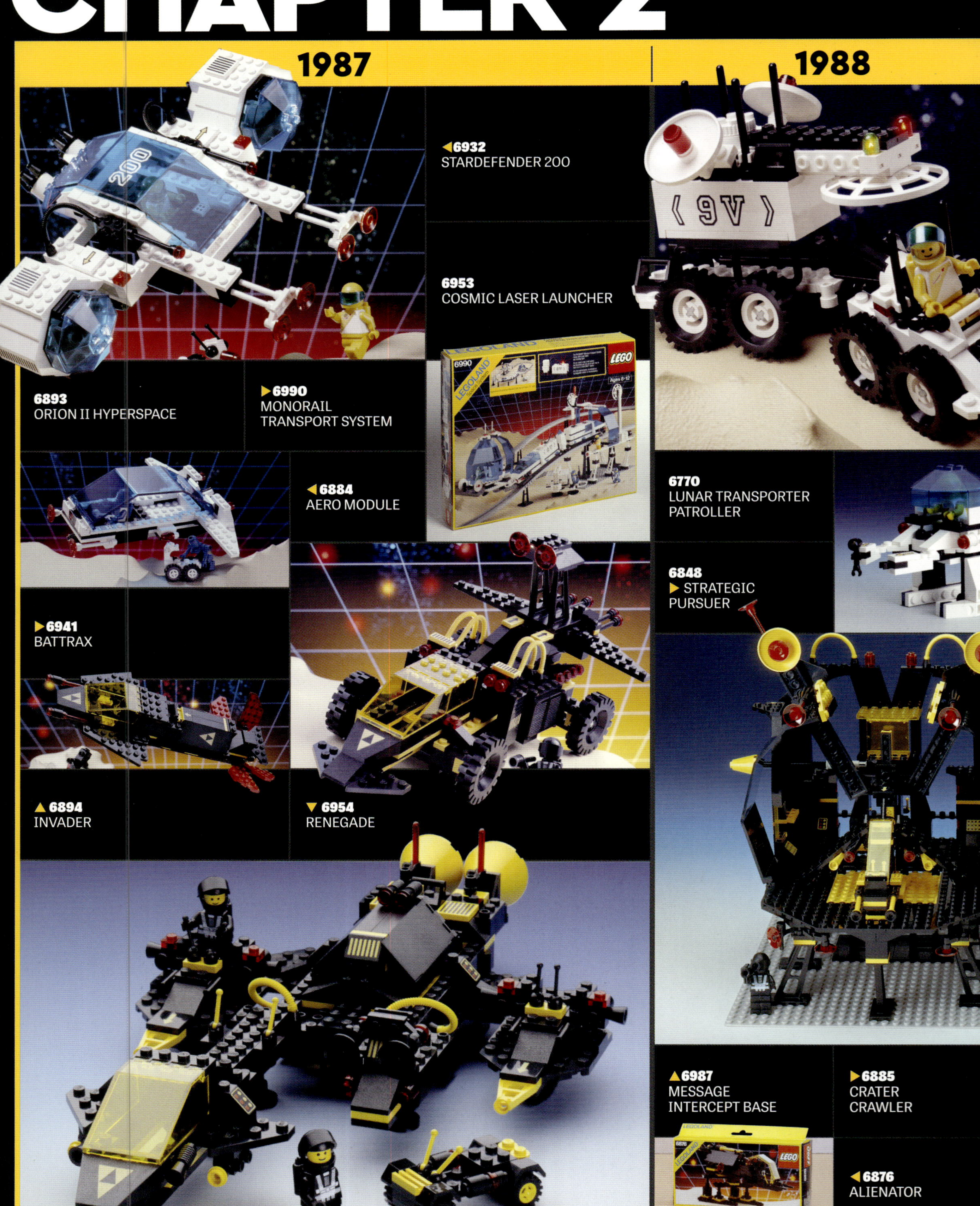

◀6932
STARDEFENDER 200

6953
COSMIC LASER LAUNCHER

6893
ORION II HYPERSPACE

▶6990
MONORAIL
TRANSPORT SYSTEM

◀6884
AERO MODULE

▶6941
BATTRAX

▲6894
INVADER

▼6954
RENEGADE

6770
LUNAR TRANSPORTER
PATROLLER

6848
▶ STRATEGIC
PURSUER

▲6987
MESSAGE
INTERCEPT BASE

▶6885
CRATER
CRAWLER

◀6876
ALIENATOR

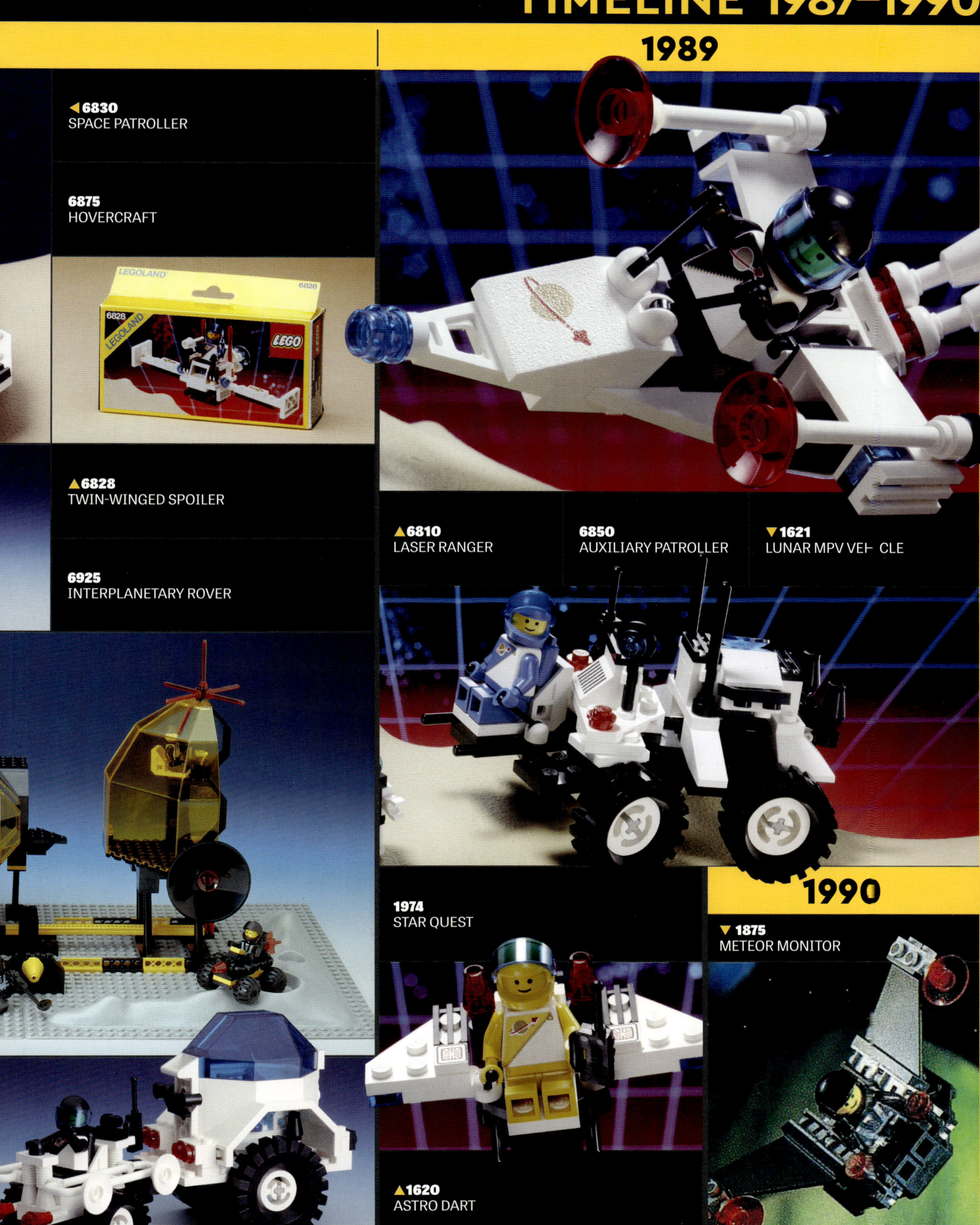

◀ **6830**
SPACE PATROLLER

6875
HOVERCRAFT

▲ **6828**
TWIN-WINGED SPOILER

6925
INTERPLANETARY ROVER

1989

▲ **6810**
LASER RANGER

6850
AUXILIARY PATROLLER

▼ **1621**
LUNAR MPV VEHICLE

1974
STAR QUEST

▲ **1620**
ASTRO DART

1990

▼ **1875**
METEOR MONITOR

CHAPTER 2

FUTURON AND BLACKTRON

1987–1990

Nine years into LEGO® Space, the product assortment was refreshed to keep it original and eye catching to customers and toy retailers. Eight final sets in what would become known as the "Classic Space" era were released in 1987, but alongside them in US catalogs were three black models bearing the thrilling names Invader, Battrax, and Renegade. Conversely, catalogs in other countries introduced a range of white models with a gleaming, futuristic look.

In 1987, the names Blacktron and Futuron were not used. "It is fair to say that there was a somewhat blurry transition from Classic Space to Futuron and the themes that followed," says Bjarne Panduro Tveskov, who joined the LEGO Space design team in 1985, at the time these were being designed. The style of these two subthemes, however, was clearly defined.

▼ A variety of Futuron models, including monorail elements, coexist peacefully with the Blacktron 6987 Message Intercept Base in a promotional image from 1988.

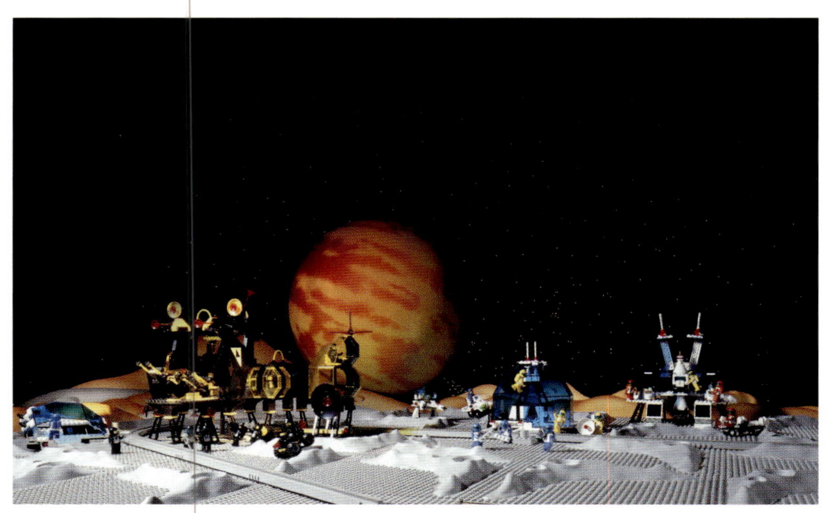

"I WANTED SOME SILVER ELEMENTS, AND MAYBE THE BLACK AND YELLOW COLOR SCHEME; LATER, I USED THESE COLORS ON BLACKTRON."

—JENS NYGAARD KNUDSEN, LEGO SPACE CREATIVE LEAD, THE LEGO GROUP, 1979 TO 1992

Blacktron sported a dramatic color scheme, one which creative lead Jens Nygaard Knudsen had been considering for many years. When the LEGO Space line was first refreshed in 1982, white with transparent blue was chosen, but another of the palettes on the shortlist was black with transparent yellow. The Futuron color palette also had precedent. The Classic Space Light & Sound System sets from 1986 (see page 83) are white with transparent blue windscreens and have a thin black line running through the fuselage. These design features were retained and amplified for the new style applied to LEGO Space for 1987.

Both of these ranges became available across all markets in 1988, and from that point, marketing began to use the names Futuron and Blacktron. US advertising hinted that they were opposing factions, such as a catalog proclaiming, "The Valor Squadron—FUTURON—Energizes

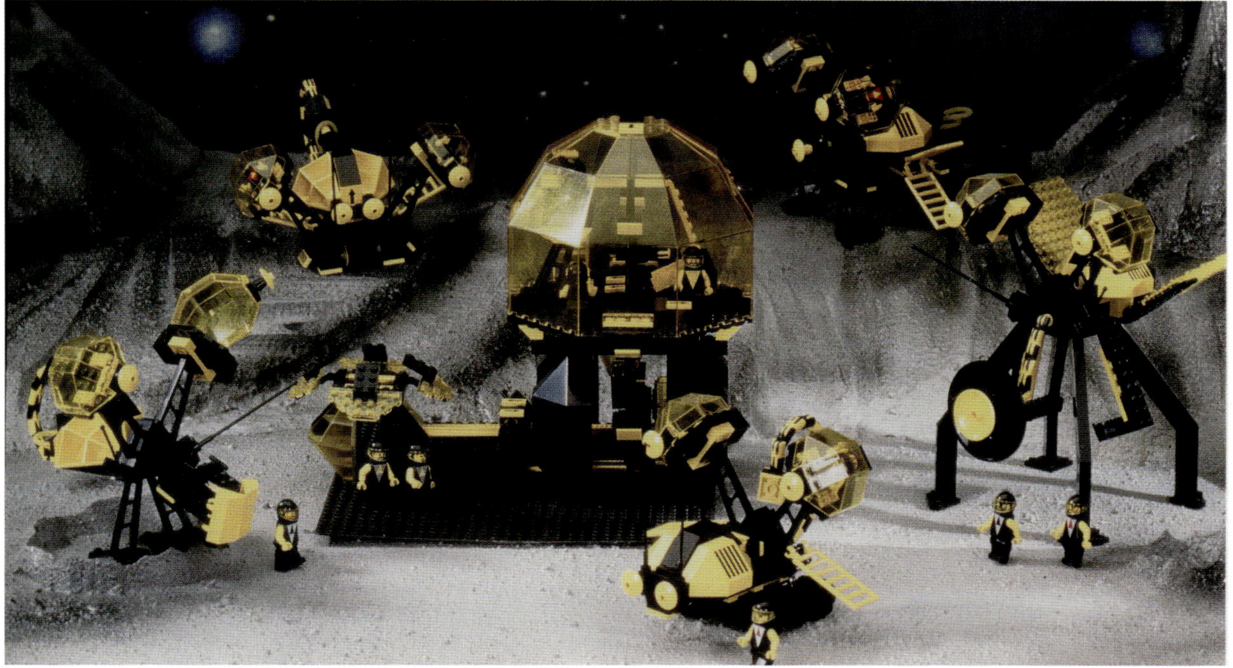

This concept model for Blacktron from 1985 includes insect-like craft and minifigures with different faces and chest insignia which were never used. Some features, such as the octagonal pod canopies, appeared later in 1991's Blacktron Future Generation.

the Galaxy as the Forces of BLACKTRON™ Continue Their Pursuit!" Marketing elsewhere in 1988, however, left the nature of their relationship more open.

The new subthemes were not only defined by their colors; element choice also played an important role. Blacktron retained the sleeker, more aerodynamic shaping of Classic Space vehicles and used LEGO® Technic connectors in a consistent manner so that segments from different products could be connected together easily. Meanwhile, Futuron made use of new elements with angled curves that were based upon octagons, although it was some other new elements that made the greatest impact. "We made a lot of new elements and colors for the Futuron theme," Jens Nygaard Knudsen recalled in a 2008 interview, "but when we tested with children, they were unimpressed: it was the minifigures' new diagonal print, still with the LEGO Space logo on it, that all of them were impressed with. Children like this kind of detail."

ALL ABOARD THE LEGO® MONORAIL

The 9 V electrical system that initially appeared in Light & Sound System sets in 1986 was first used for propulsion

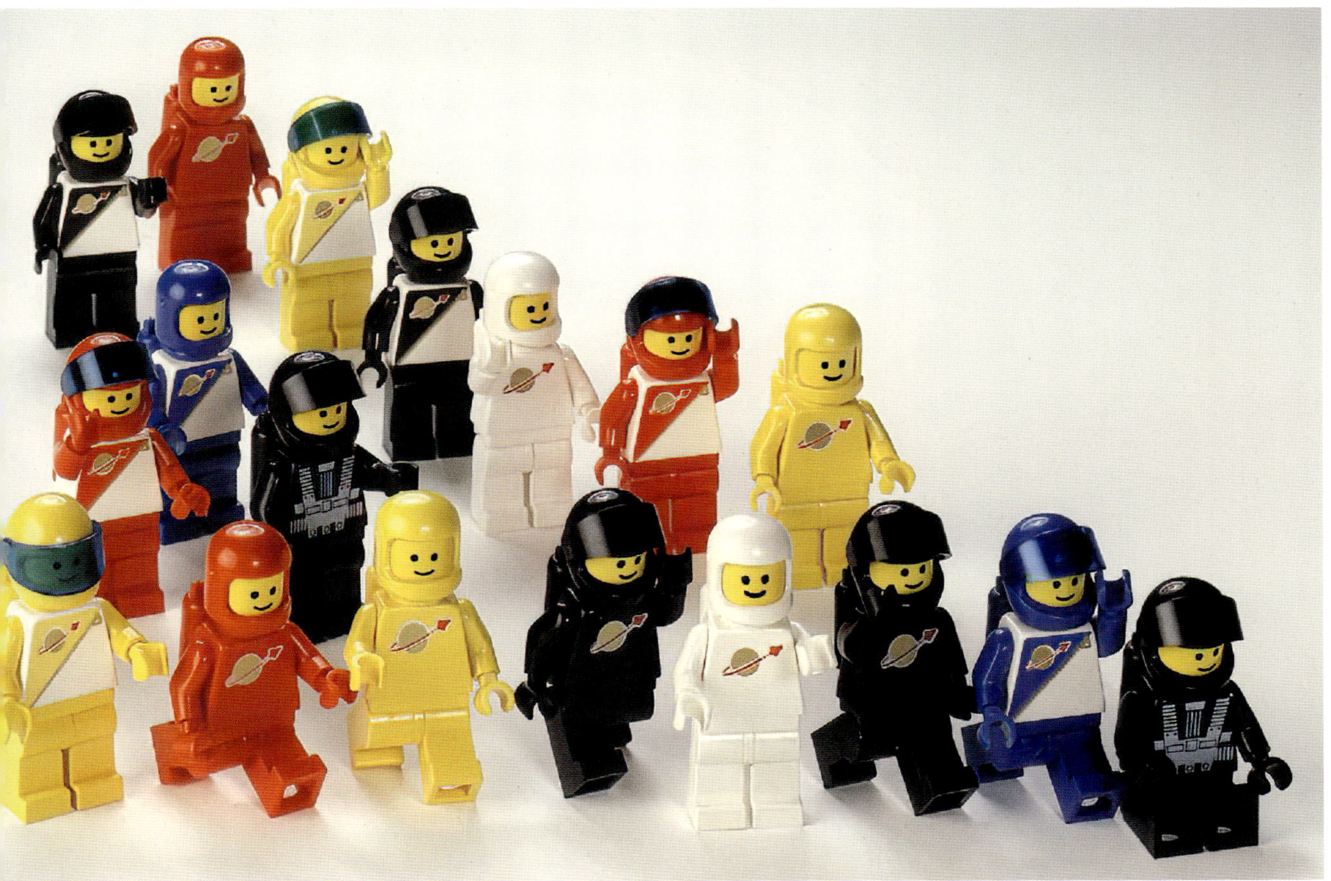

New crew: The original red, white, yellow, blue, and black astronauts were joined in 1987 by the two-tone Futuron design, in red, yellow, blue, and black, with a new visor element in transparent blue. Blacktron introduced a black-spacesuited minifigure with a harness decoration on the chest and an opaque black visor.

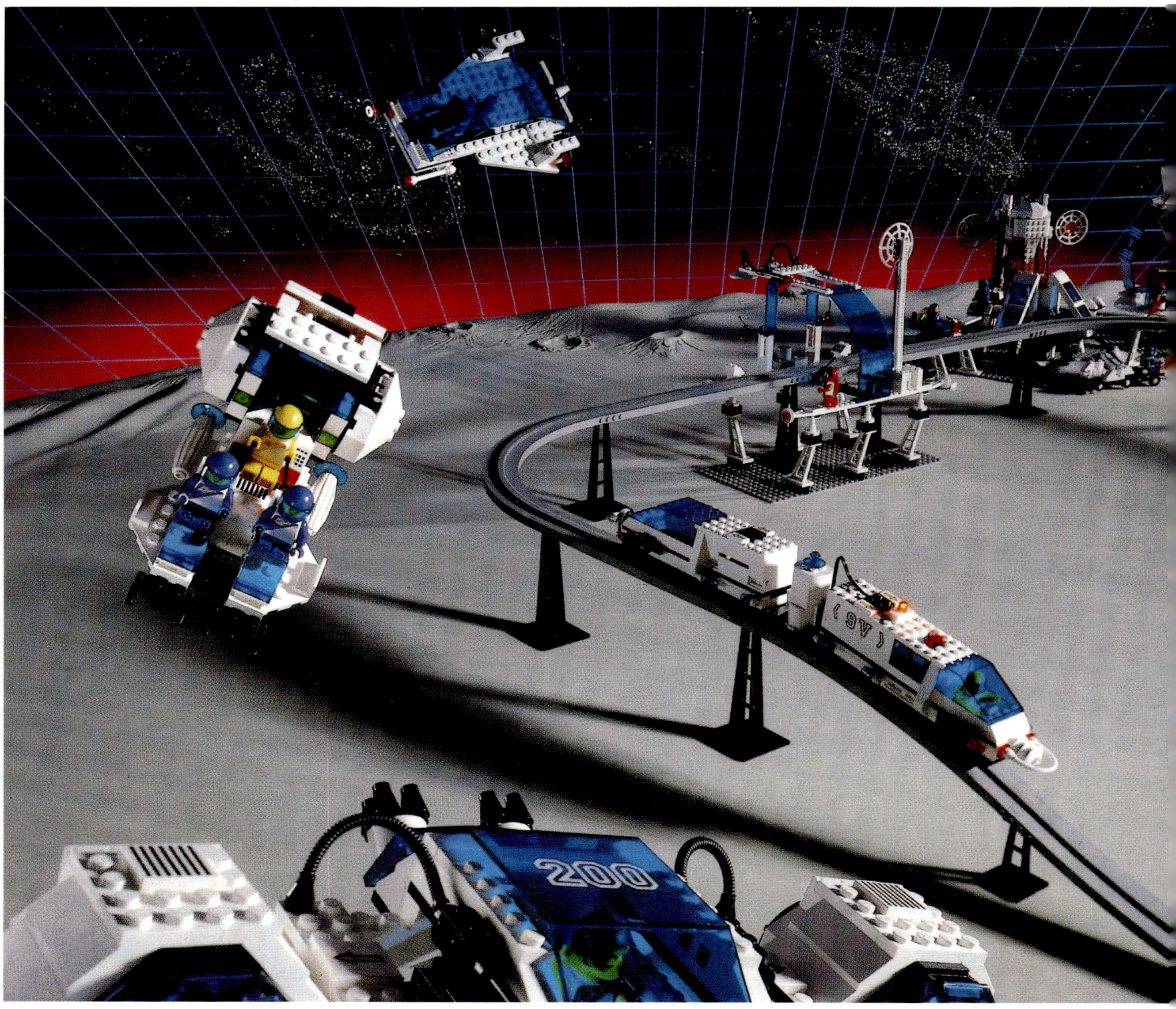

▶ In 1987, Futuron was the first subtheme to use the exciting new monorail system. In the 1990s it reappeared in Unitron, as well as the LEGO® Town line.

"THE FIRST SET I REMEMBER STARING AT FOR HOURS IN THE LEGO CATALOG AS A CHILD WAS THE MONORAIL. NOT ONE I EVER RECEIVED, BUT I WAS JUST FASCINATED BY HOW IT COULD WORK, HOW THE TRACKS WERE BUILT, AND IMAGINING HOW AWESOME IT WOULD BE TO HAVE RIDING AROUND MY ROOM."

—WILLIAM THOROGOOD, VICE PRESIDENT OF DESIGN, NEW BUSINESS, THE LEGO GROUP, 2004 TO PRESENT

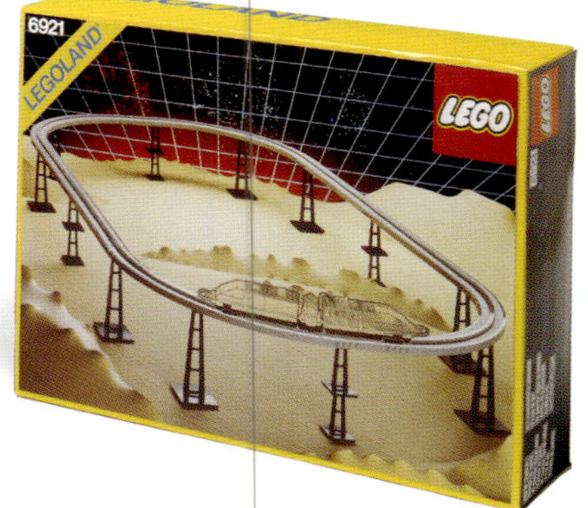

▼ Original mockup of the box for Futuron set 6921 Monorail Accessory Track.

in a 1987 Futuron set, years before the 9 V system was rolled out to LEGO® Trains and LEGO® Technic products, which used both 4.5 V and 12 V electrical systems. The monorail was a complex technological concept which took Knudsen and the LEGO Town team many years to bring to shelves. Aside from the new motors, which were developed with Peter Bolli and his team in Switzerland, the LEGO team, including Niels Milan Pedersen, designed well over a dozen new elements for the chassis, bogies, and tracks. A centrally housed motor propelled the train via gear racks on the single track, and ramp pieces enabled different levels of track. Sensors at the stations could stop or reverse the train's movement.

LEGO products using the monorail system remained available for a decade: 6990 Monorail Transport System and its supplementary product, 6921 Monorail Accessory Track, remained available until 1991; LEGO Town released an airport-themed set in 1990; and in 1994, LEGO® Space released an even larger monorail set within the Unitron subtheme.

INTERVIEW WITH

BJARNE PANDURO TVESKOV

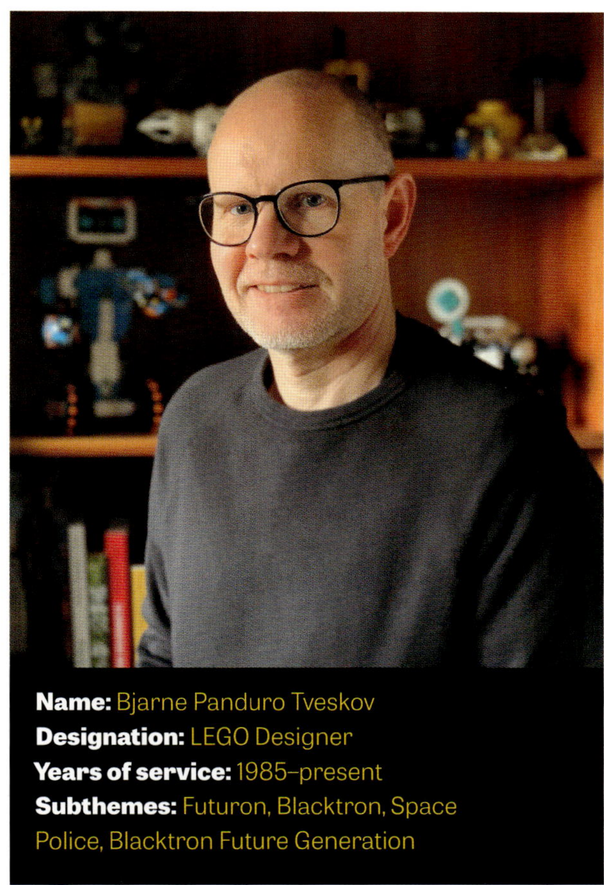

Name: Bjarne Panduro Tveskov
Designation: LEGO Designer
Years of service: 1985–present
Subthemes: Futuron, Blacktron, Space Police, Blacktron Future Generation

DID YOU OWN LEGO® SPACE SETS BEFORE YOU BECAME A LEGO DESIGNER?

I loved LEGO® toys as a child, but by age twelve, when LEGO® Space came out, I was moving on to other interests. Still, I remember being interested that the LEGO Group were doing space models. I was fascinated by science fiction, but there wasn't much merchandise around. You kind of had to grab whatever you could get; it was such a limited supply.

WHAT WAS YOUR FIRST IMPRESSION OF JENS NYGAARD KNUDSEN?

It was Jens who was running the show, for sure! I remember I asked him about those alternative models on the back of the boxes. He jumped up, and we went into this room with a lot of models, and he just grabbed something and went into a long song and dance about their meaning. He was very animated!

I was seventeen when I was interviewed for the job of designer, and had just turned eighteen when I started. I quit high school to join; I had hardly even had a job in my spare time before that. It felt like some sort of freelance gig at first; it being a full-time day job was a pretty weird thought to me. But in a way it felt like "coming home"—seeing that there were other people with creative interests who actually made a living from it. Having spent my time in school mostly drawing and thinking about stuff, it was fantastic to see there was a place for me in this world.

"IT WAS FANTASTIC TO SEE THERE WAS A PLACE FOR ME IN THIS WORLD."

DID YOU ONLY WORK ON LEGO SPACE?

Most people worked across multiple themes, but I was one of the first people hired to do only LEGO Space. When Jens and Daniel Krentz started [around 1970], the design department could pretty much fit in one room. Growth was very rapid. When I joined, it was run in a more structured way.

We were based upstairs in the old gray Systemhouse and in Ole Kirk's House [nicknamed Løvehuset for the two lion statues flanking the doorway]. The LEGO® Town team was downstairs, and we had the LEGO® Trains people next door. So, we felt that we were very close to the original LEGO company that was started in some of those buildings, and where Godtfred Kirk Christiansen had his office in the 1950s. There was always this strong sense of history, which is still the case today. It's so important that it's grown out of that building. You can see that every thing in this strong history builds upon something that came before. It means a lot, when you look back.

WHAT WAS KNUDSEN LIKE TO WORK WITH?

I've never met a person like him in my life. I'm a bit of an introvert, so seeing a person who would just broadcast everything from the inside to the outside was just mind blowing and very inspiring. Jens really had no "filter," and I mean that in a good way, as that was really freeing. He had a very special personality;

▼ A transporter carrying an articulated vehicle, designed by Bjarne Panduro Tveskov at age seventeen for his interview with the LEGO Group in 1985. He was inspired by *Space: 1999*, one of the few science-fiction programs shown on Danish TV during his youth.

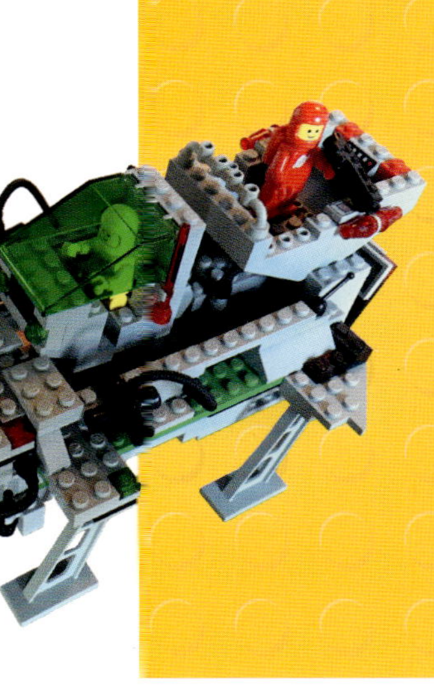

Design/ Kreativitet

Små børn/grovmotorik

Til udarbejdelse af forslag, til elementer og modeller til børn 2 - 5 år, søger vi to kreative medarbejdere.

En passende baggrund kunne være kunsthåndværker, med god sans for form, farve og design, eller lignende, kombineret med en lyst og forståelse for børns leg.

Rumfartsprogram

Til videreudvikling af vort Rumfartsprogram søger vi en fantasirig medarbejder, med god sans for former, farver og børns leg.

En passende baggrund kunne være egne eksperimenter med fremstilling af Science Fiction lignende modeller eller tegninger.

Ansøgning: Skriftlig ansøgning mrk. "Modeldesigner" bedes sendt til Personaleafdelingen.

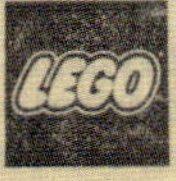

Virksomheden: LEGO Futura ApS varetager LEGO Gruppens internationale marketingplanlægning og produktudvikling. LEGO Gruppen beskæftiger internationalt omkring 4.200 medarbejdere. Vi er ca. 2.400 medarbejdere i Danmark - heraf omkring 120 i LEGO Futura ApS's tre afdelinger i henholdsvis Billund og København.

LEGO Futura ApS
7190 Billund

▲ The advertisement to which Bjarne Panduro Tveskov responded in 1985: "We are looking for an imaginative employee with a good sense of shapes, colors, and children's play."

he enjoyed putting energy into the world. He was always pushing ideas in new directions inside the company, and doing really well at it, but there were some challenges with that also. Sometimes he would go too far, back off a bit, and then try again—but that, too, was inspiring. At that age you really take it in, and it kind of builds your personality also.

Jens had a lot to handle. He was running both LEGO® Castle and LEGO Space, and also LEGO® Pirates was coming onto the radar, which was a big deal with lots of new elements and so on. Most of the LEGO assortment was reality based, but he was running all the fantasy lines. Even LEGO Castle was a bit more on the reality side, but LEGO Space is basically just imagination—rooted in NASA technology but more free flowing. This was perfect for the way Jens worked.

He would always talk about other ideas during our daily coffee break. For example, he had some great ideas on how to turn downtown Billund into some kind of theme park, with fake façades on the houses to make it look like some kind of old-time Danish village, and also building sidewalks with escalators! Of course, you had to ignore a lot of his ideas, but there would be nuggets here and there. I guess that was the same as the process of developing LEGO products; you have to explore a lot of weird ideas to get some good ideas too!

THE APPRENTICE

WHAT WAS THE PROCESS OF DEVELOPING MODELS?

Before starting at the LEGO Group, I had no idea how this work would be done—would we sit around a table and build the model together? But it was actually a case of having your own office. Model design was a very individual process then. Each person was handling a specific product, and if you know the different designers, you can see their personality a little bit in the design.

We didn't sit down for a briefing. It was more like, "Let's start building," and then Jens would come in from time to time, looking and saying, "Yes, this is good," or "Throw that away!" Thinking back, I feel it was more like an apprenticeship—a learning-by-doing, practice-based approach. You guide people without really instructing them too much.

Design was an iterative process. Our models went through phases with the model committee and the building instruction process, to make sure that they could be built. The model committee—kind of a team of model coaches, if you will, from different departments—would build the model and give feedback to the designer in a verbal meeting. For example, they might have said, "If you push down here on the model, it's going to break," and you had to find a way of binding it together. As a new designer, it was interesting to go through that process, and it was also a way of learning.

The building instructions team had a good sense of what children could do. Once a year we went to Copenhagen to see if children could build from the instructions. Some of the elements we introduced were not easy for children to build with—for instance, large shell pieces that shape the overall design are a bit unstable until you lock them together. If we went too far out on a tangent, we got feedback in terms of buildability and would pull back.

"YOU CAN SEE THAT THIS STRONG HISTORY BUILDS UPON SOMETHING THAT CAME BEFORE, AND IT IS QUITE UNIQUE TO THE LEGO GROUP."

It was also about including as many of the new elements into products as possible, and as a designer you always want to find new ways of using the elements! You hardly even had them in your hands before you began to turn them over and see how they looked from different angles, to imagine interesting shapes. That's how it works to this day: using elements in new ways and learning from other people.

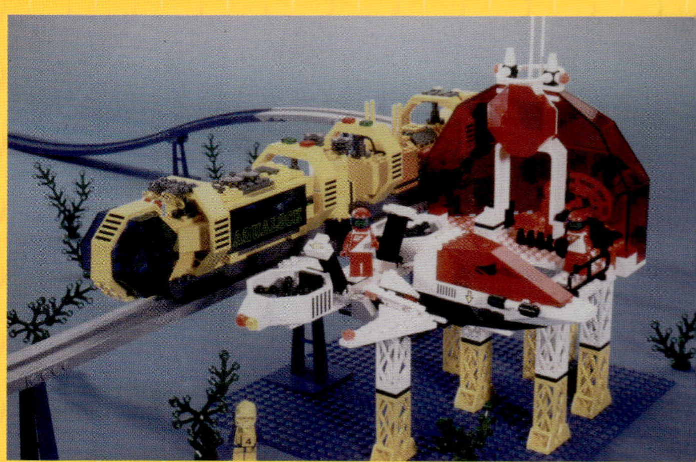

► This unreleased model by Bjarne Panduro Tveskov, dating from around 1990, is one of many early concepts for a LEGO Space theme called Seatron. The monorail was intended to serve as a link between the underwater world and a Futuron-like world above water. The design team even went on a research trip to Monaco and Paris to look at real submarines.

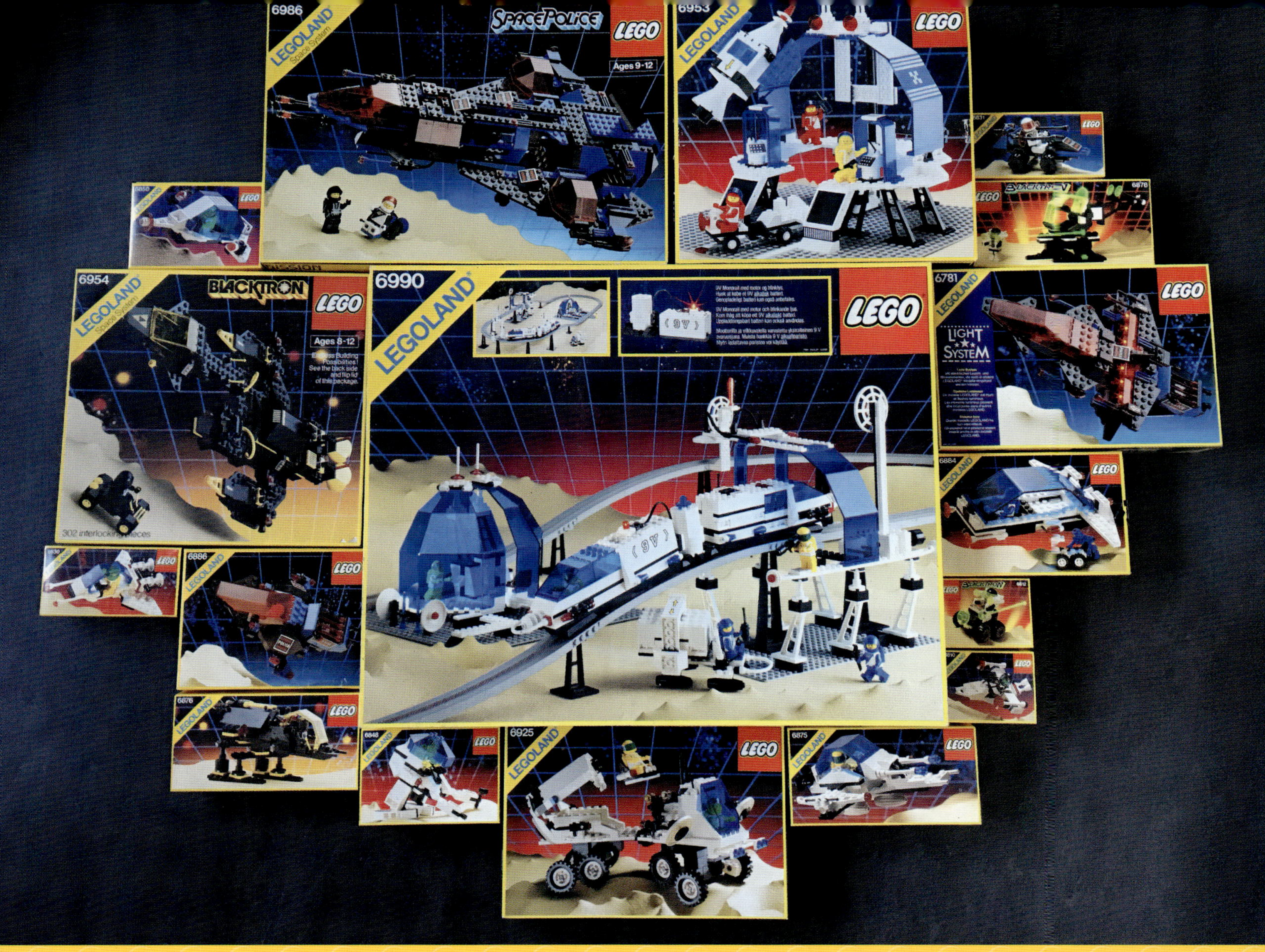

▲ A selection of LEGO products for which Bjarne Panduro Tveskov was a lead designer.

DID KNUDSEN STILL DO A LOT OF THE DESIGN WORK?

Jens would do a lot of things himself, not just the model development but also a lot of the element development. He would cut pieces up, file them, and glue them together. We had one technical draftsperson who would draw the elements. She didn't use computers; she used big pieces of paper. She would translate all these wild ideas into drawings, and then prototype elements would be made. There was a whole department that could build these rapid, short-run elements so that we could try them out, and then maybe make changes.

It's important to get good quantities of elements produced, to keep the price down. It was understood that you would use, say, a new cockpit across a whole line. However, with Blacktron we were each using differently decorated elements for each model, because element choice was very much down to the individual designer. Some of those elements are relatively rare because they have only been used in one product. This became more structured over time.

"YOU HAVE TO EXPLORE A LOT OF WEIRD IDEAS TO GET SOME GOOD IDEAS TOO."

WAS THERE STRUCTURE AROUND MINIFIGURE CHOICE? FOR EXAMPLE, DID THEIR DIFFERING COLORS HAVE MEANING?

It was only a few years ago that I heard this quote about the astronauts being color coded—each having different roles and so on. I was never informed of this. We just used whatever colors we thought looked good with the model. I often used yellow minifigures because I thought they looked great inside the transparent blue cockpits, whereas the red ones appear to be black when viewed through the blue plastic.

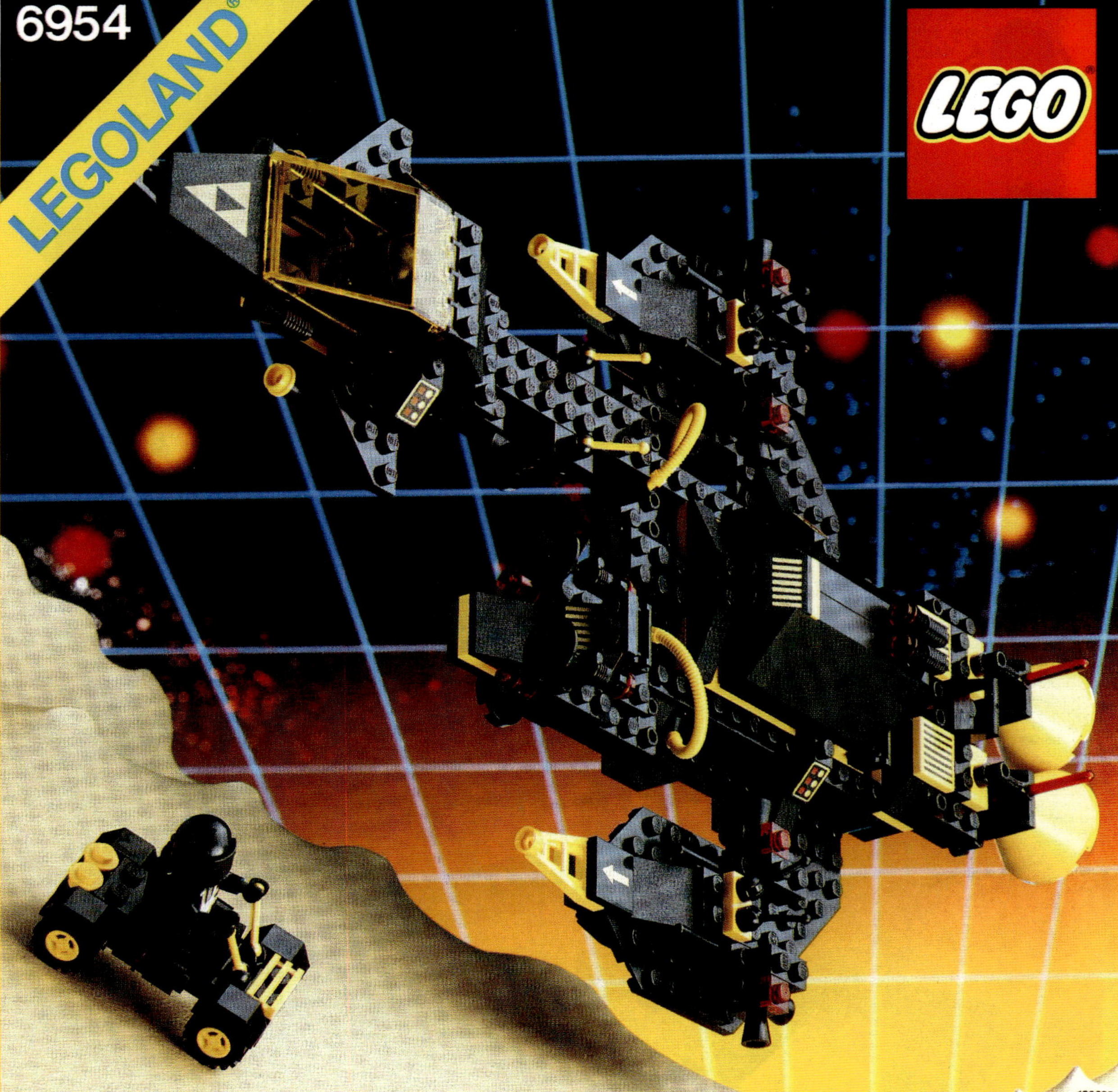

▶ For the Blacktron Renegade Tveskov designed an unusual asymmetrical spaceship.

"YOU HARDLY EVEN HAD THE NEW ELEMENTS IN YOUR HANDS BEFORE YOU BEGAN TO IMAGINE INTERESTING SHAPES."

HOW DID THE COMPLETED CONCEPT MODELS PROGRESS TO BEING CHOSEN AS PRODUCTS?

There was an "input meeting," where all the different departments showed ideas for new products, using good old-fashioned slideshows! We would prepare hundreds of slides with new ideas, and Jens would present them—for hours.

It was an annual meeting, as I recall, but I think it picked up pace in later years. It was a big affair—a lot of designers sat in, because it was a chance to see what the other groups were doing. It was mind blowing to see so much packed into a day, and with Jens presenting, it was just like a whirlwind.

WHAT OTHER KINDS OF TASKS DID YOU HAVE?

With the fixed assortment, it was easier, as you knew what you were working with, and then it was a nice break to do the alternative models. Someone would hand out bags of elements, and you would build them within a day. It was more freeform—there weren't the same demands on model stability and so on—which for me was great.

There were always odd jobs coming in, too. Sometimes you would build a big model for an exhibit or a little model for a cross-promotion. Also, for work anniversaries, LEGO employees received a gift from the company. We often had to go to product photography and TV commercial shoots, to make sure everything was in place. Sometimes those required extra models, like a big background, but that was mostly handled by other departments.

ON TO THE FUTURE

FUTURON WAS THE START OF SUBTHEMES BECOMING THE NORM FOR LEGO® SPACE. DID IT FEEL THAT WAY AT THE TIME?

There was the sense with Futuron that it was a new start. If you look at the first LEGO® Space models, there's clear inspiration from real space technology, whereas you could say Futuron is coming from more of a fantasy basis—that optimistic vision of the future, like the EPCOT theme park with the monorail at the Walt Disney World Resort.

At the time I started with the LEGO Group, the monorail was already there in prototype form, but we were still trying to find its place in LEGO products—getting the right choice of elements and scaling it down to something that could actually be produced. The designer Carsten Michaelsen and I worked on the first monorail set together. Carsten liked having lines going through the models, like the black lines that went all the way through a Futuron model in an elegant way. I was inspired by the very clean design sense he had. That was something I picked up on quite early, just by looking at what he did.

I think all designers pick up little things they see and like from other designs—both from other LEGO designers and elsewhere. Then it is often the case of refining and developing your own style. One of the many great things about the LEGO® System in Play is that even though we all work from the same array of elements, there is still plenty of room for having a personal design style.

BLACKTRON BROUGHT SUGGESTIONS OF CONFLICT TO LEGO SPACE. WAS THIS A CHALLENGE TO INTRODUCE TO CONSUMERS?

Blacktron sets were tested in Europe with consumers, and some were just too scared by the Renegade spaceship [set 6954; see page 130], so that never came out in Europe. Also, we couldn't show the Blacktron people with their visors down on many illustrations, because that might seem scary. They had to show their smiling faces and not appear to be too menacing.

I remember Jens had an idea for another theme that would be military but without weapons. They would be like UN soldiers, and they would have various sorts of tech, but there would be no weapons, to respect the LEGO Group's antiviolence stance. It's kind of a strange concept that I don't think would ever fly, but Jens would always try to find new, creative angles. ■

▼ Alternative models shown on the boxes of Futuron and Blacktron sets 6954, 6875, and 6885.

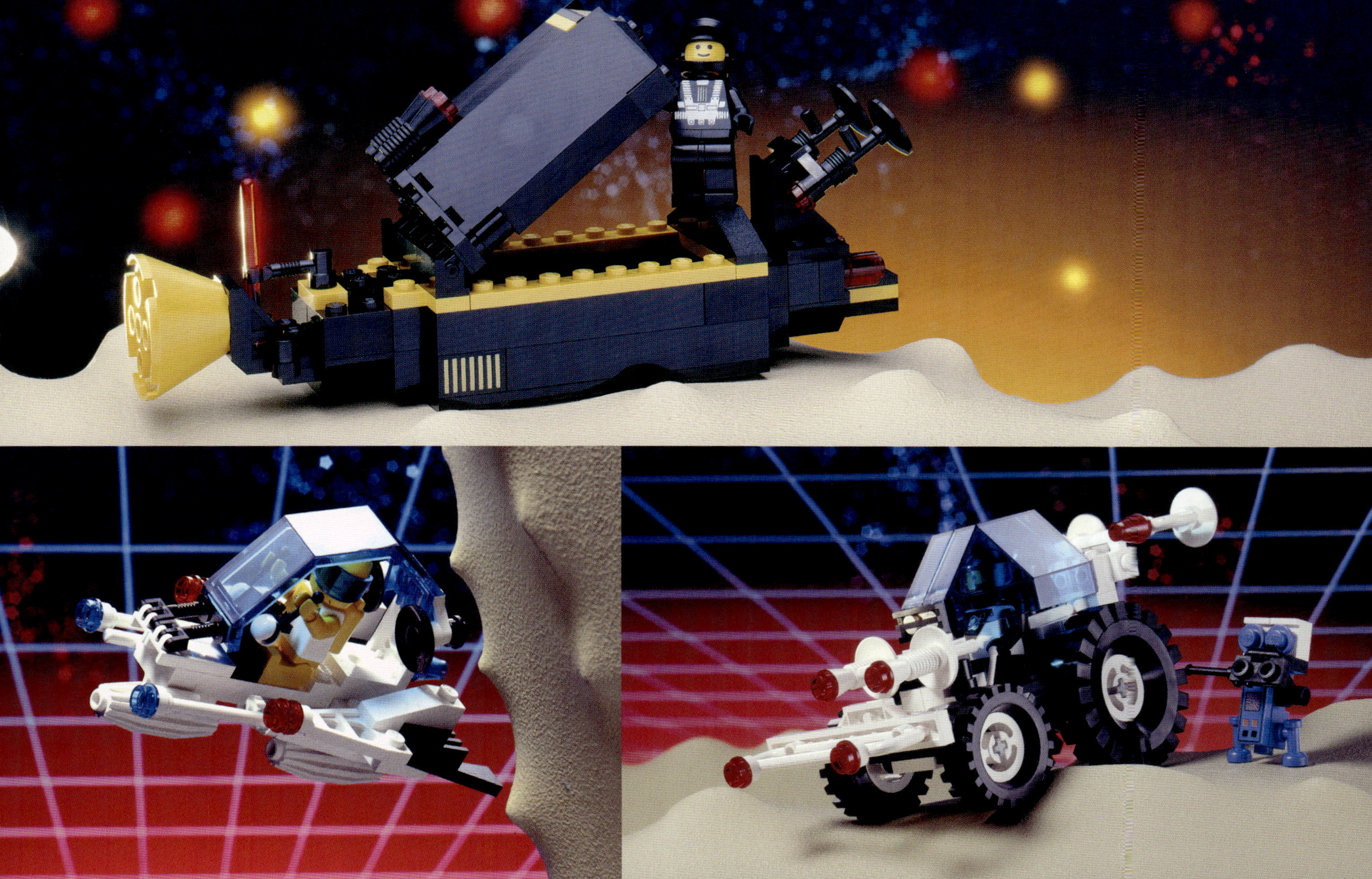

FUTURON

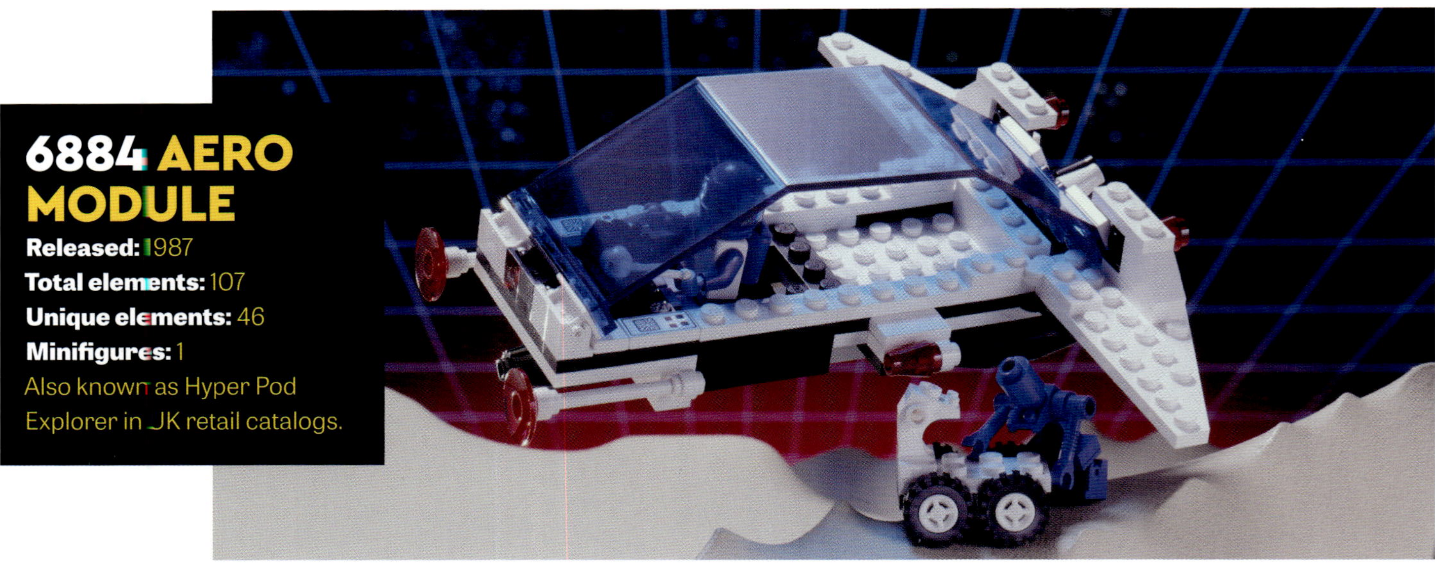

6884 AERO MODULE

Released: 1987
Total elements: 107
Unique elements: 46
Minifigures: 1
Also known as Hyper Pod Explorer in UK retail catalogs.

The Futuron aesthetic shifted away from streamlined spacecraft, and the design of this transporter could not be more different to what had come before. The windshield is an angled wall panel, creating a large canopy over the shallow rectangular body, which has exhaust cones attached at the rear. The compact rover and droid fit inside the cargo bay behind the pilot, and the hinged wings can swing forward.

▼ Original mockup of box.

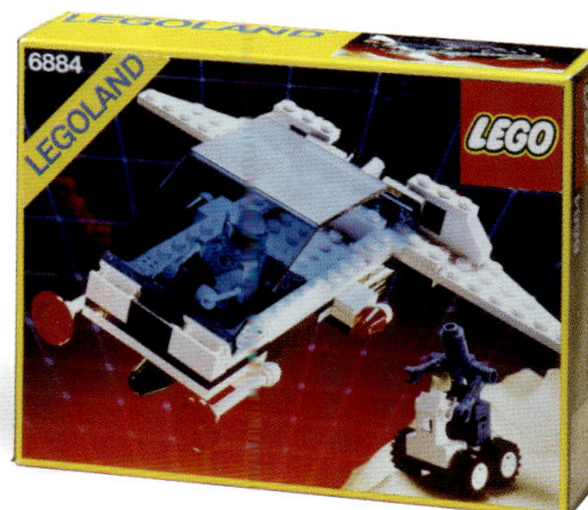

The wings folded forward after takeoff, and they soon reached cruising speed. "Fire the negative-mass rays," the droid instructed. "Otte-84, copy that," replied the pilot as she pressed a button to her side. Cone-shaped beams of concentric pulses emitted from the front, and a bright ring of red light formed ahead. "Wormhole active. Power up the plasma drive," instructed Otte-84. With the briefest of flashes, they traveled instantly through hyperspace, emerging directly above the surface of Skye, more than four hundred light years away from Messier 54. As the pod touched down on the barren moon, the protective kyan-mullite canopy sprung wide open, and Otte-84 lifted the buggy out in preparation for the survey.

*A handful of angled panels were added to the LEGO® Space inventory in 1987, including the **panel 10 x 6 x 11**, which could be aligned with the large quarter-dome element but is here utilized as a minimalist canopy. Many sets included decorated versions, and the element was used until 1996.*

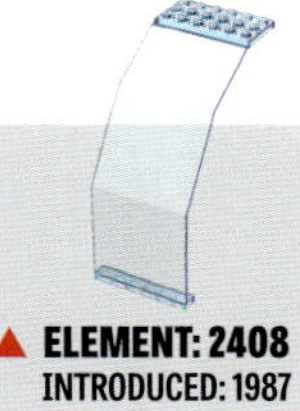

▲ **ELEMENT: 2408**
INTRODUCED: 1987

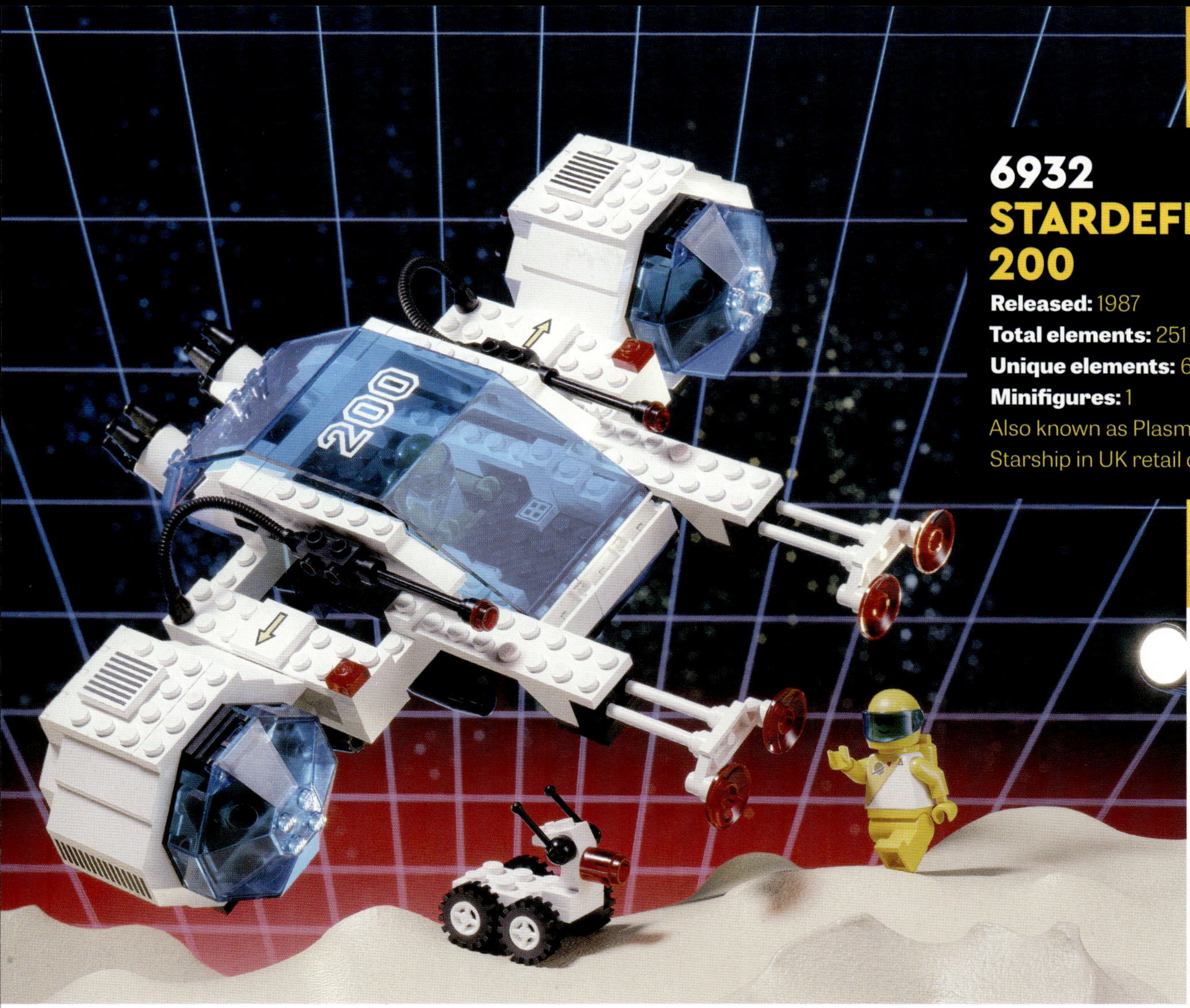

6932 STARDEFENDER 200

Released: 1987
Total elements: 251
Unique elements: 67
Minifigures: 1
Also known as Plasma-Drive Starship in UK retail catalogs.

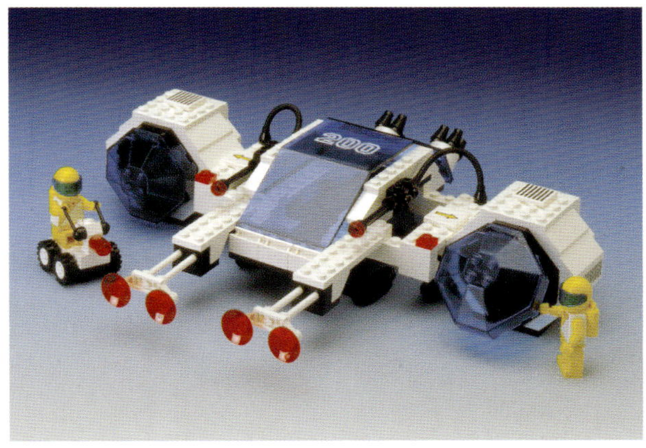

Unusually for LEGO® Space, this craft is wider than it is long. Like set 6884, an angled wall panel creates the windshield, and exhaust cones are attached at the rear, but here the fuselage is enclosed by bricks on the sides. The twin pods flip open at the front and on top, and can be detached and combined together into an unusual figure-eight design. They have no visible controls of their own.

The "200" range designation was assigned to plasma-drive starships, which proved perfect for reassignment as defensive craft following the attacks on the near-complete Star Base on Epsilon Gb. Its detachable pods used an octagonal design: many times cheaper to construct than a circle and more efficient than a square at minimizing heat gain during the zeptosecond-long journey through hyperspace. Whenever one of the strange black ships appeared, a Stardefender would deploy and then dock its two pods together, shooting them through a wormhole to a position just behind the enemy. There, the pods would divide to create triangulation points behind the hostile craft and, with a fire of the negative-mass cannons, send these unknown assailants to a random location in the cosmos.

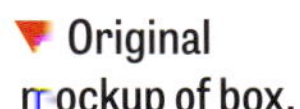
▼ Original mockup of box.

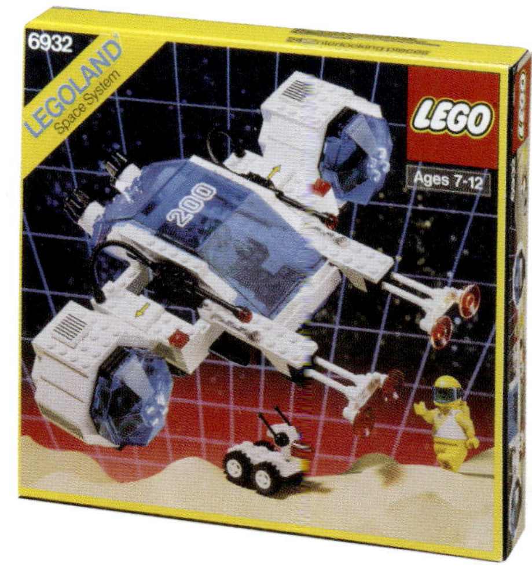

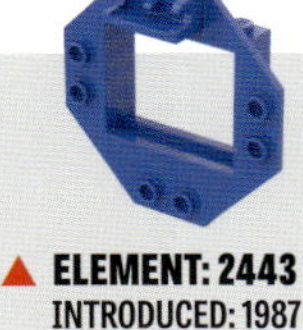
▲ **ELEMENT: 2443**
INTRODUCED: 1987

*The unusual **hinged window frame 1 x 4 x 3 with octagonal panel** was designed to pair with the octagonal canopy element. It takes the form of a standard window with a large octagonal frame that has sideways-facing studs and a hinge base. It appeared in six sets before being discontinued in 1996.*

6953 COSMIC LASER LAUNCHER

Released: 1987
Total elements: 210
Unique elements: 83
Minifigures: 2
Also known as Star Base One in UK retail catalogs.

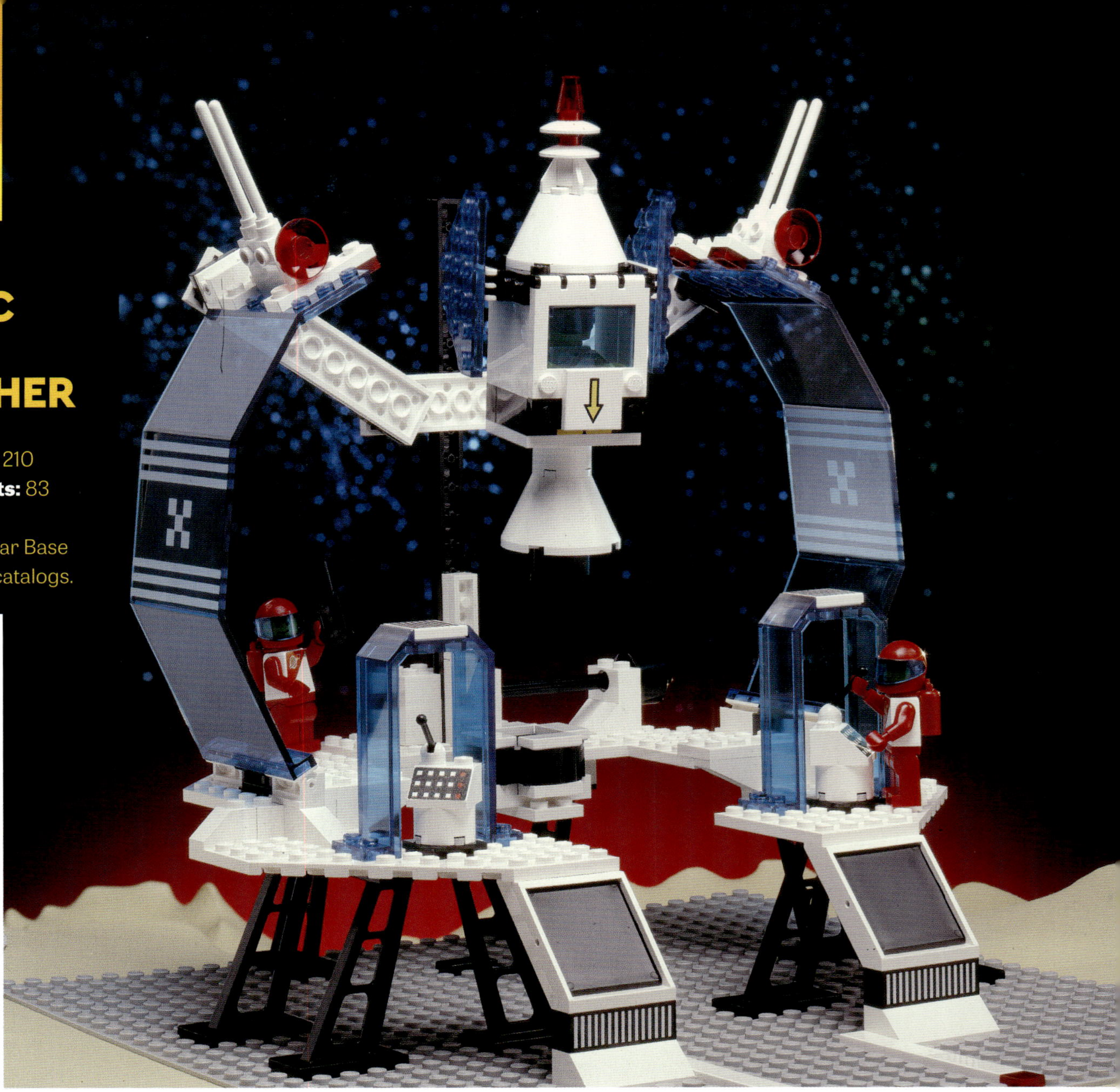

▼ Original mockup of box.

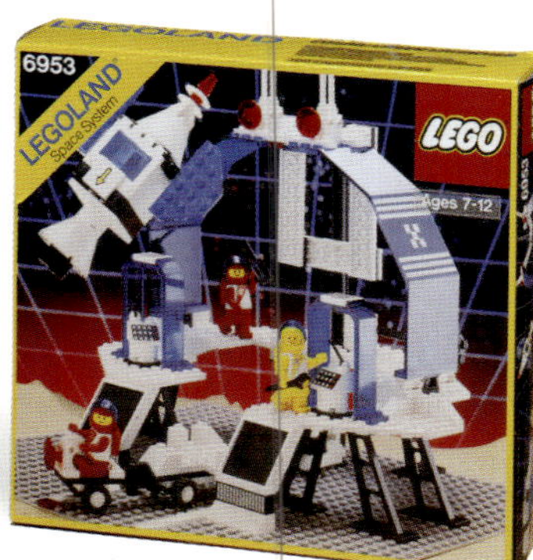

LEGO® Space bases often suggested a larger structure through the use of a sparse framework; in this case, the ring-shaped platform and two angled panels hint at a dome. The design printed on the large blue transparent panels was inspired by a series of icons designed by artist Ron Cobb for the 1979 movie *Alien*. A minifigure fits inside the satellite, which is delivered to the launch pad by a long articulated vehicle. A simple rack-and-pinion mechanism is operated at the rear, opening the two angled panels outward in unison. This was the only Futuron set to contain red minifigures, although they were also available in the supplementary pack 6703, and the torso came in LEGO® DACTA™ set 9355.

With the first Star Base now complete, the miner drove the Coronal Mass Extractor into the center of the launch zone and climbed aboard. The technicians opened the bay for launch, closing it again as soon as the payload was clear. Once they had powered up the hyoid platform beneath their feet, ultraphotovoltaic energy flowed to the negative-mass cannons on top of the base, and from her command post, the senior technician opened comms. "SB-1 to CME-1. Prepare for star lifting." She fired the cannons, creating a massive wormhole directly in front of the Coronal Mass Extractor. Beyond it, framed perfectly by the ring, lay the blue supergiant star Tycho. From the safety of his position some four hundred light years away, the miner fired his plasma laser through hyperspace to heat a pinpoint in the star's atmosphere. Soon, a solar flare erupted, blasting plasma toward Skye.

The transparent blue shields covering the two command posts are made from one of the more unusual angled panels to be introduced in this era: ***panel 3 x 2 x 5⅔****, which is effectively a tall bracket. It was also used prominently in the monorail set, and in the Blacktron base as a shuttle craft windscreen. In the 1990s, it occasionally appeared in LEGO® Town and LEGO DACTA sets.*

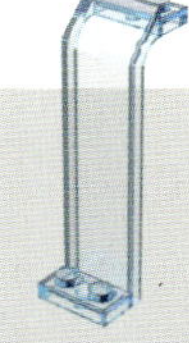

▲ **ELEMENT: 2448**
INTRODUCED: 1987

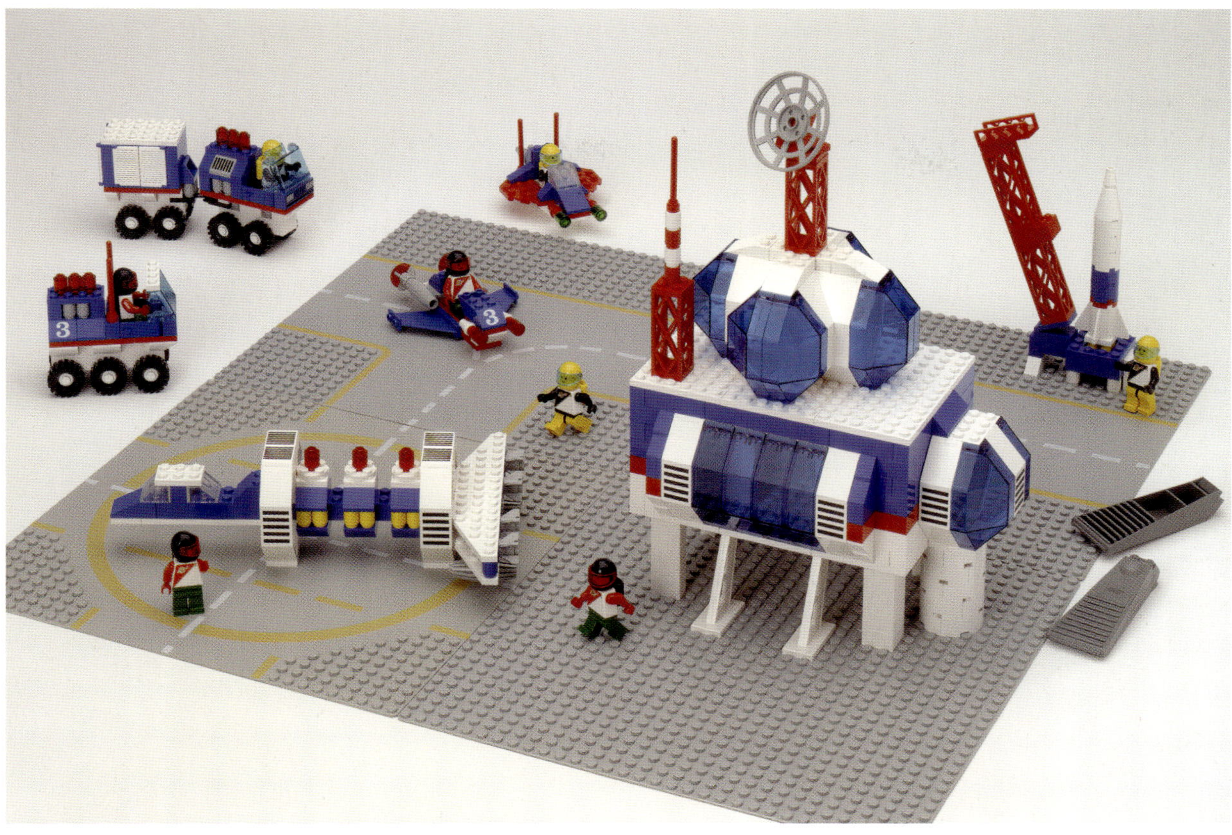

The red torso element from set 6953 also appeared in the 1991 space-themed set 9355, released by LEGO DACTA (now LEGO® Education).

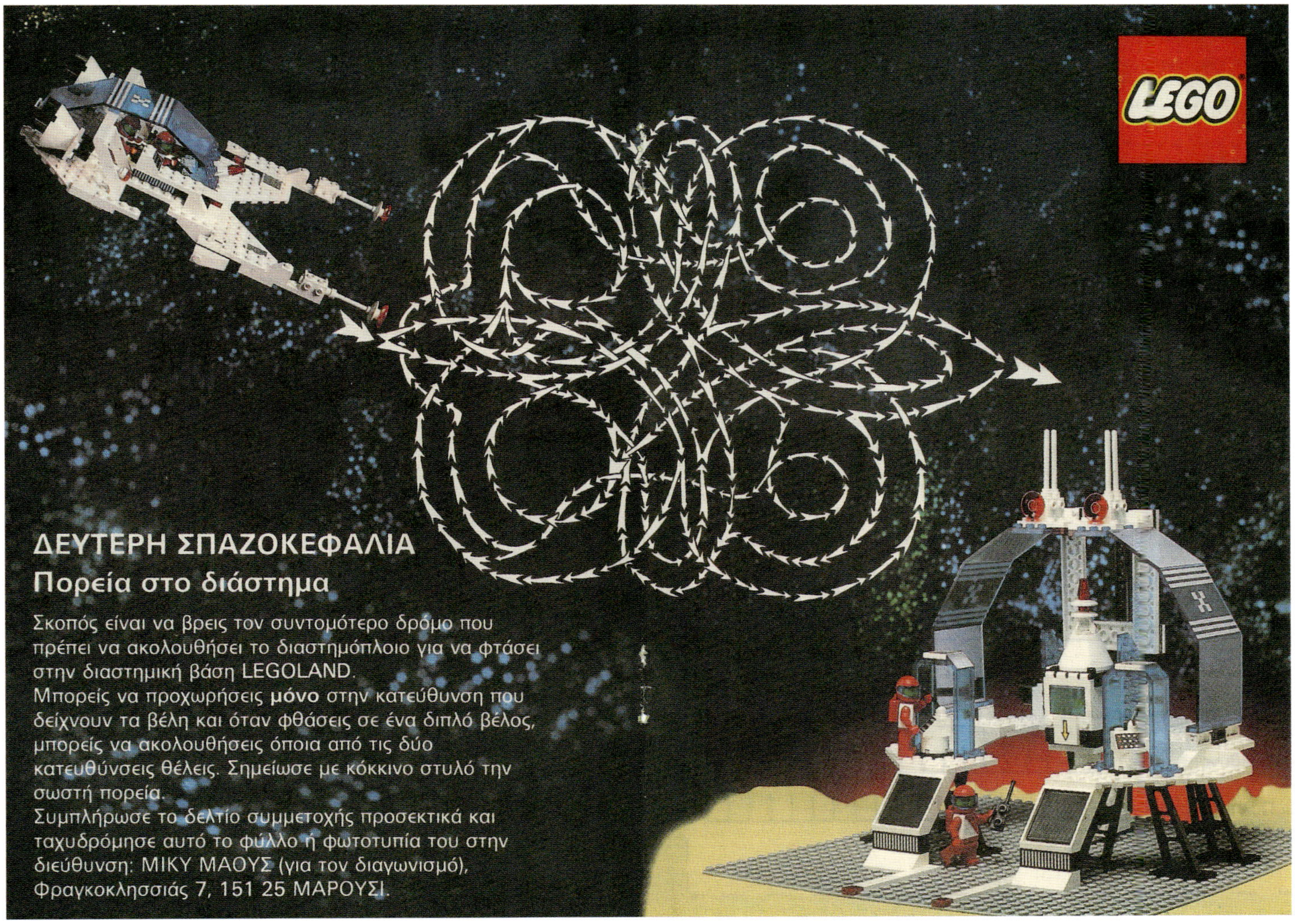

LEGO

ΔΕΥΤΕΡΗ ΣΠΑΖΟΚΕΦΑΛΙΑ

Πορεία στο διάστημα

Σκοπός είναι να βρεις τον συντομότερο δρόμο που πρέπει να ακολουθήσει το διαστημόπλοιο για να φτάσει στην διαστημική βάση LEGOLAND.
Μπορείς να προχωρήσεις **μόνο** στην κατεύθυνση που δείχνουν τα βέλη και όταν φθάσεις σε ένα διπλό βέλος, μπορείς να ακολουθήσεις όποια από τις δύο κατευθύνσεις θέλεις. Σημείωσε με κόκκινο στυλό την σωστή πορεία.
Συμπλήρωσε το δελτίο συμμετοχής προσεκτικά και ταχυδρόμησε αυτό το φύλλο ή φωτοτυπία του στην διεύθυνση: ΜΙΚΥ ΜΑΟΥΣ (για τον διαγωνισμό), Φραγκοκλησσιάς 7, 151 25 ΜΑΡΟΥΣΙ.

This competition in a Greek *Mickey Mouse* magazine from 1987 features set 6953 and one of its alternative models from the back of the box.

6990 MONORAIL TRANSPORT SYSTEM

Released: 1987
Total elements: 731
Unique elements: 162
Minifigures: 5
Also known as Space Trak Centre in UK retail catalogs.

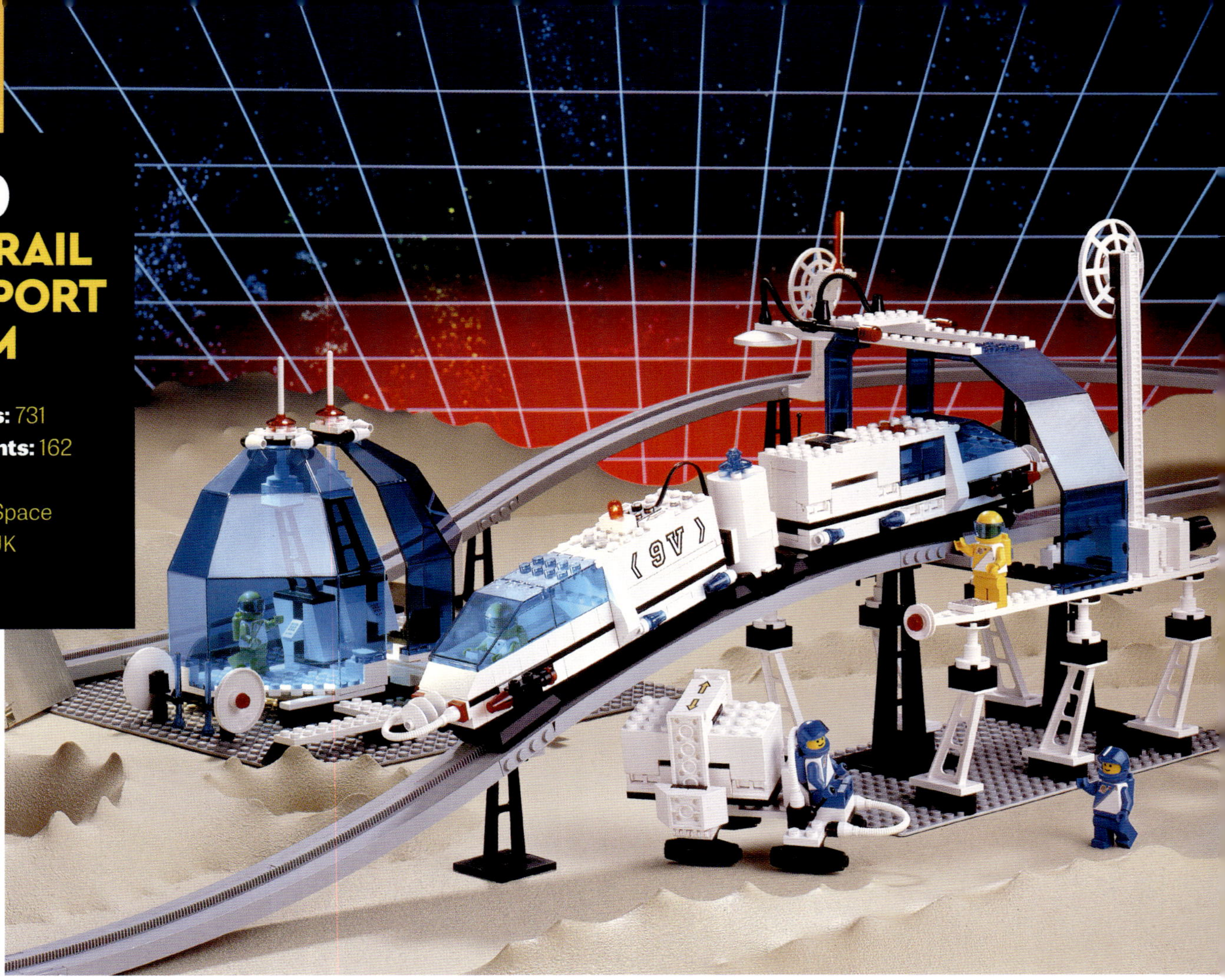

▶ The model comes with five minifigures, more than any other Futuron set. The tiny scooter fits inside the cargo containers.

The first-ever LEGO® monorail runs between a domed base and a raised platform. At each station a round knob printed with an arrow is rotated to either control the direction of movement or to stop the train as it arrives. The train comprises twin cabs, a central motor, and two cargo carriages: one holds the 9 V battery box, while the other has two detachable containers that can be unloaded in different ways at each station. At the domed base, turning a dial shifts and rotates the quarter-dome elements, bringing the whole base close to the track to receive a container. At the raised station, a rack controls a platform for moving cargo vertically. A shuttle craft is also provided, which can clasp a container.

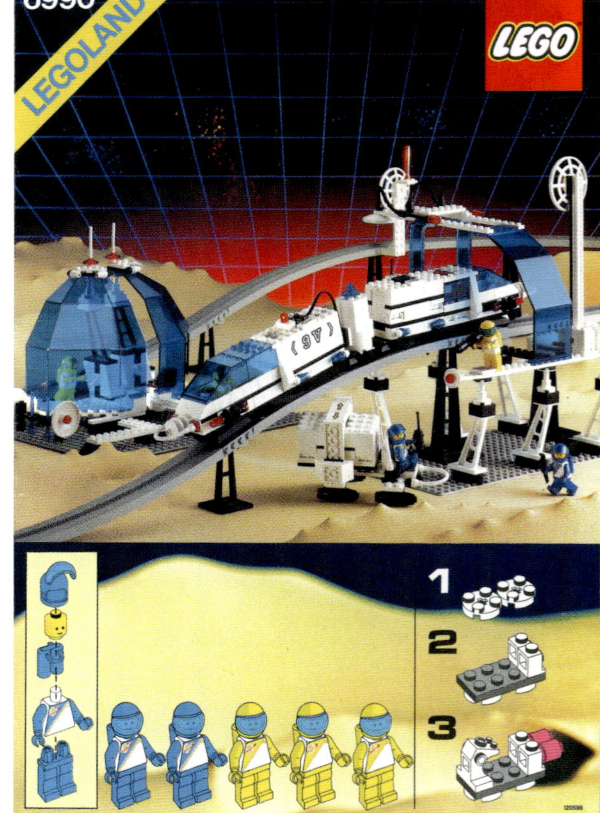

▼ Original mockup of box.

With shipments from Skye coming thick and fast, the plasma processing system was proving highly efficient and was ripe for scaling up. The cargo shuttle brought individual containers to the freight station elevator, which purified the plasma as it rose to the railway. Along its frictionless journey to the lab, the train passed through a compression fusion arch, reducing the volume of the plasma inside each container to the size of a zamosphere. Care was taken at the lab station to move the unstable cargo completely away from the track before activating the ogdoöspheric insulation panels, allowing the technician to safely remove and store the compressed plasma. They could fit enough in a single container to power an entire Futuron colony for a generation.

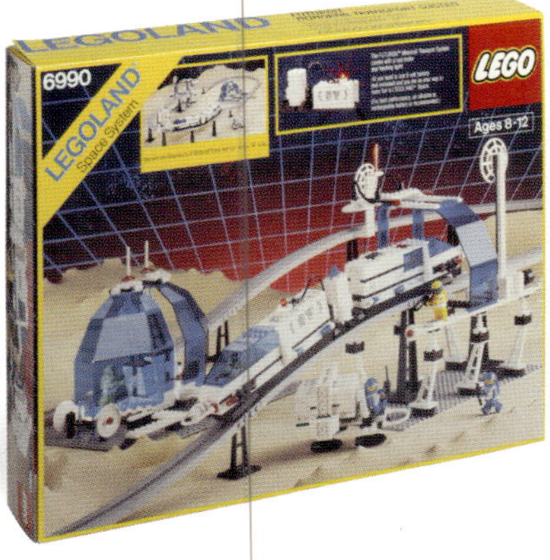

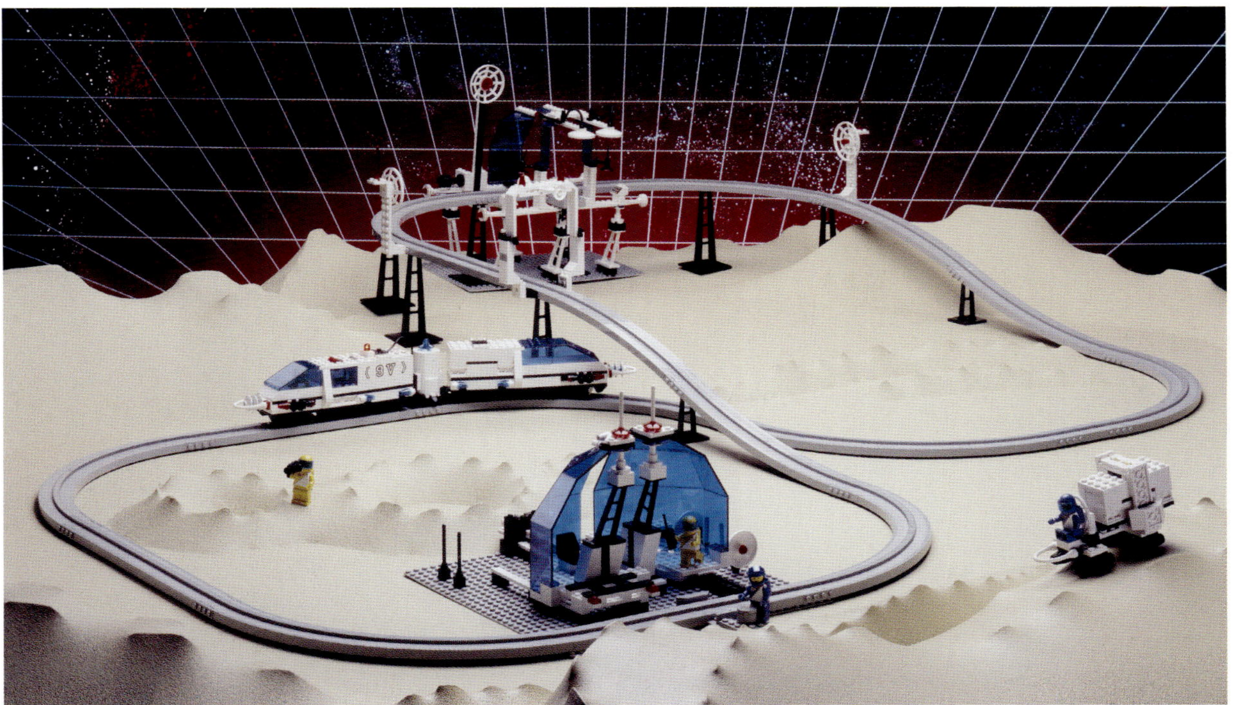

The monorail track could be nearly doubled in length by adding the supplementary set 6921.

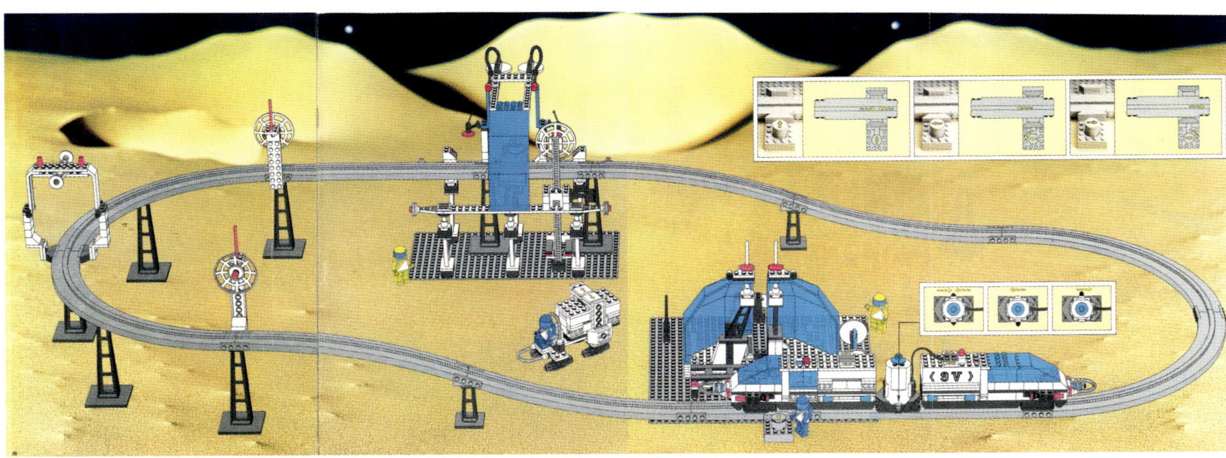

The final step of the building instructions includes visual explanations of how the directional controls work.

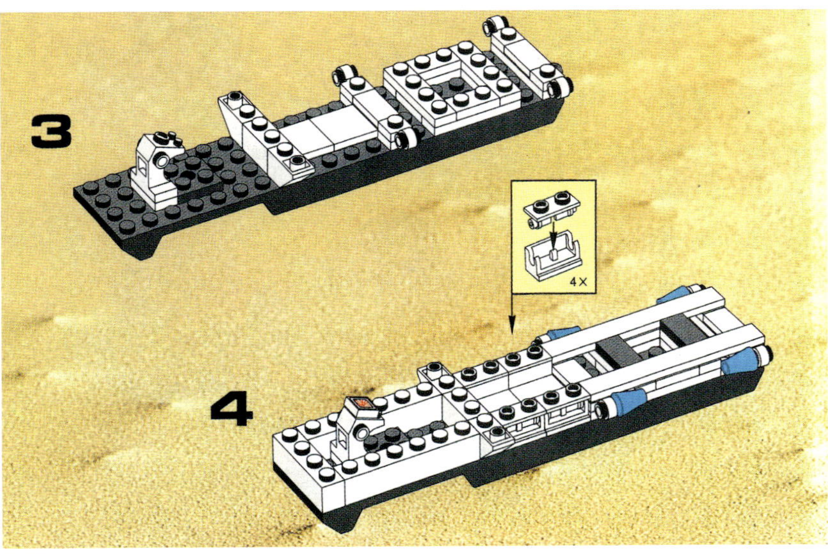

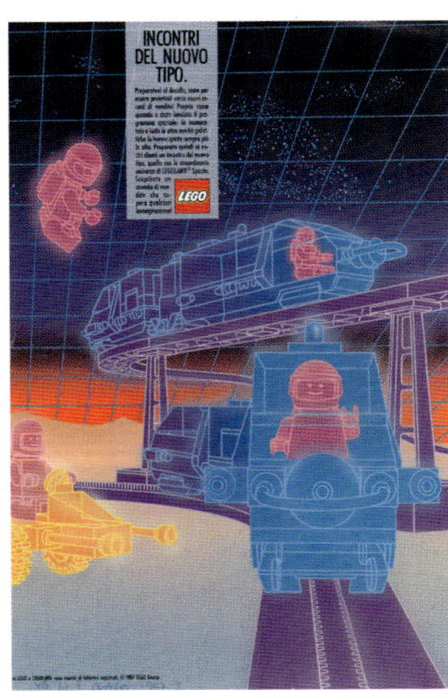

This unusual 1987 ad featuring the new monorail was aimed at retailers, suggesting that they offer customers "an encounter of the new kind . . . You will discover a world of sales beyond any imagination!"

The sheer size and thrilling shape of the ***quarter-dome panel 10 x 10 x 12*** *make it one of the most iconic elements ever to be introduced for LEGO® Space. It was used in seven different bases up until 1998 and was also used as a canopy for the massive M:Tron vehicle 6989 Mega Core Magnetizer.*

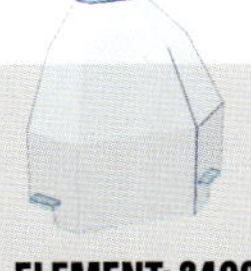

ELEMENT: 2409
INTRODUCED: 1987

6893
ORION II HYPERSPACE

Released: 1987
Total elements: 162
Unique elements: 65
Minifigures: 2
Also known as Roboprobe Transporter in UK retail catalogs.

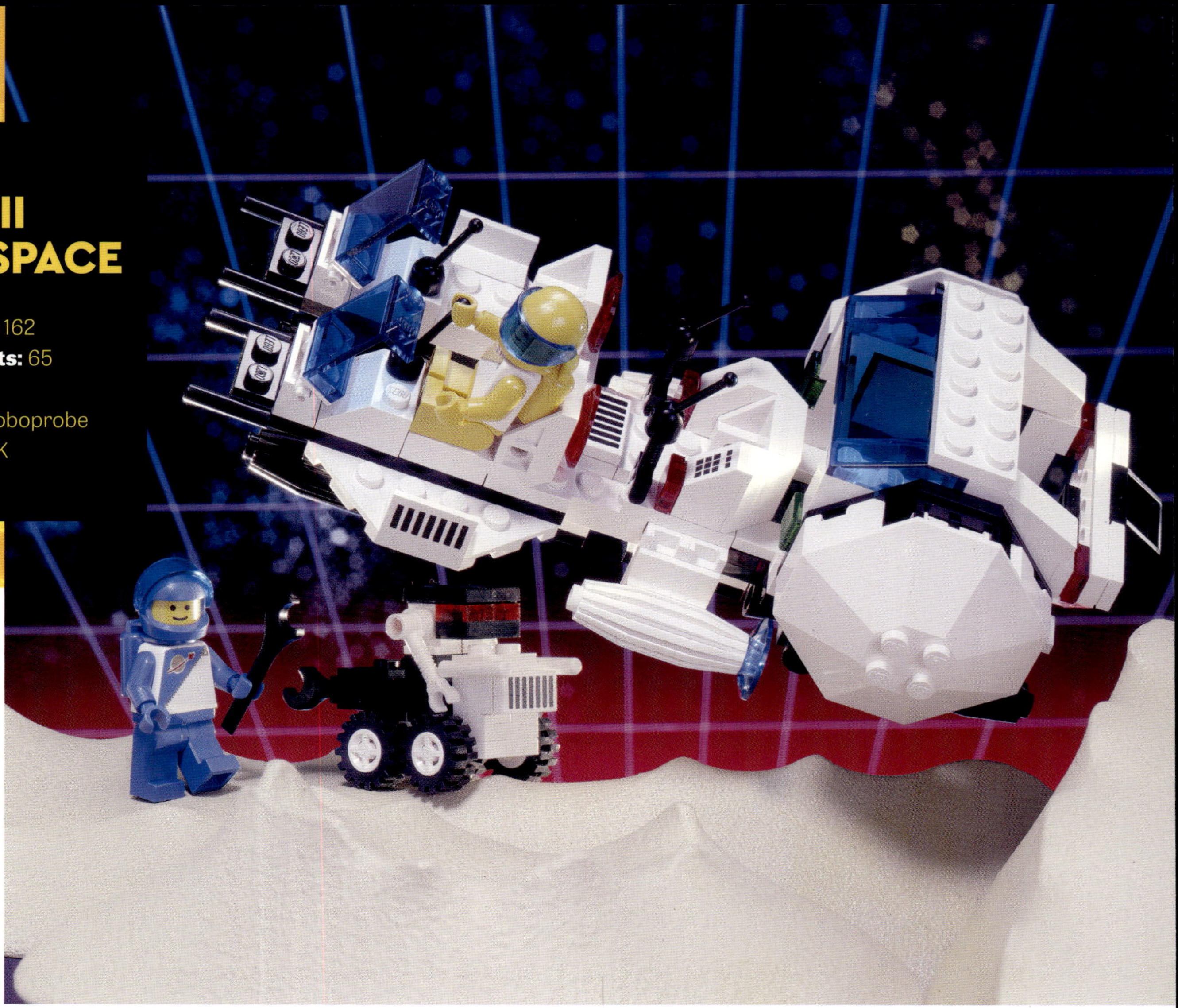

▶ The final page of the building instructions includes a photograph highlighting the modular structure of the model.

▼ Original mockup of box.

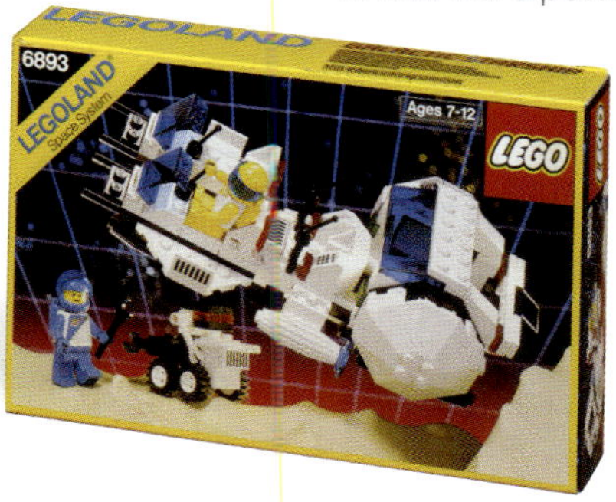

This three-seater ship splits into three segments, revealing the centrally housed hover chair. The droid in this set is integrated into the rover itself. The rear cargo pod can fit the rover via a fold-down ramp, and its side-mounted octagonal pods swing upward.

The Orion series comprised propulsion units designed specifically for fuel efficiency in hyperspace transport. Their kyan-mullite windscreens offered sufficient protection from the hypermatter, and the compact plasma drives were powerful enough to carry heavy freight—in this case, a hover chair and roboprobe laboratory. As soon as they had traversed the wormhole, the spectrometrist scrambled back to his hover chair, and the pilot ejected him, along with the cargo pod. Landing safely on Skye, he checked in with the autonomous research unit as it drove down the lab's ramp.

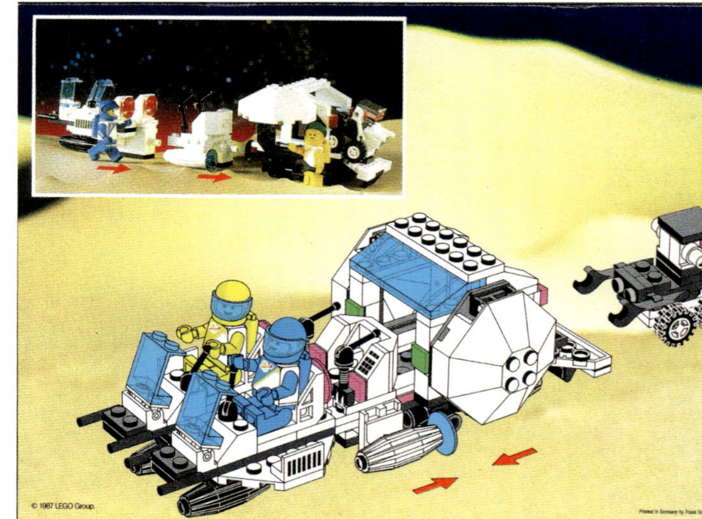

"My systems are fully functional, thank you," replied Otte-93, as it opened the pod canopies to protect itself from the solar winds.

*LEGO® astronauts had somehow survived the vacuum of space without needing visors for eight Earth years. At last, LEGO® Space introduced the **minifigure visor**—transparent blue for Futuron, and opaque black for Blacktron—and it appeared in LEGO® Town sets the following year. The same design is still in use today.*

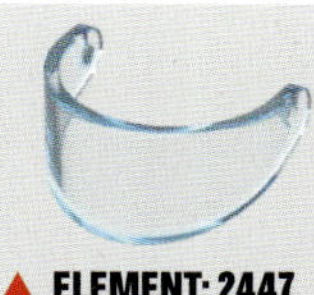

▲ **ELEMENT: 2447**
INTRODUCED: 1987

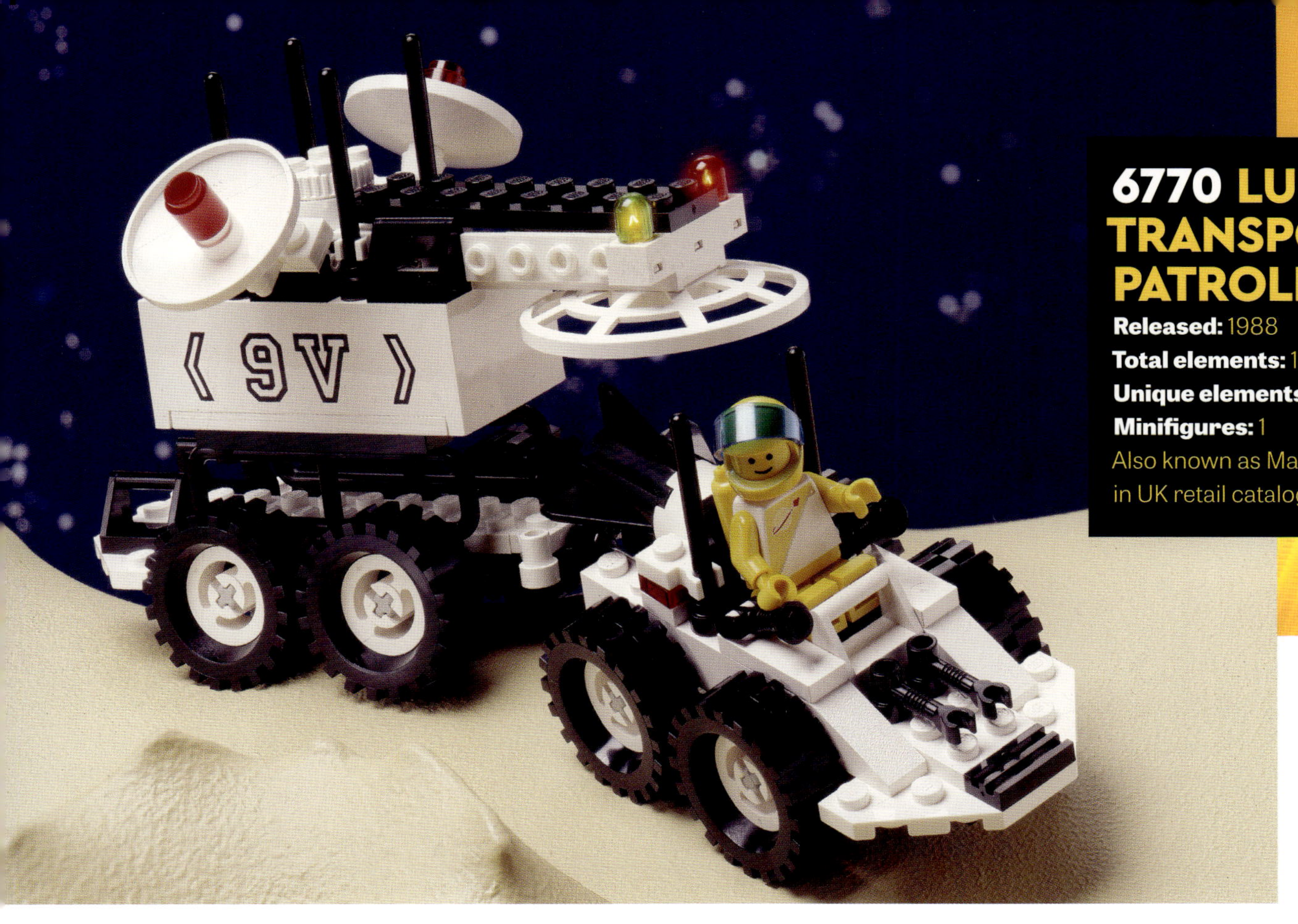

6770 LUNAR TRANSPORTER PATROLLER

Released: 1988
Total elements: 117
Unique elements: 48
Minifigures: 1
Also known as Magma Carrier in UK retail catalogs.

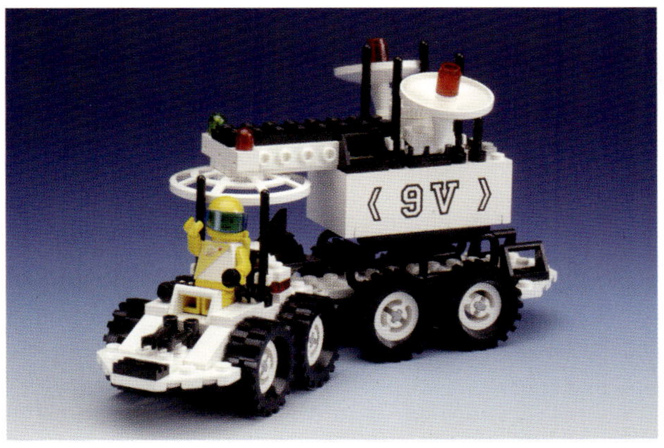

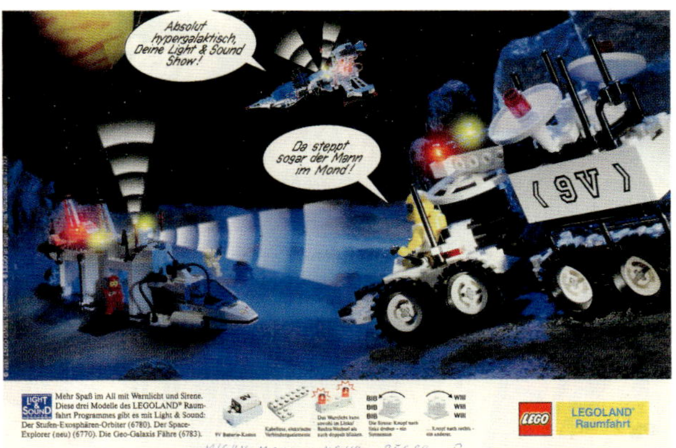

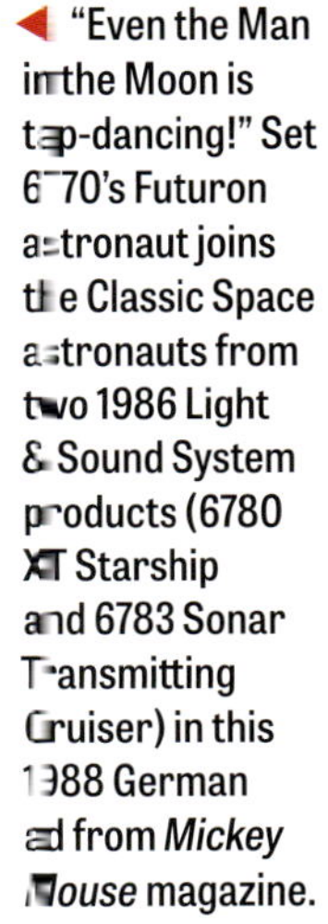

◀ "Even the Man in the Moon is tap-dancing!" Set 6770's Futuron astronaut joins the Classic Space astronauts from two 1986 Light & Sound System products (6780 XT Starship and 6783 Sonar Transmitting Cruiser) in this 1988 German ad from *Mickey Mouse* magazine.

The sole 1988 product utilizing the Light & Sound System was an eight-wheeled articulated transport vehicle. The prime mover, when separated, has oversized rear exhaust and tires, giving it a powerful look. The trailer cargo is some kind of industrial equipment, composed of a 9 V battery box with a projecting arm and various dishes and antennas.

The seismologist drove across the barren landscape of Skye slowly enough to allow time for the low-frequency sound waves coming from the infrasonic transducer behind them to obtain readings from the moon's core. As they entered the B7 quadrant, the sound frequencies rose to within the range of human hearing, becoming a piercing wail. Definitely an oncoming eruption. They parked and decoupled, providing unfettered access for the dish, before clambering on top to replace the red transducers with the green sonar interpolators. This was ancient tech from the Milky Way civilization, but it just might bring under control the damage that their solar lifting was causing. As they sped back to base on the turbobuggy, a dark shape with flecks of yellow appeared in the skies above the abandoned equipment.

▼ Original mockup of box.

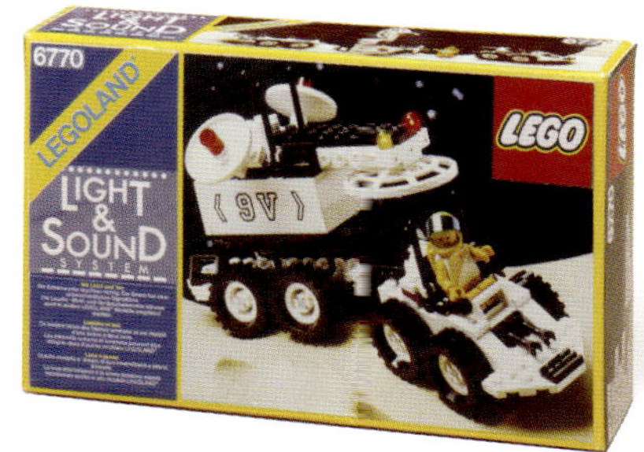

Bracketed elements, allowing sideways building, were introduced by LEGO® Space in 1979. A challenge with using them is that they can quickly throw construction "off grid," because LEGO elements are not perfect cubes. ***Bracket 1 x 2 / 1 x 4*** *was the first of a new style that featured a thinner vertical section, which helps keep construction within the LEGO grid in many situations. There are now more than a dozen different LEGO brackets using this style.*

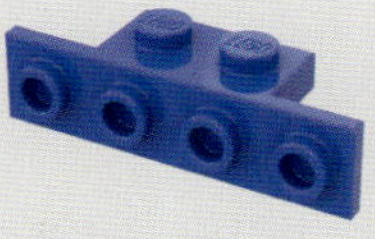

▲ **ELEMENT: 2436**
INTRODUCED: 1987

6828 TWIN-WINGED SPOILER

Released: 1988
Total elements: 57
Unique elements: 27
Minifigures: 1
Also known as Sky Walker in UK retail catalogs.

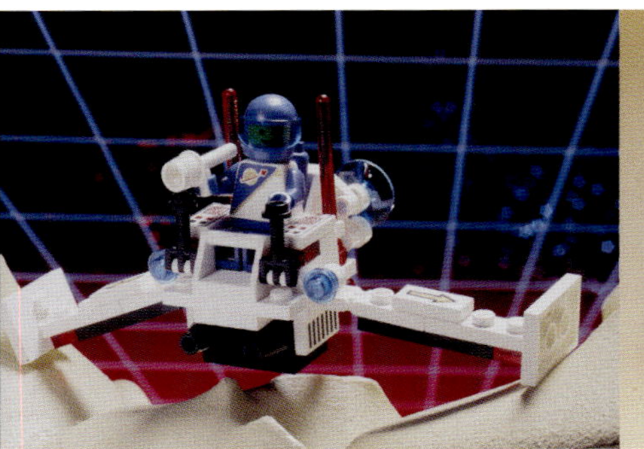

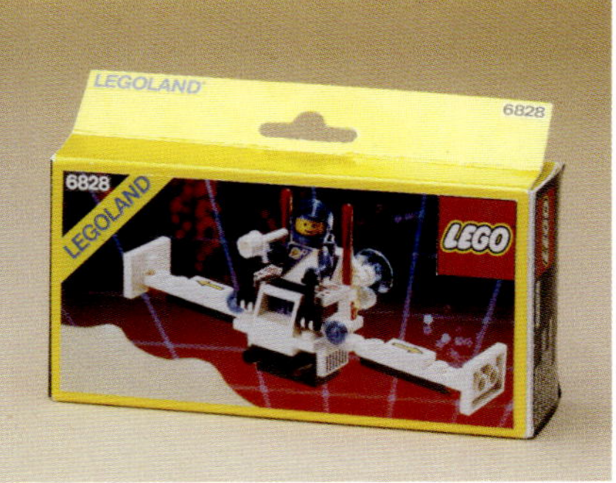

This simple but ingenious design features wings that swing down to become landing gear—or perhaps the legs of a walking mech. The black levers can act as controls but also look great facing forward in tandem with the black intake cones.

The pilot skimmed low above the lava flow, scanning for signs of any remaining hyperconverged storage units in Skye's B7 quadrant. A strong reading came in, and she straightened up as she lowered the wings down. The craft descended lightly onto the bubbling red mass: the cobalt gravity manipulators kept it stable, while her bio-suit kept her icy cool. Walking gently through the lava, she soon located one of the units, and picking up her handheld gravity winch, she began to draw it out of the lava, hoping its iridium casing would hold out just a little longer.

__Bracket 2 x 3/1 x 3__ was designed to be a signal stand and was used exclusively within the LEGO® Trains system—until its final appearance, when it arrived in LEGO® Space. Its angled design provided strength for its ingenious utilization here as both wings with spoilers and feet for a mech.

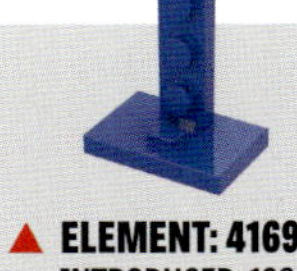

▲ **ELEMENT: 4169**
INTRODUCED: 1980

6848 STRATEGIC PURSUER

Released: 1988
Total elements: 62
Unique elements: 31
Minifigures: 1
Also known as Robo-Rider in UK retail catalogs. Included in multipack 1530 Space Value Pack.

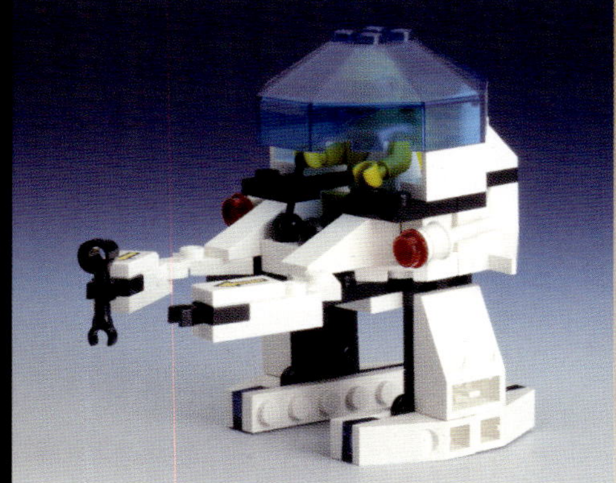

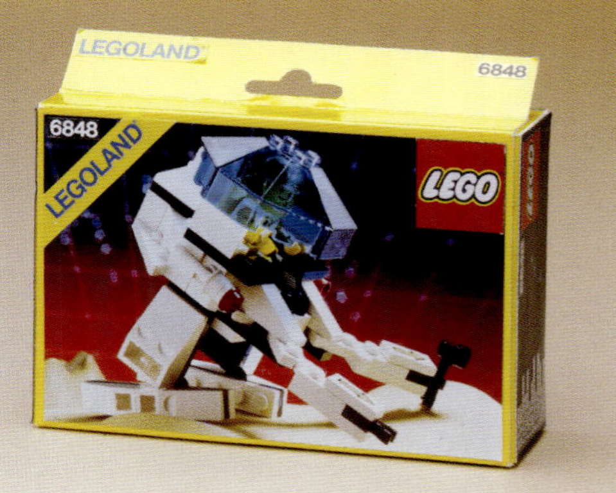

There is generous space inside the bubble canopy of this walker mech. The angled forms so common in Futuron are achieved in the legs by using inverted slopes attached sideways. Hinges allow it to tilt forward and bend its arms inward.

The seismologist worked her way slowly across the igneous surface of the B7 quadrant, the autonomous Roborytter exosuit guiding her strategically toward each vent in turn. Arriving at a fresh site, she angled the twin fissure pluggers to fill the vast hole before bending to seal the edges with the sonar interpolator. It was painstaking and important work, but dangerous too. The enemy were attacking frequently, and given that the Futuron hadn't required any military technology for dozens of millennia now, they were ill equipped to defend themselves—and certainly would not fight back.

The __octagonal canopy windscreen 6 x 6 x 2__ is strongly associated with Futuron, although it appeared in a couple of Classic Space sets beforehand and frequently within Blacktron Future Generation later on. Its angled dome shape gives a pod-like form to ships and vehicles, and the element also found use in LEGO® Castle as a tent canopy.

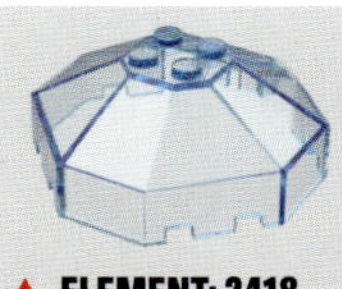

▲ **ELEMENT: 2418**
INTRODUCED: 1987

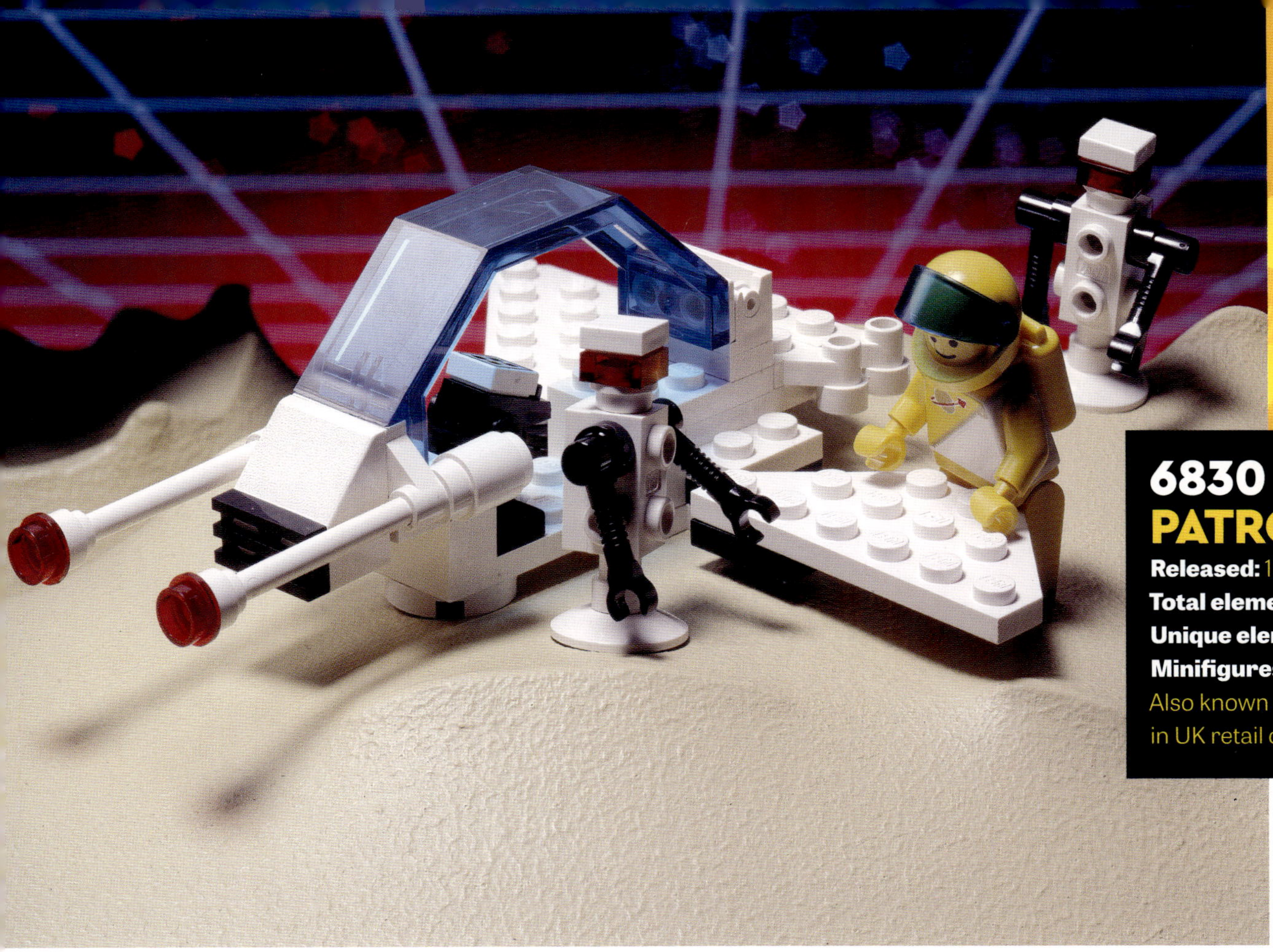

6830 SPACE PATROLLER

Released: 1988
Total elements: 51
Unique elements: 20
Minifigures: 1
Also known as Sky-Streak in UK retail catalogs.

◀ The rear exhausts could be detached to become helpful droids, here assisting with wing reconfiguration.

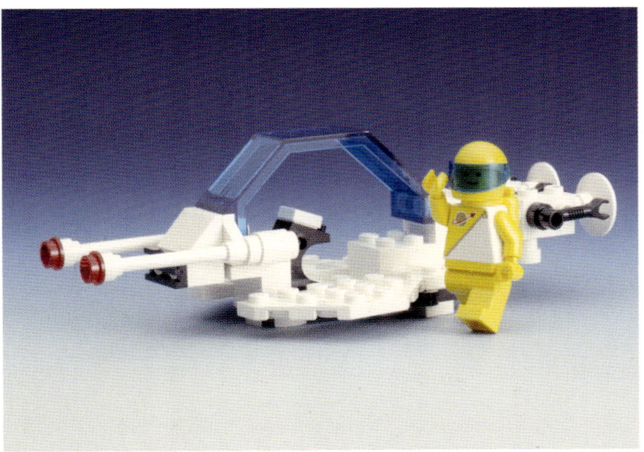

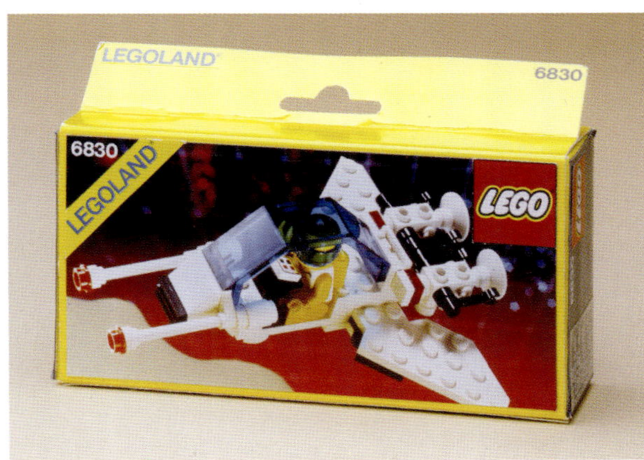

This compact shuttle neatly incorporates Futuron's octagonal design aesthetic, and the wings can swing backward to create a 45° angle. Surprisingly for a small set, there are two droids provided; however, the reason soon becomes clear: they can be attached as rear thrusters.

"I have visual," said the audiologist as he spotted a faint yellow triangle glinting in the blackness beyond the Mercator grid. "Activating interception mode," said Otte-30-Alpha. The wings folded inward as the plasma drive fired up and created a wormhole in their path. Otte-30-Beta detached itself from the rear in the moment before the ship streaked through hyperspace, and a split second later Otte-30-Alpha did the same as the patroller reappeared on the far side of the hostile craft. The two units quickly faced one another and initiated the phase shift oscillator. The black ship soon diverted its course to escape the pulse, speeding away unharmed. "Interception successful. Do not pursue," confirmed Otte-30-Alpha as the two droids reconnected to the craft. "Communications exfiltrated. Playing now." A voice could be heard, loud and clear. It was a human voice! "Assimilator to Blacktron base. Come in."

*The black intake vent at the front of this craft is **modified tile 1 x 2 with grille**, an element which proves that a specialized design can be immensely flexible in its purpose, even when small. Its thin grooves add textured detailing to models, and this effect is heightened when it is attached to an element of a different color. It has appeared in well over three thousand LEGO® sets.*

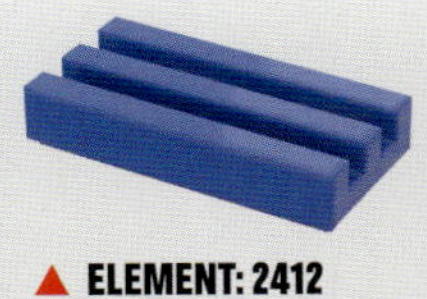

▲ **ELEMENT: 2412**
INTRODUCED: 1987

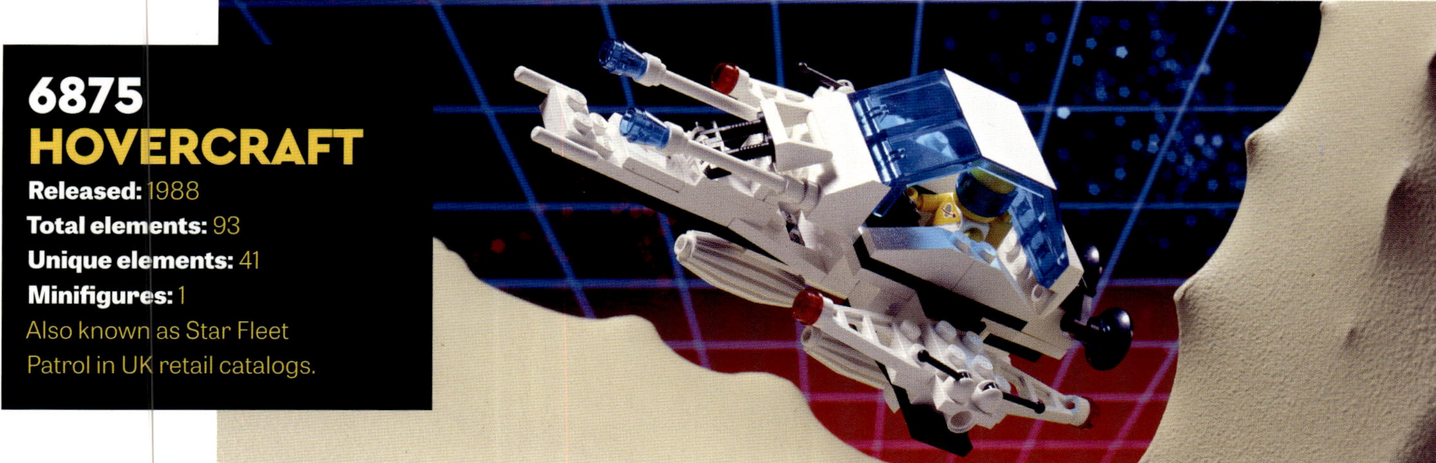

6875 HOVERCRAFT

Released: 1988
Total elements: 93
Unique elements: 41
Minifigures: 1
Also known as Star Fleet Patrol in UK retail catalogs.

This small craft has a couple of unusual design features. The wings are two detachable craft that can be joined into one, and the main canopy has a locking mechanism: the black levers at the front must first be lowered before the transparent blue panel can be hinged backward.

*The modular approach was used increasingly within LEGO® Space sets, and the introduction of the **modified plate 2 x 2 with pinhole** offered the functionality of a LEGO® Technic brick within a more compact space. Here, it is prominent on the sides of the detachable wings, providing not only function but also an interesting industrial form.*

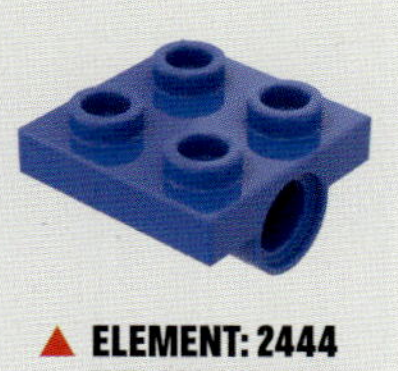

▲ **ELEMENT: 2444**
INTRODUCED: 1987

After countless millennia of peace, a new Star Fleet division had been hastily formed to protect the Futuron from the ongoing attacks. Mining craft which had originally been designed for star lifting were instead reassigned to patrol duties. The P-class hovercraft, with their detachable triangulation satellites for high-energy electron mapping, were easily repurposed to detect plasma anomalies instead. If Blacktron craft came within range, the cobalt gravity manipulators could hold them in place long enough for a Stardefender to jump to assist.

6885 CRATER CRAWLER

Released: 1988
Total elements: 101
Unique elements: 48
Minifigures: 1

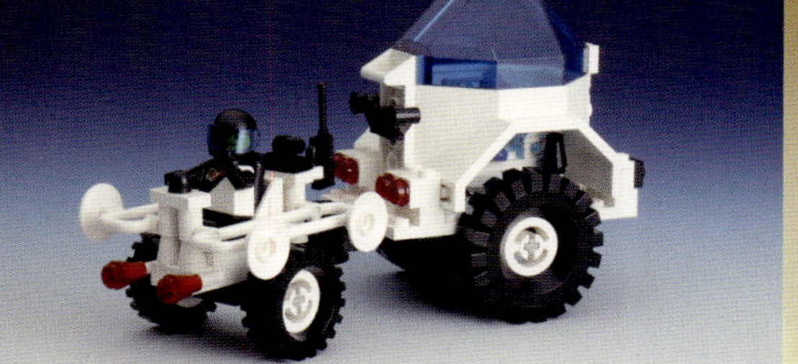

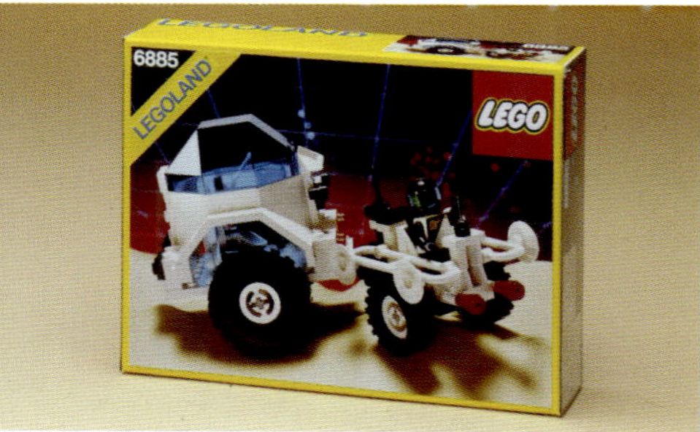

The trailer of this highly characterful articulated vehicle features a covered docking station containing a tiny craft piloted by a stocky blue droid. Different-sized tires give it a tractor-like appearance, and the cab's distinctive bodywork is made of four space guns, to which two of the new white fence bar elements are attached sideways.

"Main team to base. Come in, Saturn," said Otte-85 as they trundled away from the incident zone. There was no reply, just the squeal of a signal scrambled by the increasing ionization levels. "Driver, stop and power up the electromagnetic shielding," commanded the droid, and soon the two pairs of zinc generator coils surrounding the astronaut were humming gently. The solar canopy unbuckled, and as the pod opened, Otte-85 rose swiftly upward and forward to the safety of the shield column above the driver. Now their connection to Saturn base was as clear as day, but it wasn't the Futuron who answered.

*Of the family of six-brick-tall angled panels introduced in 1988, **concave corner panel 4 x 4 x 6** is perhaps the strangest looking. It enabled the creation of octagonal corridors and large stanchions; however, its utilization here is perhaps the most ingenious among the seven sets it appeared in. Two are attached sideways to create not only the central octagonal cargo bay but also the wheel arches.*

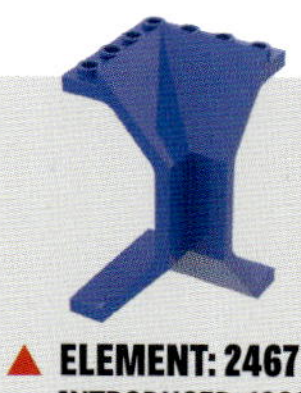

▲ **ELEMENT: 2467**
INTRODUCED: 1988

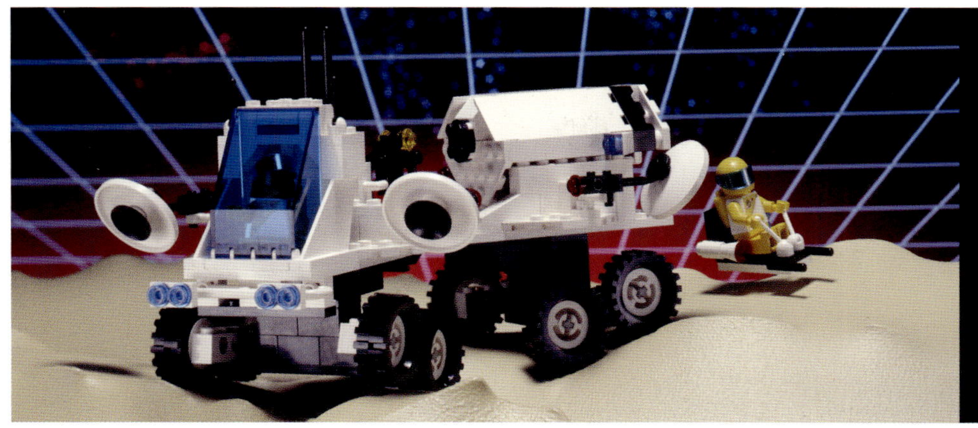

6925 INTERPLANETARY ROVER

Released: 1988
Total elements: 211
Unique elements: 64
Minifigures: 2
Also known as Galaxy Rover in UK retail catalogs.

This large eight-wheeled vehicle with suspension features a vertically mounted canopy, snuggled in between two tail elements to create an interesting profile. The headlights are attached to a fence element placed sideways. The rear pod hinges open to reveal a small craft piloted by a second minifigure, and two identical droids are also provided.

The shuttle pilot parked in the cargo pod just as the geophysicist was powering down the cobalt gravity manipulators. "The droids will be back soon—let's get packed up," he shouted. They were so focused upon packaging the astatine safely for the jump, they didn't notice the robots' return until they coupled to the vehicle. "Wait . . . did you give those droids a makeover?" the geophysicist laughed. "They've turned black and yellow!" All four of the rover's negative-mass dishes suddenly fired up. "Ni-25, desist!" the pilot screamed as she leapt aboard. A huge red ring flashed around the vehicle as the planet disappeared, replaced by a mountainous range lit by two crescent moons. "Where did you jump us?" she shrieked. "Welcome to the Blacktron," answered a dark figure looming in the inky blackness. "Although, I believe your ancestors once knew mine as Sort Elektronika."

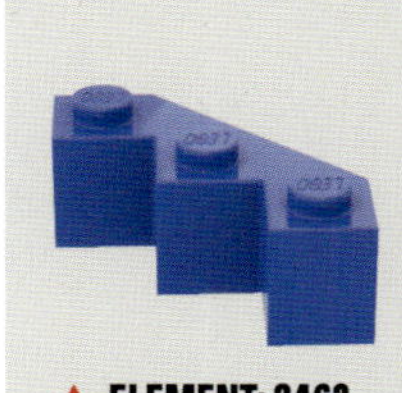

▲ **ELEMENT: 2462**
INTRODUCED: 1988

Although it is perhaps more readily associated with LEGO® Pirates *and* LEGO® Castle *models, the* ***modified facet brick 3 x 3*** *matched the angled look of* Futuron *perfectly and soon reappeared in* Blacktron Future Generation. *The family of faceted bricks expanded in later years to include 2 x 2, 4 x 4, and 5 x 5 versions, as well as matching quarter-dome elements which first appeared in* Space Police 2.

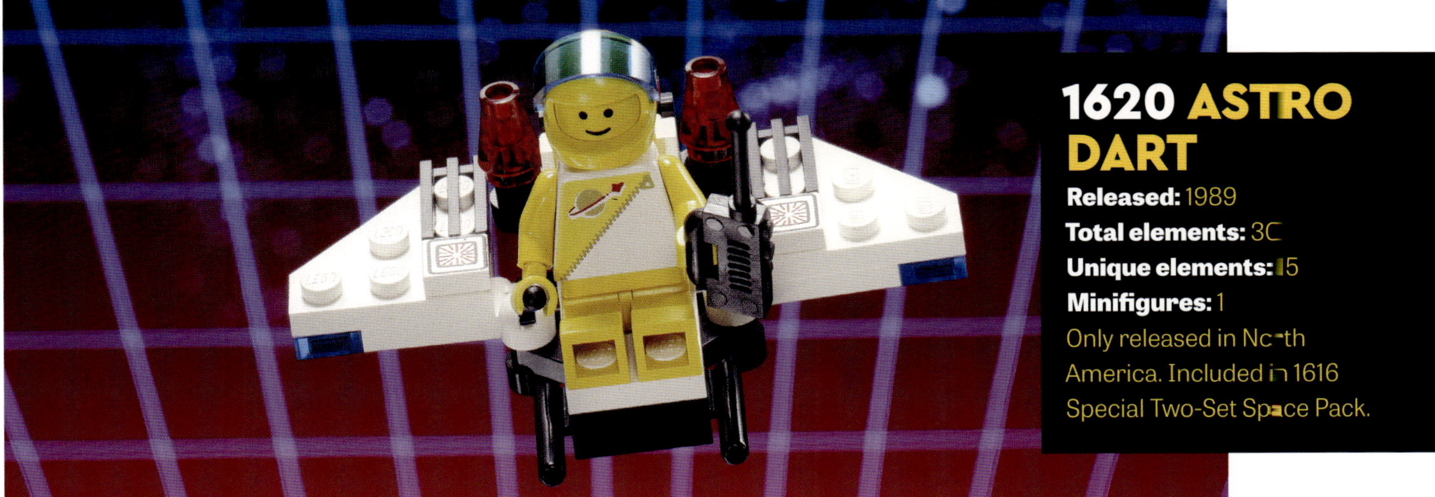

1620 ASTRO DART

Released: 1989
Total elements: 30
Unique elements: 15
Minifigures: 1
Only released in North America. Included in 1616 Special Two-Set Space Pack.

This small glider makes effective use of two 3 x 3 corner wedge plates, attached with hinges so that they can be angled downward. The rear thruster unit can also be tilted so that the transparent red cones point either upward or forward.

Finding herself skittering just within the limits of the Mercator grid, the agent angled the inverse thrusters to bring her trajectory down close to the surface. Within moments, she had a Blacktron Interceptor on her tail. Steering with one hand, she snatched her intergalactic comms unit with the other and tapped its scramble button. "Dart Silver-9 to Saturn base. Confirmed, the Blacktron have our hyperspace tech." She lurched as the Interceptor entered firing range, placing both hands momentarily on the flight sticks to angle the wings and spin quickly out of harm's way. "But I know their location; it's in the Triangulum galaxy. Stand by."

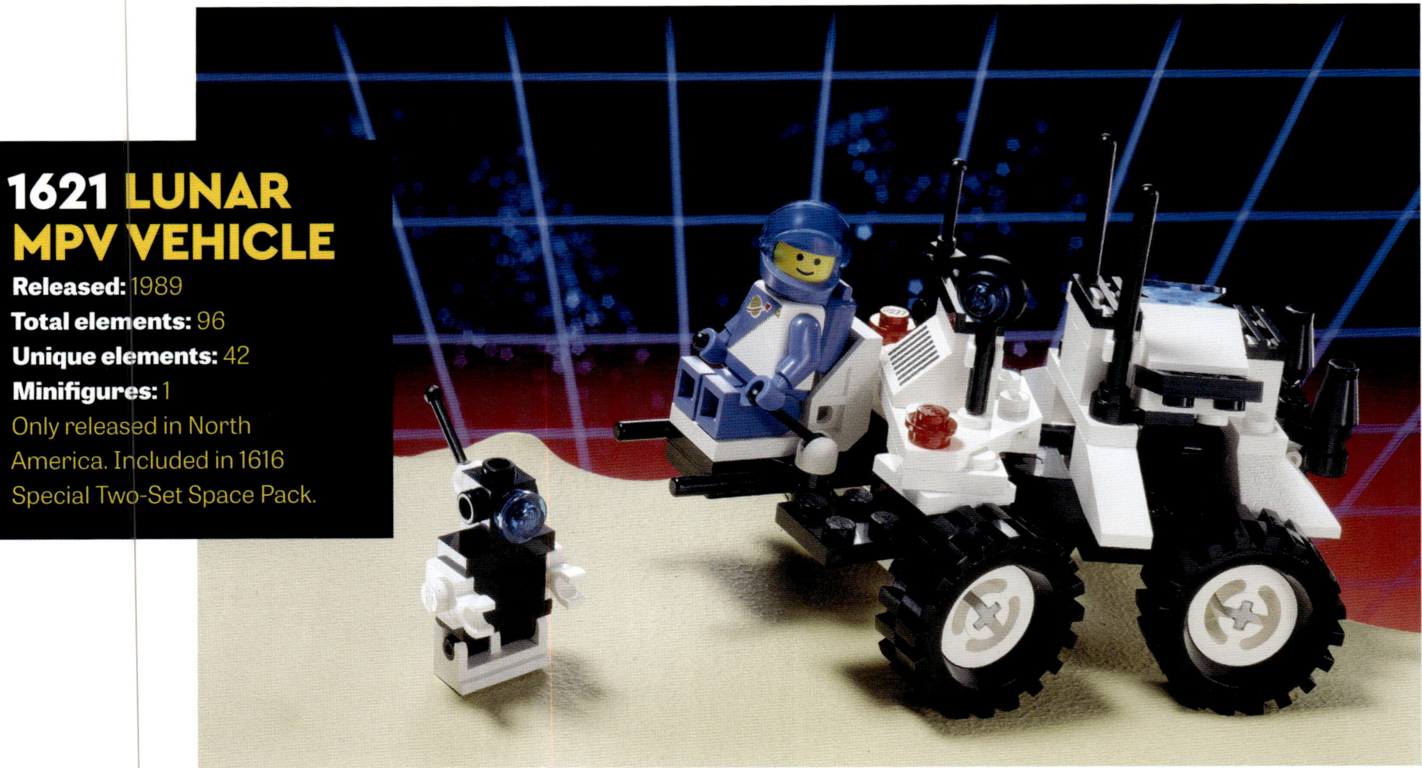

1621 LUNAR MPV VEHICLE

Released: 1989
Total elements: 96
Unique elements: 42
Minifigures: 1
Only released in North America. Included in 1616 Special Two-Set Space Pack.

Although small, this model includes functional steering. It uses a simple and clever mechanism controlled by tilting the central section of the trailer, first developed for Space Police set 6895 Spy Trak 1 the previous year. The twin cab is asymmetrical, providing the droid with a small platform alongside the driver.

"Hey, that's my seat," said the geologist as the droid sat quietly in the command chair. "I am the mission senior," replied Six-21, correctly. "Sure, but your seat is too small for me!" he replied, smiling at the cheeky robot. After a short pause, Six-21 flew silently to the left-hand chair. They set off, using the zircon antennas to detect any nearby plasma anomalies, which the cobalt gravity dish would then tilt to latch on to. This would steer the vehicle to the source location, where they would angle the twin warp cannons to annul the potential threat.

6850 AUXILIARY PATROLLER

Released: 1989
Total elements: 46
Unique elements: 24
Minifigures: 1
Also known as Space Pod in UK retail catalogs.

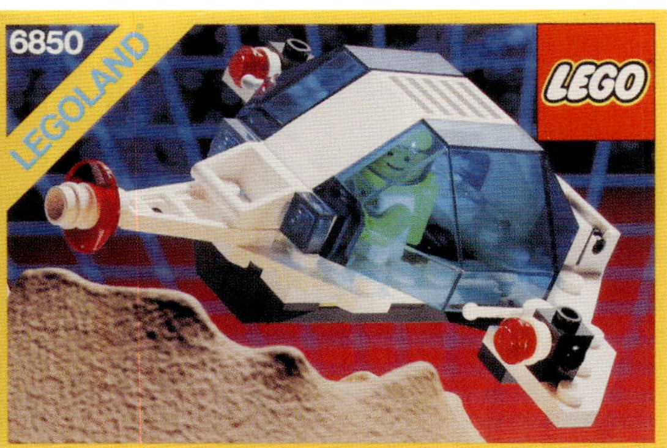

Despite being one of Futuron's smaller sets, this bug-like craft incorporates many design hallmarks of the theme, including an octagonal planform with bubble canopy, minimalist printed elements, and tapered diagonal shaping at the front of the canopy. When opened, the swing wings can be folded inside the cockpit, fitting snugly around the pilot. Interestingly, there are no visible means of propulsion, and the pilot's forward view is seemingly obscured.

The scout swung the negative-mass cannons out of the pod and pulled the canopy shut with a clunk. Within moments, he had jumped to Tycho. "Omega Patrol 50 to Star Base 7. I am in position and monitoring for the Blacktron." These long shifts were tiresome, but staring at a star at close range for hours was oddly calming. Thankfully, he had scored one of the second-gen craft today, with the additional glare protection afforded by the white diamond spine on their striated canopies. The spectral analyzer blinked on the holographic display in front of him, revealing unstable temperature fluctuations on the other side of the star. Probably nothing, but it gave him something to do. With a swipe of the controls, the cannons began to power up.

The clean design of ***modified tile 1 x 2 with bar handle*** *is generic enough to be used to represent many different things, and its functionality fulfills multiple purposes. As well as acting as a handle for both minifigures and humans, it has a rectangular space on its rear side that can loosely clasp two studs—as seen in this model.*

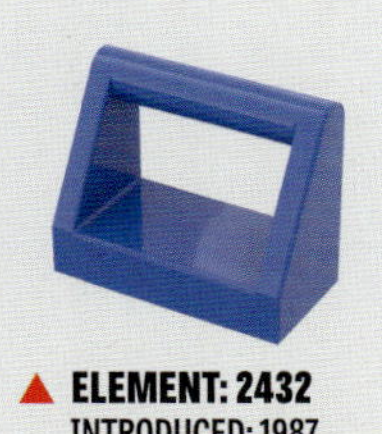

▲ **ELEMENT: 2432**
INTRODUCED: 1987

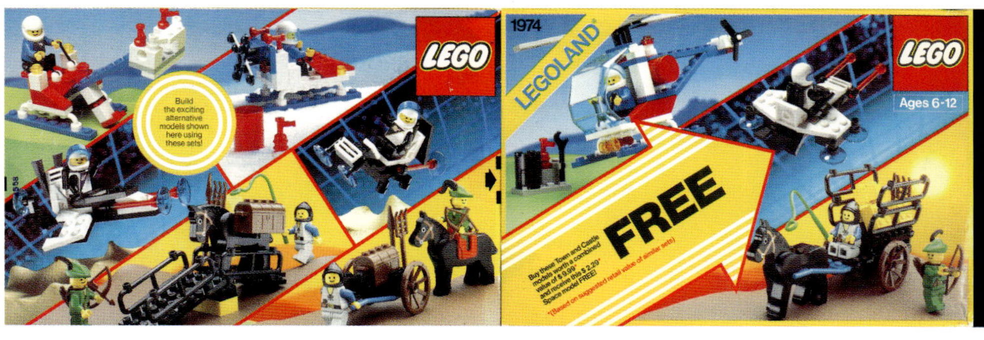

1974 STAR QUEST

Released: 1989
Total elements: 37
Unique elements: 17
Minifigures: 1
Sold as part of a triple pack with a LEGO® Town and a LEGO® Castle set.

The white wings on this small ship can be angled downward, and the four transparent blue dishes appear to be its only method of propulsion. This is one of the few Futuron craft to bear the LEGO® Space logo and the only one to contain a minifigure wearing a white helmet.

With Blacktron Predators closing in around her, the pilot took a desperate course of action. She turned toward Tycho, diving straight toward a solar flare belching from its fiery surface. The enemy craft dispersed, unable to weather the intense heat. With a touch of a control at her side, she rotated so that the underside of her craft faced the starburst and fired up the cobalt gravity manipulators, bracing herself for impact. The plasma struck and blasted her far away from her pursuers as she rode the wave, delicately rotating the handlebars in order to remain balanced.

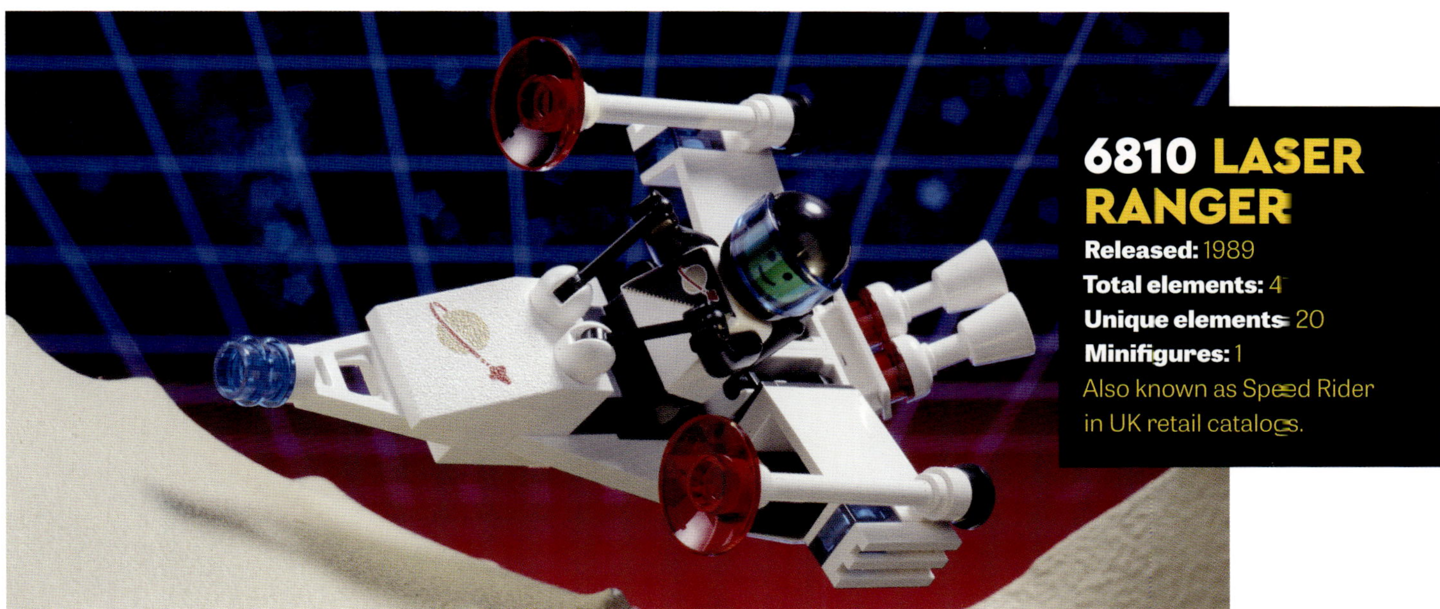

6810 LASER RANGER

Released: 1989
Total elements: 4[illegible]
Unique elements: 20
Minifigures: 1
Also known as Speed Rider in UK retail catalogs.

The planform of this small shuttle is closer to the triangular forms of Classic Space craft than Futuron's octagons, and its bodywork includes some inventive element usage: white window panels are used for the hinged wings, and two minifigure goblets act as rear thrusters.

The pilot sped into the asteroid belt, tugging deftly on the flight sticks to manipulate the wing alignment and thruster angle, sending the Ranger into sharp curves around each lump of icy rock. With no hope of keeping up, the Blacktron craft began blasting whole sections of the belt into dust. The pilot sped to safety, but then suddenly jerked the controls and flipped the ship to face the dark assailant directly. The cobalt gravity manipulator shot a beam to grip the Blacktron craft in place as the twin plasma lasers powered up. "Control to Ranger 10, you are not authorized to engage" came the order over comms. "That star-lifting laser will obliterate them! Desist." The pilot smiled resolutely as she replied. "Negative, Control. I reckon it's time some of us did a bit of policing."

▼ Original mockup of box.

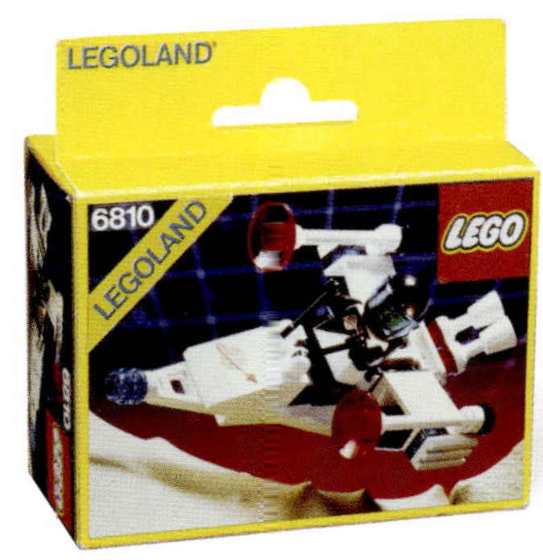

*To accompany the 1 x 4 x 3 version from 1980, LEGO® Town introduced the **panel 1 x 2 x 2** in 1985, and this useful element found its way into space two years later.*

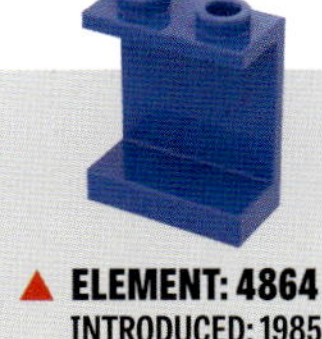

▲ **ELEMENT: 4864**
INTRODUCED: 1985

They are usually utilized as windows or wall segments, but here they create arched wings.

BLACKTRON

6894 INVADER

Released: 1987
Total elements: 159
Unique elements: 66
Minifigures: 1
Also known as Blacktron Cruiser in UK retail catalogs.

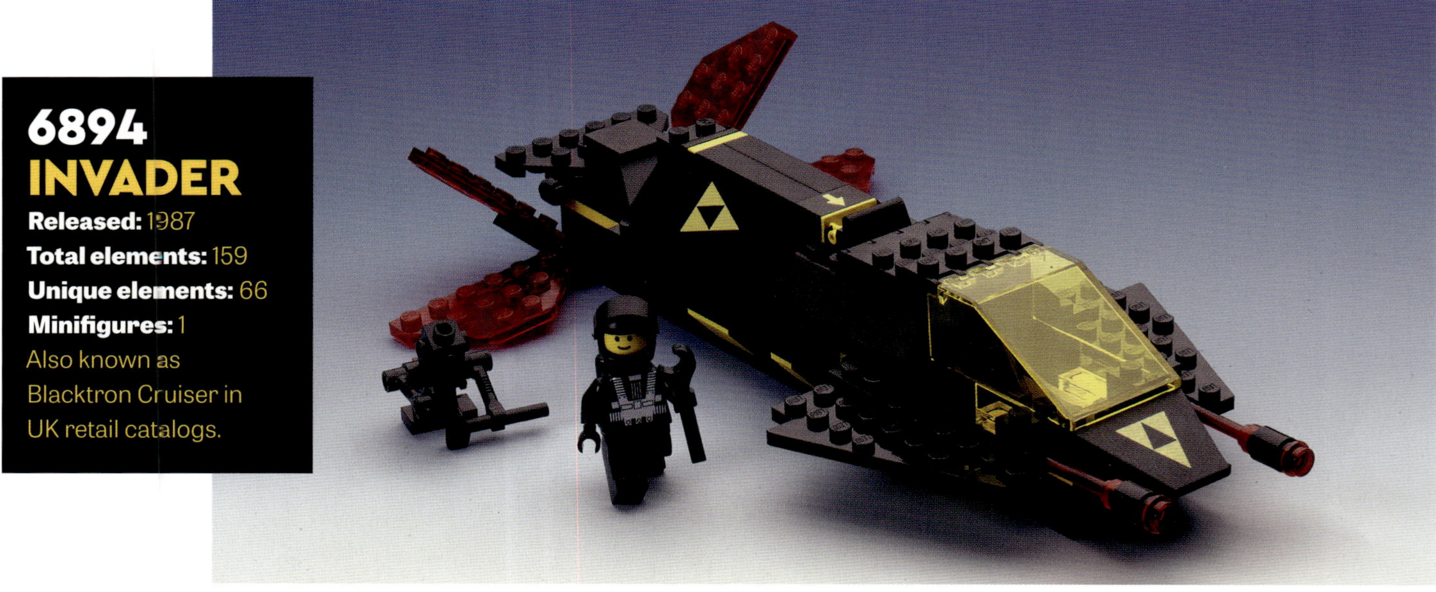

This sleek freighter consists of three detachable modules. The cab detaches to become a shuttle of delta planform, and its wings can also swing outward. Its canopy is double jointed with transparent yellow and opaque black sections, providing access to a storage compartment behind the pilot. The model's angular central section is a spacious cargo pod that hinges upward to reveal the droid and some tools. With a small empennage and transparent wings that can form octagons or an insect-like X formation, the detached rear section could be a satellite.

▼ Original mockup of box.

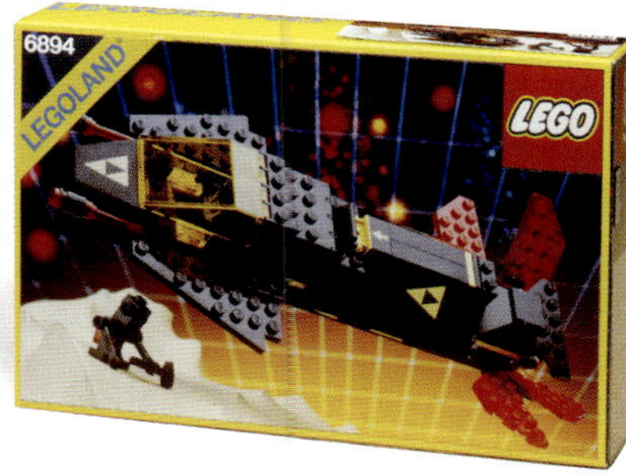

"Warp speed disengaged. Deploying orbital surveillance weaponry," said the pilot. The red blur at the rear of the craft slowed, revealing four oscillating wings which soon ground to a halt, forming two octagonal radars. The entire aft section detached and entered an orbit around Epsilon Gb. "Ejecting Trojan capsule." The remaining ship split in two, sending the black pod and its droid hurtling down through the Mercator grid toward the glittering white Futuron settlement below. After changing from cruise mode to covert, he switched the communication channels to speak to the capsule. "Otte-94, I am reconnoitering the base. Let me know when you have infiltrated."

*The **tapered wedge 4 x 4** was introduced for LEGO® Town sets, and then appeared in LEGO® Space sets with printed decorations. It looked smart decorated with the LEGO Space logo in sets 6783 and 1499, and then its trapezoidal facet proved perfect for the insignia of Blacktron that appeared in three sets.*

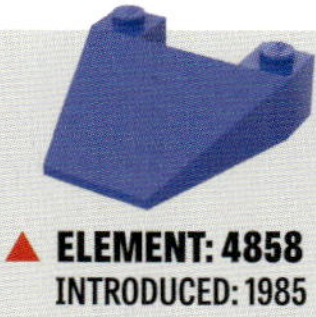

▲ **ELEMENT: 4858**
INTRODUCED: 1985

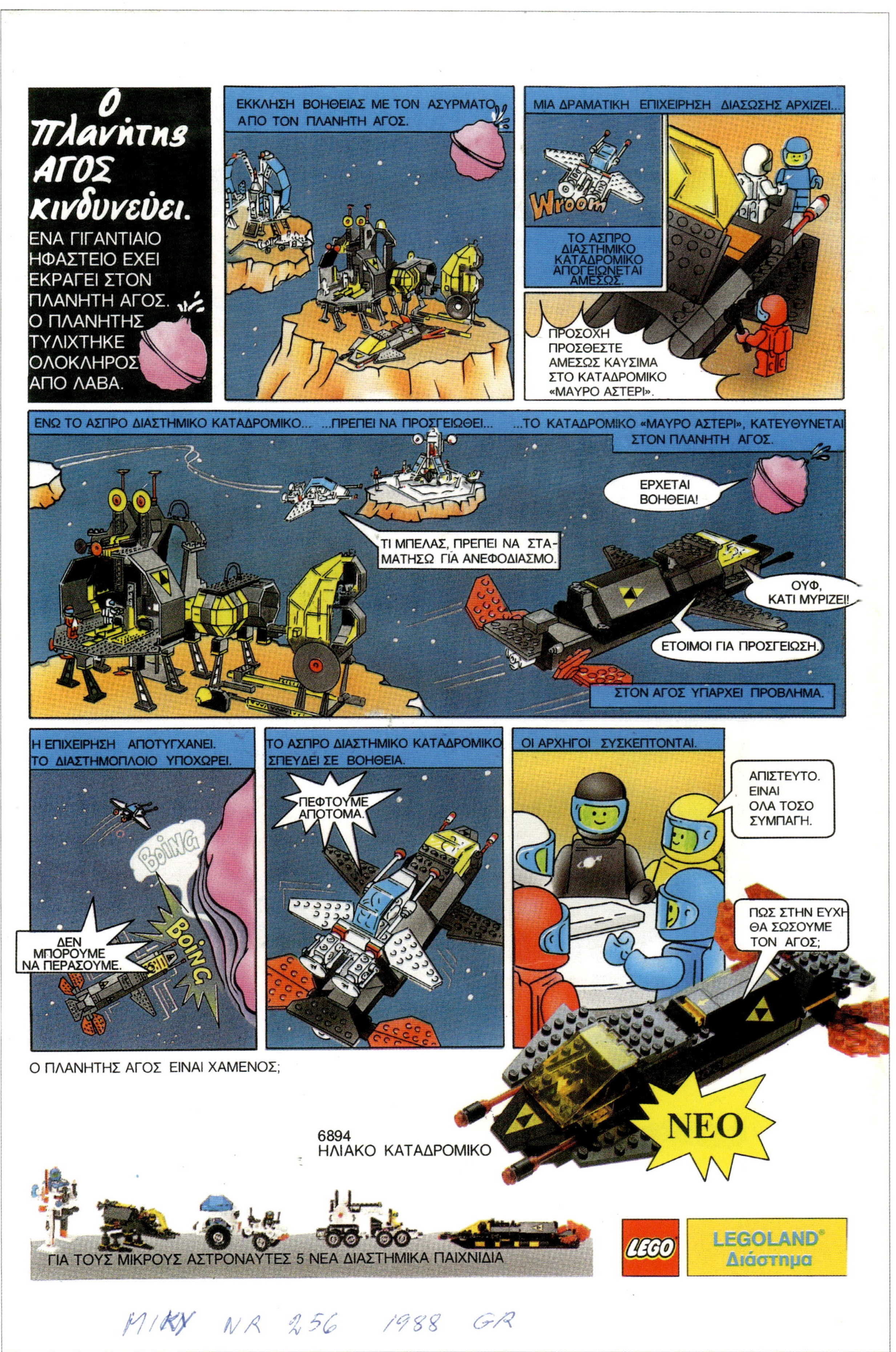

In this Greek comic from 1988 featuring a mix of Classic Space, Futuron, and Blacktron products, the Invader comes to the rescue of the planet Agos, which has been consumed by lava. However, the ending of the story is left to the reader's imagination: "Is planet Agos lost?"

6954
RENEGADE

Released: 1987
Total elements: 315
Unique elements: 95
Minifigures: 2
Only released in North America.

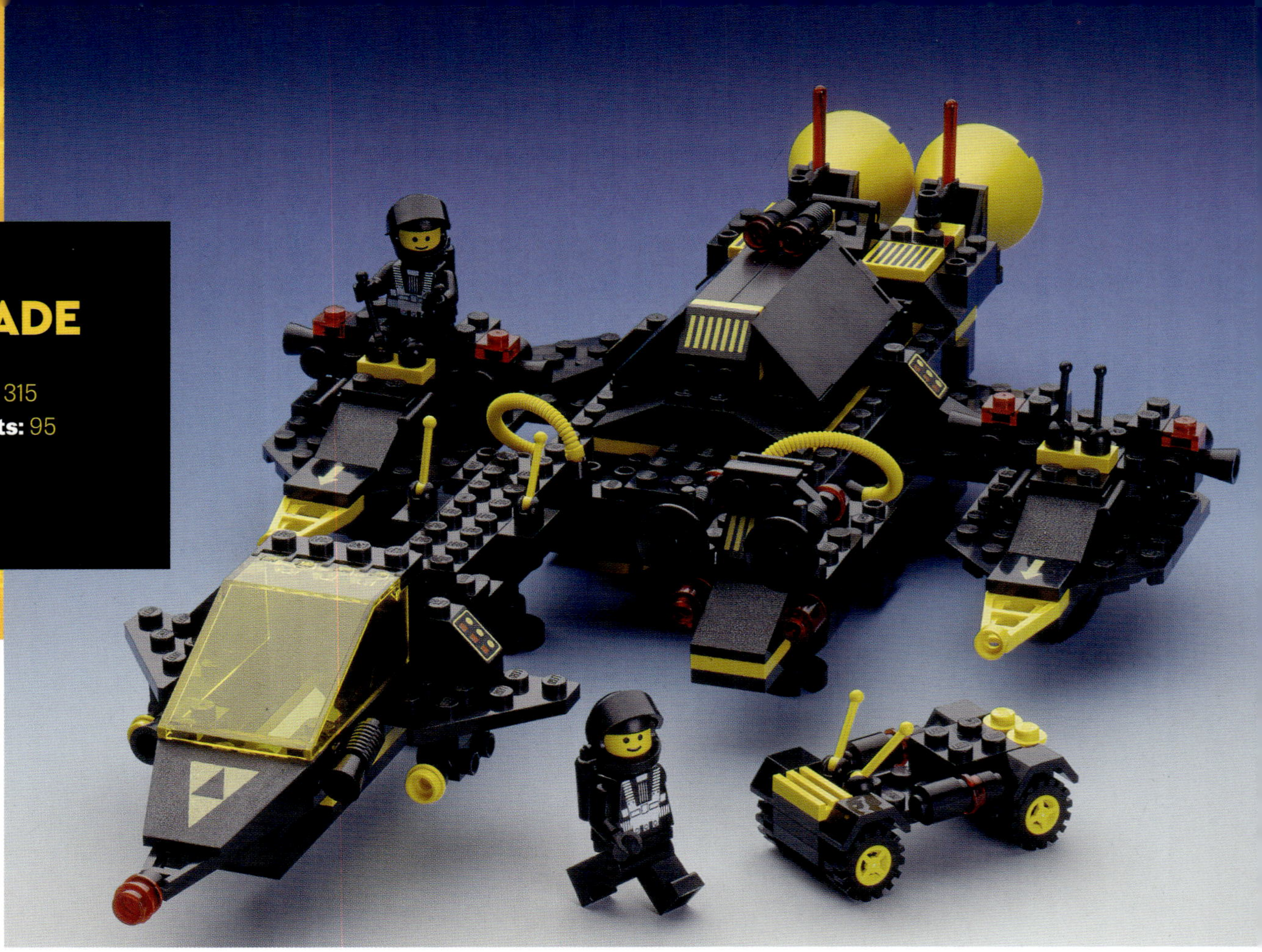

▶ This slick buggy fits inside the cargo pod. It was the first LEGO Space set since 1979 to use the angled mudguard element.

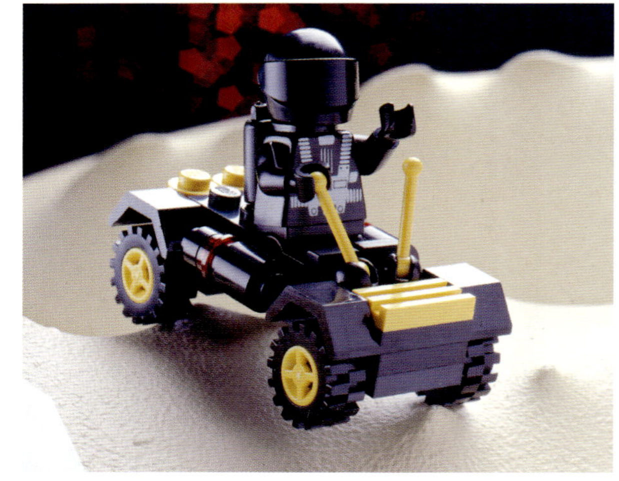

The defining feature of this dramatic ship is its asymmetrical forked planform; nothing like this had ever been seen in LEGO® Space. It has five detachable modular sections, providing a myriad of possibilities for rearrangement, including the rear thrusters, which not only swing open to release the cargo container but also detach individually. The turrets on the wings also detach to create scout ships but are not modular, as they are connected by bars rather than LEGO® Technic pins.

▼ Original mockup of box.

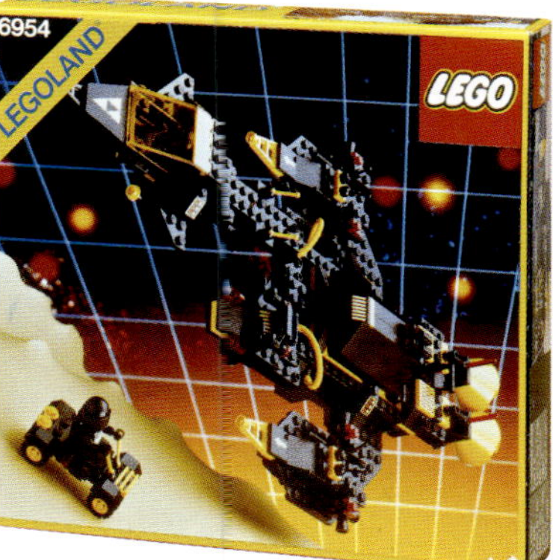

"Battrax to Renegade," said the captain, "we are in retreat. The Futuron have the droid. Take out the base." Beneath his impenetrable visor, the sergeant smiled coldly. "We won't leave a trace, Captain. Command must never know of this." The reconfiguration sequence commenced as soon as the Renegade entered the atmosphere of Epsilon Gb, fragmenting the ship. The twin scout craft lifted away from the wings as the cockpit and auxiliary command unit separated from the front. At the rear, both thrusters detached as the empennage swung open to release the cargo pod. One thruster reconnected to the auxiliary command while the other joined the rear of the cargo pod, just as the sergeant reversed the cockpit onto the front. As each new craft sped a safe distance away, the remaining hulk circled downward, hurtling unstoppably toward Star Base 1. "Initiate self-destruct," the sergeant commanded from her shuttle, "in three, two, one . . ."

*The **wedge plate 10 x 10 with 4 x 4 cutout and cut corner** does not necessarily look like a LEGO Space element, but that's the only theme it was used for in its first year of production, appearing here as well as in sets 6990 and 6953. Its use in the Renegade is particularly inventive, providing a strong structure for the forked wings.*

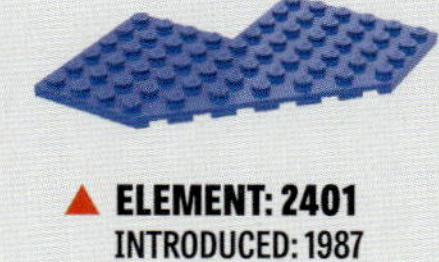

▲ **ELEMENT: 2401**
INTRODUCED: 1987

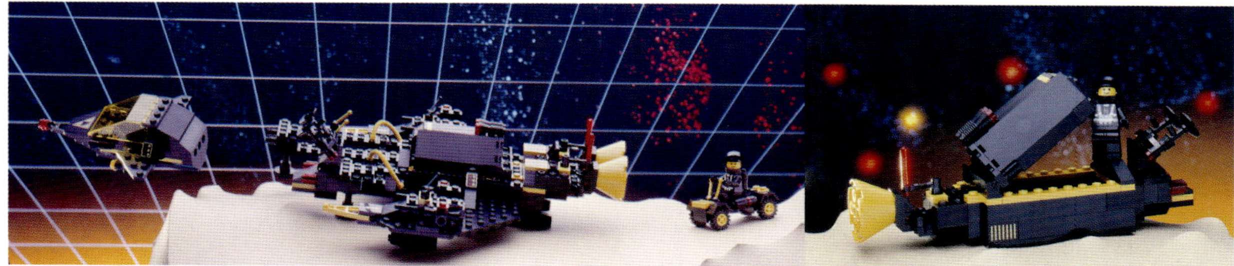

◀ Two of the many possible combinations of modules.

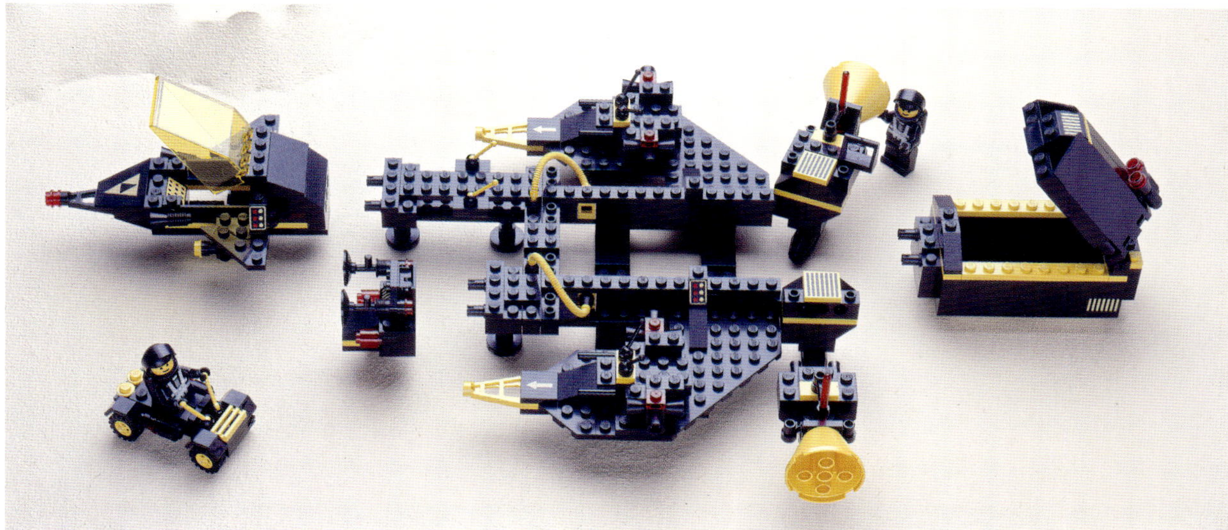

◀ The back of the box revealed all of the modular sections that composed the Renegade.

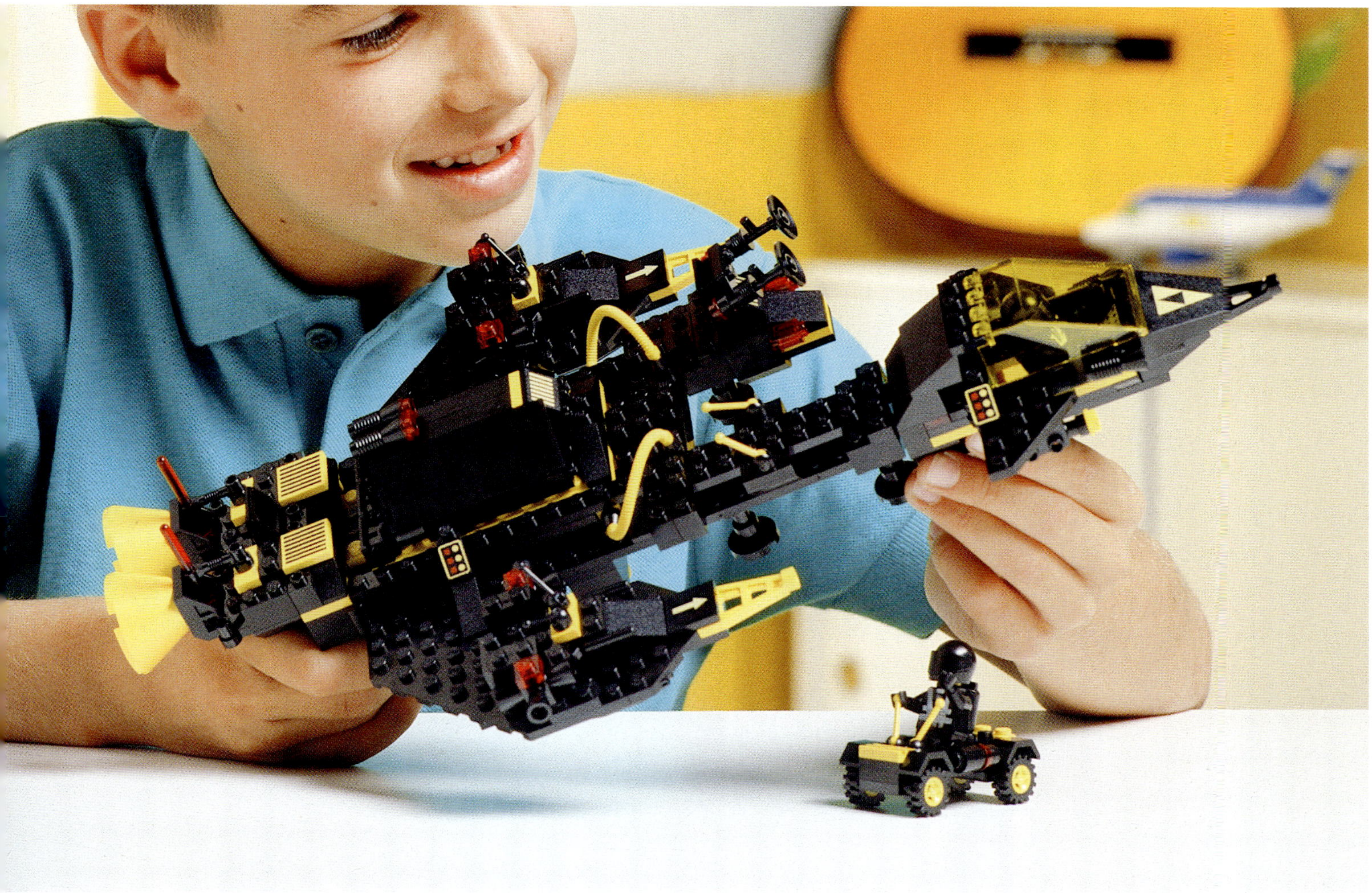

As with many sets of this period, the interior of the box lid featured a child enjoying the completed model.

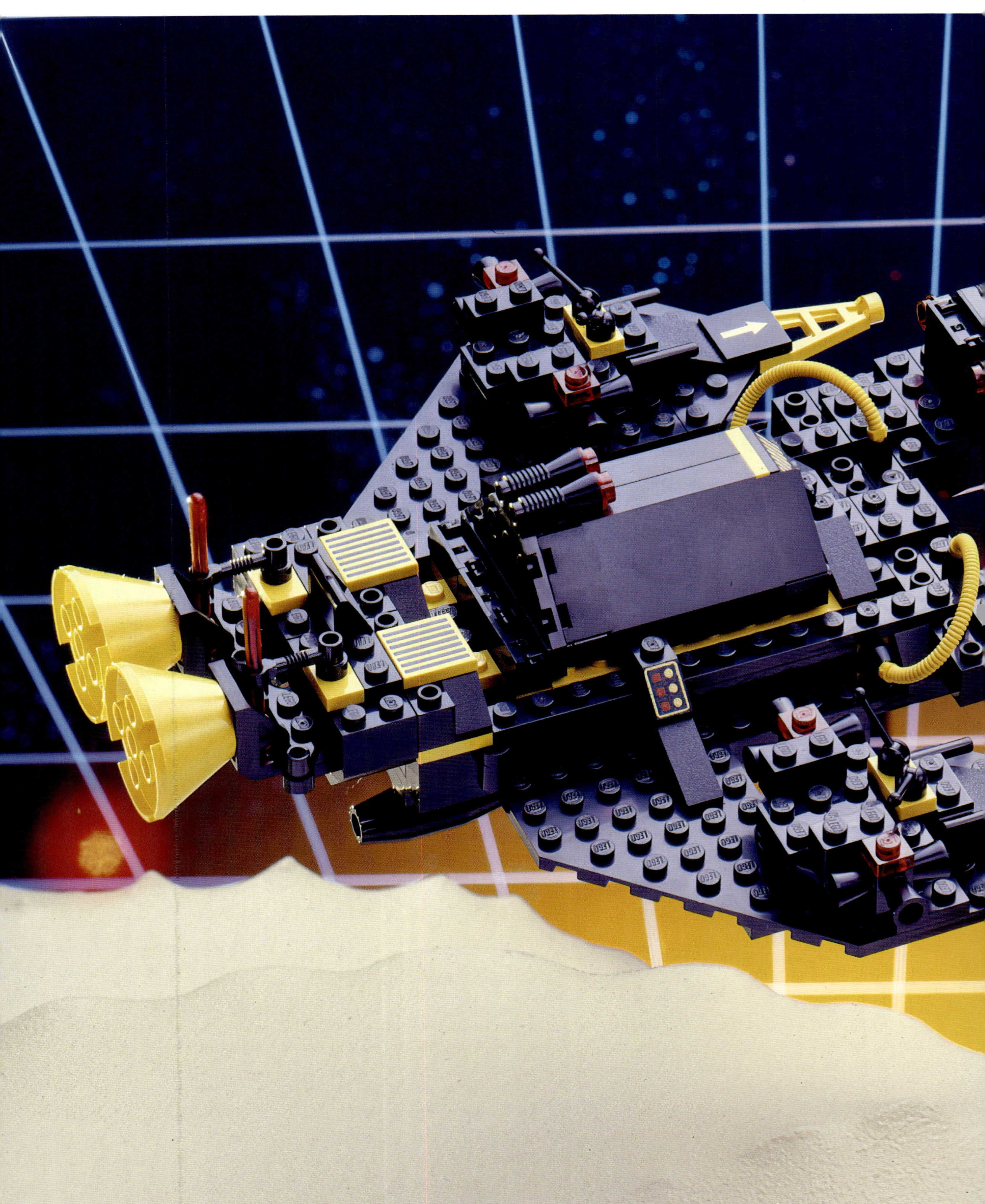

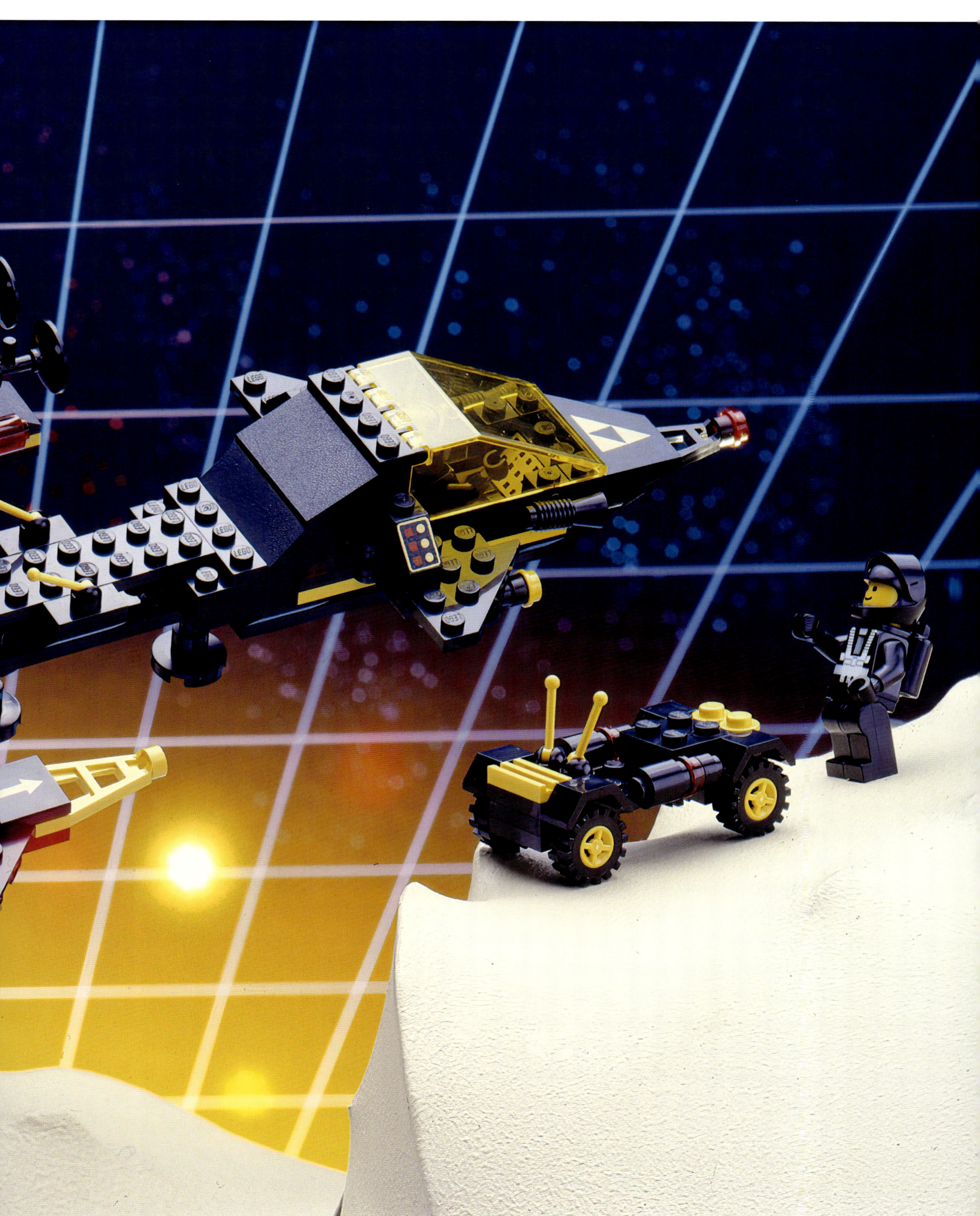

6987 MESSAGE INTERCEPT BASE

Released: 1988
Total elements: 575
Unique elements: 146
Minifigures: 5
Also known as Blacktron Star Base in UK retail catalogs.

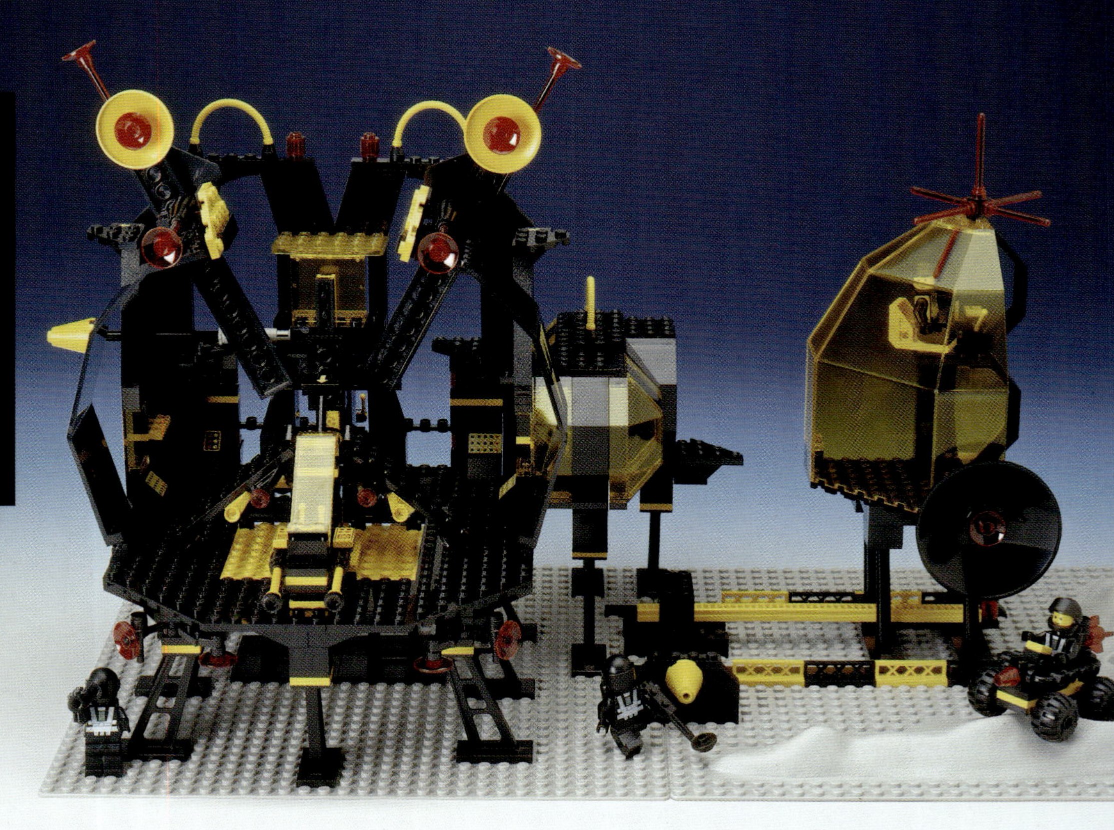

▶ The landing bay in its closed position. Note the large arched bricks above the roof; this is the only time they've been used in a LEGO Space set to date.

Aside from the monorail, this was the largest LEGO® Space set to date, exceeding 1983's 6980 Galaxy Commander by more than one hundred elements. The base comprises a cavernous, detailed landing bay connected by a winding octagonal corridor to a domed control tower platform. The rack-and-pinion mechanism seen in 6953 Cosmic Laser Launcher is used twice here: to open the clamshell launch pad and to slide the entire control tower dome to join the corridor. Delightfully, this dome is topped by a studded brick covered with transparent red antennas. The shuttle craft looks suitably menacing with its extreme droop wing, and the buggy has a futuristic design using long curved bar elements. This was the final set to include the gray crater baseplate; in 1996 a decorated light blue version appeared in a LEGO® Aquazone set.

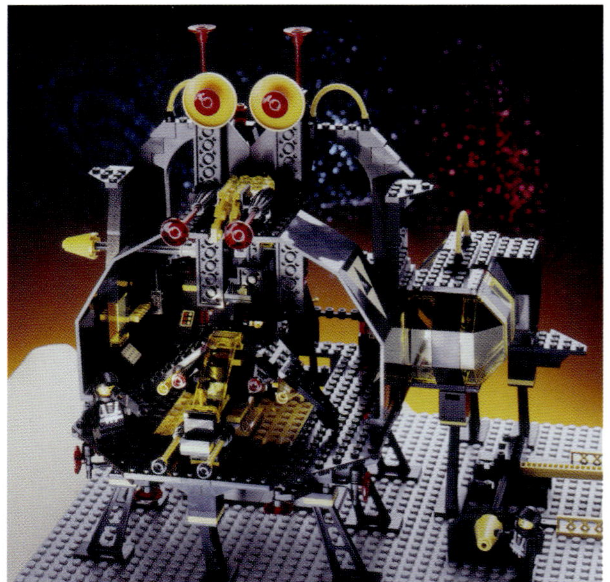

▼ Original mockup of box.

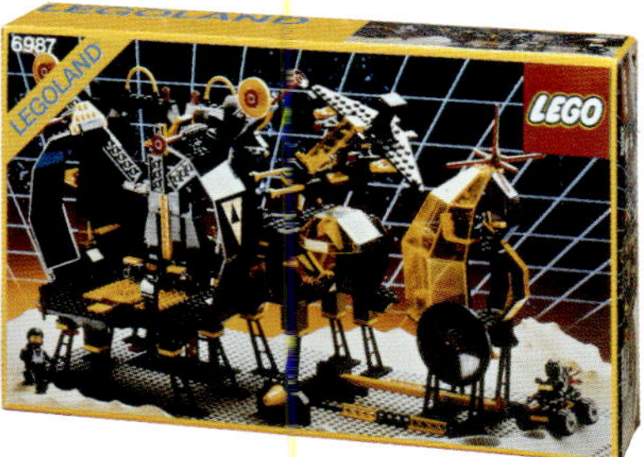

The craft lowered its wings as it approached the base, steering toward the dome that was expanding before it like two open palms. "Infiltrator to Star Base. Futuron hypernetwork has been activated. Initiate interception." The data scientist descended from the bay control unit and slammed down their suppression visor as they ran through the winding corridor to the security tower. They activated the ogdoösphere, waiting for the ground crew below to fully separate it to its protective position before initiating the isotropic antenna. "Commencing data exfiltration," they said as the gamma rods glowed red above them.

*The full gamut of angled panels introduced in 1988 are used in the architecture here, including the **convex corner panel 3 x 3 x 6** as a prominent corner window within a corridor wall that hinges open for play access. In other sets, two were combined to form octagonal dome canopies.*

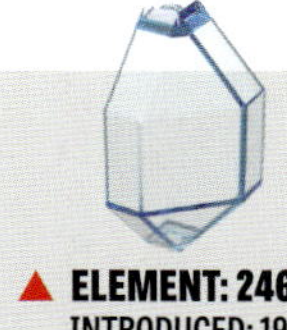

▲ **ELEMENT: 2468**
INTRODUCED: 1988

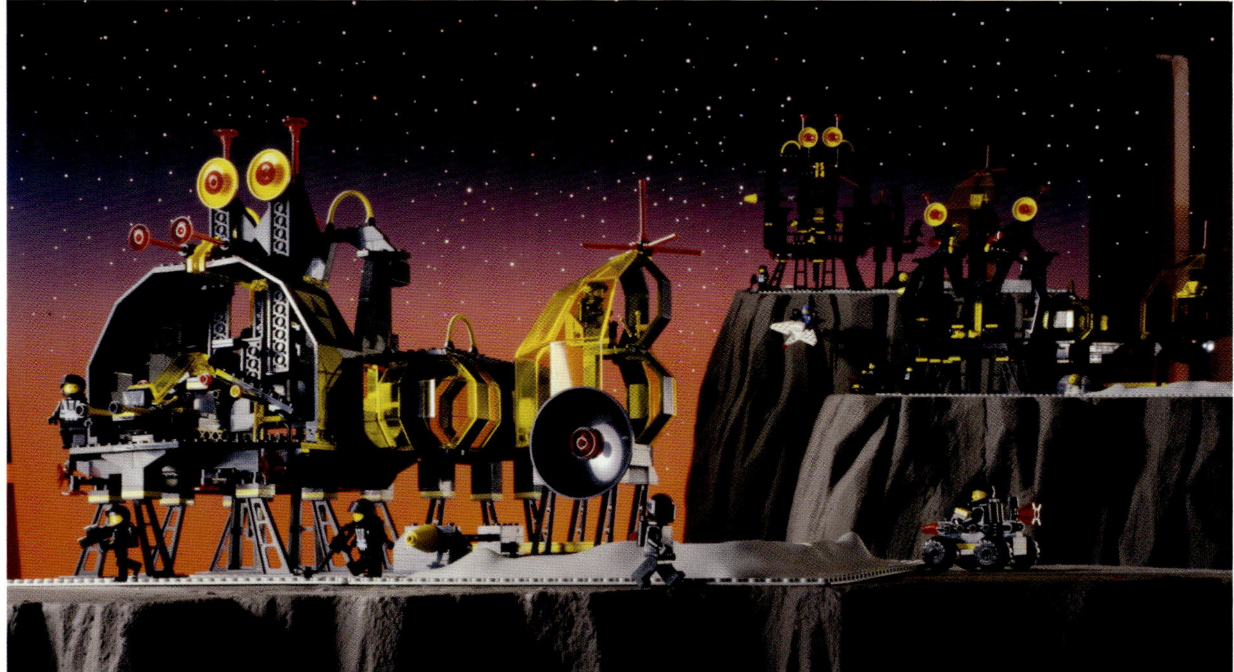

◀ This thrilling promotional photograph features not one but three Message Intercept Bases, heavily guarded by Blacktron crew, who all appear oblivious to the lone scout ship performing reconnaissance.

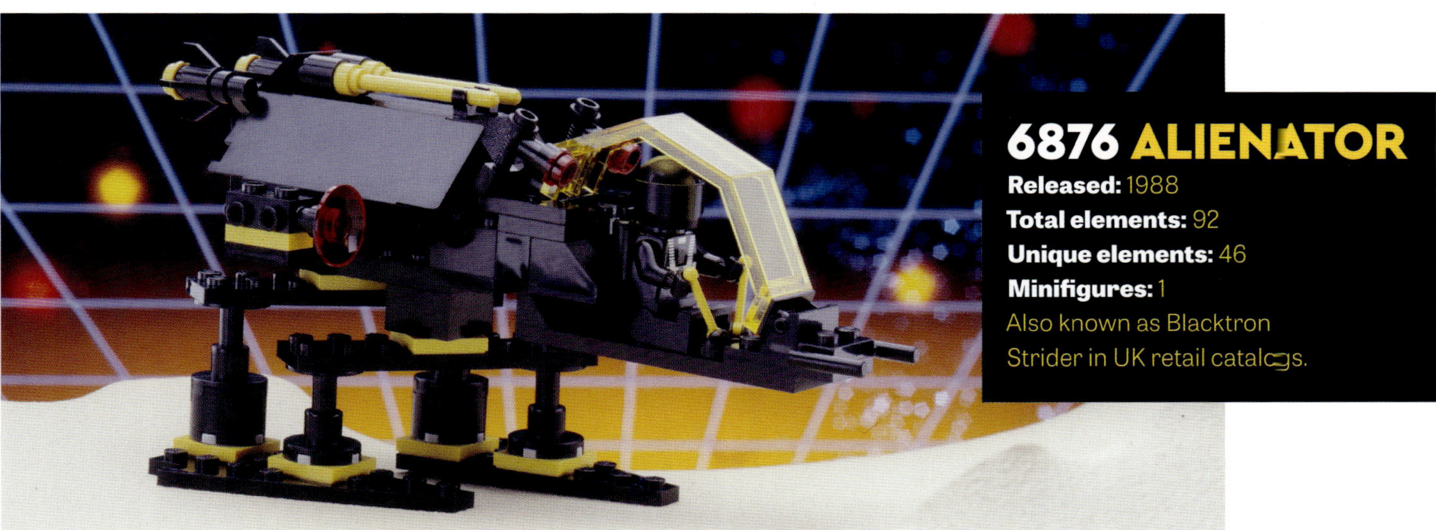

6876 ALIENATOR

Released: 1988

Total elements: 92

Unique elements: 46

Minifigures: 1

Also known as Blacktron Strider in UK retail catalogs.

Although this set does not contain a large quantity of elements, its clever design and the number of large elements included make it an impressive walker. The wedge-shaped feet and staggered heights of the two pairs of legs create an illusion of forced perspective, as though the Alienator is bearing down upon you. The windscreen is held closed by the two black levers behind it, and the whole front cab detaches to provide modularity with other Blacktron sets.

The walker moved inexorably forward with mechanical grunts, striding through the radioactive diamond dust of Epsilon Gb. Something glistening white caught the wrecker's eye, and she stopped for a closer look. Releasing the magnolevers behind her head, she detached the cockpit to swoop down and lifted it out using the front-mounted gravity prongs. Eureka! She immediately contacted the mother ship. "Futuron HSU located. Pinging it to you now." She opened the teleport pod, angling its yellow quantum extrapolator rods toward the mother ship, and climbed in. After shifting the hyperconverged storage unit across with her handheld gravity winch, she fired up the quantum laser and watched as the cube dematerialized with a loud "ping!"

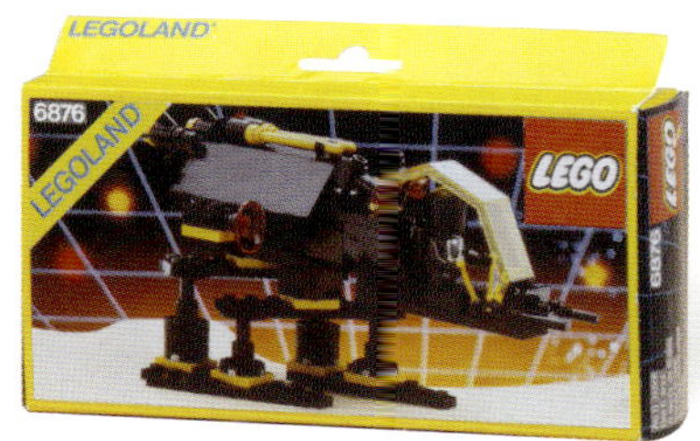

▼ Original mockup of box.

*Following the large angled panels that premiered in 1987, a family of smaller, six-brick-high angled panels were introduced the next year. The **panel 3 x 2 x 6** could be placed alongside the matching corner elements but was more frequently used as a canopy, as seen here.*

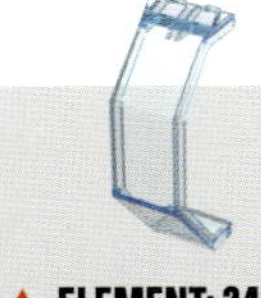

▲ **ELEMENT: 2466**
INTRODUCED: 1988

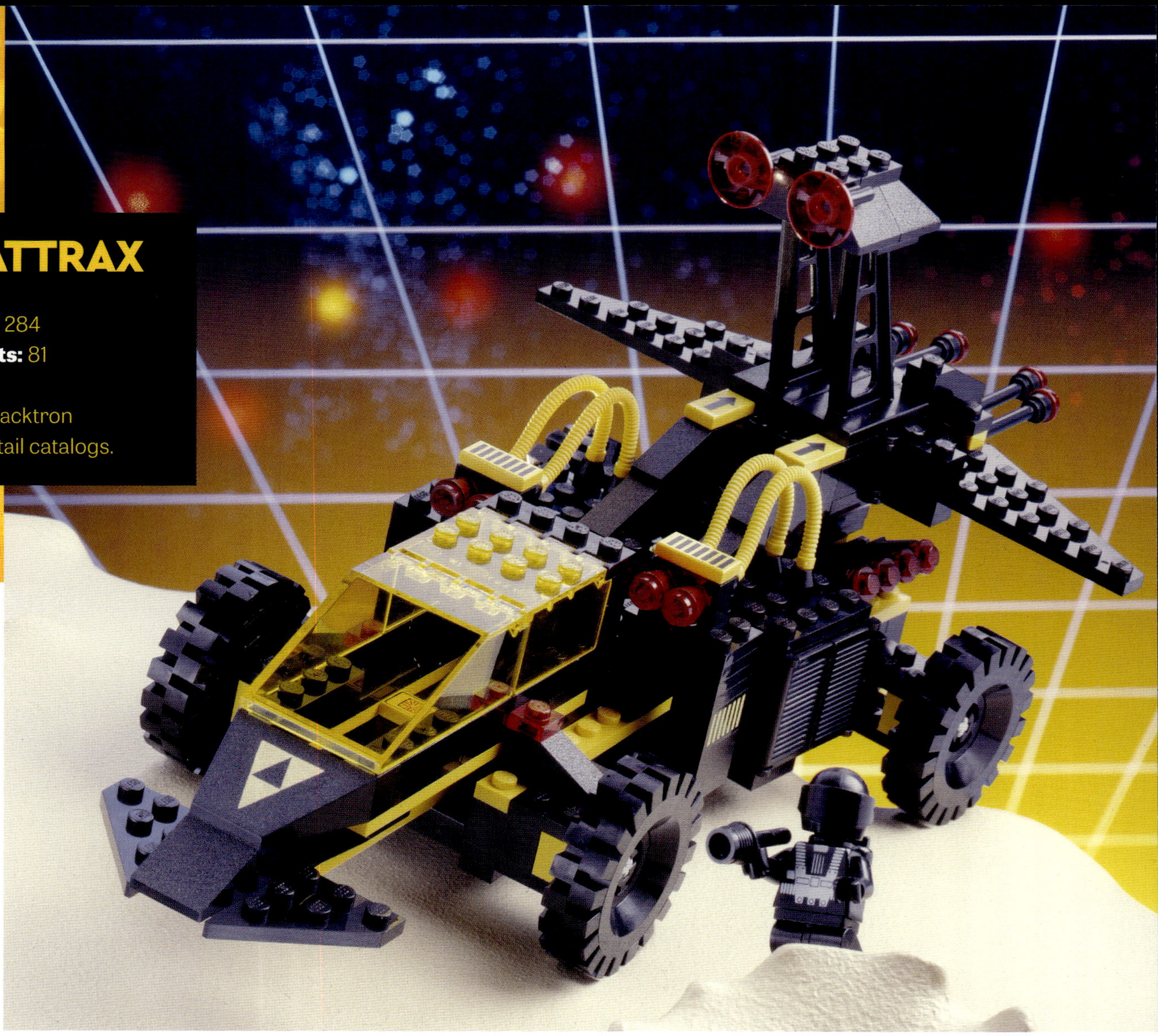

6941 BATTRAX

Released: 1987
Total elements: 284
Unique elements: 81
Minifigures: 1
Also known as Blacktron Prowler in UK retail catalogs.

▼ Original mockup of box.

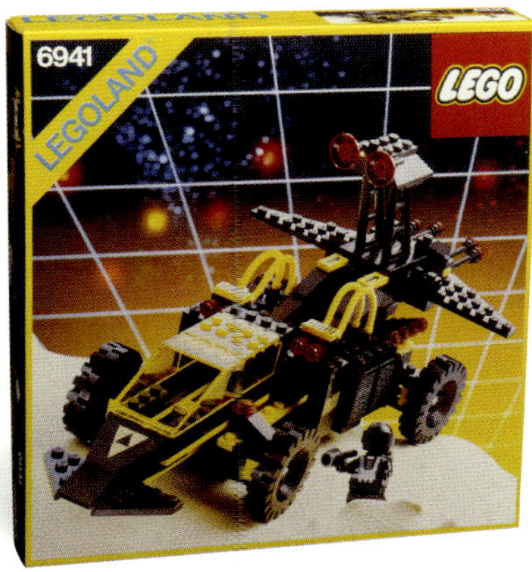

This articulated ground vehicle with functional steering brings a whole new level of excitement to LEGO® Space, with features resembling a rear-engine dragster! The spoiler with vertical stabilizer is a detachable shuttle that can be piloted on its own. The front module, which consists solely of the long cockpit, has a double-hinged canopy with space for two minifigures. It too can be operated on its own but looks better when combined with the rear module. Although the remaining central module lacks a driving seat, it still looks great on its own. Its central pivot point is simply a vertical LEGO® Technic axle, and the prominent containers with shutters contain a variety of tools.

As the captain traversed the dunes surrounding Futuron Star Base 1, an unexpected message came in. "Invader to Battrax. Urgent assistance required. Craft down, copilot wounded, droid captured. Sending location." He disengaged prowler mode and fired up the rear thrusters. The Battrax shot across the landscape, diamond dust flying in its wake. Upon reaching the Invader, he and the pilot lifted their injured colleague into the rear seat of the Battrax as they formulated a plan. "The capsule is salvageable. We can couple it to your shuttle," proposed the pilot. "Good plan," replied the captain as he headed to the rear of the Battrax to decouple. "I'll drive to the droid's location; you cover me from above. The droid takes the capsule, and you and I reconnect for maximum firepower. Anything goes wrong, we obliterate all resistance—and the droid."

*Not only does the **windscreen canopy extender 4 x 4 x 2** augment the existing canopy from 1983 to give a fresh look, but it also increases play functionality by adding a joint within the windscreen. It was also used in the monorail trains, and, as new four-module-wide canopy elements were introduced into LEGO Space in the following years, the canopy extender continued to be combined with those.*

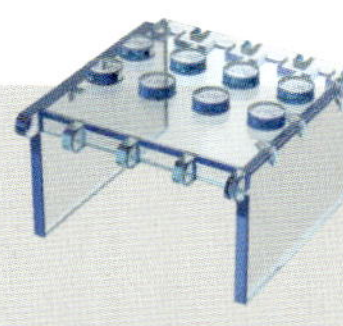

▲ **ELEMENT: 2337**
INTRODUCED: 1987

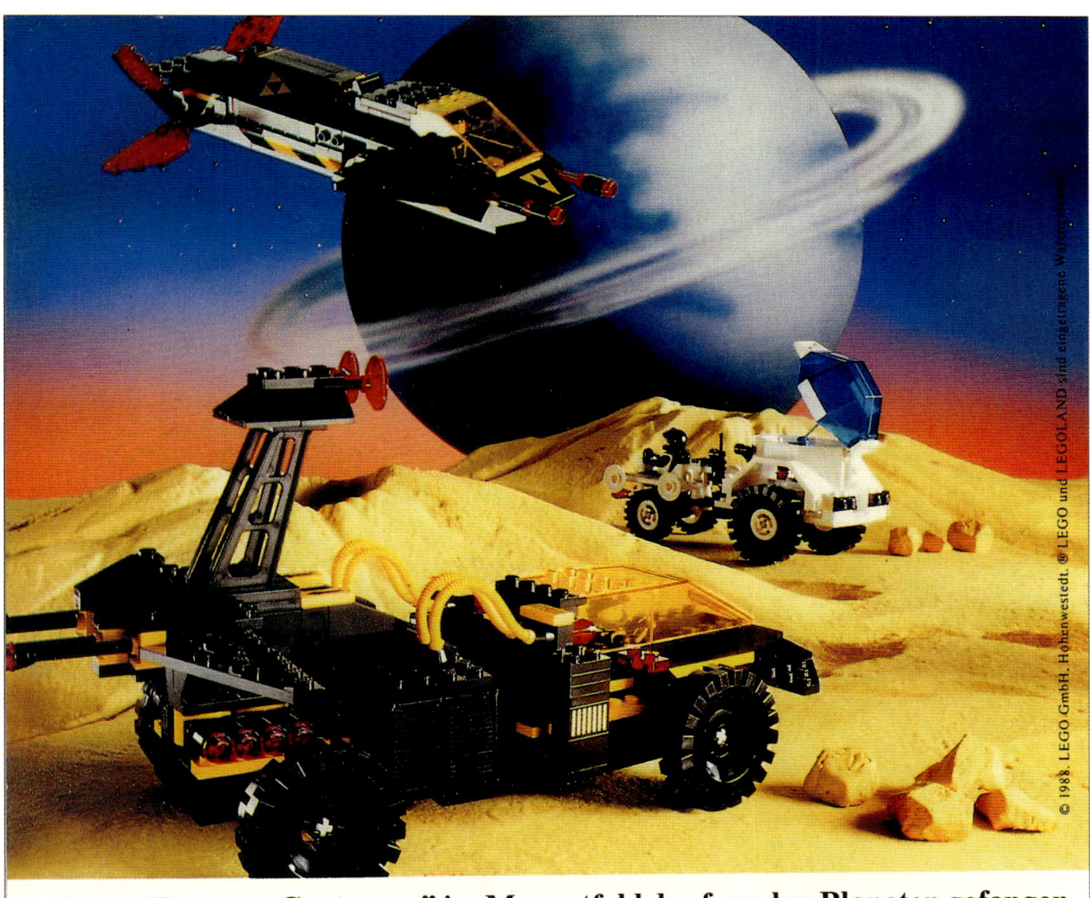

The Battrax and the Invader (here named "Centaurus") join the Futuron set 6885 Crater Crawler in this German ad of 1988. No factional conflict is suggested within the copy.

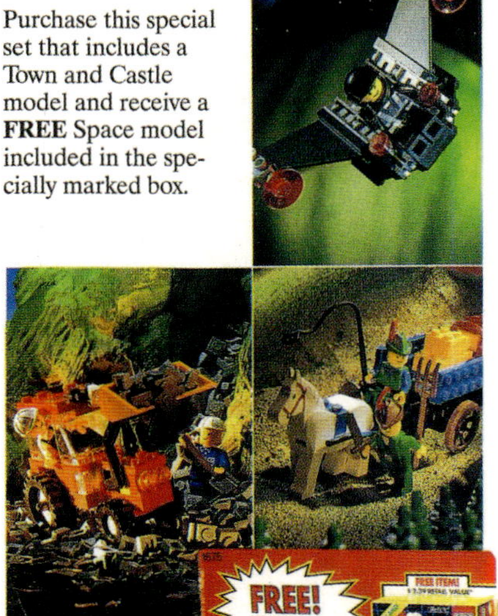

A bit of a "space oddity," this limited-retail set was released two years after the previous Blacktron set and used white elements in place of yellow, similar to Futuron—and also Blacktron Future Generation, which would be released the following year. An interesting craft is created using just a few elements, including the large tail elements as folding wings, as in 6845 Cosmic Charger in 1986.

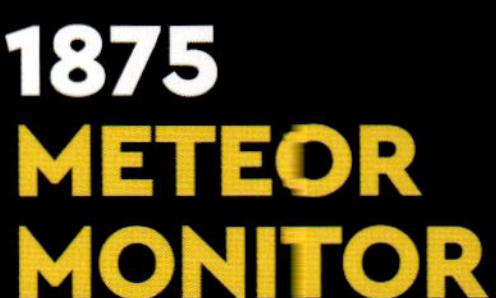

1875 METEOR MONITOR

Released: 1990
Total elements: 32
Unique elements: 12
Minifigures: 1
Only released in North America. Included in 1675 Three Set Bonus Pack.

The soldier ran through the wreckage of the Futuron city, searching for a flight-worthy craft. All he could find was a tiny monitor craft, intended for clearing meteor debris from interstellar flight paths. It would have to do. As he rose higher, he could see the full extent of the destruction his compatriots had wrought. And for what? His hopelessness was replaced by a sudden conviction. He would find the others and convince them to jump far away. Somewhere the Blacktron would never, ever find them. A place where future generations could live in peace.

◀ Meteor Monitor was included as part of a 1990 offer including LEGO® Town and LEGO® Castle sets.

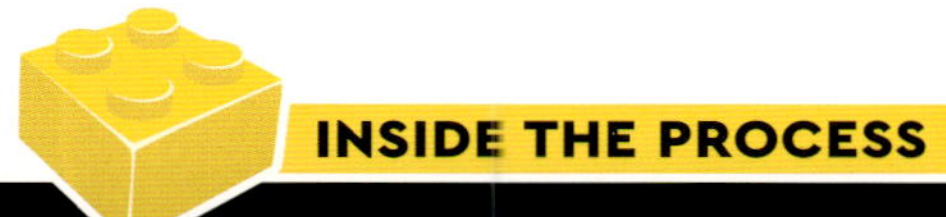

INSIDE THE PROCESS

SHOOTING STARS

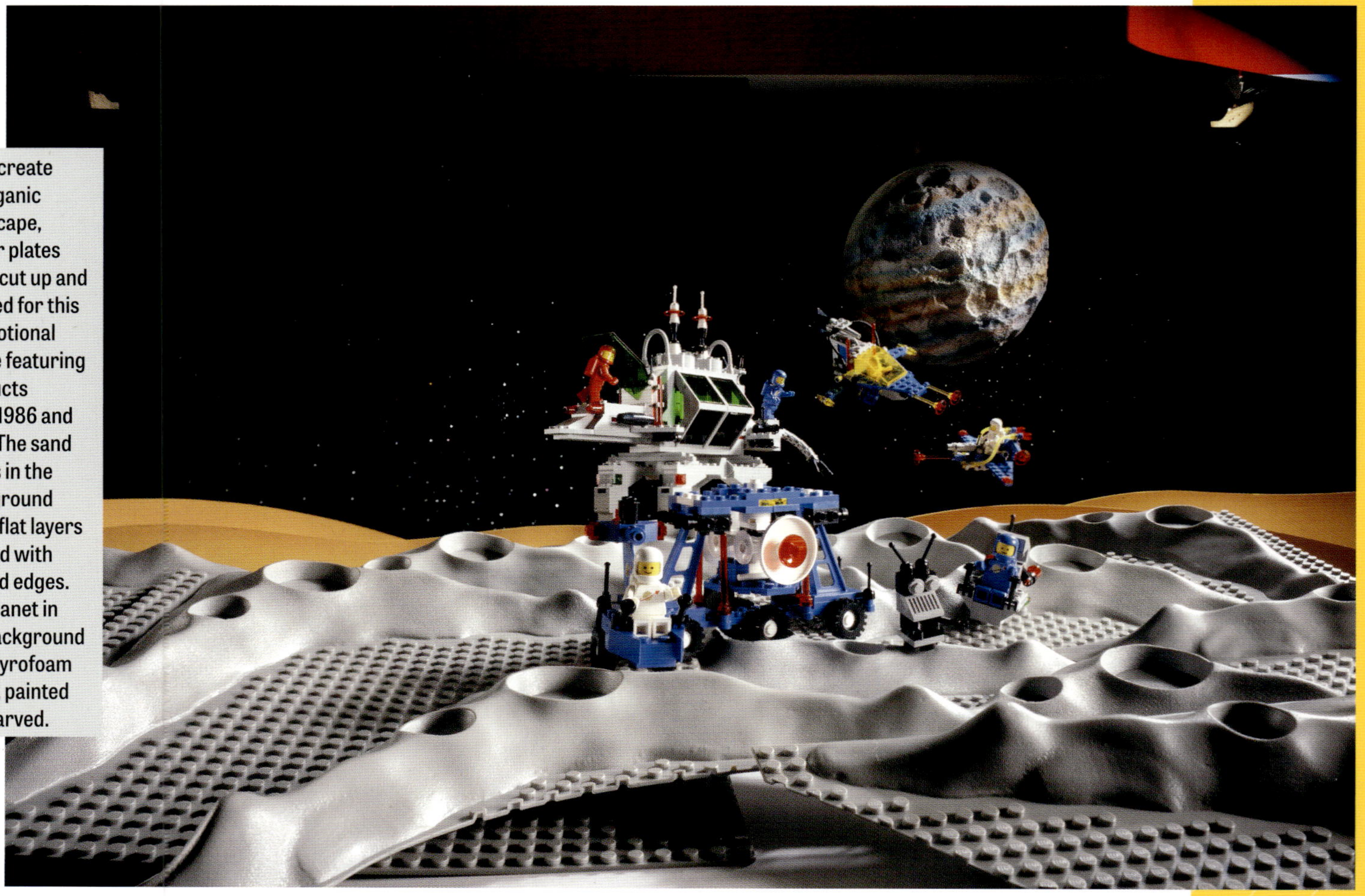

▶ To create an organic landscape, crater plates were cut up and layered for this promotional image featuring products from 1986 and 1987. The sand dunes in the background were flat layers of card with curved edges. The planet in the background is a Styrofoam globe, painted and carved.

Just as the design of LEGO® models evolved over the years, so too did the photographs adorning the boxes, catalogs, and advertisements. In the era before digital image editing software, as much work as possible was done in-camera. These original open matte photographs from the LEGO archives provide a glimpse into the artistry involved.

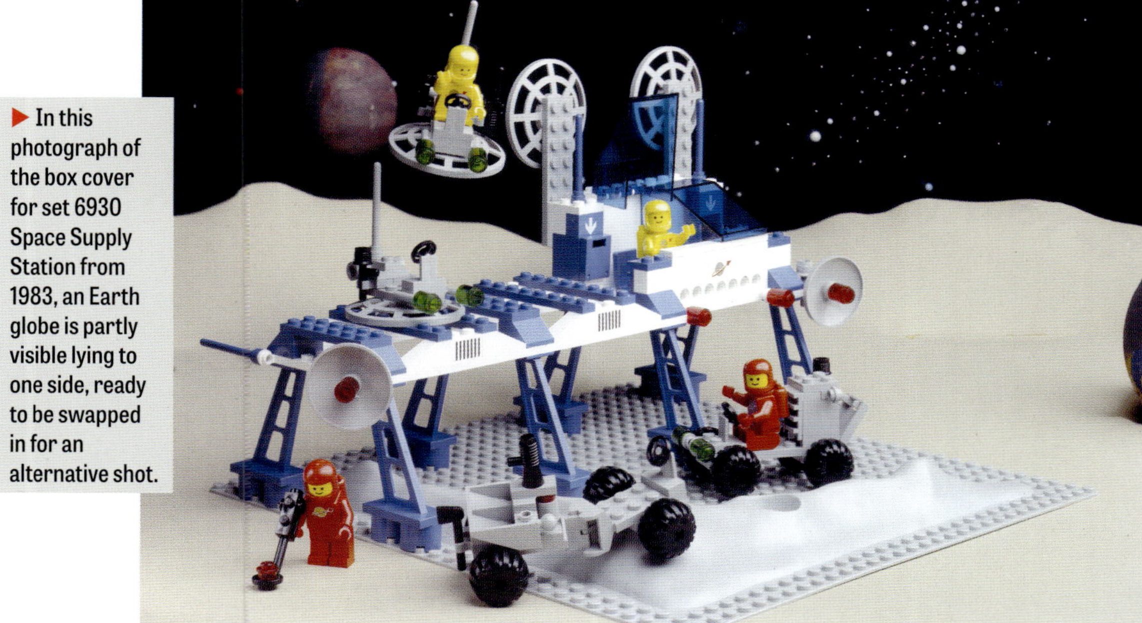

▶ In this photograph of the box cover for set 6930 Space Supply Station from 1983, an Earth globe is partly visible lying to one side, ready to be swapped in for an alternative shot.

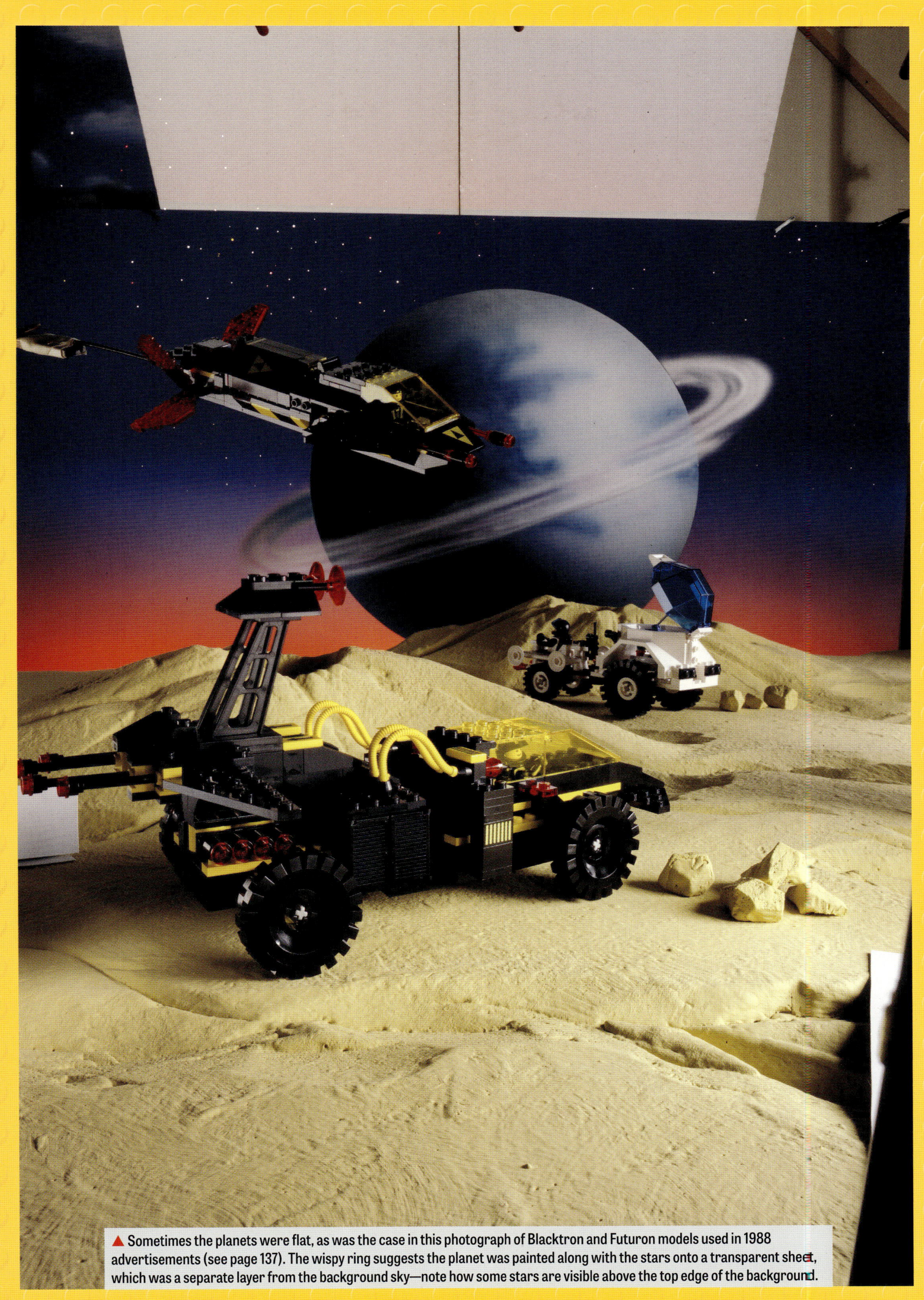

▲ Sometimes the planets were flat, as was the case in this photograph of Blacktron and Futuron models used in 1988 advertisements (see page 137). The wispy ring suggests the planet was painted along with the stars onto a transparent sheet, which was a separate layer from the background sky—note how some stars are visible above the top edge of the background.

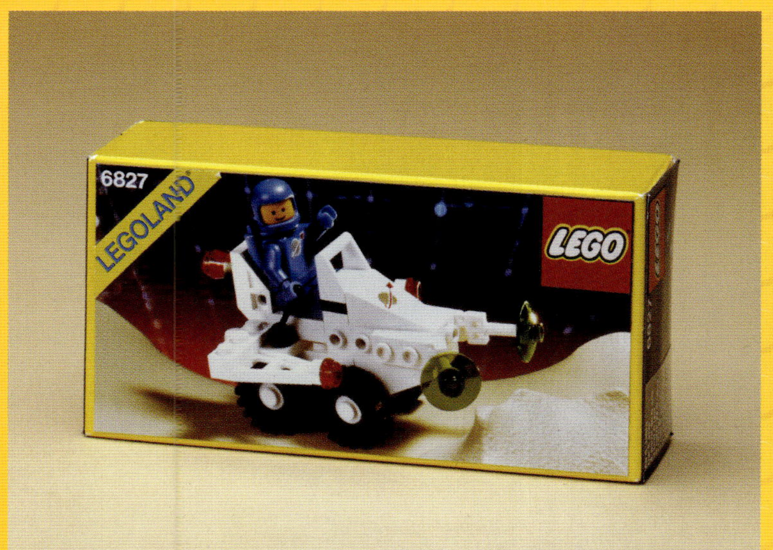

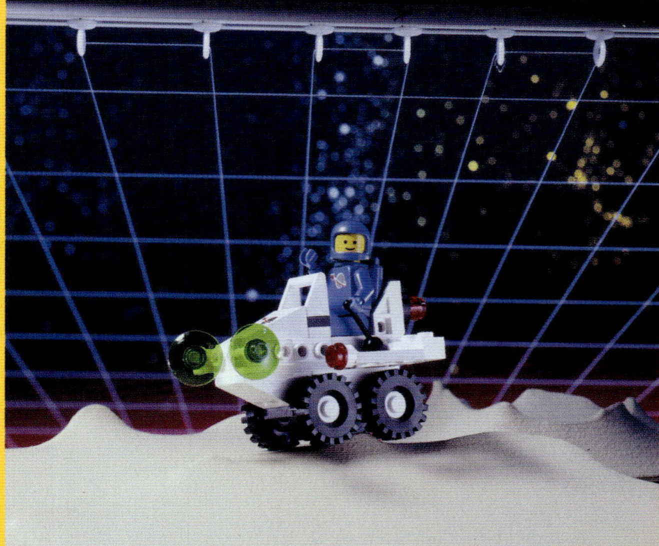

▶ From 1987, a three-dimensional grid was added to the background. This, too, was a physical object, consisting of wires threaded between adjustable hoops spaced along a frame made of metal rails. The illusion of depth was created by spacing the hoops at different distances along each rail: wider at the top and narrower at the bottom.

"I REMEMBER SEEING THAT BACKGROUND FOR FUTURON MODELS FOR THE FIRST TIME, AND IT WAS PRETTY MIND BLOWING . . . IT'S REAL— IT'S A PHYSICAL THING!"

—BJARNE PANDURO TVESKOV, DESIGNER, THE LEGO GROUP, 1985 TO PRESENT

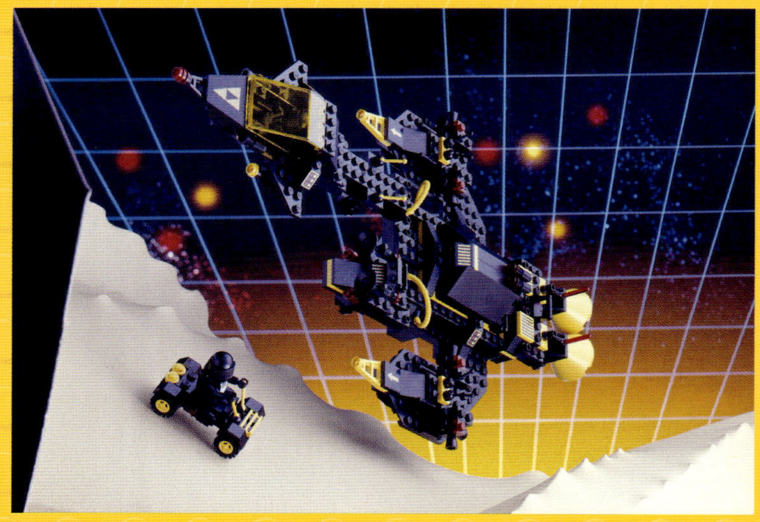

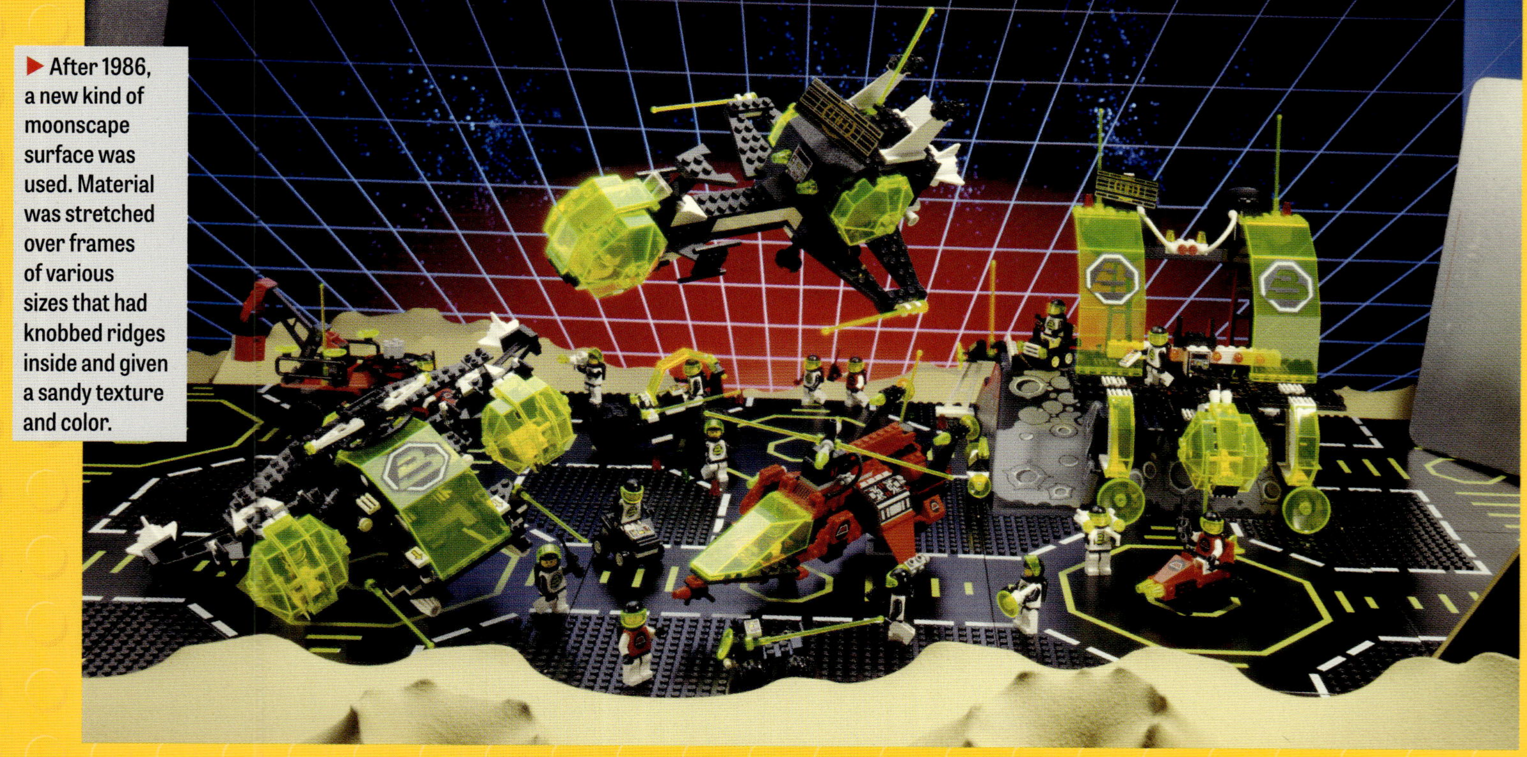

▶ After 1986, a new kind of moonscape surface was used. Material was stretched over frames of various sizes that had knobbed ridges inside and given a sandy texture and color.

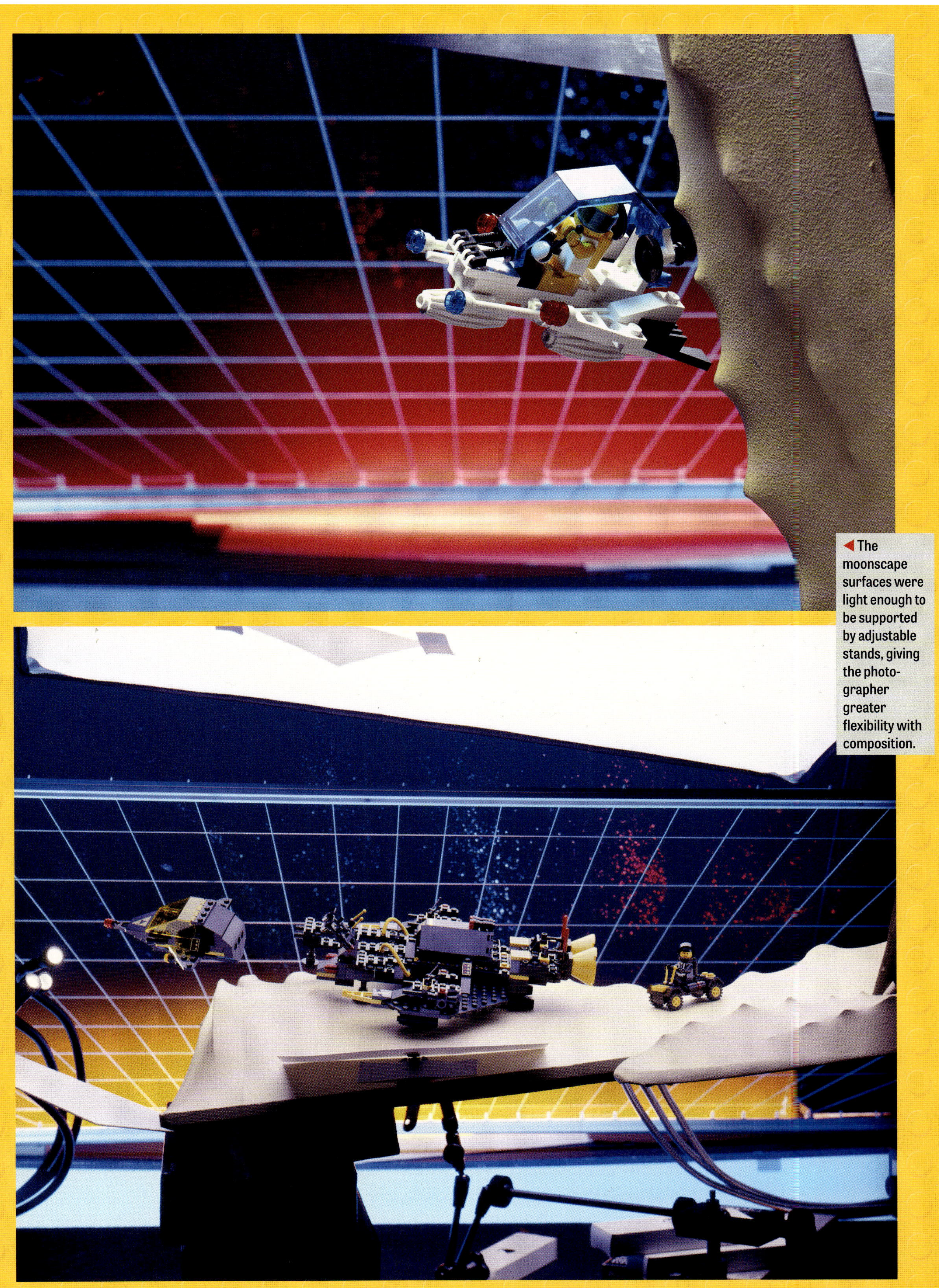

◀ The moonscape surfaces were light enough to be supported by adjustable stands, giving the photographer greater flexibility with composition.

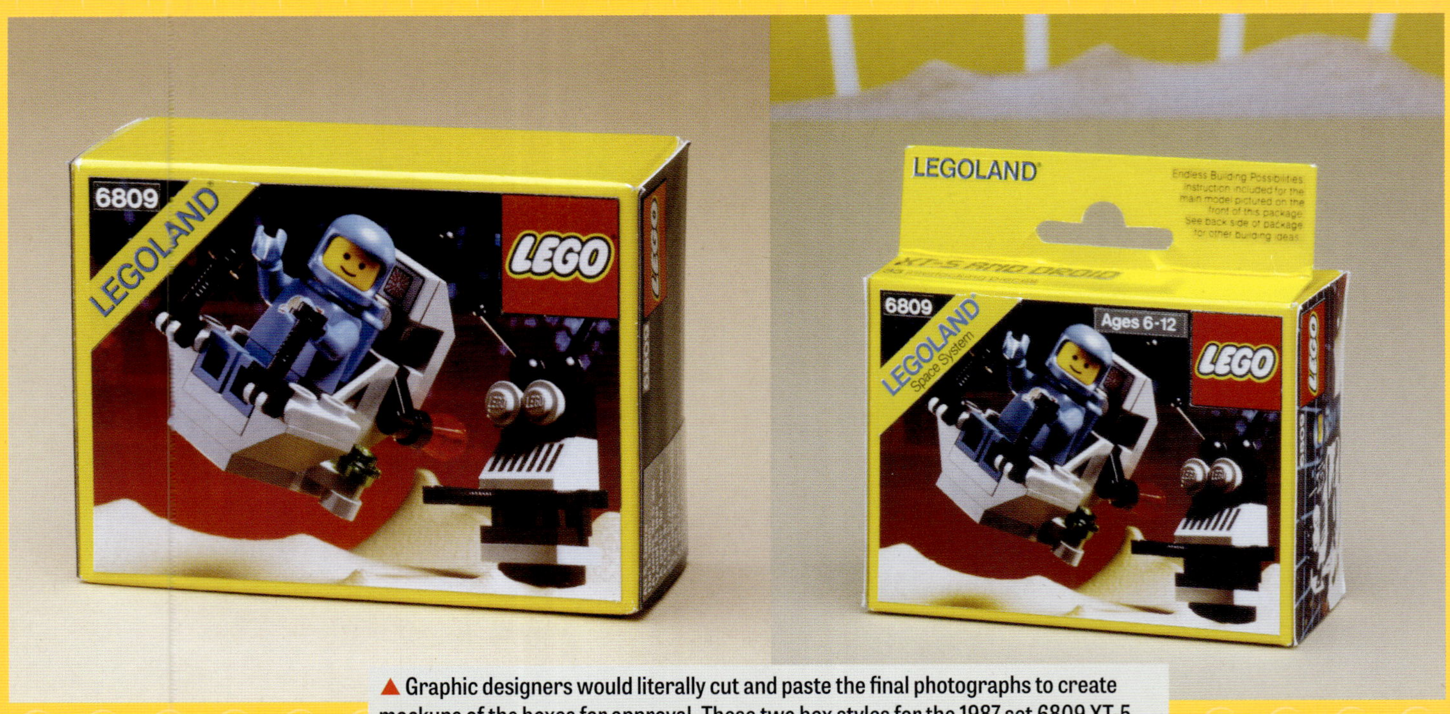

▲ Graphic designers would literally cut and paste the final photographs to create mockups of the boxes for approval. These two box styles for the 1987 set 6809 XT-5 and Droid reveal differences in US packaging, including the age mark and the legend "Space system" shown on the front and the name of the set and piece count on the top.

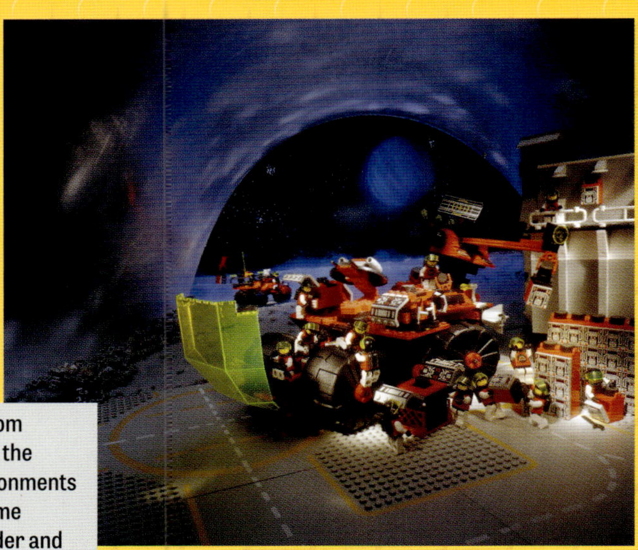

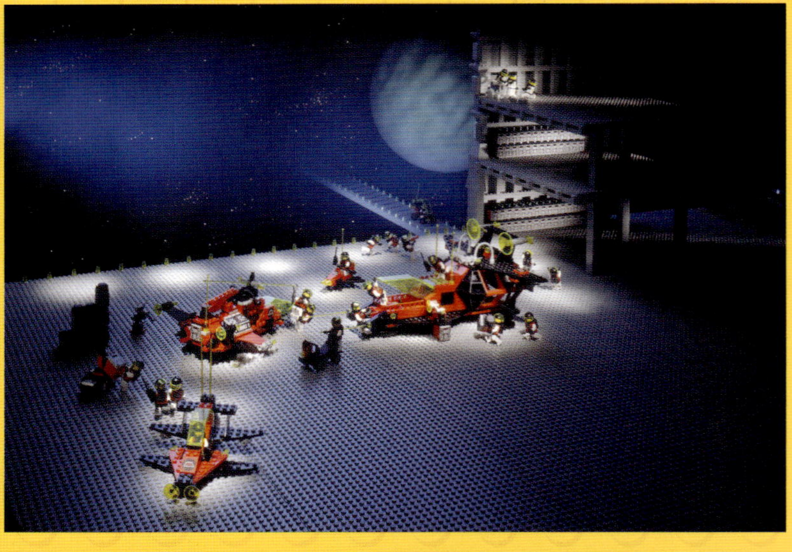

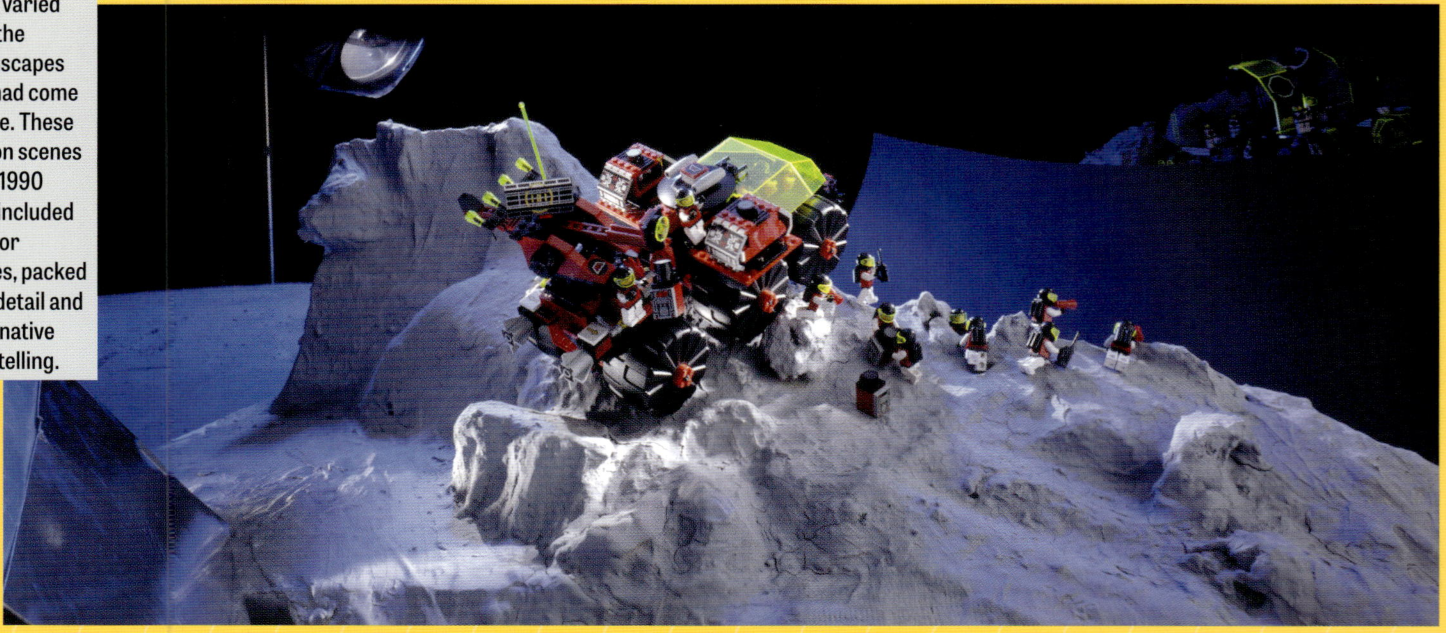

▶ From 1989, the environments became grander and more varied than the moonscapes that had come before. These M:Tron scenes from 1990 even included interior spaces, packed with detail and imaginative storytelling.

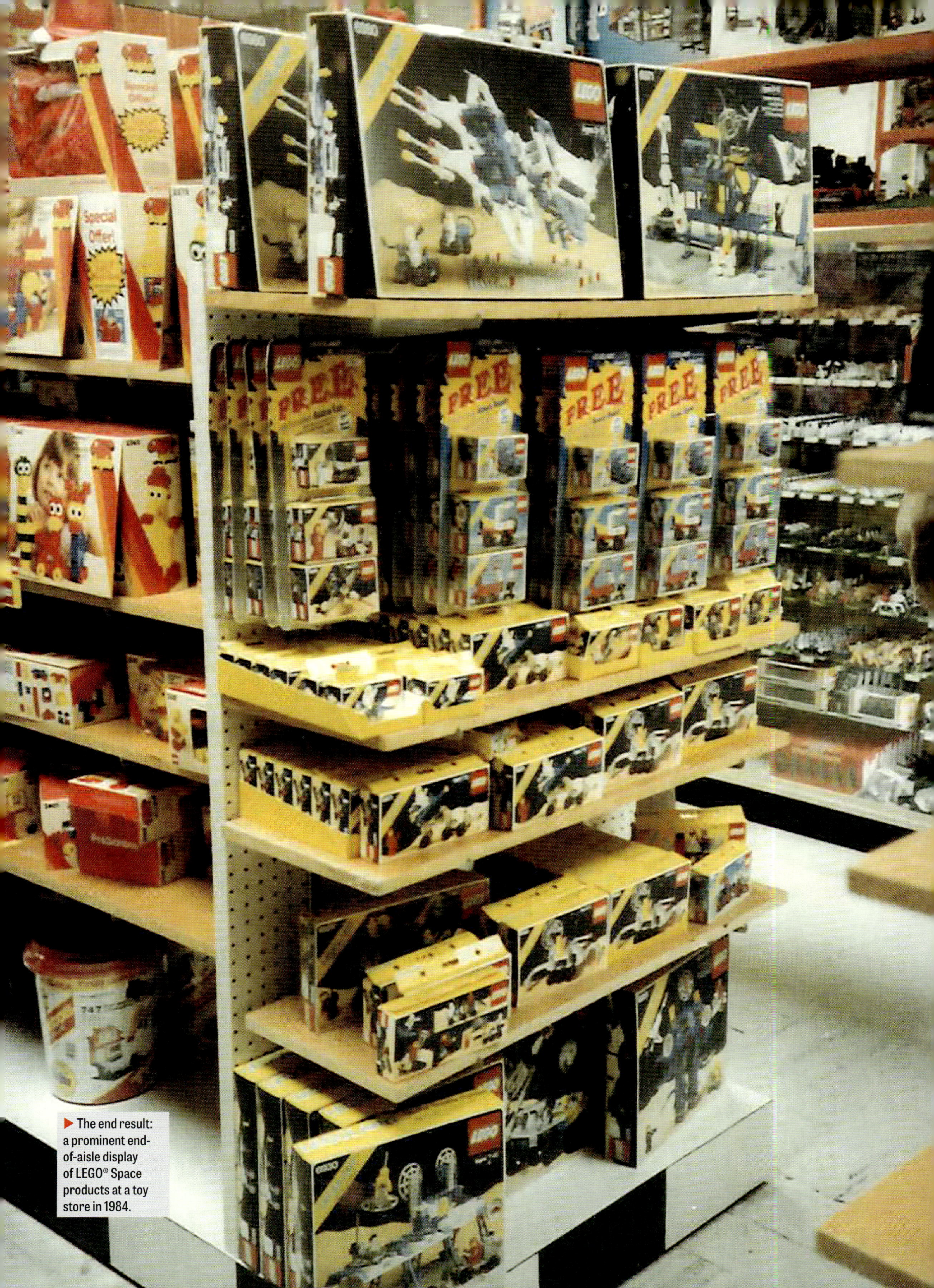

▶ The end result: a prominent end-of-aisle display of LEGO® Space products at a toy store in 1984.

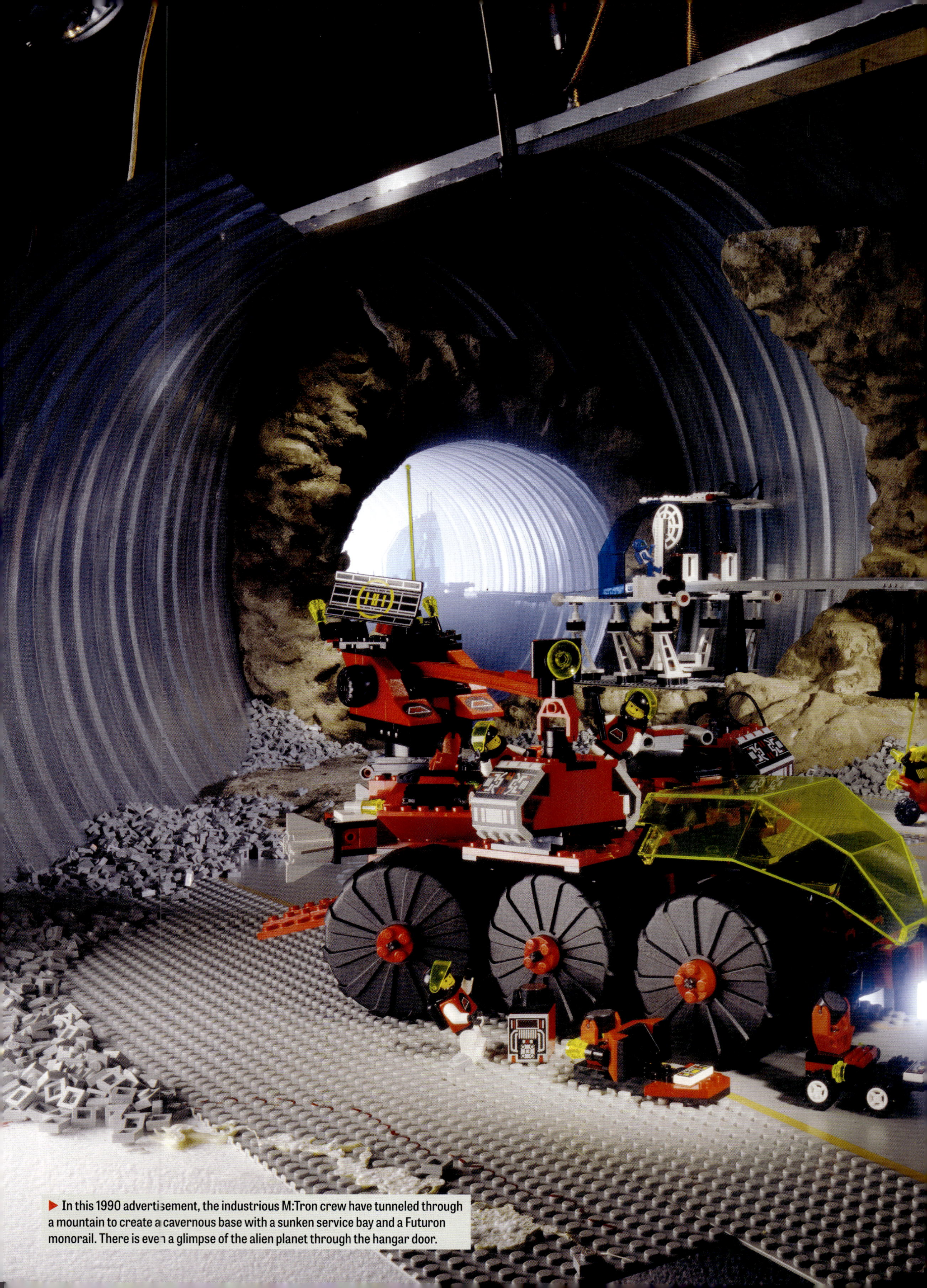

▶ In this 1990 advertisement, the industrious M:Tron crew have tunneled through a mountain to create a cavernous base with a sunken service bay and a Futuron monorail. There is even a glimpse of the alien planet through the hangar door.

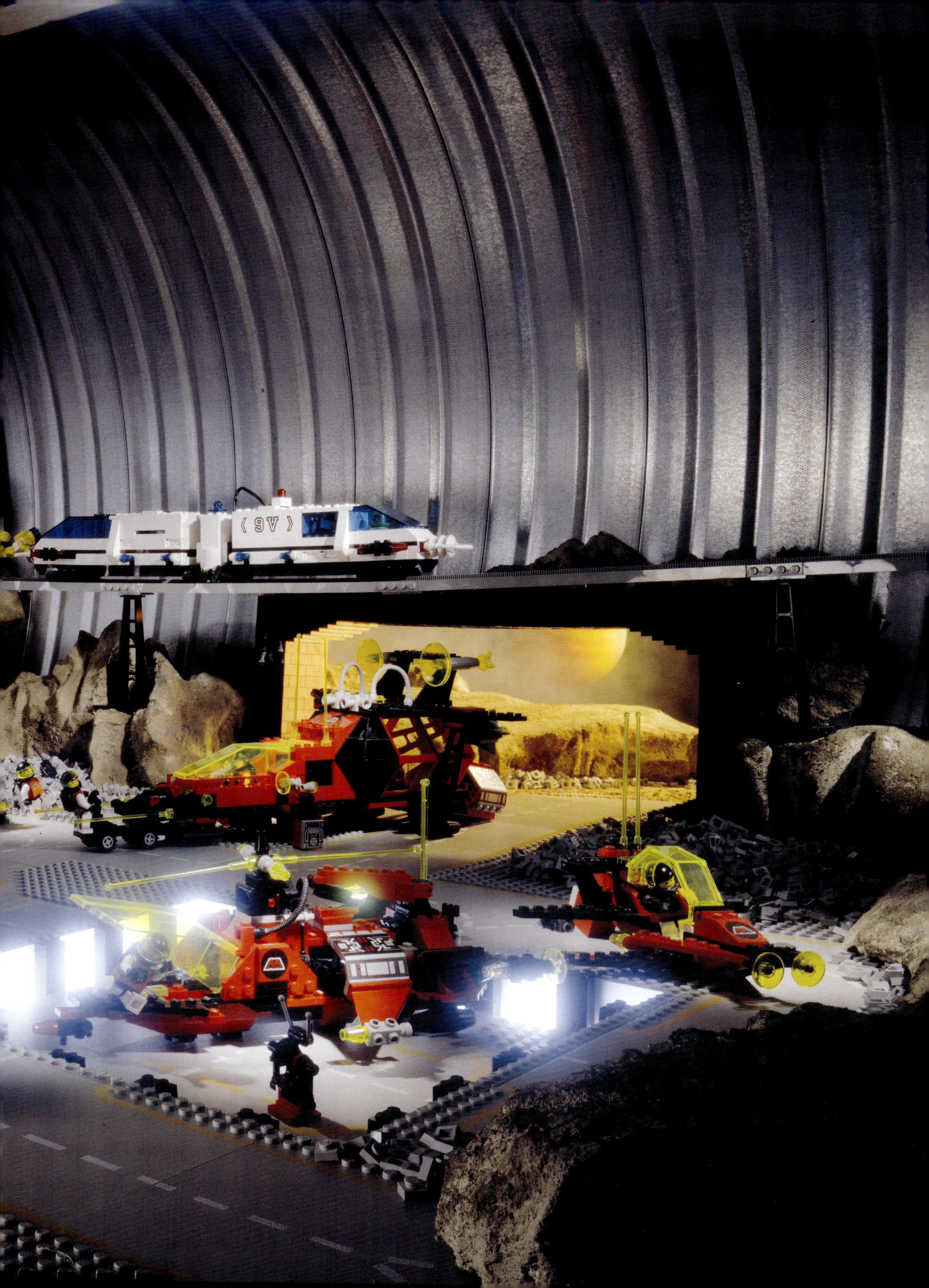
9V

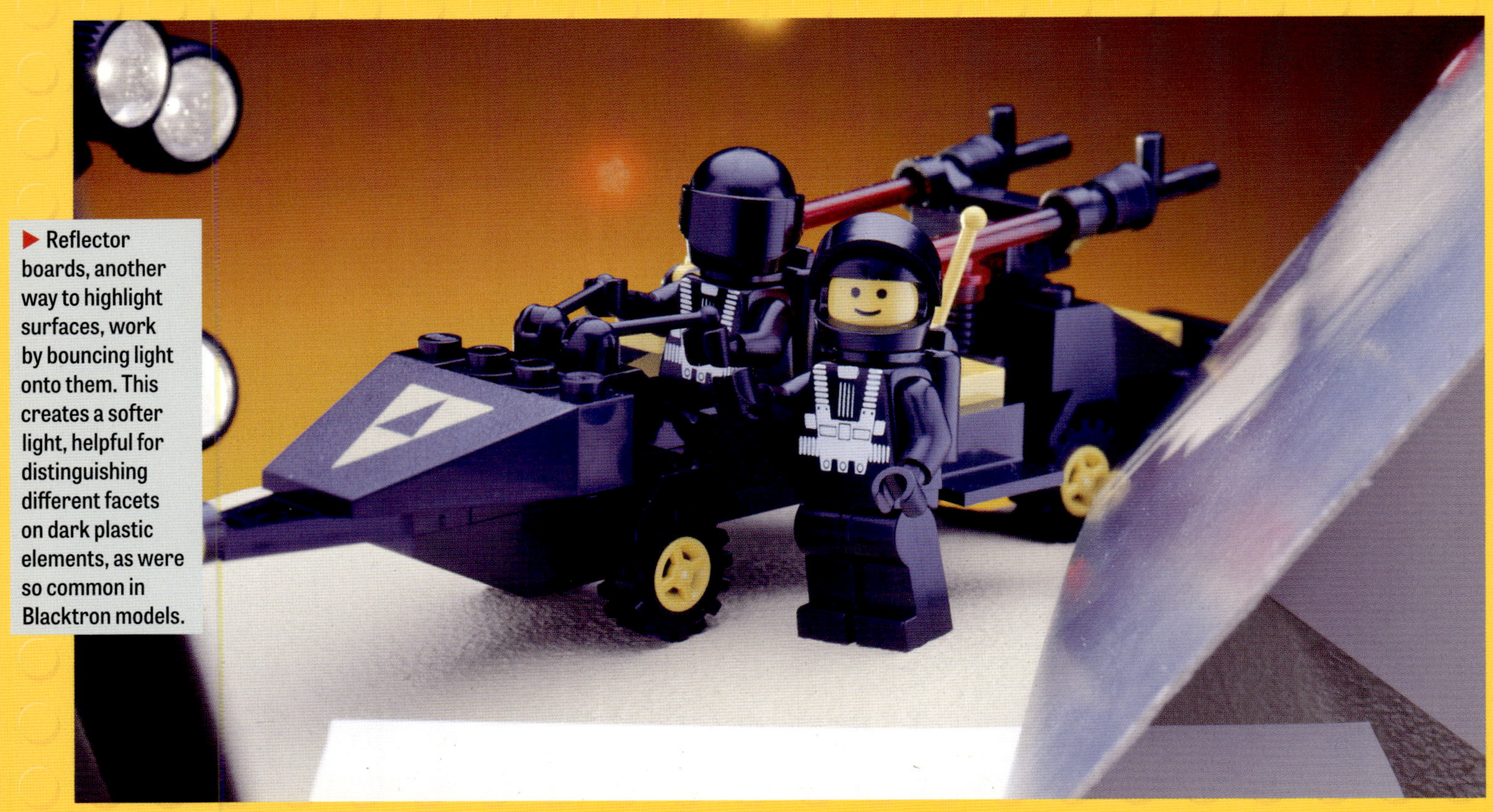

▶ Reflector boards, another way to highlight surfaces, work by bouncing light onto them. This creates a softer light, helpful for distinguishing different facets on dark plastic elements, as were so common in Blacktron models.

"BLACKTRON WAS VERY HARD—PHOTOGRAPHING BLACK MODELS ON A BLACK BACKGROUND—BUT THEY DID A FANTASTIC JOB."

—BJARNE PANDURO TVESKOV, DESIGNER, THE LEGO GROUP, 1985 TO PRESENT

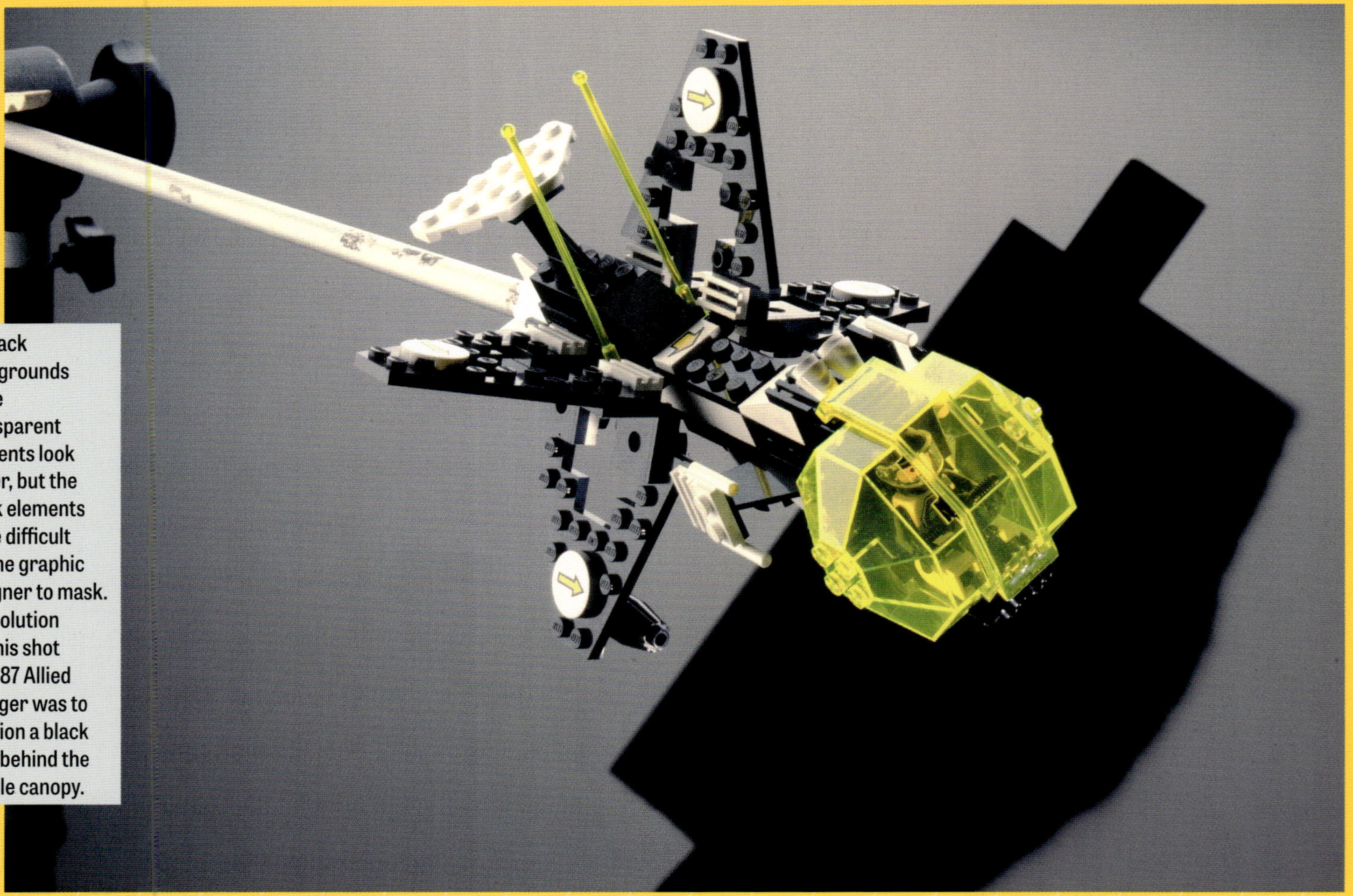

▶ Black backgrounds make transparent elements look better, but the black elements were difficult for the graphic designer to mask. The solution for this shot of 6887 Allied Avenger was to position a black card behind the bubble canopy.

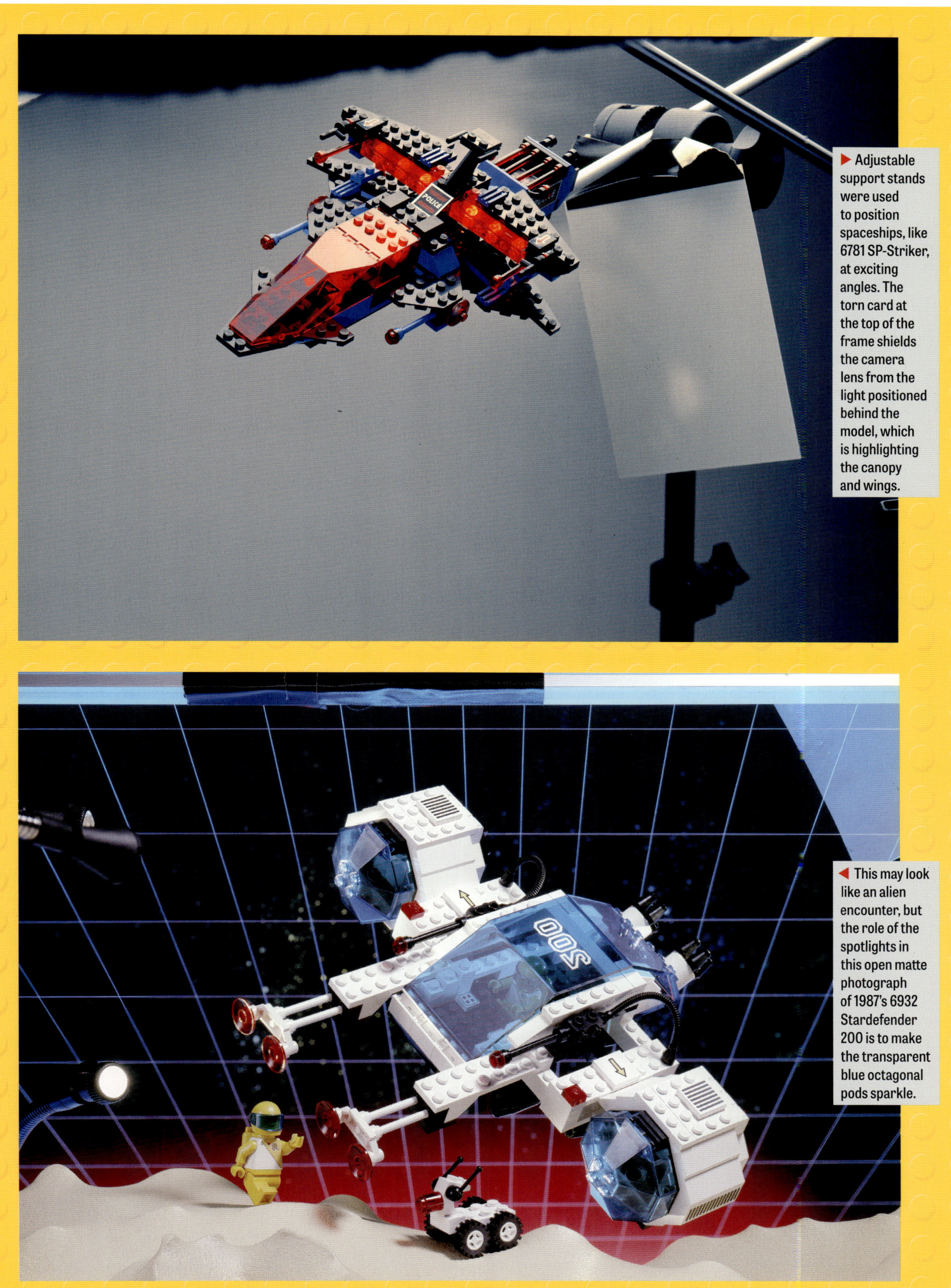

▶ Adjustable support stands were used to position spaceships, like 6781 SP-Striker, at exciting angles. The torn card at the top of the frame shields the camera lens from the light positioned behind the model, which is highlighting the canopy and wings.

◀ This may look like an alien encounter, but the role of the spotlights in this open matte photograph of 1987's 6932 Stardefender 200 is to make the transparent blue octagonal pods sparkle.

CHAPTER 3

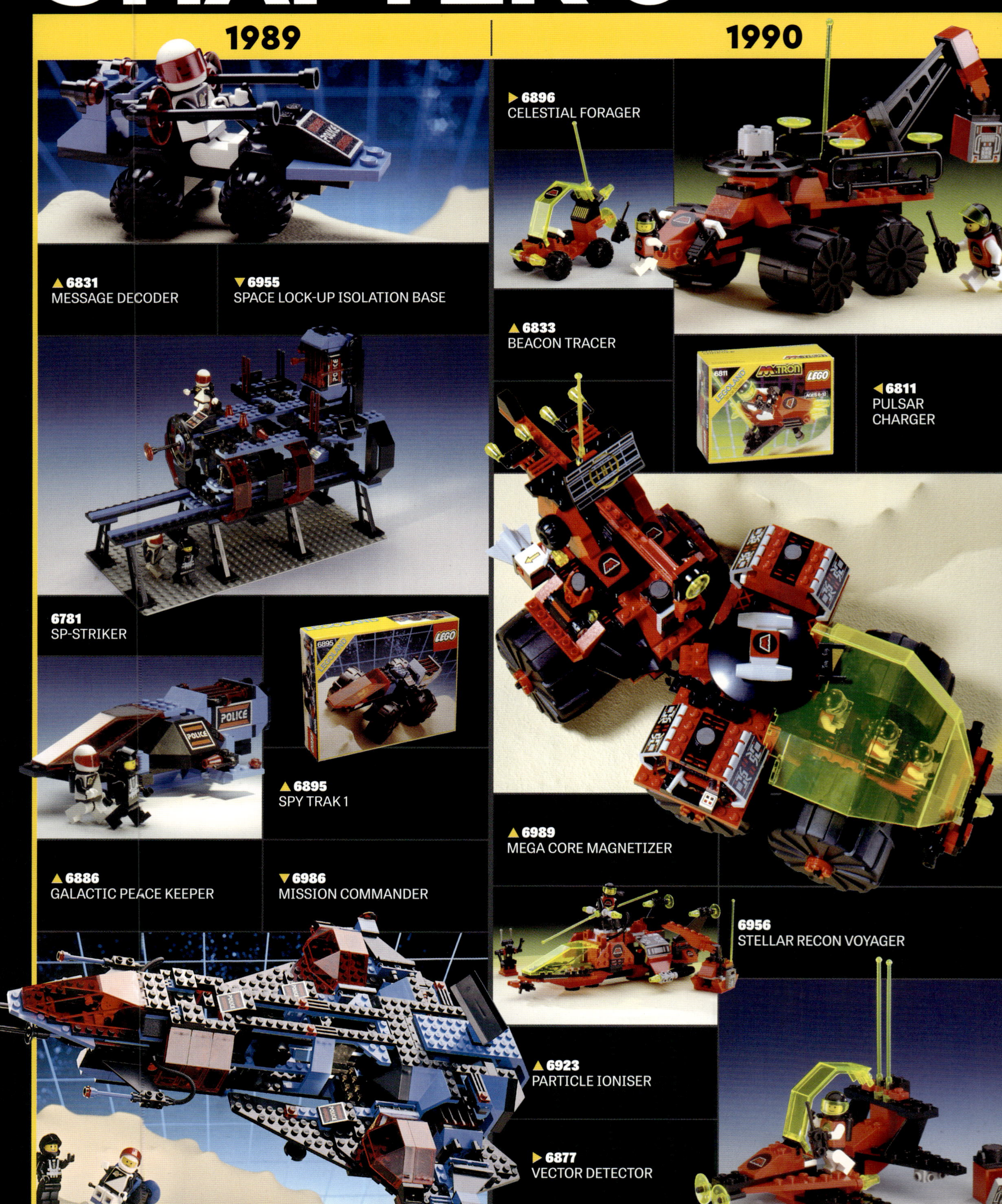

▲ 6831
MESSAGE DECODER

▼ 6955
SPACE LOCK-UP ISOLATION BASE

6781
SP-STRIKER

▲ 6895
SPY TRAK 1

▲ 6886
GALACTIC PEACE KEEPER

▼ 6986
MISSION COMMANDER

▶ 6896
CELESTIAL FORAGER

▲ 6833
BEACON TRACER

◀ 6811
PULSAR CHARGER

▲ 6989
MEGA CORE MAGNETIZER

6956
STELLAR RECON VOYAGER

▲ 6923
PARTICLE IONISER

▶ 6877
VECTOR DETECTOR

1991

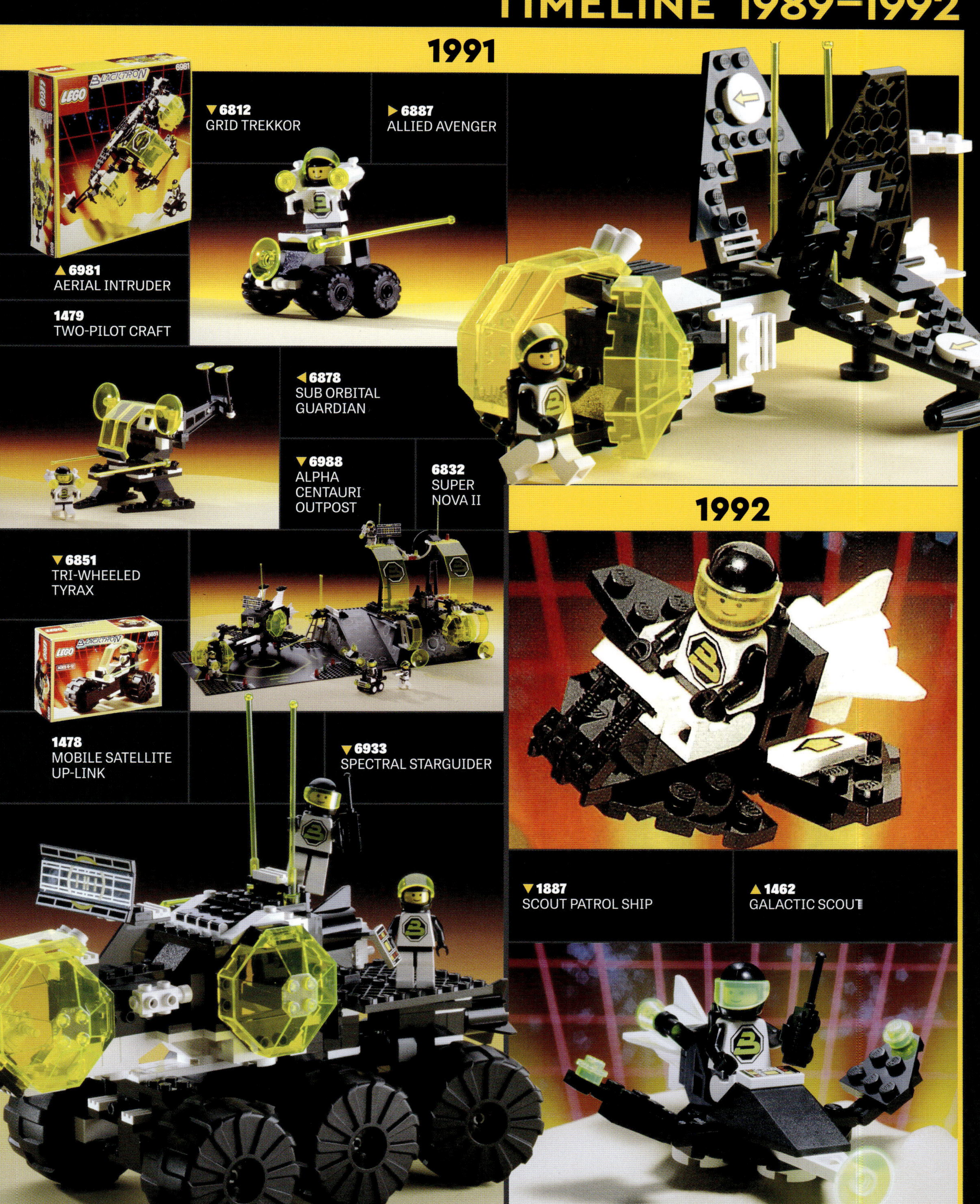

▲6981
AERIAL INTRUDER

1479
TWO-PILOT CRAFT

▼6812
GRID TREKKOR

▶6887
ALLIED AVENGER

◀6878
SUB ORBITAL
GUARDIAN

▼6988
ALPHA
CENTAURI
OUTPOST

6832
SUPER
NOVA II

▼6851
TRI-WHEELED
TYRAX

1478
MOBILE SATELLITE
UP-LINK

▼6933
SPECTRAL STARGUIDER

1992

▼1887
SCOUT PATROL SHIP

▲1462
GALACTIC SCOUT

CHAPTER 3

SPACE POLICE, M:TRON, AND BLACKTRON FUTURE GENERATION

1989–1992

Following the success of Blacktron, exciting new worlds of play continued to expand the universe of LEGO® Space year after year. The 1989 subtheme Space Police was followed in 1990 by M:Tron, and 1991 brought a second wave of Blacktron sets (sometimes marketed as Blacktron Future Generation, as we will distinguish it here).

Segmenting the LEGO Space line into different factions every year presented marketing challenges as to how these would be differentiated on boxes and in catalogs and advertising. The delineation of Futuron and the original Blacktron as "range designations," as they were called internally, was first proposed by the US marketing team in late 1986. The idea was well received in Billund but raised complex considerations for a multinational company which needed to retain some consistency across all markets. Should names such as Blacktron be translated to local languages? Would adding the names of subthemes to boxes confuse customers? What would the impact be upon internal processes, such as product development and budget allocations?

"YOU CAN'T ESCAPE THE LEGO® SPACE POLICE . . . NO MATTER HOW MANY LEGS YOU'VE GOT."

—US TV AD, 1989

Blacktron, with its specially designed and trademarked logo, was considered a special case, and initially there was no intention to introduce further range designations. However, after the Futuron and Blacktron range designations were successfully trialed in US test

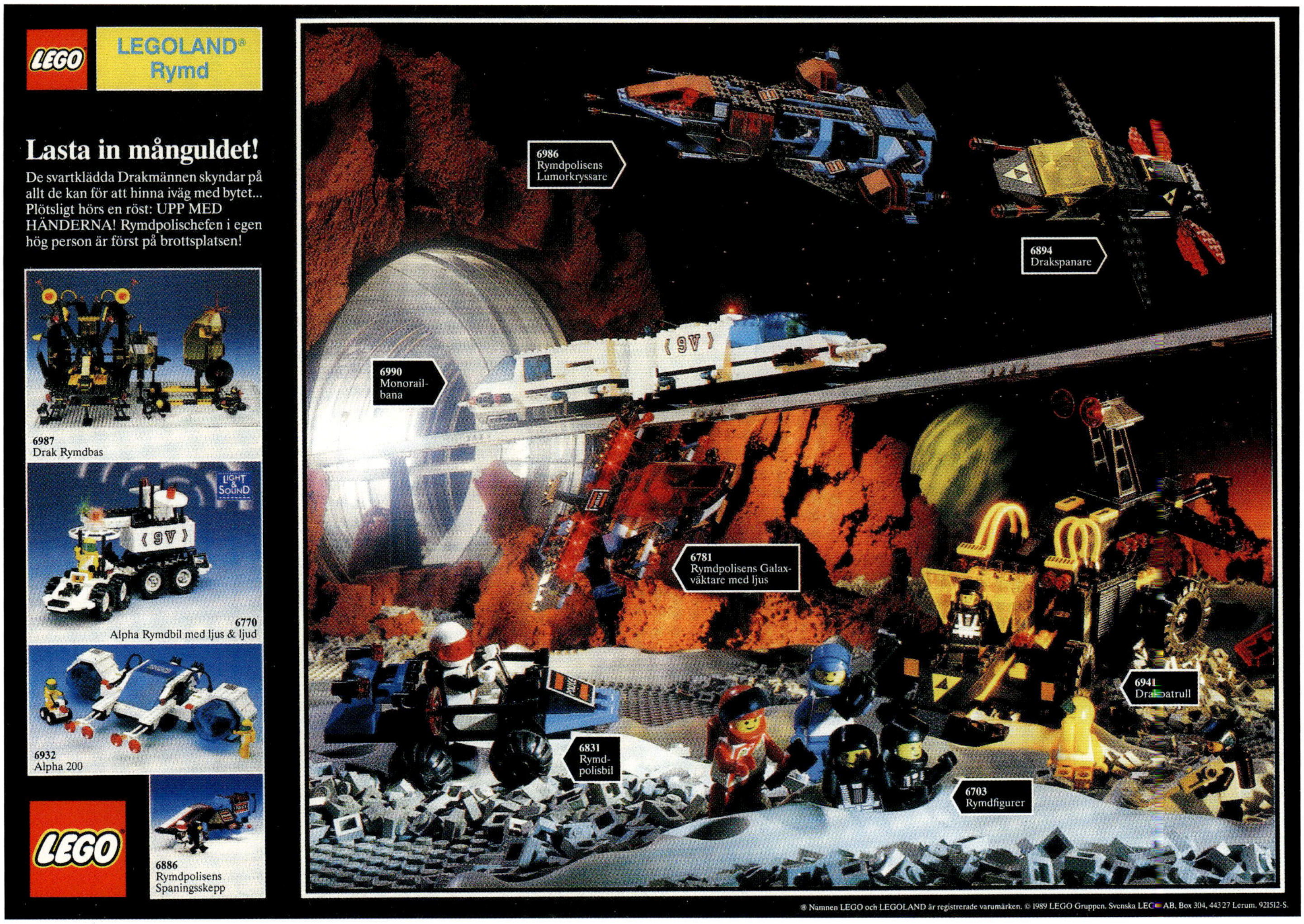

LEGO

LEGOLAND®
Rymd

Lasta in månguldet!

De svartklädda Drakmännen skyndar på allt de kan för att hinna iväg med bytet... Plötsligt hörs en röst: UPP MED HÄNDERNA! Rymdpolischefen i egen hög person är först på brottsplatsen!

6987
Drak Rymdbas

LIGHT & SOUND

6770
Alpha Rymdbil med ljus & ljud

6932
Alpha 200

LEGO

6886
Rymdpolisens Spaningsskepp

6986
Rymdpolisens Lumorkryssare

6894
Drakspanare

6990
Monorail-bana

6781
Rymdpolisens Galax-väktare med ljus

6941
Drakpatrull

6831
Rymd-polisbil

6703
Rymdfigurer

® Namnen LEGO och LEGOLAND är registrerade varumärken. © 1989 LEGO Gruppen. Svenska LEGO AB. Box 304, 443 27 Lerum. 921512-S.

Space Police arrived on the scene in 1989, and the advertising left little doubt as to who was misbehaving. In this exciting Swedish example from 1989, Blacktron—named Dragonmen locally—are making off with Futuron "moon gold."

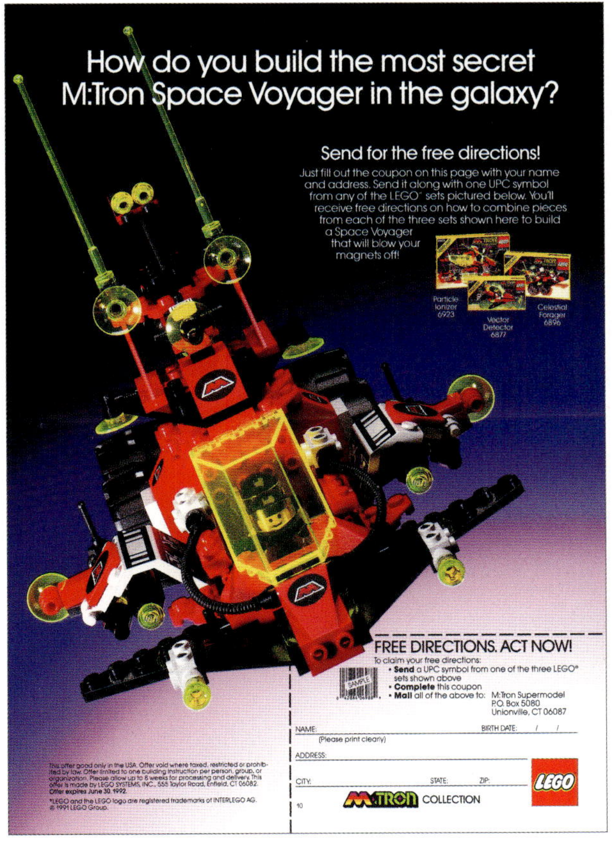

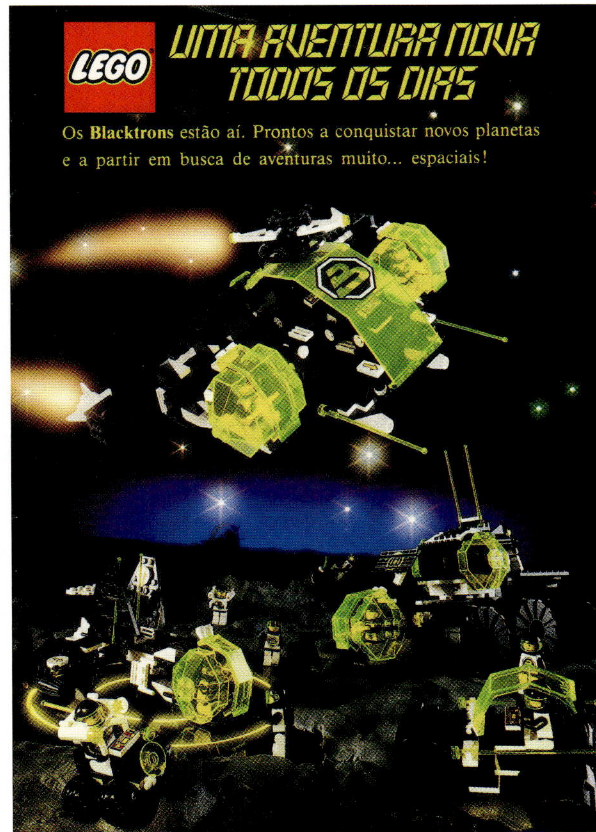

◄ A US promotion in 1991 offered special instructions to build a Space Voyager, using elements from M:Tron sets 6877, 6896, and 6923.

◄ "Ready to conquer new planets . . ." This Portuguese advertisement of 1991 highlights some of the Blacktron Future Generation vehicles with bubble canopies.

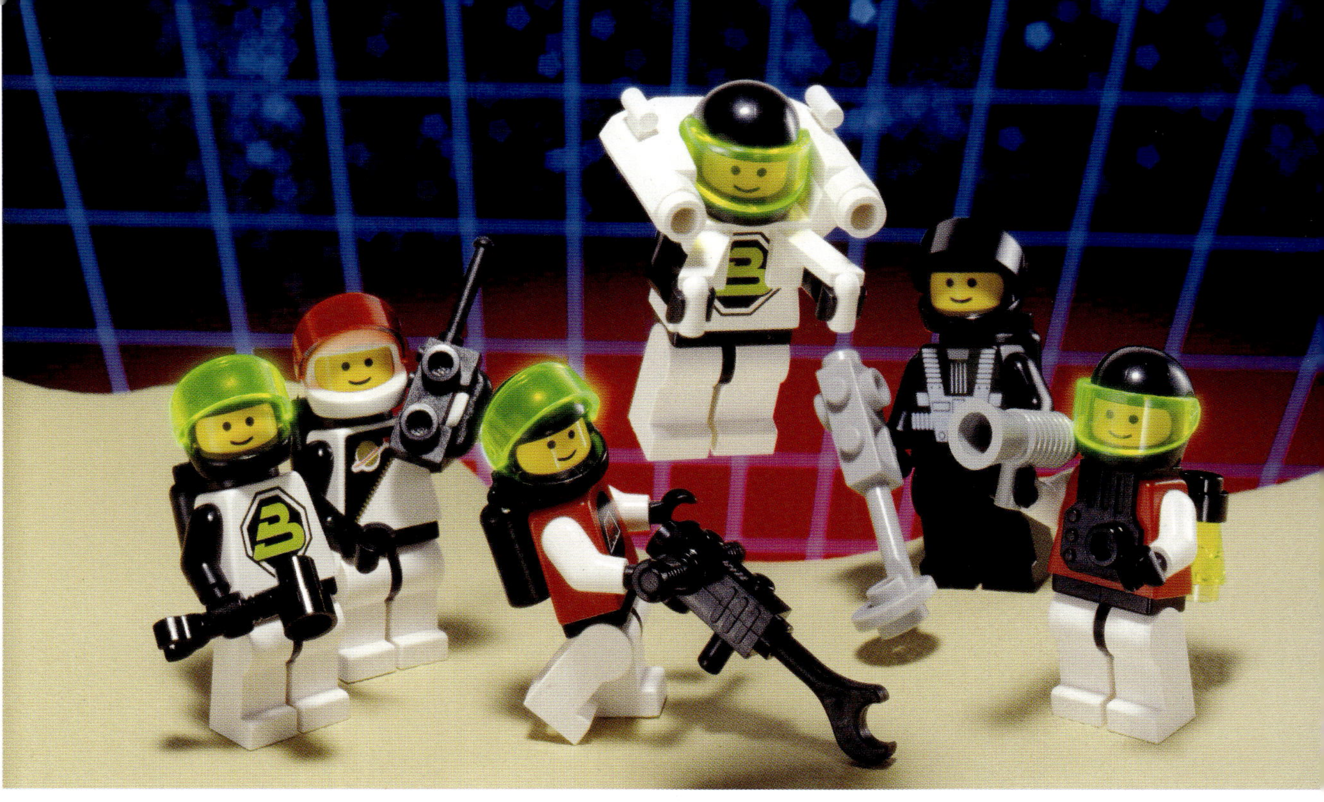

The supplementary set 6704, released in 1991, combined astronauts from four LEGO Space subthemes. An original Blacktron minifigure with chest harness and opaque black visor is included alongside two Blacktron Future Generation minifigures, updated with transparent fluorescent green visors and a "B" logo on white torsos. M:Tron minifigures are quite similar, except they have white arms and red torsos with an "M" logo. Space Police share the same torso decoration as Futuron but have white helmets, gloves, and legs and transparent red visors.

markets in 1987 and rolled out globally in 1988, Space Police also became a range designation in 1989.

The concern around confusing customers with an array of LEGO® brands, themes, and subthemes was also addressed at this time. M:Tron was the final range designation to bear the LEGOLAND® legend on the boxes. Beginning in 1991, this was replaced by a consistent "LEGO® System" brand across LEGO Space, LEGO® Town, and LEGO® Castle, with the range designation logo alongside.

▼LEGO Space boxes had always borne the diagonal LEGOLAND banner. After 1991, this was replaced with consistent LEGO® System branding, as seen on Blacktron Future Generation set 6887 Allied Avenger.

"I NEVER REALIZED BLACKTRON WERE BAD GUYS UNTIL I SAW THEM IN A SPACE POLICE PRISON!"

—FRÉDÉRIC ROLAND ANDRE, SENIOR DESIGN MANAGER, THE LEGO GROUP, 2010 TO PRESENT

EMERGENCY SERVICES

One reason for introducing new subthemes was the desire to secure the longevity of the LEGO® Space line. "We wanted to introduce some subthemes which could be recurring, like LEGO® Town had with police and fire rescue," explains designer Jørn Thomsen, who joined the LEGO Group at the time when those two ideas were being developed as Space Police and Magna-Tron (as it was initially called).

The suggestion in US advertisements from 1988 that Blacktron were the "bad guys" was made explicit worldwide by the introduction of Space Police in 1989, as the sets included a "Space Lock-Up" containing a Blacktron figure. However, some promotional images showed the two factions coexisting in harmony. Ultimately, the choice of narrative was up to the children.

Red spaceships were a wild idea, perhaps inspired by the fire rescue concept, and upon release in 1990, M:Tron were indeed advertised as being a rescue team, but not exclusively. A more exciting aspect had been introduced to the theme: five of the sets included magnets, used to move cargo containers and small vehicles.

Blacktron Future Generation brought a fresh color scheme of black and white with transparent fluorescent green, applied to both the models and the minifigures. The ship designs became more insect-like, with bulbous heads and narrow bodies.

Thank you for your rapifax of 23 October, 1986.

I have with great interest read your arguments in favour of introducing a range designation for some of the 1988 novelties. I cannot and will not argument against the items you put forward - I think they are correct seen from an LSI marketing viewpoint.

However, I find that the increased use of range designations raises the question whether or not we are complicating our consumer message. We would then say to the consumer:

"Go and buy":

the Monorail item 6990

from

the Whitetron range

from

the Space system

from

the LEGOLAND product line

from

LEGO Co.

This is an extremely complicated message and I find that perhaps the problem is primarily the LEGOLAND designation, which does not communicate anything and secondarily, to some extent, the Space system (generic). Logically, we should then discontinue the LEGOLAND and the Space system designations.

- 2 -

An internal communication of 1986 responds to the US marketing team's suggestion that Blacktron and Futuron—at this time known as Whitetron—become specific "range designations" within the LEGO Space line. It was observed that the existing LEGOLAND designation now held little meaning for consumers.

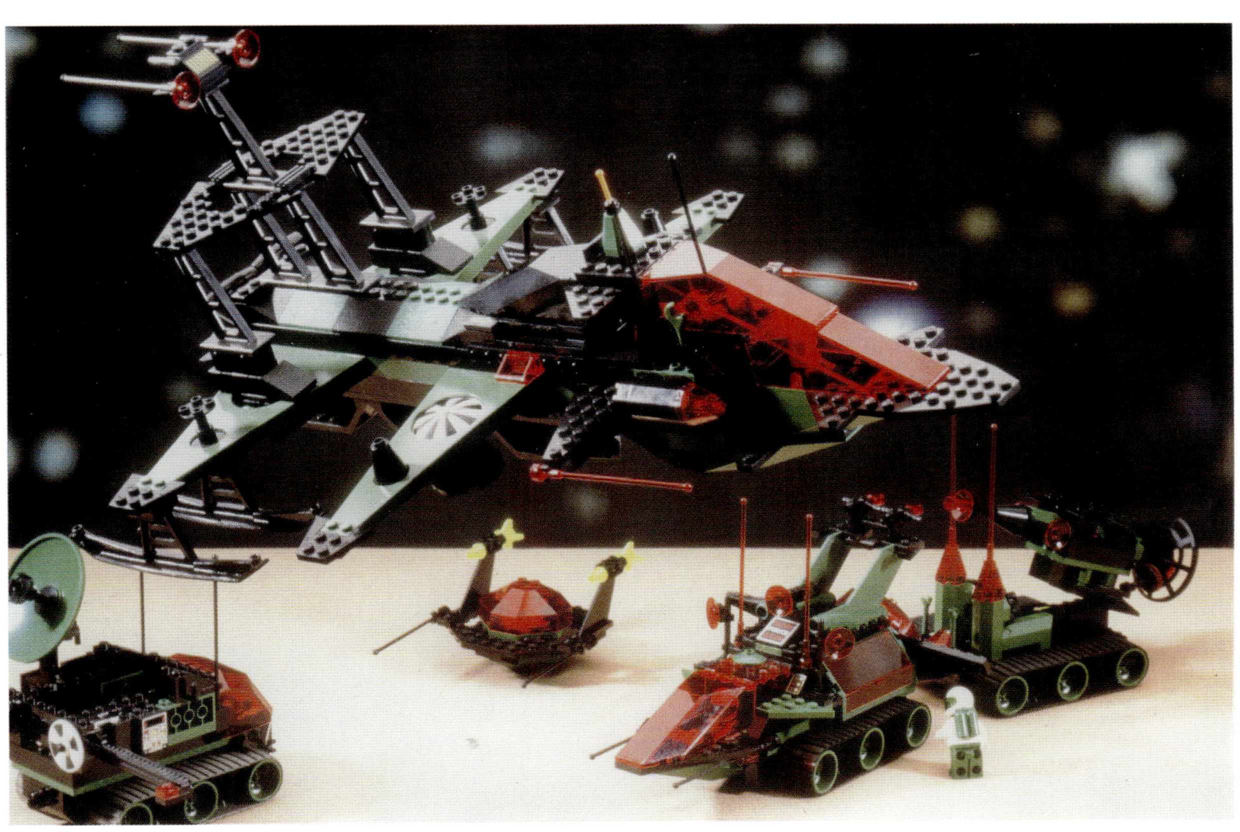

An early concept for the Space Police subtheme from around 1986, which includes some wing and tail elements that were never released. However, the family of octagonal bricks visible on the rocket launcher were eventually released in the LEGO® Aquazone theme in 1995.

"THE SECOND BLACKTRON COLOR SCHEME IS, TO THIS DAY, ONE OF THE BEST OF ANY LEGO LINES, IN MY OPINION."

—WILLIAM THOROGOOD, VICE PRESIDENT OF DESIGN, NEW BUSINESS, THE LEGO GROUP, 2004 TO PRESENT

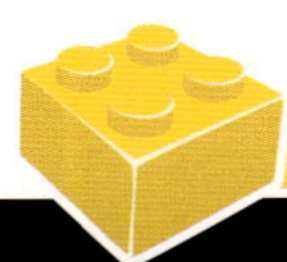

INTERVIEW WITH

BJARNE PANDURO TVESKOV

WHAT DROVE THE CHANGE TO BRING IN SO MANY NEW SUBTHEMES?

Even without the levels of customer involvement and user-centered design that we have today, the LEGO® brand was growing like crazy, and it was becoming a more multinational company. The US had been a fairly small LEGO market, but suddenly it really started taking off.

Also, the toy business became more story driven during the 1980s, and there was a sense that you needed to be able to "see" what was new. There was a bigger need for novelty coming from the markets. For instance, with Blacktron doing well, it was very clear why this was: something was happening—there were "bad guys." Hence this annual cycle developed of having something new happening every year—and I also think that's why it became more distinct: to have a new line every year, a new color, a new backstory, and so on.

WHAT ARE YOUR RECOLLECTIONS OF DEVELOPING SPACE POLICE?

I remember Space Police was supposed to be white with transparent green, to be like Futuron, but that looked very much like the color of German police cars, so it was changed to a whole different color scheme of blue, black, and transparent red.

The prison cell from Space Police, with its red transparent laser bars and a Blacktron minifigure inside, lends itself to role-play and is a good modular unit for combining two or more sets, since the cells can be interchanged between the different Space Police models.

My favorite model that I ever designed is also from that subtheme: 6781 SP-Striker. It has a little monitor inside that lights up with a schematic of the model itself. The model does not include the sound brick, as it would have been more expensive to include both, and as Jens Nygaard Knudsen remarked at one point, "Lights are much more important, since the kids can always make their own sounds with their mouths!"

We also worked with some external companies on having hologram stickers for Space Police—that was a big 1980s fad, to have stickers with depth in them—but it fell through. I think their surface would scratch and did not live up to quality requirements.

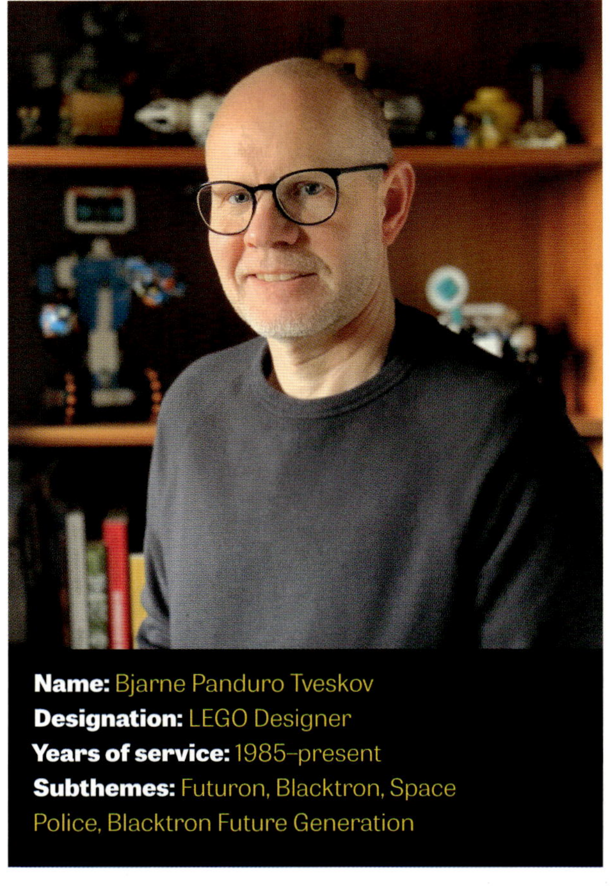

Name: Bjarne Panduro Tveskov
Designation: LEGO Designer
Years of service: 1985–present
Subthemes: Futuron, Blacktron, Space Police, Blacktron Future Generation

DID YOU WORK ON M:TRON AND BLACKTRON FUTURE GENERATION?

I worked on the concept for M:Tron, but I didn't do any of the final models. We started using color more boldly, with the red. That color had really been outside of our radar; LEGO® Space was white, gray, blue, and some black. And then we thought, red spaceships? Maybe that could work.

It was also exciting to get a new fluorescent material in the transparent pieces from the German company Bayer AG. This was called LISA [short for *Licht Sammler*, meaning "light collector"]. We used LISA in green in M:Tron and Blacktron Future Generation but also had it in some other colors, like orange and purple.

I did design a few of the Blacktron Future Generation sets. There was a brief to do a new round of Blacktron, but to make the models somewhat less dark, and I built many color combinations to test this. One that I quite liked had little brick-built rainbows integrated into the designs, with fluorescent orange cockpits. ■

▶ A wide variety of M:Tron development models from 1987, prior to magnets being introduced into the concept.

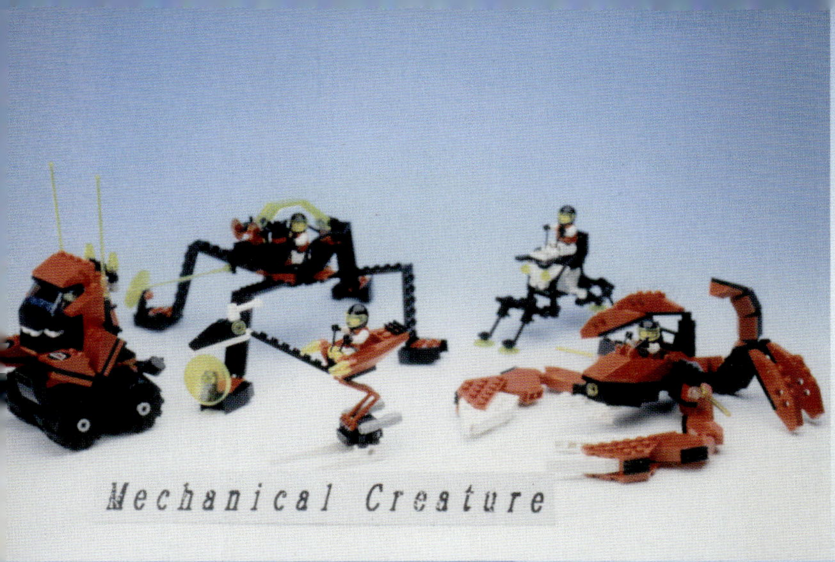

Mechanical Creature

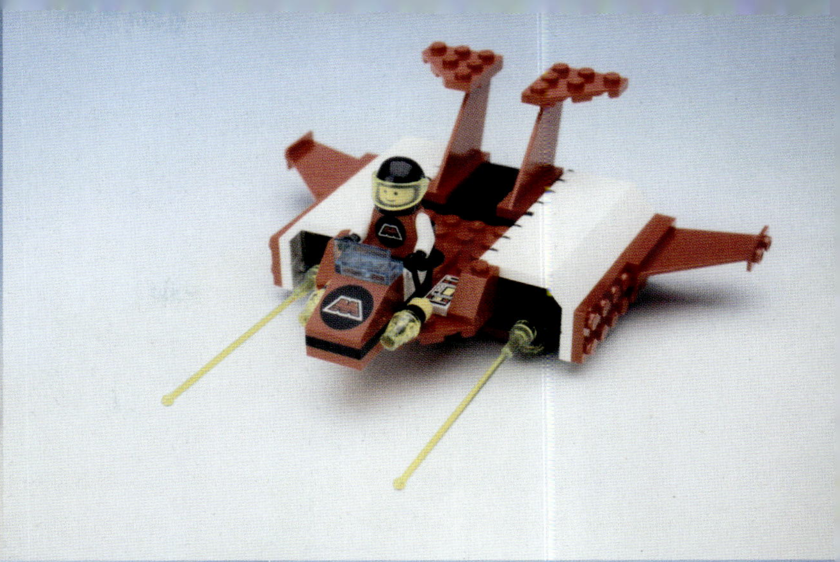

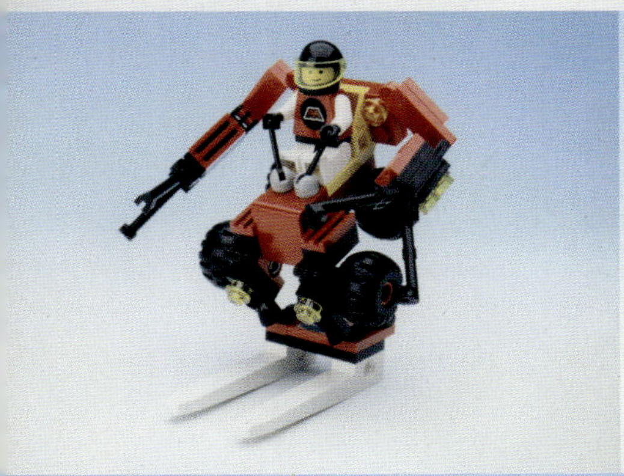

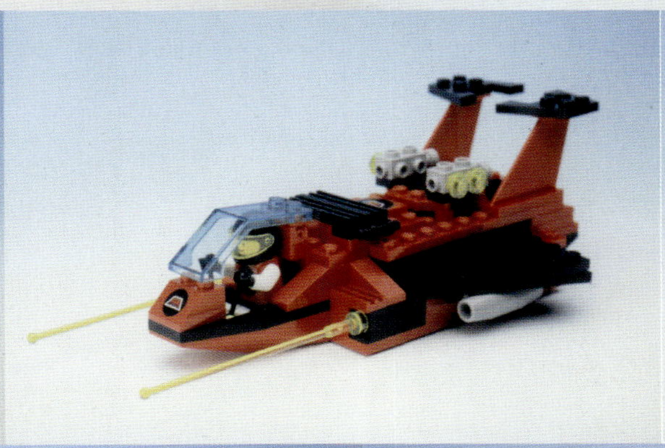

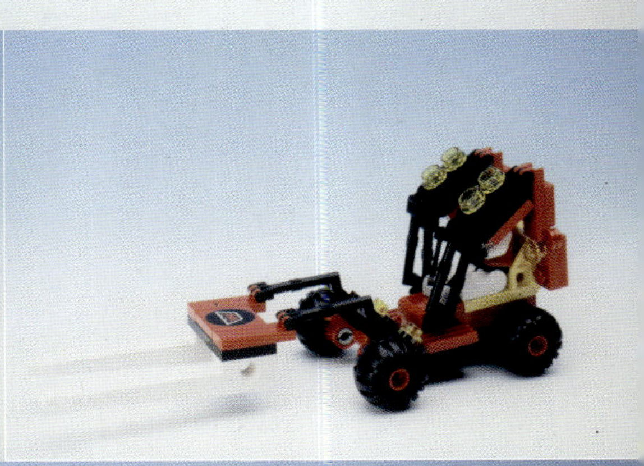

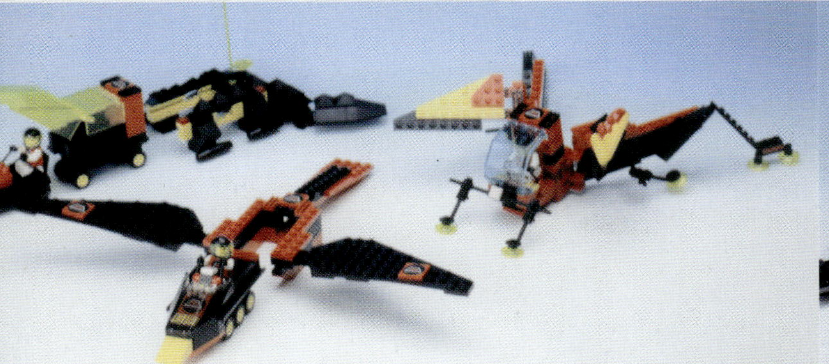

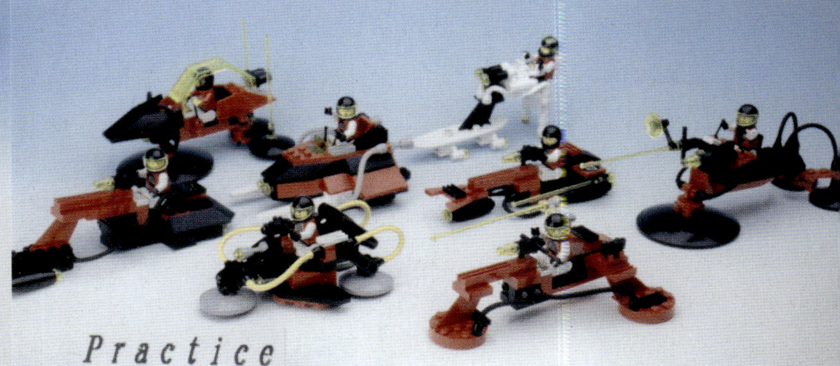

Practice

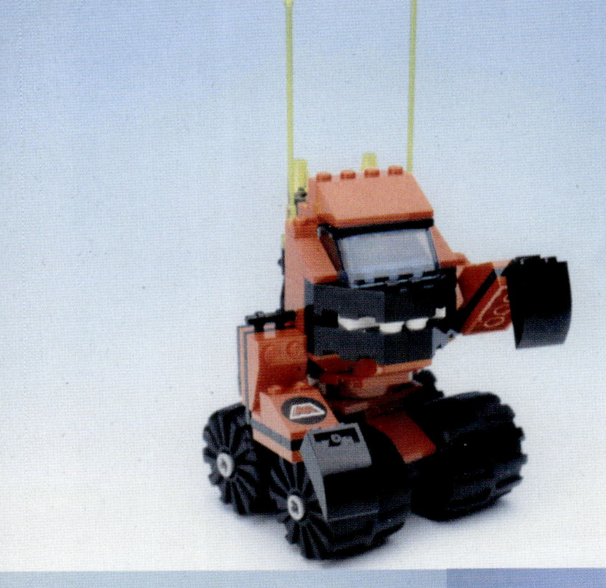

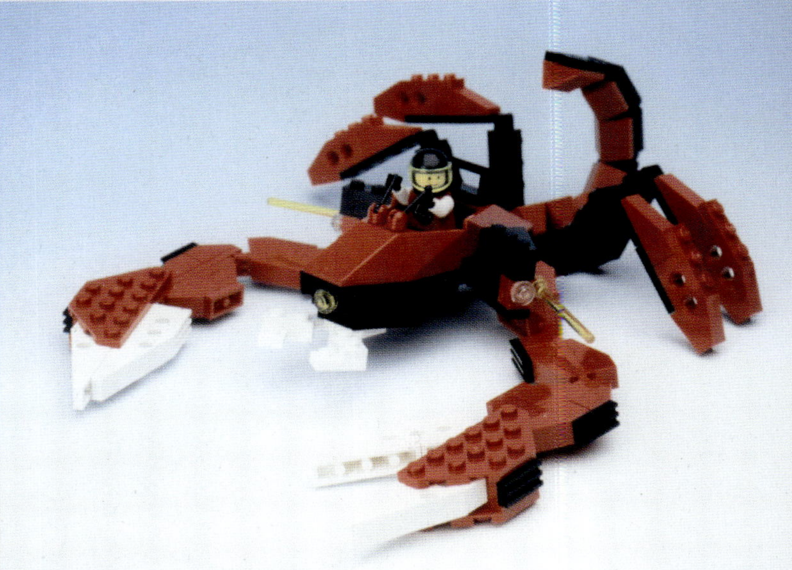

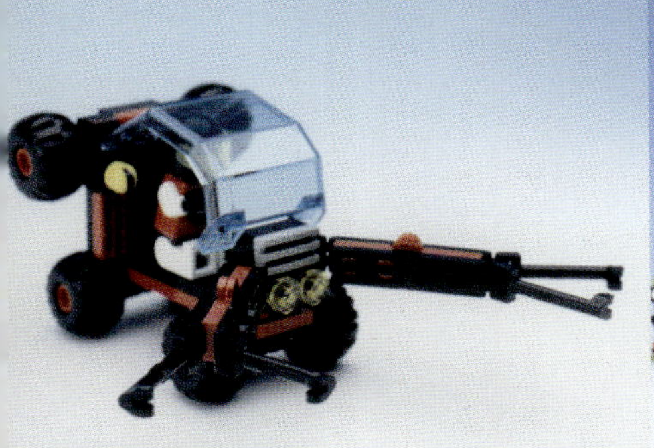

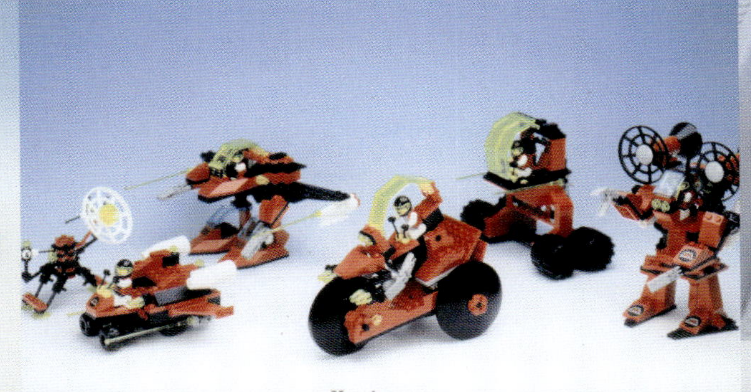

Union

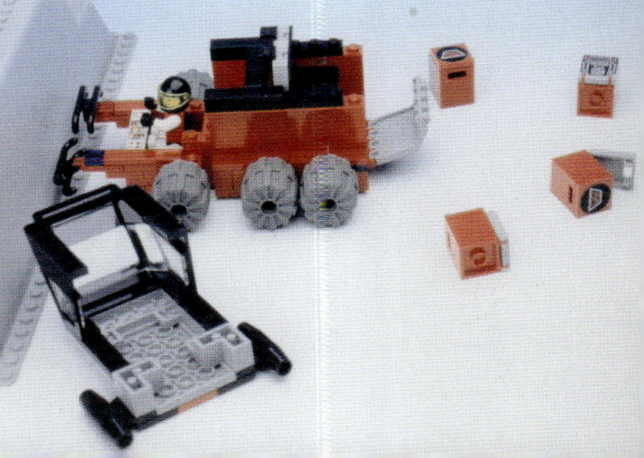

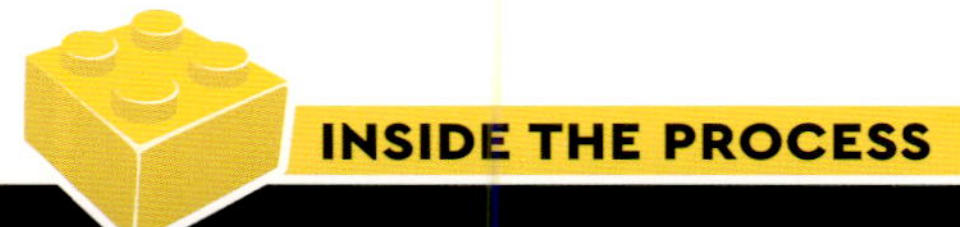

INSIDE THE PROCESS

ALTERNATE UNIVERSE

More than 150 LEGO® Space products were released between 1978 and 1992, but hundreds of ideas never made it to market: concepts, sketch models, element prototypes, novel color schemes, and even entire subthemes. Here is a selection from the archives of the LEGO Group, some published for the first time.

Many of these were never intended for sale. For example, some models were built only to show a potential new color scheme or the building possibilities provided by new elements. This was to protect patents; the images would be sent to an external notary to document the existence of the new ideas.

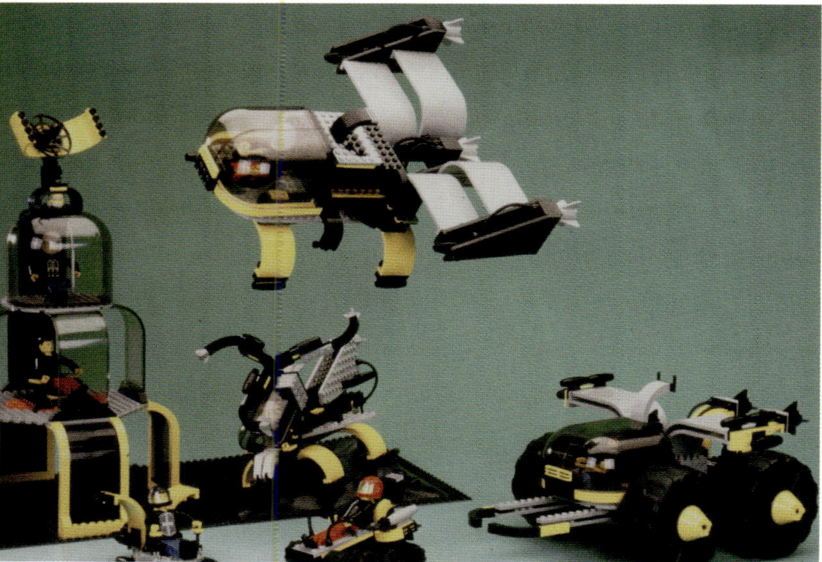

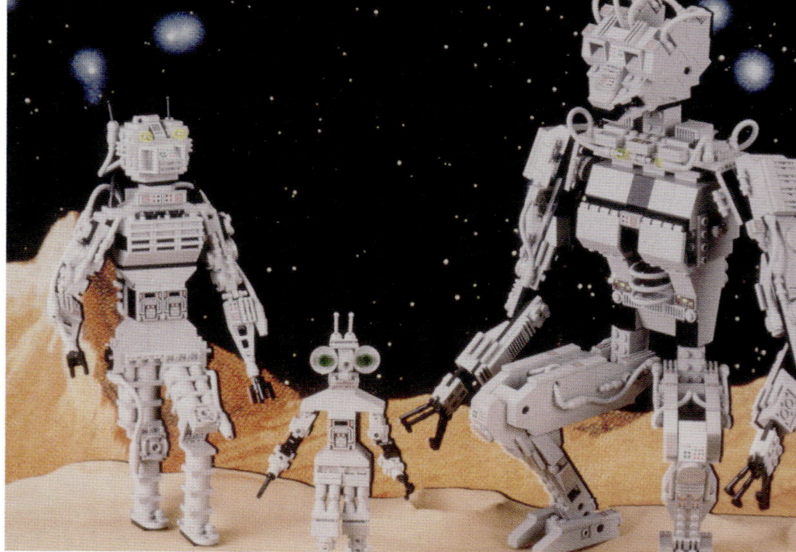

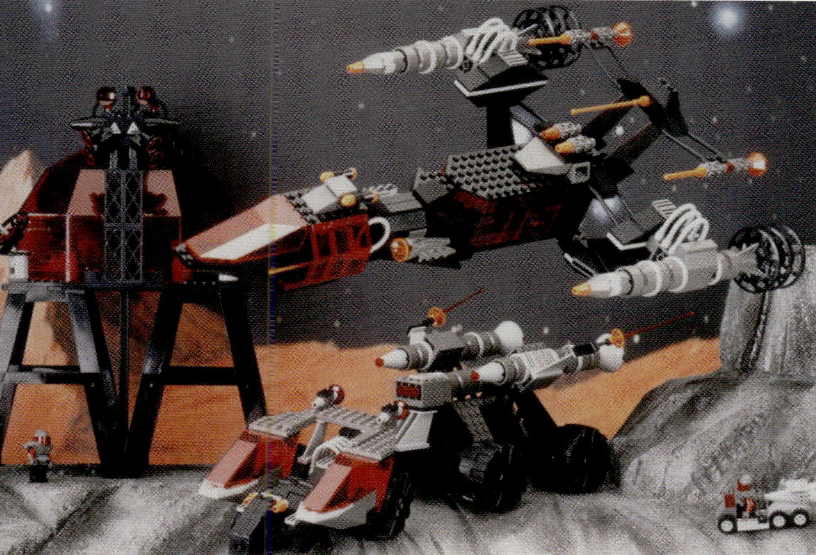

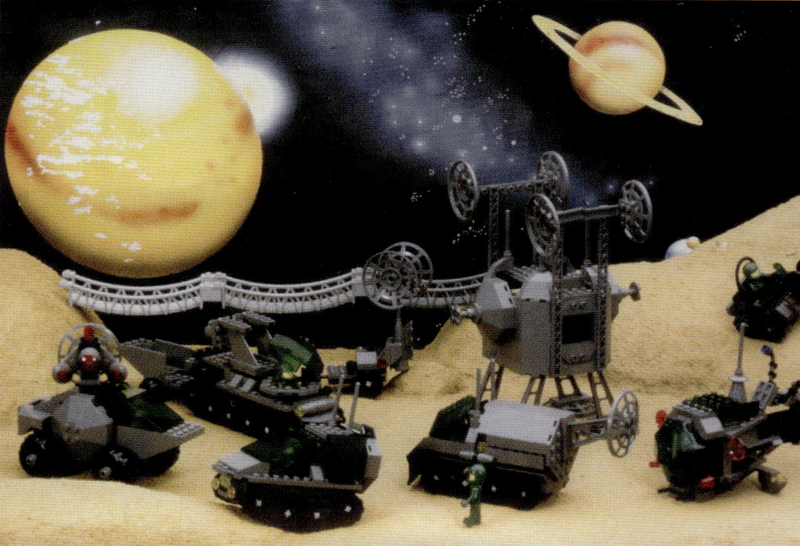

"WE WOULD OFTEN MAKE PROTOTYPE ELEMENTS IN A RANGE OF COLORS, SO IT WAS ALWAYS AN ATTRACTIVE PROSPECT TO TRY OUT NEW AND UNUSUAL COLOR COMBINATIONS—SOMETHING QUITE UNIQUE TO THE LEGO SPACE LINE."

—BJARNE PANDURO TVESKOV, DESIGNER, THE LEGO GROUP, 1985 TO PRESENT

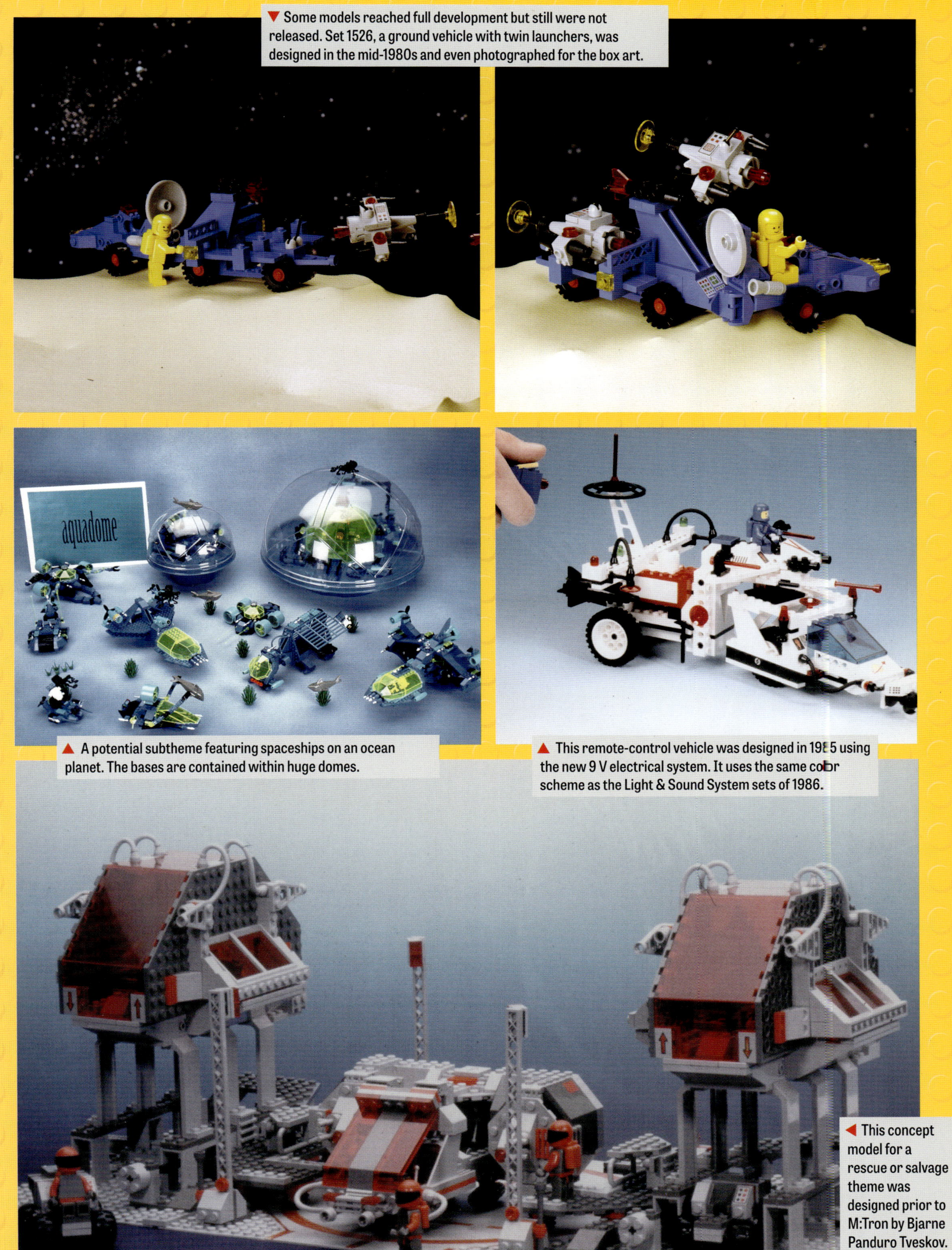

▼ Some models reached full development but still were not released. Set 1526, a ground vehicle with twin launchers, was designed in the mid-1980s and even photographed for the box art.

▲ A potential subtheme featuring spaceships on an ocean planet. The bases are contained within huge domes.

▲ This remote-control vehicle was designed in 1985 using the new 9 V electrical system. It uses the same color scheme as the Light & Sound System sets of 1986.

◀ This concept model for a rescue or salvage theme was designed prior to M:Tron by Bjarne Panduro Tveskov.

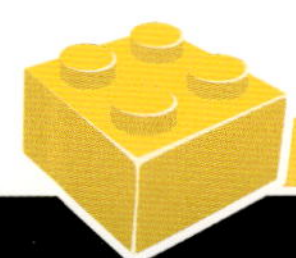

INTERVIEW WITH

JØRN THOMSEN

DID YOU GROW UP PLAYING WITH LEGO® PRODUCTS?

I had a lot as a child. I always got small LEGO® boxes when I went shopping in a nearby big city, or as a reward, such as visiting the dentist. I built all the things that I thought were fascinating: trains, trucks, tractors, buildings, and motorcycles. I used rubber bands to transfer power from one wheel to another, because at that time there were no gears in the LEGO range. The smaller LEGO elements were also exciting to my parents. I remember they used them to build the floor plan of their new house.

WHAT LED YOU TO JOIN THE LEGO GROUP?

I trained as an electrician and worked as one for thirteen years. One day I saw an unusual job advertisement in the Sunday paper. It said that the LEGO Group was looking for "an astronaut who would like to fly around in his own spaceship." It was my dream job, and I had a lot of ideas, so I set about applying for the position.

I sent in an application and got an interview with Jens Nygaard Knudsen. I brought a large space vessel with me. It was completely red, as I didn't have a very large selection of LEGO bricks, but even with a limited assortment, you can create play and a story. My idea was a floating planet, so everything happened underwater.

Jens was very interested, and I was told that I would be sent some LEGO elements the following week. I then had a week to build a space theme, photograph it, and create a story. Along with the LEGO elements, I also received tape and other things so that I could send the model back securely. However, we chose to drive for an hour to Billund to deliver it in person, so that I could be sure that it would arrive whole!

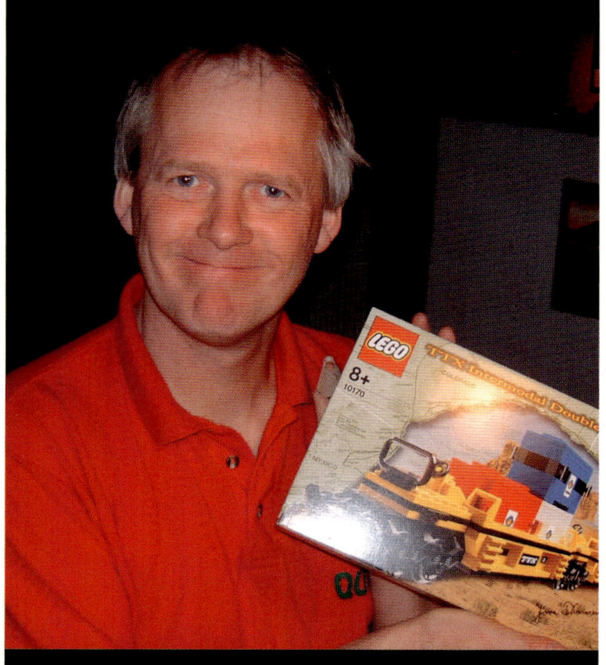

Name: Jørn Thomsen
Designation: LEGO Designer
Years of service: 1987–2023
Subthemes: Futuron, Space Police, M:Tron, Blacktron Future Generation, Space Police 2, Ice Planet 2002, Unitron, Spyrius, Life on Mars

WHAT WAS STARTING THE JOB LIKE?

I was given a welcoming reception where I was shown my working space with lots of LEGO elements, and there was a lot of display space. There were shelves in the hallways from floor to ceiling covered with small models; these were thoroughly cleaned by Jens several times a year!

I was excited to meet the designers from LEGO® Pirates, LEGO® Castle, and LEGO® Space over coffee that first morning. I thought that this meeting was happening because I had started, but it turned out there was one every morning in Jens's group. There would be pleasant conversation, and everyone was kept informed about what was being worked on. On

BEFORE 1989, LEGO® SPACE WAS ABOUT ROCKET TRANSPORTATION, SETTLEMENT ON ALIEN PLANETS, AND EXPLORATION. BLACKTRON AND FUTURON THEN INTRODUCED THE CONCEPT OF "GOODIES AND BADDIES," SO IT WAS A NATURAL NEXT STEP TO COME UP WITH SPACE POLICE.

top of the daily coffee meetings, Jens came around once a week to see what we were working on.

There were times when I helped with functions or alternate models for LEGO Castle and LEGO Pirates, but mostly I built for LEGO Space. The very same week that I was hired I started building the latest models, which turned out to be Space Police.

NEW IDEAS

WHAT ARE YOUR RECOLLECTIONS OF CREATING SPACE POLICE?

Before 1989, LEGO® Space was about rocket transportation, settlement on alien planets, and exploration. Blacktron and Futuron then introduced the concept of "goodies and baddies," so it was a natural next step to come up with Space Police.

My first model was a vehicle where all four wheels could be steered by a rocking mechanism. There was a policy at the time that we could use only a limited range of LEGO® Technic elements; otherwise, they thought the models might look too much like one another. Gears were among the reserved elements, and so to create the steering function I used a cross shaft instead, with two sloped bricks that push against it when you tilt the central section of the model, causing all four wheels to turn.

DID KNUDSEN STILL DESIGN MODELS DURING YOUR TIME?

Although Jens was our manager, he did also build models himself and came up with suggestions for new themes. He also had many ideas for new elements and made prototypes of them using pliers, hobby knives, and glue. They weren't pretty, but you could clearly see the idea! The element workshop would then make an actual prototype, which could take a week, but often when they would then show it to Jens, this instantly sparked further ideas. Jens constantly had new ideas! So often the pliers would come out again and the prototype was modified. Soon the procedure was improved, and the workshop did not spend as much time creating the first prototype. It would therefore be a little rougher, so that when they showed it to Jens, it was more acceptable if he changed it.

Consequently, we were the group that had the most new elements that had been designed but not yet launched. We also had many new colors of existing elements, as Jens was very insistent that we had the right elements to build with. As a result, we had continual visits from the other designers to see what new elements were available!

HOW DID YOU COME UP WITH NEW IDEAS? WAS THIS DONE IN CONJUNCTION WITH THE MARKETING TEAM?

Today, marketing is an integral part of the design team but at that time, they sat in a building somewhat far away and came to visit. Marketing regularly tested different themes with children and often concluded that a certain theme had "won" the test. However, Jens liked all themes, and his opinion was that even if only some of the children liked a certain theme, he still felt that it should be produced.

Jens often spoke of "free development," where new ideas could flow. Jens was always bubbling with new ideas, and all ideas had to be explored! For

For his first set, 6895 Spy Trak 1, Thomsen developed an ingenious yet simple steering mechanism. It was adapted for use in several other sets.

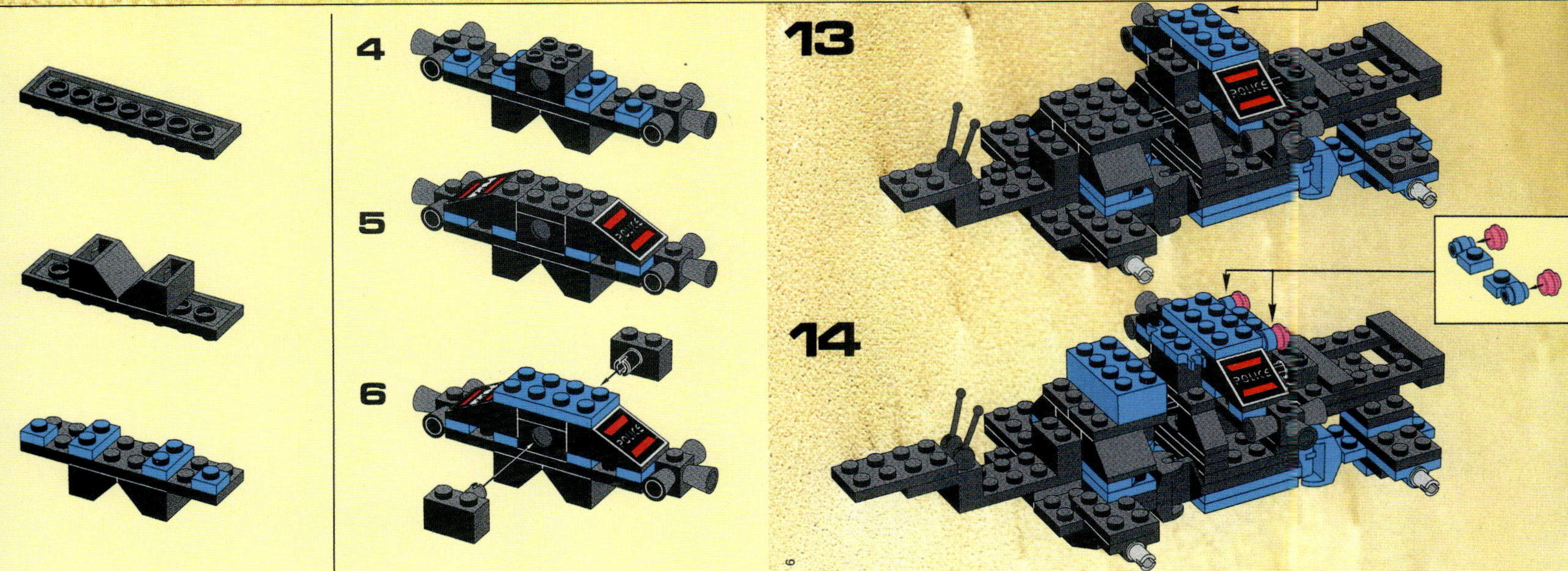

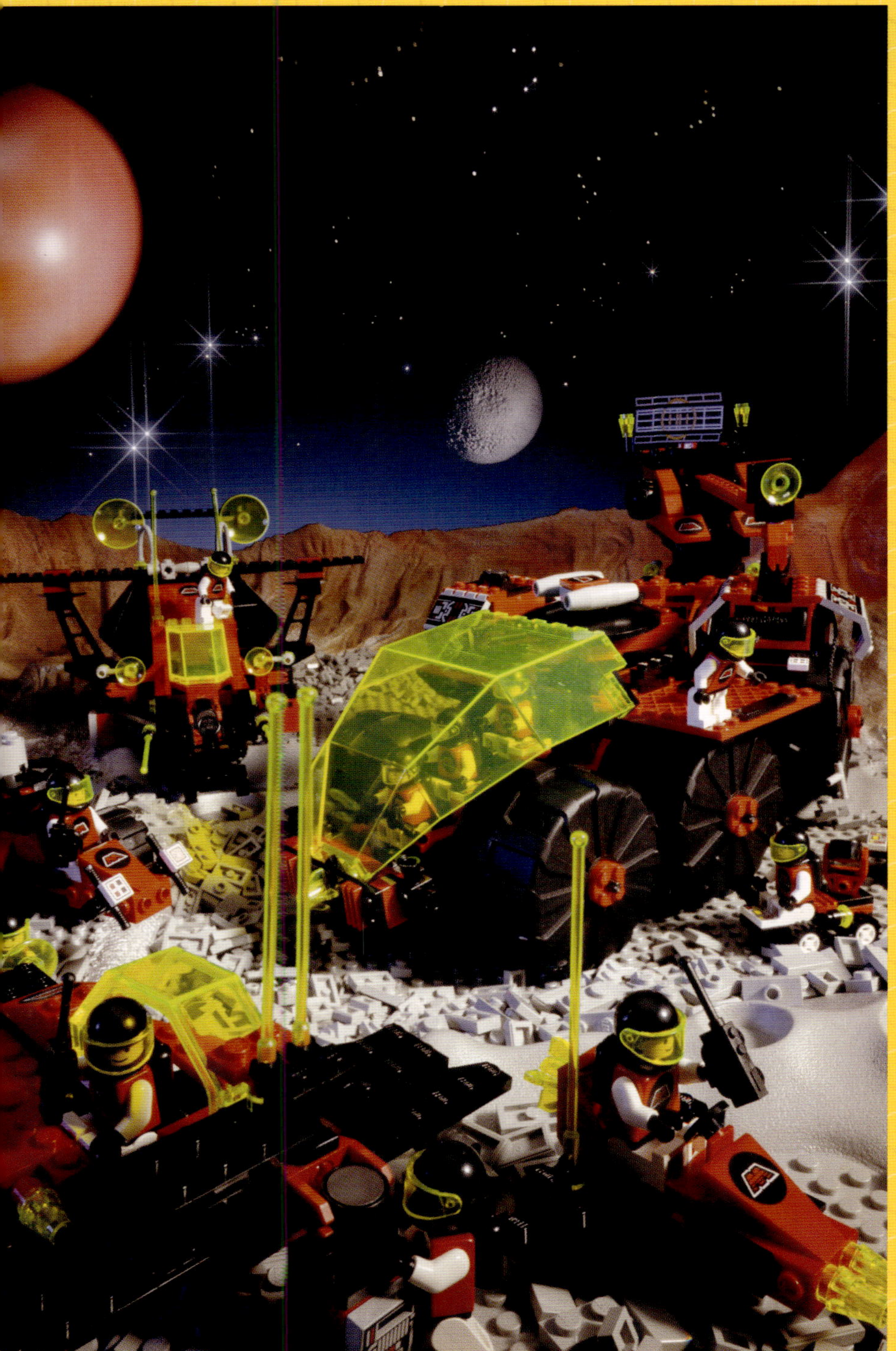

▲ M:Tron vehicles assemble below a starry sky for this promotional image of 1990.

example I worked on mirrors that could be used for periscopes, electromagnets, colored liquid inside LEGO® elements, new functions for the LEGO monorail, and much more. We even tried a 9 V electrical system on the vacuum-formed baseplates, in the form of an applied decoration that conducted the current. And I remember in 1991 we built a LEGO Space theme with a skeletal horse! That element did not come onto the market until 2007 [in LEGO Castle sets, not LEGO Space].

THE MAGIC OF M:TRON

WHAT WAS THE THINKING BEHIND INTRODUCING MAGNETS TO M:TRON?

Magnets are fun and magical. The magnets that had been in LEGO® Trains had always fascinated me; it is such a clever way to connect the carriages together. We felt they were something that would fit well with LEGO Space.

We already had magnets on LEGO trains, so it was an obvious step to develop two elements for holding the magnets: one for a crane, and one for something to be lifted with. They were first used in M:Tron, and then later on for the Ice Planet and Aquazone subthemes.

Rather than including a base in the M:Tron range, we talked in terms of the large vehicle [6989 Mega Core Magnetizer; see page 174] being the base. So the vehicle had to be really big! The large plastic wheel from Space Police had been a success, so Jens developed a smaller and a larger one. My colleagues had hoped that an even bigger wheel could have been developed; however, that proved not to be an option because the connecting elements—the LEGO® Technic snap and cross axle—were not sufficiently strong for such a huge wheel. The LEGO Space line had developed a lot of large wall panel elements, usually used in buildings, but I used the [quarter dome panel 10 x 10 x 12] as a huge cockpit! It went well with the big wheels.

The fluorescent material called LISA was launched in transparent green for M:Tron, but we could only use it sparingly, so there was a long debate about which elements provided the strongest visual effect when this new material was used. We developed a new eight-module-long antenna that looked really good, as well as using LISA for the existing cockpit, visor, and decorative elements.

BLACKTRON FUTURE GENERATION THEN INTRODUCED MORE ELEMENTS USING LISA BUT DID NOT RETAIN THE M:TRON CANOPY?

There was a strong desire for cockpit elements to be changed regularly, and we had used the same cockpit in different colors for both Space Police and M:Tron. So for the new large Blacktron craft in 1991, we built a bubble-shaped cockpit design using existing elements, including the octagonal canopy.

To create the bubble effect we developed a new element that allowed sideways building. It looked like two regular 2 x 4 bricks on top of each other, but with studs on two of the sides. Nowadays, it's not unusual to have studs on the side of elements, but in 1991, it was still thought that it might cause building challenges for children if you opened up the possibility of building sideways. ■

SPACE POLICE

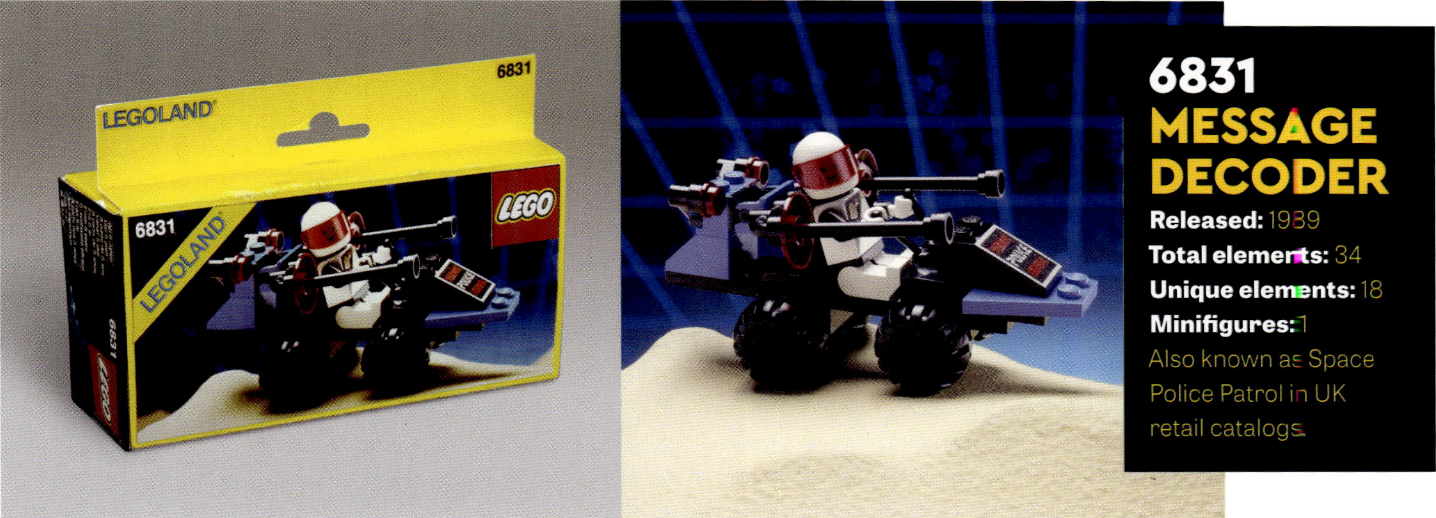

6831 MESSAGE DECODER

Released: 1989
Total elements: 34
Unique elements: 18
Minifigures: 1
Also known as Space Police Patrol in UK retail catalogs.

◀ Original mockup of box.

This little buggy has some distinctive features. Its chassis is long and slopes downward at the front, and some unusual equipment is mounted at the rear: a U-shaped antenna attached to a hinge brick, so that it can either surround the driver or be angled skyward. As it is a small model, this is the only Space Police set not to include a Space Lock-Up.

After nearly half a day of peering over the dune, the intelligence specialist finally saw the glistening yellow ogdoösphere of the Blacktron Star Base sliding open. They leapt into action, angling the dipole aerial at their side directly toward the base, and shifted a control lever to initiate the exfiltration. At last! They had finally found the perfect vantage point, and in the split second when the Blacktron equipment was exposed, the data flowed instantaneously. Smiling with relief, they reversed quickly back down the dune and stood up to reposition the antenna, pointing to the Mission Commander high above. The cipher dials glowed red as the precious data was decrypted and transmitted.

*The **vehicle mudguard 2 x 4 with angled arch** is strongly associated with the LEGO® Space ground vehicles of 1979, where it appeared in gray. Thereafter it appeared in only two more LEGO Space sets: this one and 6954 Renegade from 1987, both times in black. Its octagonal arch suits the aesthetic of late-1980s LEGO Space themes well.*

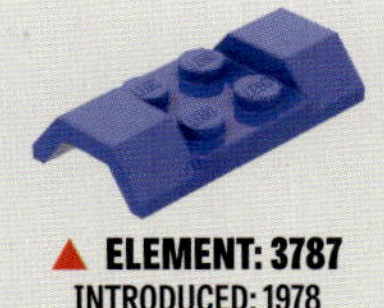

▲ **ELEMENT: 3787**
INTRODUCED: 1978

6886 GALACTIC PEACE KEEPER

Released: 1989
Total elements: 121
Unique elements: 51
Minifigures: 2
Also known as Space Police Hunter in UK retail catalogs.

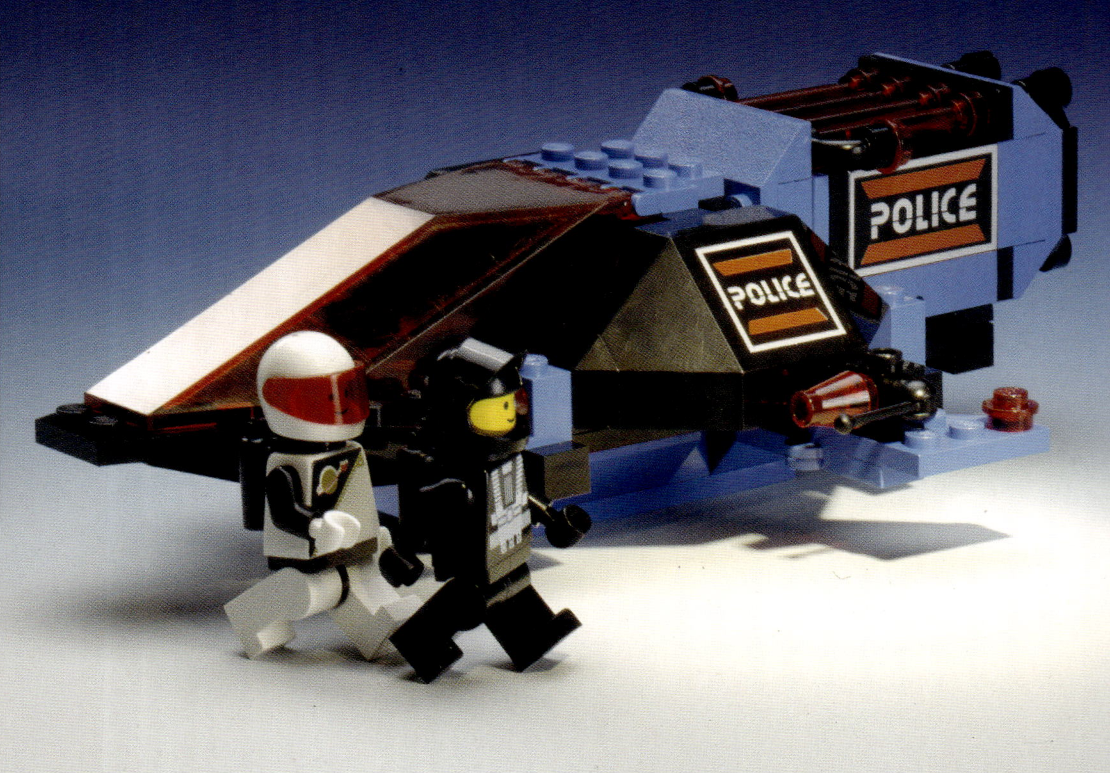

This patrol craft is more compact than other Space Police ships, and yet it still manages to include the canopy element and Space Lock-Up module. It has a bulbous fuselage, formed by two black convex corner panels positioned at either side of the cockpit. This is the only set to provide these panels with the iconic "Police" logo decoration. Two angled winglets are connected by a series of hinge plates, which allows them to be folded under and then into the panels. One Space Police and one Blacktron minifigure are provided.

▶ An alternate build featured on the back of the box.

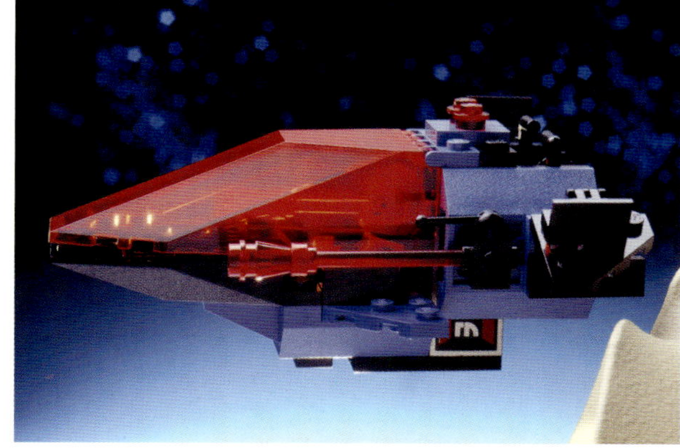

▼ Original mockup of box.

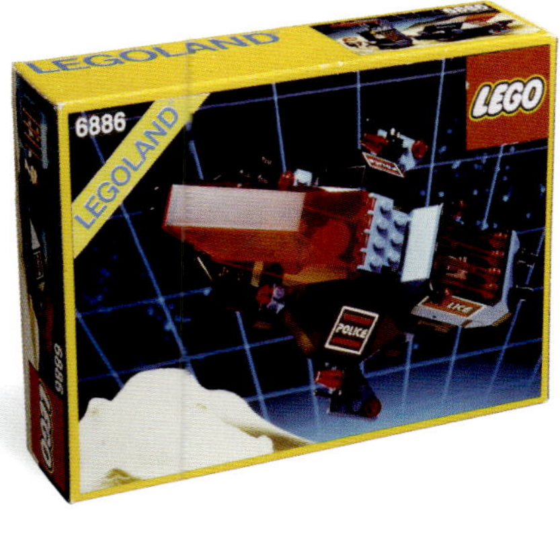

The distant stars drifted slowly past the glowing rods that surrounded him. He lay in the lockup with a calm smile on his face. "The stars are beautiful," he thought. "It's nice here." A familiar noise briefly broke him from his trance: the winglets deploying. They must be approaching enemy craft? With that thought, memories flooded back: his discovery and attempted arrest of the Blacktron scout, the ensuing combat, and his capture and imprisonment within his own lockup. The hunter becomes the hunted! He had to get out . . . somehow reach the cockpit . . . suddenly the red glow became brighter. "But space is cold," he thought. "This lockup is warm. It's okay that I am in here." The officer stared blankly as the stars slowed to a standstill.

The striking appearance of the four largest Space Police ships and vehicles is due in part to the transparent red ***windscreen canopy 10 x 4 x 2⅓****, with its distinctive elongated diagonal facets. It soon returned in transparent fluorescent green in M:Tron and made occasional appearances in other LEGO® Space themes throughout the 1990s.*

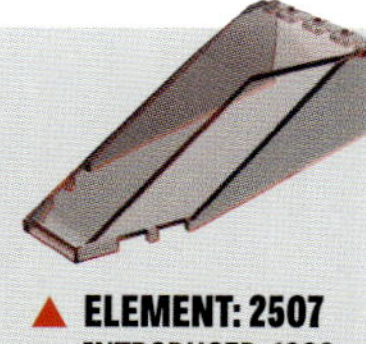

▲ **ELEMENT: 2507**
INTRODUCED: 1989

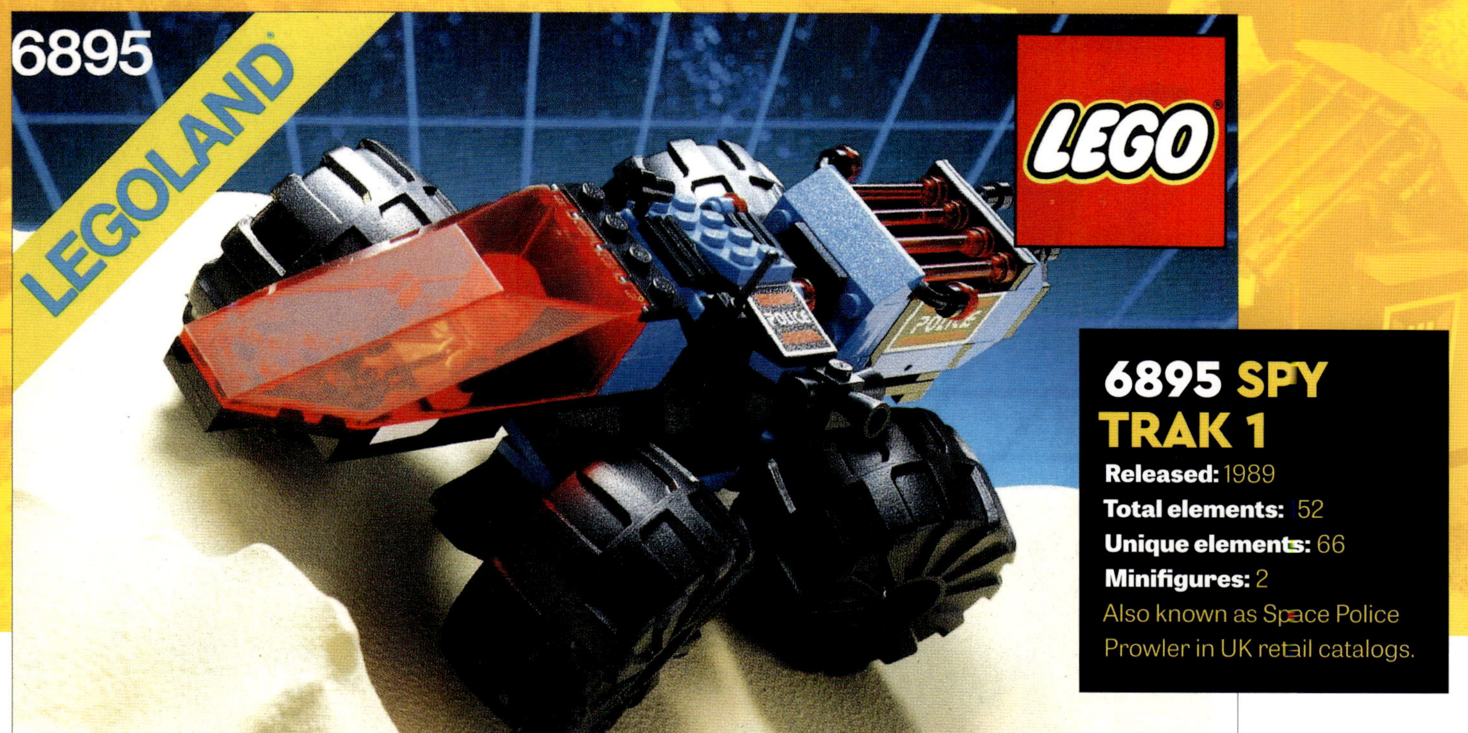

6895 SPY TRAK 1

Released: 1989
Total elements: 52
Unique elements: 66
Minifigures: 2
Also known as Space Police Prowler in UK retail catalogs.

"IT HAS SO MANY COOL FUNCTIONS CRAMMED INTO A RELATIVELY SMALL SET: FROM THE AWESOME COCKPIT, CHUNKY OFF-ROAD WHEELS, WORKING STEERING, AND MODULAR LASER PRISON IN THE BACK TO THE CLEVER STEERING ACTUATED BY THE ROCKING PANEL ON TOP. AS A KID, IT BLEW MY MIND HOW SOMEONE HAD COME UP WITH THAT!"

—WILLIAM THOROGOOD, VICE PRESIDENT OF DESIGN, NEW BUSINESS, THE LEGO GROUP, 2004 TO PRESENT

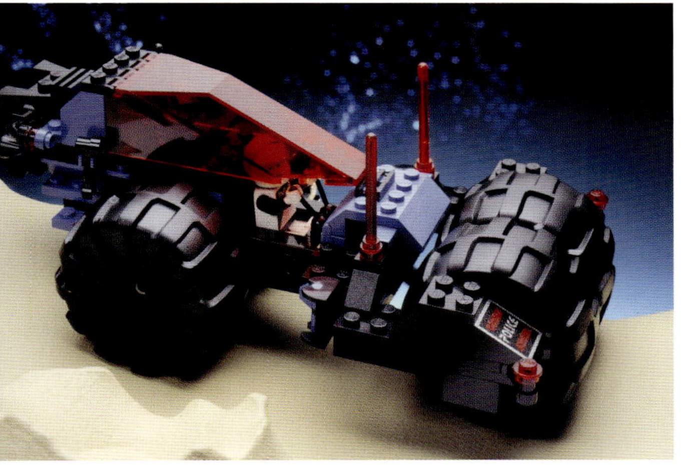

◄ An alternative model shown on the back of the box.

At the heart of this chunky ground vehicle lies an ingenious mechanism (see page 159). The chassis is articulated at two joints, with a single LEGO® Technic axle binding the front and rear sections only. In the central section, a rocking mechanism can be tilted by pushing one of the decorated slopes, which causes an inverted slope underneath to push against the axle. The front and rear wheels rotate to different angles, and the vehicle turns. The set includes both a Space Police and a Blacktron minifigure.

"Engage self-drive," said the sergeant. Gravity pulses began to shoot out from the cobalt nav sensors at the center of the vehicle, steering it effortlessly through the mountainous terrain. "Prowler to iso base," she said over the comms. "No sign of the Blacktron yet; they've strayed a long way from their craft." Glancing up through the windshield, she spotted a dark figure scaling the cliff face ahead. "Disengage self-drive," she shouted, slamming on both levers as she shot straight toward the rockface. The moment before impact, she engaged the gravity cushions and swiftly turned upward. The Blacktron spy froze in helplessness as the vehicle thundered up the vertical cliff face toward him, its lockup already opening in anticipation.

*A new type of wheel rolled its way into LEGO® Space in the form of the **hard plastic large wheel**, measuring over two inches (54 mm) in diameter. Like the small balloon wheel of 1982, it was not made of rubber. Its interior is hollow, which earned it the nickname "the yogurt cup" among designers at the time.*

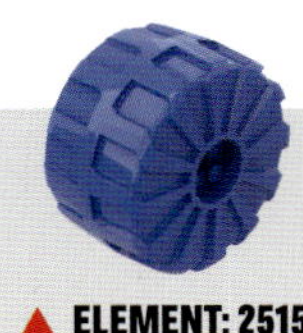

▲ ELEMENT: 2515
INTRODUCED: 1989

6955 SPACE LOCK-UP ISOLATION BASE

Released: 1989
Total elements: 252
Unique elements: 91
Minifigures: 3

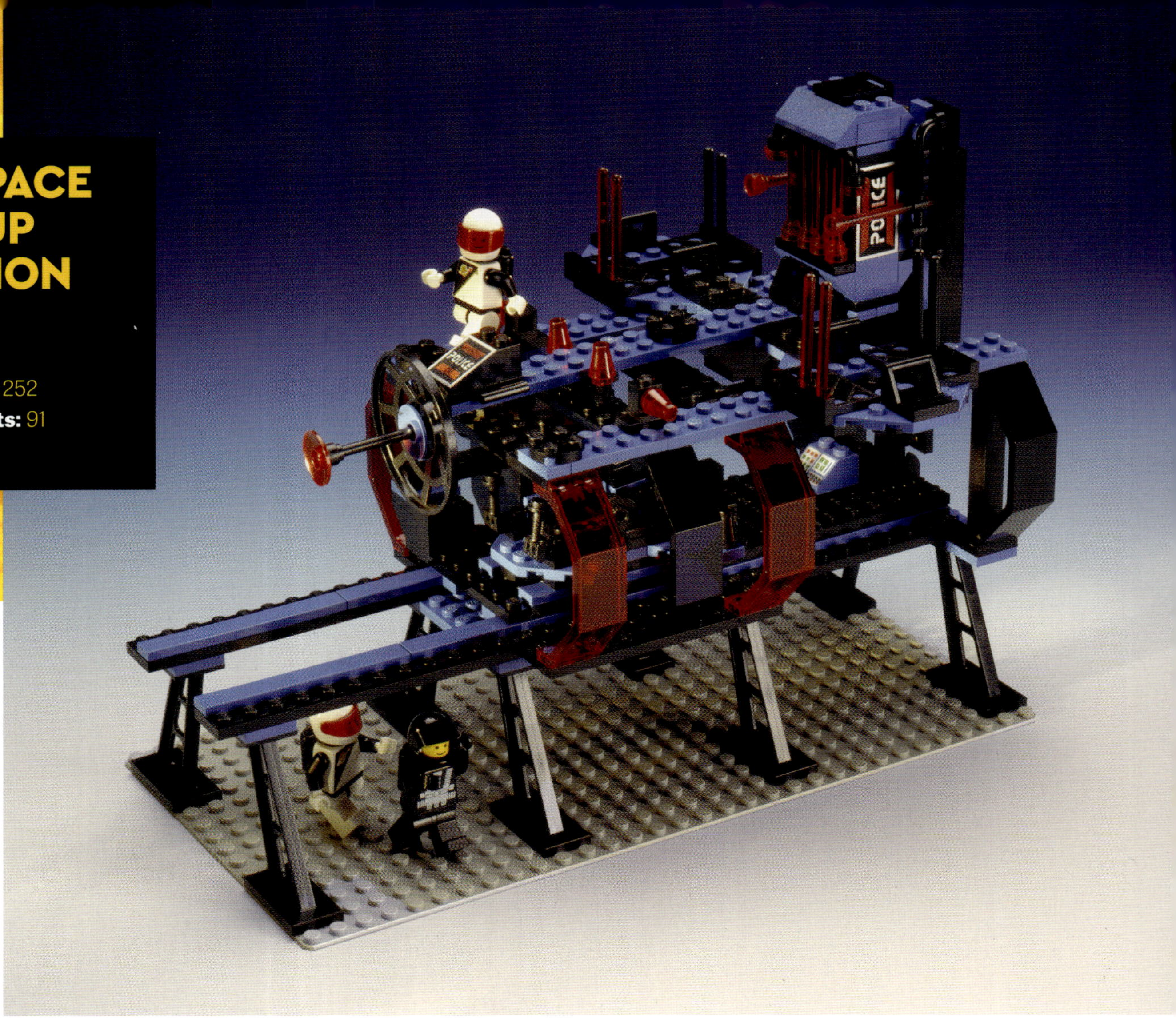

▼ Original mockup of box.

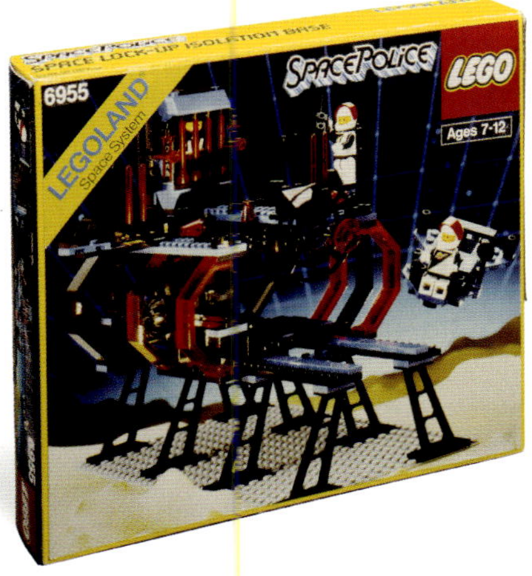

This two-tiered station with a shuttle is smaller than the previous two space bases yet delivers one of the grandest play features of any LEGO® Space set. Turning the wheel on the side operates a rack-and-pinion mechanism that simultaneously raises the radar dish high above the station, rotates the Space Lock-Up module down into the hangar, and slides the internal launch pad forward to release the shuttle. The lattice-like structure and profusion of tech equipment gives the design a utilitarian feel, as befits an isolation base. A Blacktron prisoner is provided, along with two Space Police minifigures.

The approaching shuttle ejected the lockup and reversed into the bay below. The cell guided itself into the stasis field projector where the sheriff stood, bringing him face to face with the Blacktron general inside. "Finally." He smiled as he switched off the delta wave generator. "I got you." Life slowly returned to the general's eyes, which examined the barren outpost surrounding her. "But for how long," she said calmly. The sheriff laughed as he turned away. "Well, you've got plenty of time to answer that—we'll send someone to pick you up in a few decades." Descending to the command post in the bay, he configured the stasis program and initiated the launch sequence before jumping into the back of the waiting shuttle. As the craft moved forward for takeoff, the Faraday cage began to glow red, and the lockup descended inexorably inside. The scrambler dish rose high above the base, rendering the entire asteroid undetectable. The general's limbs slowly hardened, but she remained wide awake.

*In some applications, using a **panel 1 x 2 x 3** offers no advantage over stacking three 1 x 2 bricks instead. Its use within the Space Lock-Up modules is a good example of its benefits: the inner recess provides additional space inside, and its exterior surface can be decorated. Unusually, left- and right-hand versions of this "Police" logo were provided, so that the text is still readable when the Space Lock-Up lies horizontally.*

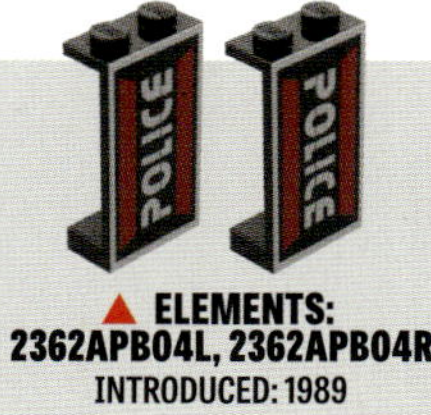

▲ **ELEMENTS: 2362APB04L, 2362APB04R**
INTRODUCED: 1989

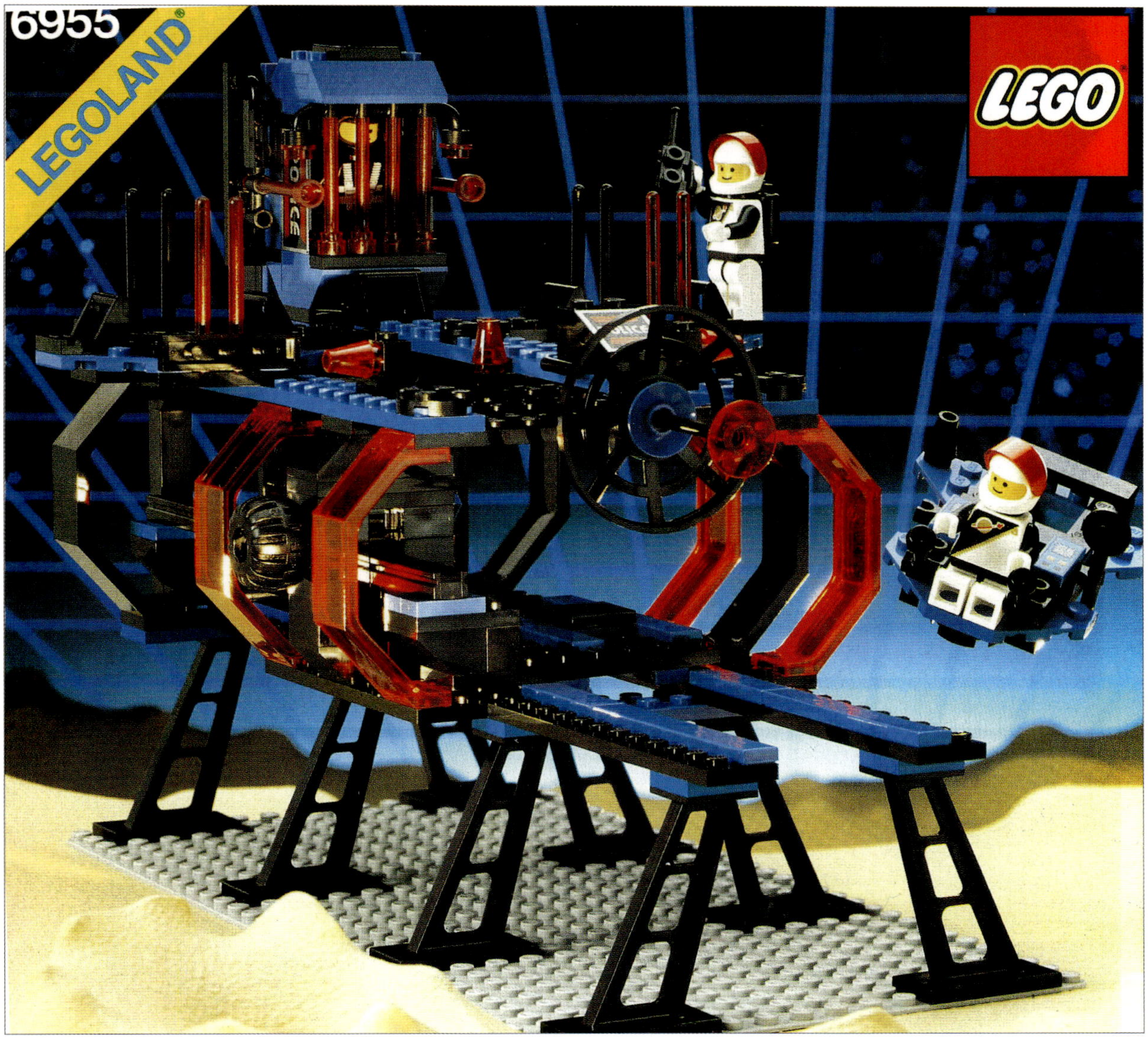

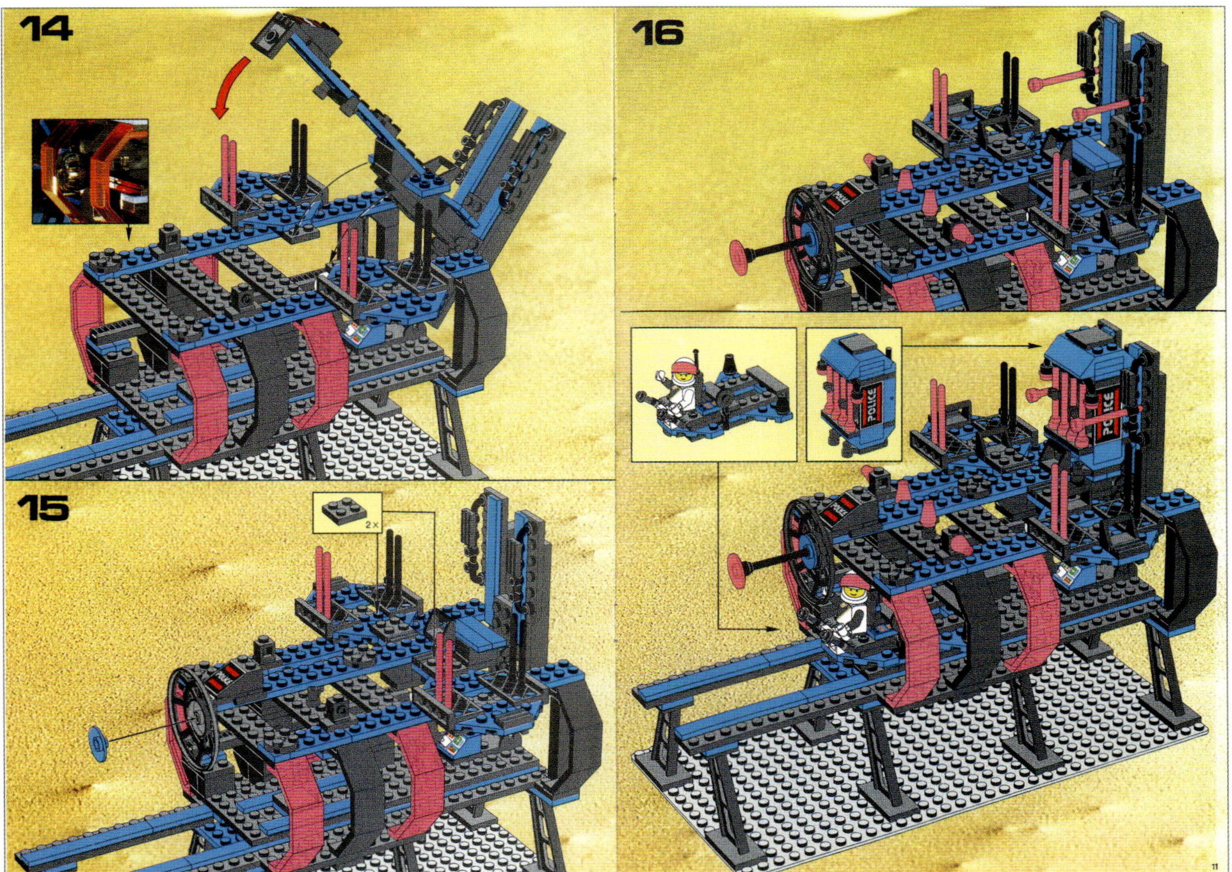

The final stages of the building instructions show the radar being lowered, positioning the Space Lock-Up and shuttle for play.

6781 SP-STRIKER

Released: 1989
Total elements: 231
Unique elements: 88
Minifigures: 2
Also known as Space Police Prisoner Transport in UK retail catalogs.

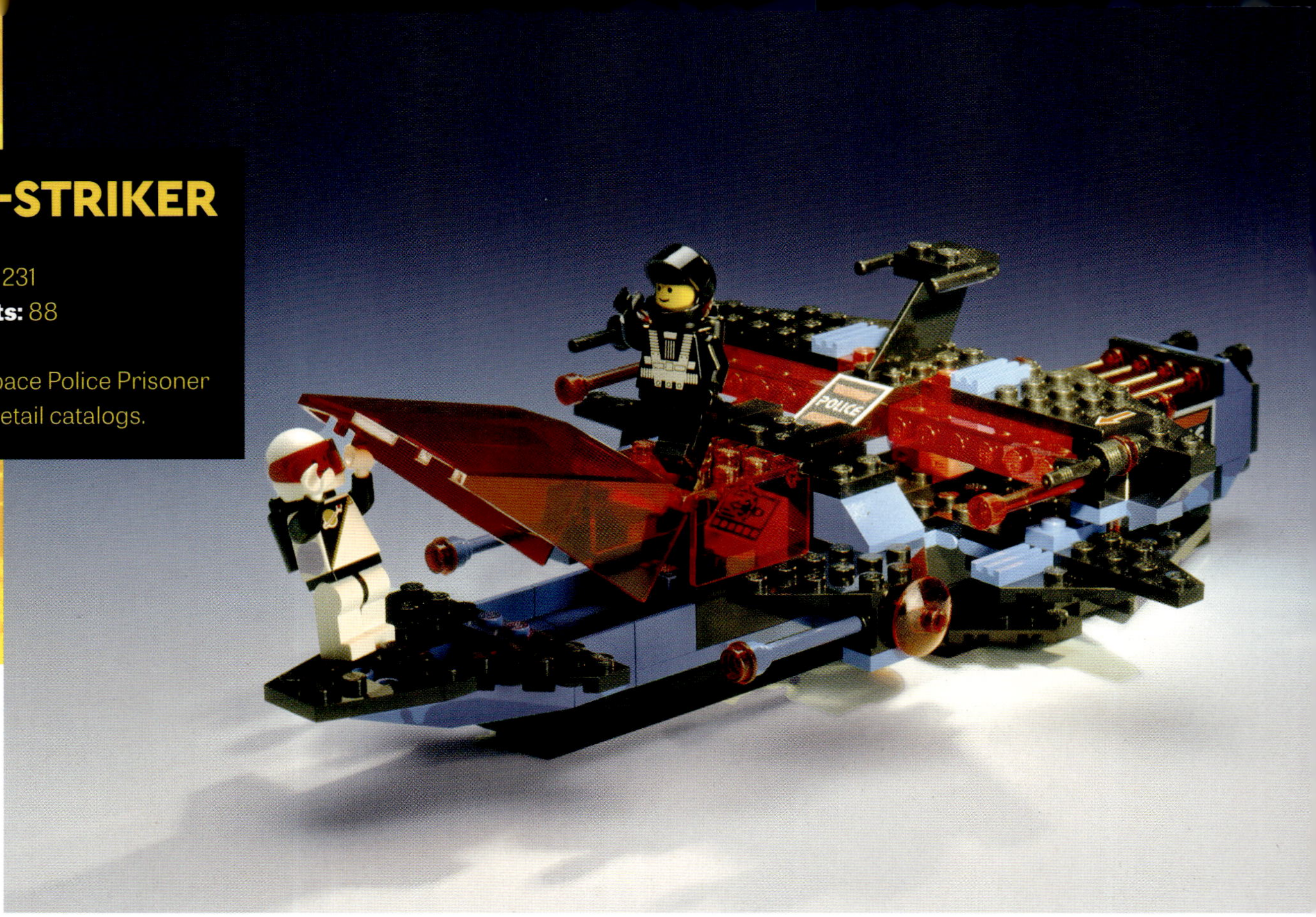

▶ The battery box and light-up console are part of a separate module that slides into the rear of the main craft, so that the console is visible at the rear of the cockpit.

This impressive craft has a multitude of wings, with the lower rear pair able to fold downward on hinges. It uses the Light & Sound System, and although it does not include a sound brick, there are two new elements: the 1 x 8 brick with three lights, and a control panel with a schematic diagram of the ship, which lights up. The 1 x 8 light bricks have transparent red 1 x 6 bricks attached sideways over the lamp covers. The Space Lock-Up module is mounted on the rear, and once detached, the upper rear half of the model slides out, for extra playability and to provide easier access to the battery. One Space Police and one Blacktron minifigure are provided.

"I enforce this galaxy, and you will comply with our laws," barked the officer. "You are under arrest for the attack on Star Base 1. You will be placed in delta pacification during transport," she said as she bundled the wriggling Blacktron sergeant into the lockup. As soon as she activated the delta wave rods, the criminal's eyes glazed over like black dots. Clambering into the cockpit, the officer checked the schematics to ensure that the life-support system was fully operational. After takeoff, she lowered the wings to exit the atmosphere quickly and turned on the sirens, ensuring that any Blacktron interceptors in her path would be drawn in by the tractor beam until close enough to strike them with the aglao-pellets.

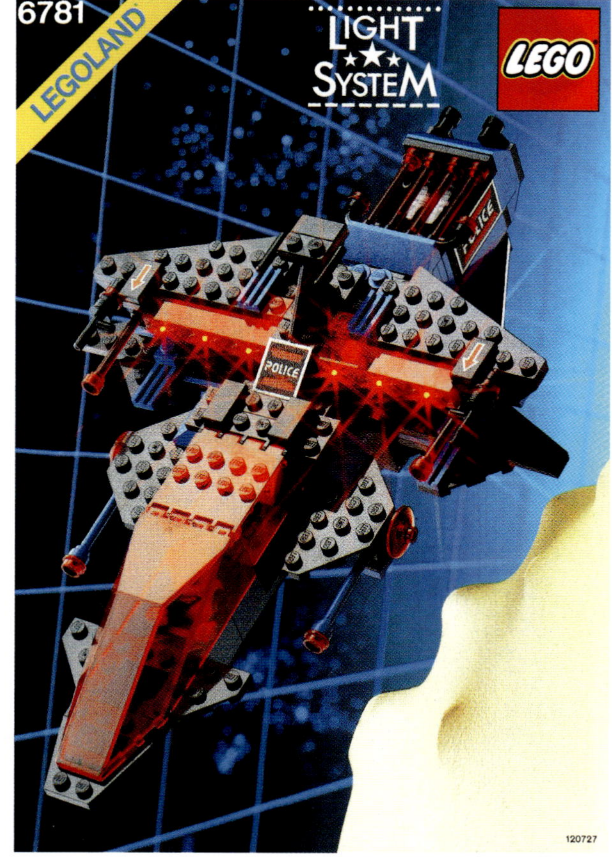

This is the only set to include the ***sloped electric light cabinet 45° 2 x 2****, which took the LEGO® Space computer panel to a whole new dimension by lighting it from within. This hollow element has a cutout section at the rear which accommodates a 1 x 2 electric light element, and a printed transparent pane clips into its angled frame.*

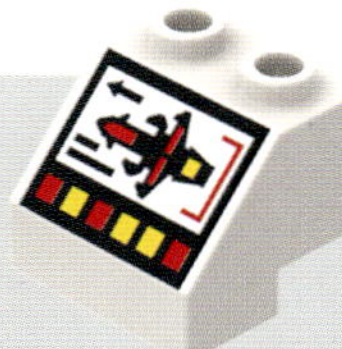

▲ INTRODUCED: 1989

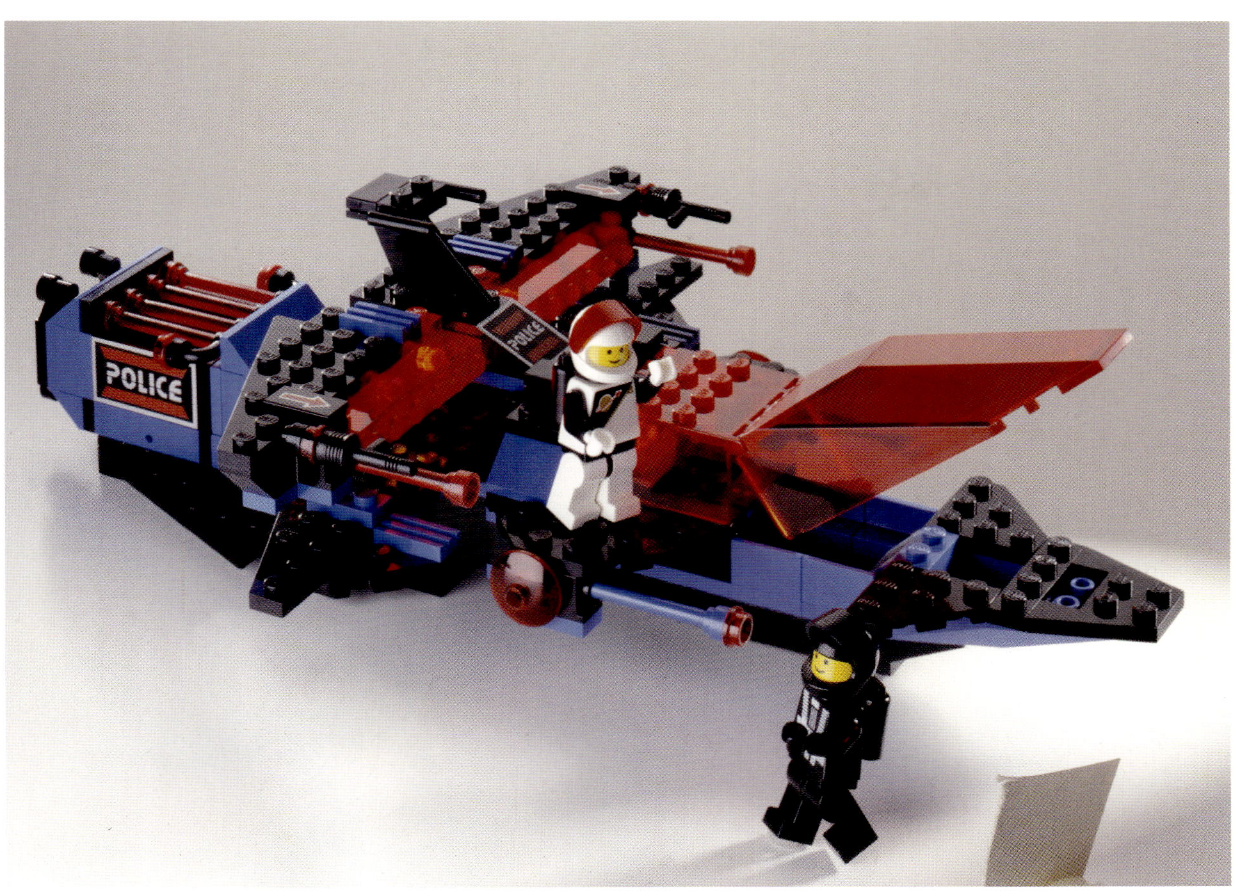

"THE LIGHT SYSTEM, MODULAR PRISON WITH LASER BARS, THE DOUBLE-HINGED COCKPIT, AND ALL OF THE AWESOME TRANSPARENT LASER CANNONS MADE THIS ONE OF THE COOLEST SPACE VEHICLES EVER TO SEVEN-YEAR-OLD WILL!"

—WILLIAM THOROGOOD, VICE PRESIDENT OF DESIGN, NEW BUSINESS, THE LEGO GROUP, 2004 TO PRESENT

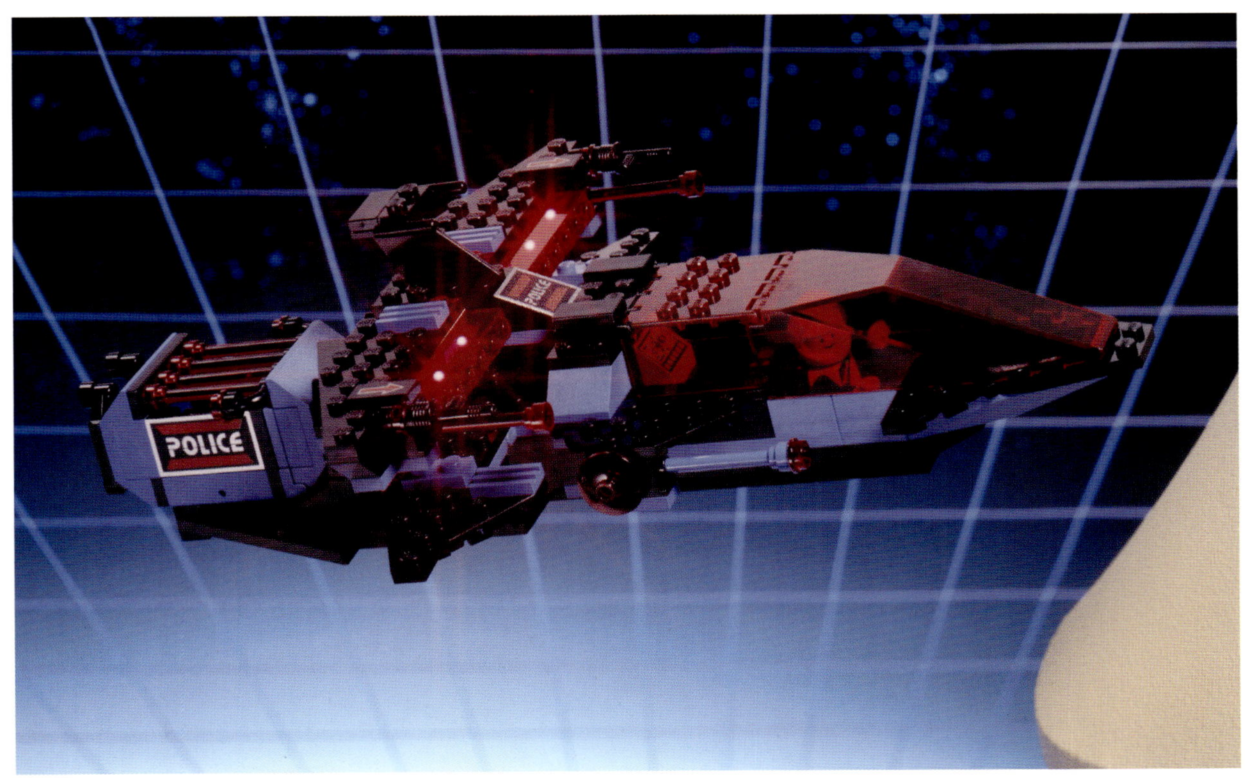

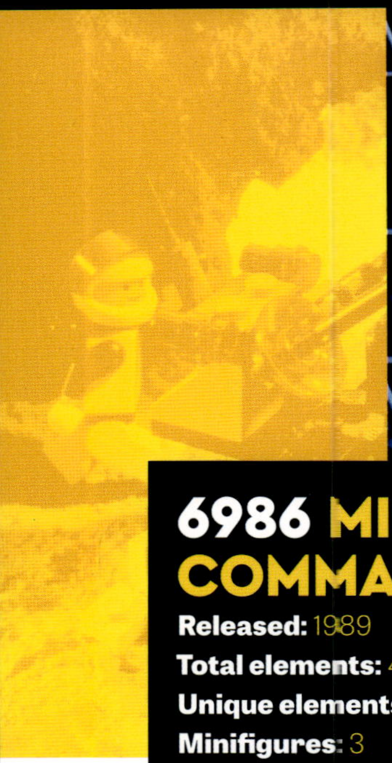

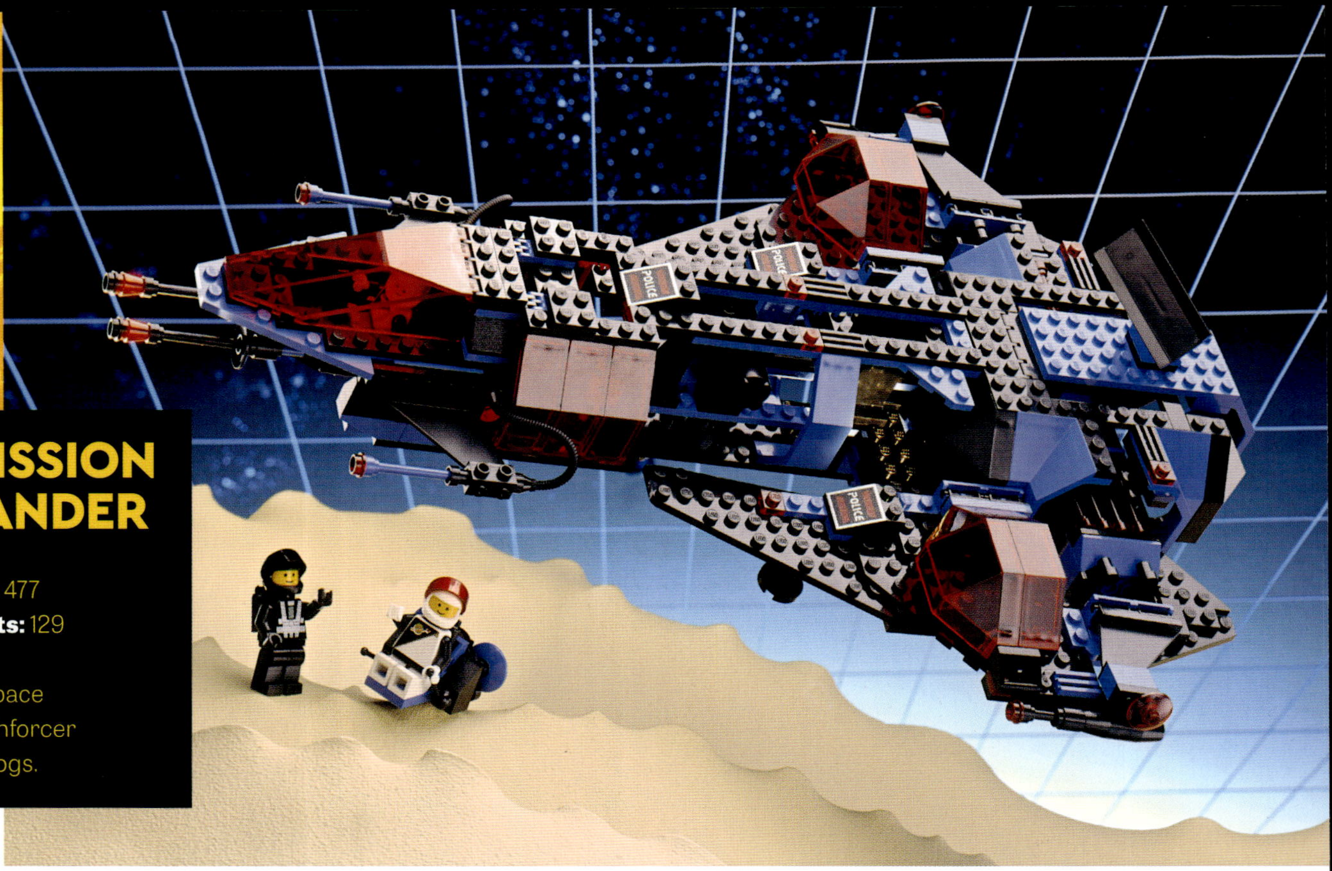

6986 MISSION COMMANDER

Released: 1989
Total elements: 477
Unique elements: 129
Minifigures: 3
Also known as Space Police Galactic Enforcer in UK retail catalogs.

The largest Space Police set has a dramatic design with a long neck and forked rear wings which each have a transparent red pod containing a patrol scout ship. In addition to being able to open the cockpit windscreen as usual, rotating a round brick on the ship's central corridor activates a mechanism that hinges the entire nose section upward while pushing the pilot's seat forward—revealing that the seat is actually a ground vehicle, long enough to carry a Space Lock-Up as cargo. The ship's side weapons can be tilted upward and then rotated inward. This unlocks the entire aft section: a cruciform module made of four concave corner panels.

As yellow streaks flew toward her, the pilot fired the negative-mass pulsors at the artillery operator on the ground, enveloping the Blacktron fire with ultra-photovoltaic energy. "I'll create a distraction," said the commander as a pod opened to release his scout ship. "Deploy prison," the pilot instructed, and the pulsors folded inward to release the aft section, which glided downward. Now freed of the additional weight, the pilot hit maximum power. "Engage autopilot! Run swoosh maneuver." The commander's scout ship blocked the Blacktron operator's escape as the giant craft bore down toward them. Suddenly, its cockpit opened like a jaw, and the pilot was ejected forward in a buggy which bounced toward them. The great ship banked upward, and the pilot and commander cornered the artillery operator at the same moment. "I enforce this galaxy, and you will comply with our laws," said the commander with a grin.

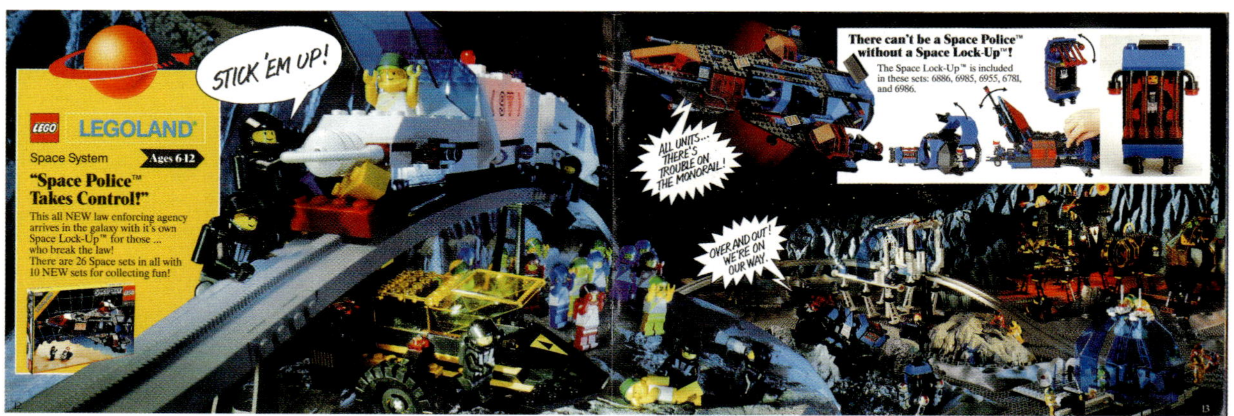

▶ The 1989 US catalog highlighted the exciting functions of the Mission Commander: the detachable aft section, the hinged front, and the Space Lock-Ups.

*The **snowplow with hinge 6 x 3** first appeared in yellow in LEGO® Town and soon found a new life in outer space as a rectangular radar. This is the first and only LEGO® Space set to include it unprinted: it soon reappeared with detailed decorations in 6989 Mega Core Magnetizer and later LEGO Space subthemes.*

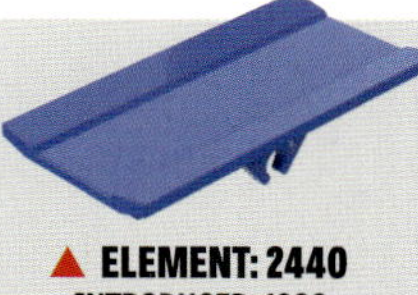

▲ **ELEMENT: 2440**
INTRODUCED: 1988

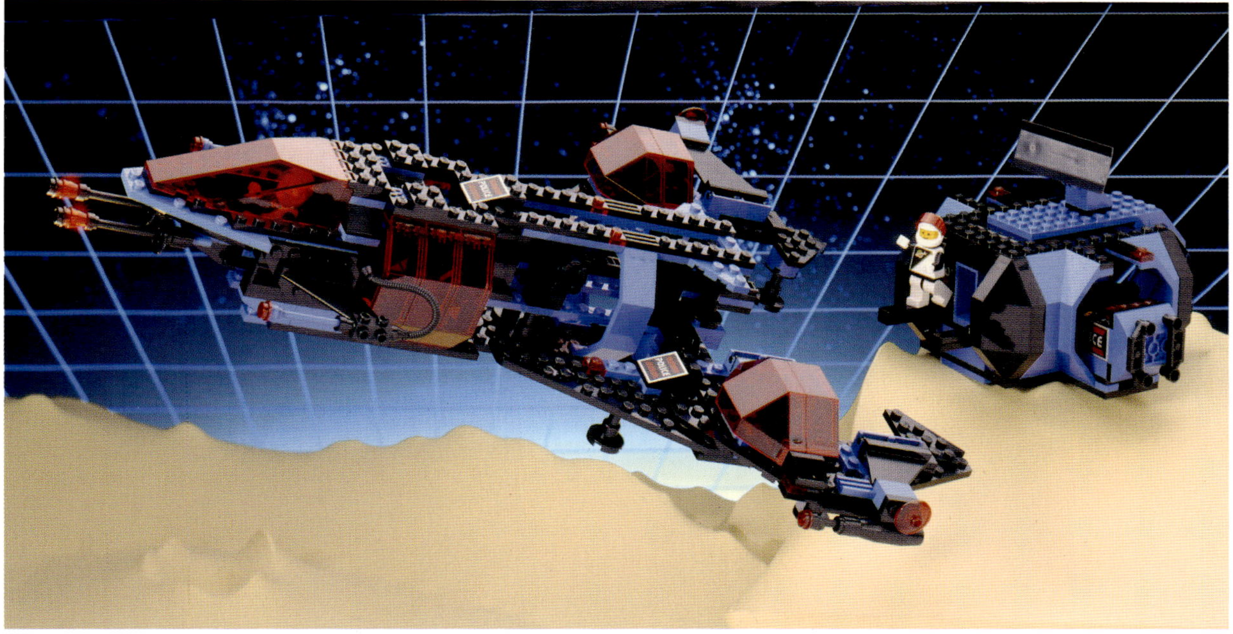

◀ The rear section of the craft is a detachable prison with room for two Space Lock-Ups.

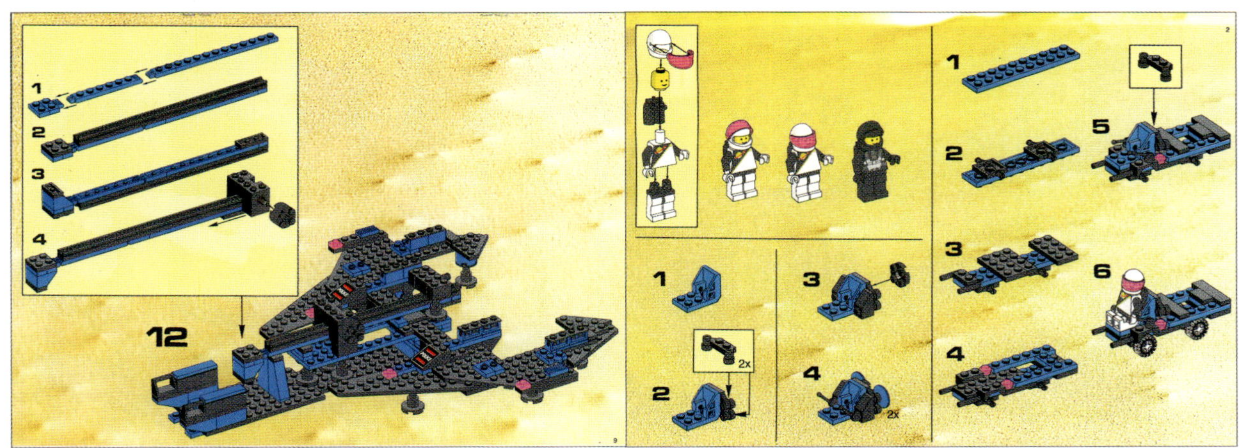

◀ The rack-and-pinion mechanism at the heart of the Mission Commander swings the entire cockpit upward.

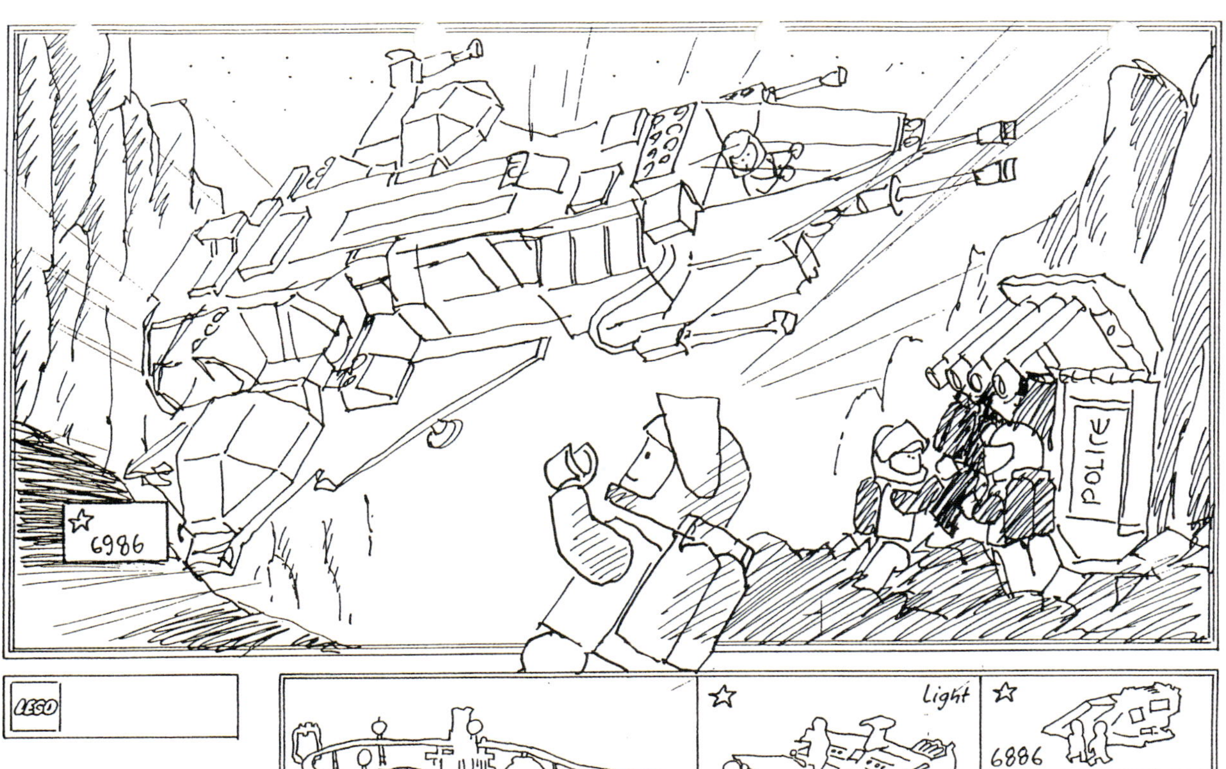

◀ Original layout sketch for an advertisement featuring the Mission Commander.

M:TRON

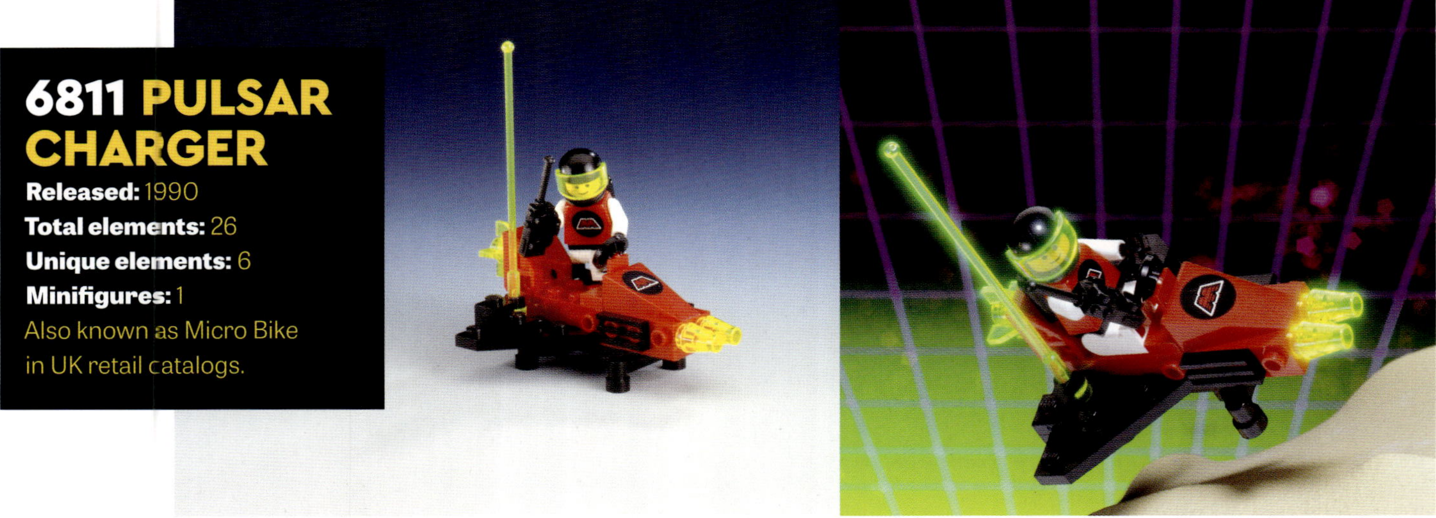

6811 PULSAR CHARGER

Released: 1990
Total elements: 26
Unique elements: 6
Minifigures: 1
Also known as Micro Bike in UK retail catalogs.

▼ Original mockup of box.

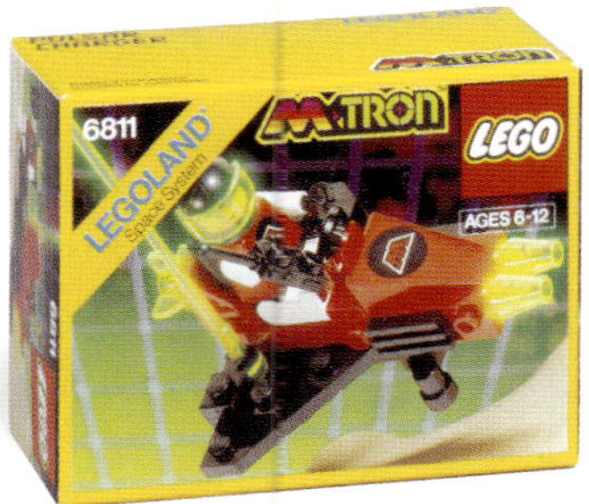

This small craft contains three elements made of the new transparent fluorescent material (see page 154), including the rear thruster. The red fuselage is formed largely by two existing elements—the space seat from 1984 and the cockpit space nose from 1986 printed with the "M" logo. The four black feet are landing gear and a possible secondary means of propulsion. Multiple controls are included: a steering wheel and two side levers.

The seismologist kept their hands steady on the nav wheel as they charged along the damaged mine shaft. An alert from the starboard geoanomaly grille brought them to a swift halt, and they switched on their cosmo-comms unit. "Captain, I'm seeing deep grooves scarring one side of the tunnel . . . signs of an explosive. The collapse was not an engineering error." Suddenly, the magnatelluric antenna bleeped, but the warning came too late: more rubble tumbled down, trapping them instantly. Knowing the shaft was still too narrow for other M:Tron craft to reach them in time, they cleared enough debris to reach the nav wheel. Firing the front-mounted maglev traction beams in short pulses, combined with small bursts on the vertical propulsors, they managed to jiggle the craft enough to gradually loosen the rocks and free themself.

*The **whip antenna** is thin, flexible, and very long and includes a thicker bar section that can connect to a minifigure hand or a clip. It is the third antenna in LEGO® Space history, and although the three have varied designs, they all adhere to a system: the whip antenna is exactly twice the length of the original antenna from 1979, which is twice the length of the small adjustable antenna from 1985.*

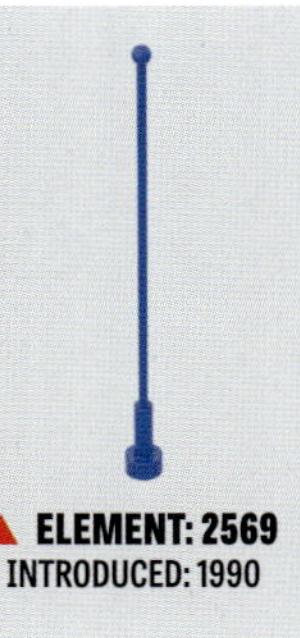

▲ **ELEMENT: 2569**
INTRODUCED: 1990

This design makes effective use of its comparatively low element count, with multiple tiers of wings, two of the new whip antennas, and a magnet at the rear to clasp a small container. An angled panel in transparent fluorescent green creates the slender windshield, which can be folded forward after rotating the two red plates with clips keeping it in place at the back.

The pararescue officer took the remaining ferrotracker out of the stasis safe and nestled it between some rocks. Lifting off to hover just above the asteroid's surface, she carefully reversed to lift the heavy cabinet with the hypermagnet. "Heading to the delta position now," she said into her cosmocomms unit to the miners trapped deep below. "Clear the area." Coming to a halt, she turned the craft to point in between the two ferrotrackers. With a swipe of the controls, the twin magna-helix antennas glowed, and a fluorescent triangular force field appeared across the landscape, linking the ferrotrackers. With another swipe, the field connected to a fourth tracker held by the miners. "Prepare for extraction," warned the officer. The twin magna-wimbles crackled into action, and the force field kept the asteroid steady as rocks loosened and tumbled vertically into space, leaving a giant triangular access shaft.

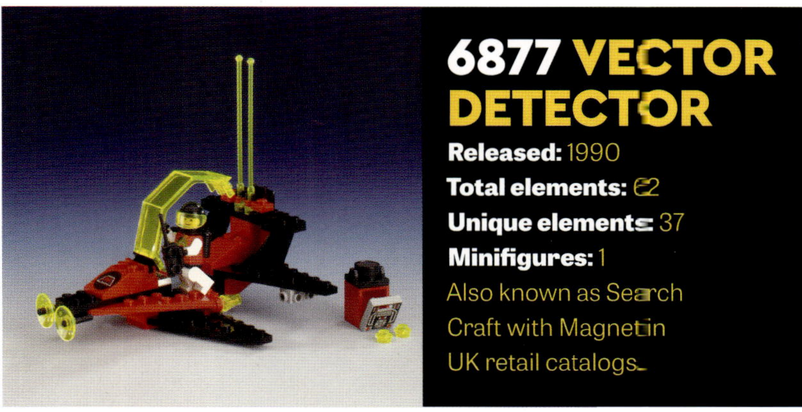

6877 VECTOR DETECTOR

Released: 1990
Total elements: 62
Unique elements: 37
Minifigures: 1
Also known as Search Craft with Magnet in UK retail catalogs.

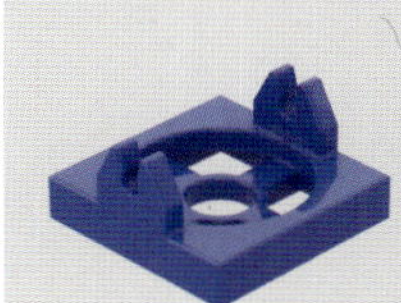

▲ ELEMENT: 2609
INTRODUCED: 1990

Magnet holder tile 2 x 2 *is one of two specialized fastenings introduced by* M:Tron. *It has a similar base to the turntable base 2 x 2, with two arms designed to clasp the existing* LEGO® Trains *magnet element from 1980. It is used by many later* LEGO® Space *subthemes, beginning with Ice Planet 2002 in 1993 and ending with Aquazone in 2007. In 1998, an inverted version was introduced to clasp magnets underneath.*

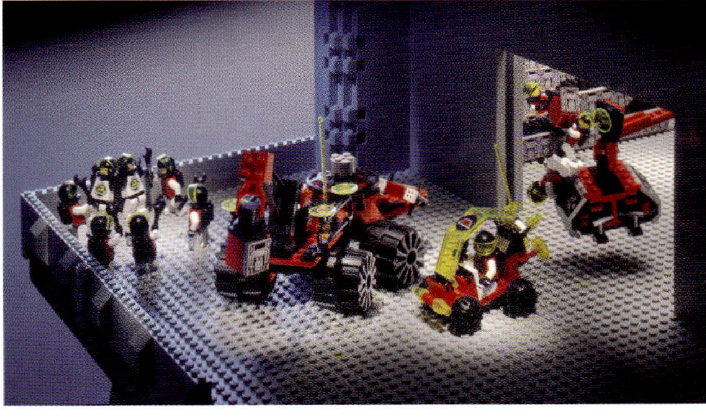

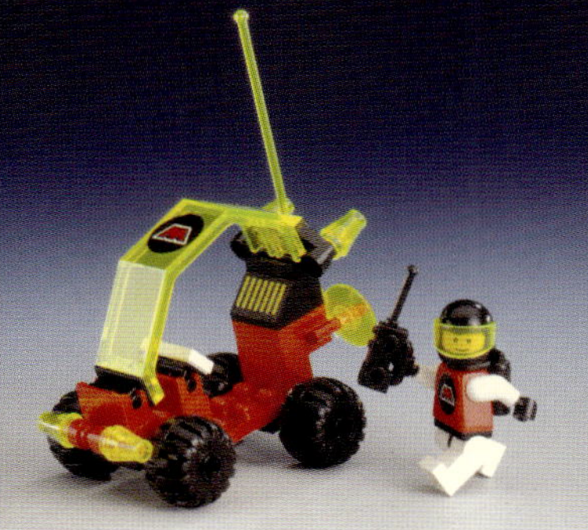

6833 BEACON TRACER

Released: 1990
Total elements: 40
Unique elements: 20
Minifigures: 1
Also known as Inspection Buggy in UK retail catalogs.

This novel buggy design has an interesting side profile, featuring a sloped chassis using a bracket 8 x 2 x 1⅓ and two hinge bricks. This small set makes extensive use of the new transparent fluorescent material, including a canopy with the "M" logo that opens forward, multiple rear thrusters, and an unusual arrangement of cones at the front. Like most M:Tron minifigures, the driver is provided with a radio to stay in touch with the team.

The dynadisks sped the buggy at lightning speed through the shafts deep below the asteroid's surface as the magnetic resonance analysis cones searched deep within the rock at either side of the tunnel for more ferrocrystals. They threw up an unusual reading, and the chief prospector reversed to the source. He unbuckled the canopy at the rear, engaged its magna-spectrometric mode, and adjusted its slider controls to obtain a detailed reading. Lithium? Surely not. Pinpointing the source, he discovered it wasn't ore but something attached to the tunnel wall. He picked up his cosmocomms unit. "Base, I've found a device. Never seen anything like it. May be some kind of beacon?"

◀ ▲ **Who needs the Space Police when you have space wrenches? The crews of M:Tron sets 6833 Beacon Tracer and 6896 Celestial Forager take on the Blacktron Future Generation in this 1990 promotional image.**

A new printed control panel, ***tile 1 x 2 on white with red and yellow controls****, appeared exclusively in* M:Tron *sets in 1990, and then* Blacktron Future Generation *the following year. The only other* LEGO® Space *subtheme it appeared in was* RoboForce *in 1997; however, its bold, geometric design was generic enough to be used by other themes, until it was discontinued in 2004.*

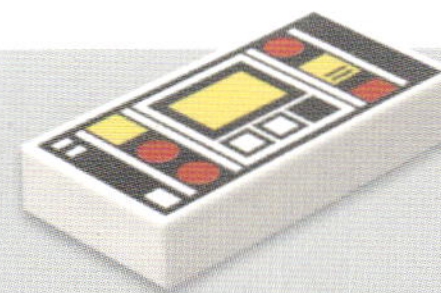

▲ ELEMENT: 3069BP68
INTRODUCED: 1990

6896 CELESTIAL FORAGER

Released: 1990
Total elements: 92
Unique elements: 47
Minifigures: 1
Also known as Astro Wrecker in UK retail catalogs.

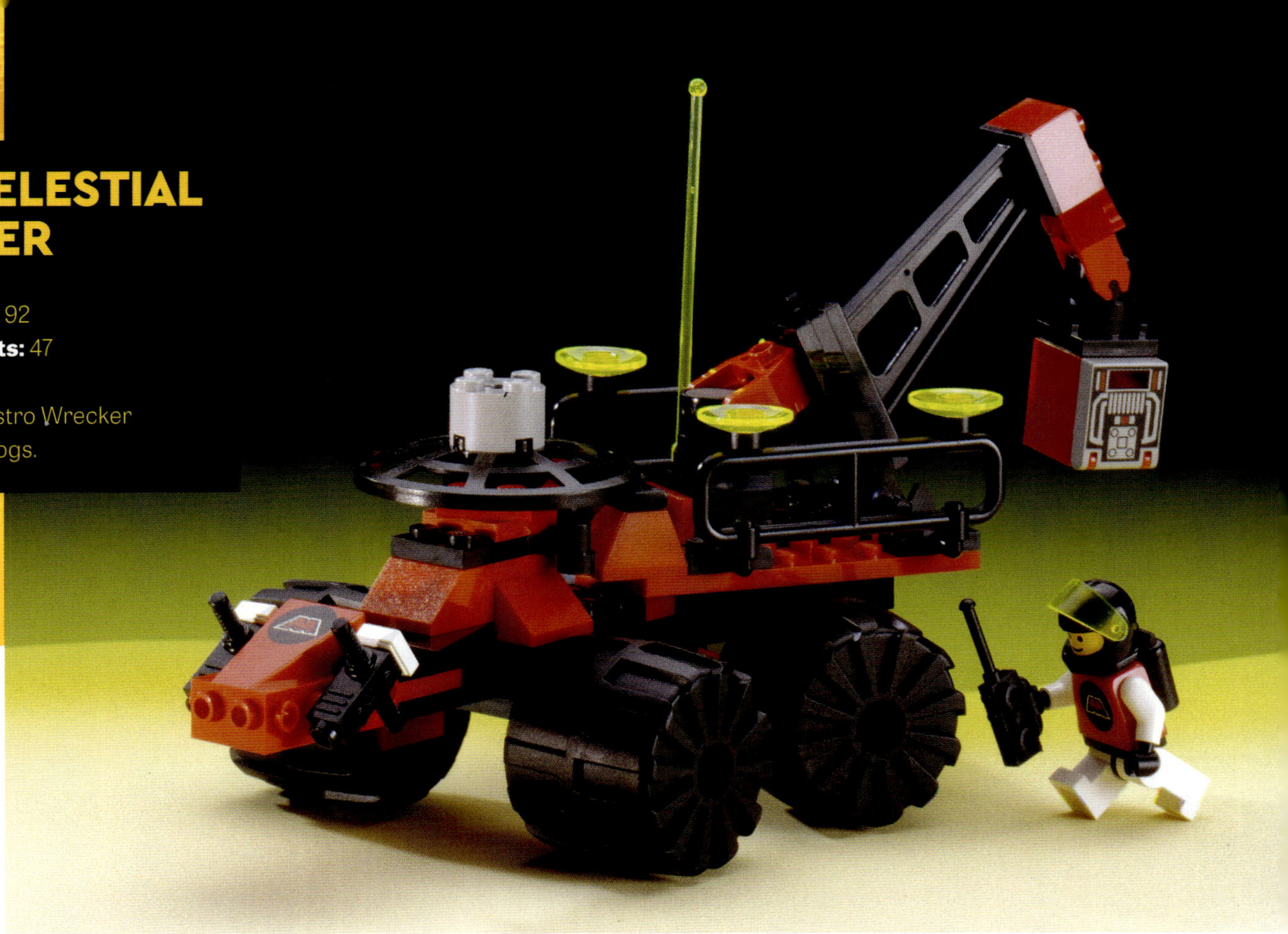

"I HAD NEVER PLAYED WITH MAGNETS BEFORE, SO THIS WAS LIKE MAGIC!"

—WILLIAM THOROGOOD, VICE PRESIDENT OF DESIGN, NEW BUSINESS, THE LEGO GROUP, 2004 TO PRESENT

Two pairs of magnets are included in this articulated truck; one holds the crane in place at its base, and once released, the crane can be lowered to collect a container using the second magnet. The large radar dish can be turned to steer the vehicle, using a refined version of the four-wheel control mechanism that designer Jørn Thomsen developed for set 6895 Spy Trak 1 (see page 163).

▼Original mockup of box.

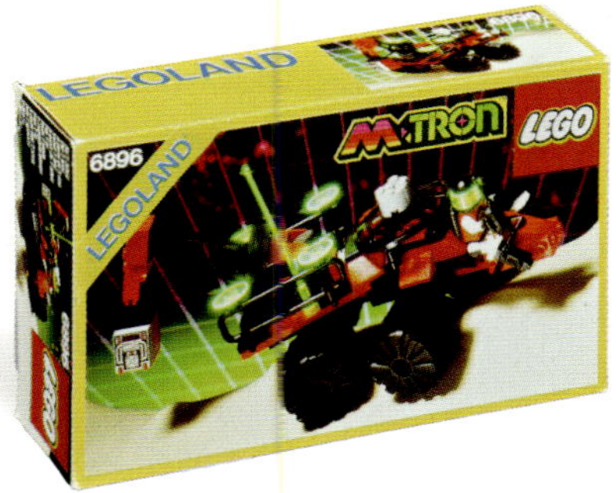

It was lucky the wreckage vehicle was on autopilot: the driver had to use all of his concentration to avoid tumbling off as they bounced across a particularly rocky patch of the asteroid. The nav saucer rotated swiftly in response to data from above, steering toward the source of ferrocrystals floating in the ring of slurry now orbiting the asteroid. Powering down, he deactivated the hypermagnet to uncouple the crane and lower the stasis safe to the surface, ready for storage. Returning to the cab, he flipped the port control arm to engage the magnaresonance disks. A square column of fluorescent light rose up to the ring above, pushing the rocks further out into space and drawing the fragments of ferrocrystal downward.

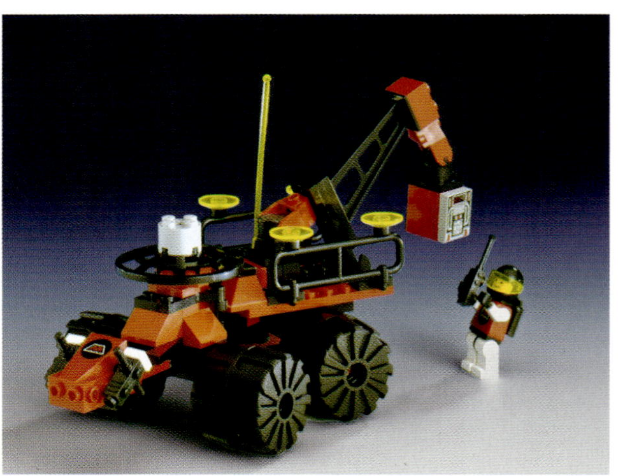

*This extraordinary element, featuring stud and bar connections on all sides, manages to be both specialized and generic in its design. Although commonly known as the **minifigure chainsaw body**, it wasn't used as such until 1993. Before then, it was utilized in a variety of ways in M:Tron and Blacktron Future Generation sets: as a tool, a weapon, and a thruster. Here, it is combined with a white tile to form a steering control.*

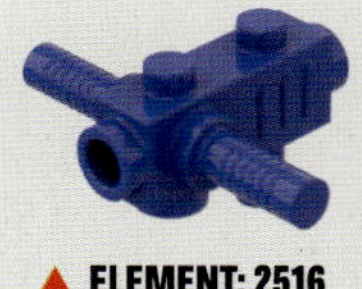

▲ ELEMENT: 2516
INTRODUCED: 1990

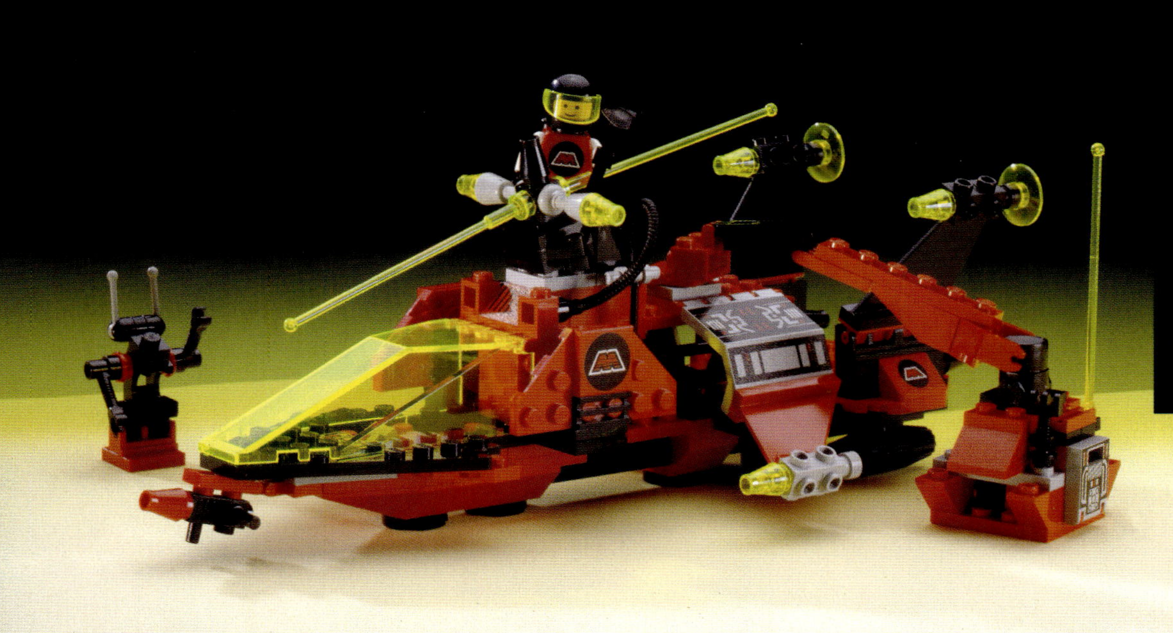

6923 PARTICLE IONIZER

Released: 1990
Total elements: 193
Unique elements: 91
Minifigures: 1
Also known as Cosmicopter in UK retail catalogs.

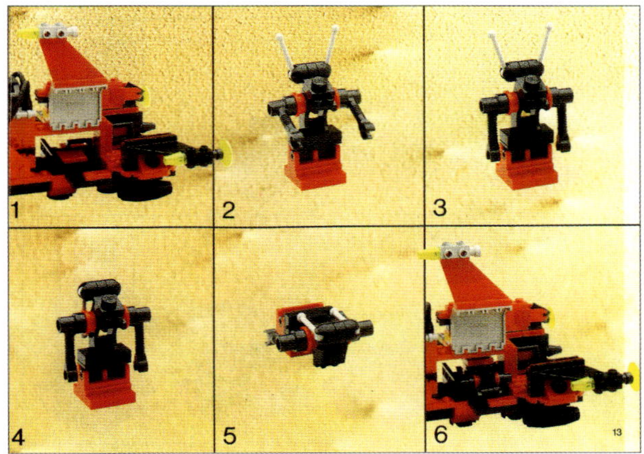

◄ The building instructions include photographic steps to show how the droid is folded to fit into the cargo area.

A helicopter-spaceship with a magnetic crane? Anything is possible in LEGO® Space, and even though this vehicle is fantastical, it includes realistic details, such as the tilted rotor with an engine fed by intake hoses. Its highly unusual wings are angled, swept forward, and of staggered height. They lift upward to reveal a droid, folded up for storage. The tails of the empennage hinge downward to become wings, unlocking the rear cargo—a large hexagonal container housing a tool rack and a transparent fluorescent round brick with fins. The crane's articulated boom concertinas outward, making it long enough to deposit the cargo to the rear of or beside the craft. The black chainsaw body element on the nose can be angled downward.

"I BUILT IT SO MANY TIMES, I DIDN'T NEED THE INSTRUCTIONS!"

—FRÉDÉRIC ROLAND ANDRE, SENIOR DESIGN MANAGER, THE LEGO GROUP, 2010 TO PRESENT

"Anubis, get the cargo onboard. Prep for urgent evac!" shouted the captain over the increasing rumble. The droid deftly steered the container into place, and as it leapt toward its charge pod, it folded itself into a neat cube to fit inside. As the insulation wings lowered, the cathode rotor began to spin. The neon lasers on the empennage began to fire on nearby trace elements of helium, creating metastable ionized molecules that were drawn into the acceleration chamber of the magnaplasmadynamic thruster. As the ground began to crumble around the ship, the hypermagnet forced spin-polarized triplet helium through the exhaust, sending the ship spinning across the cosmos at near light speed.

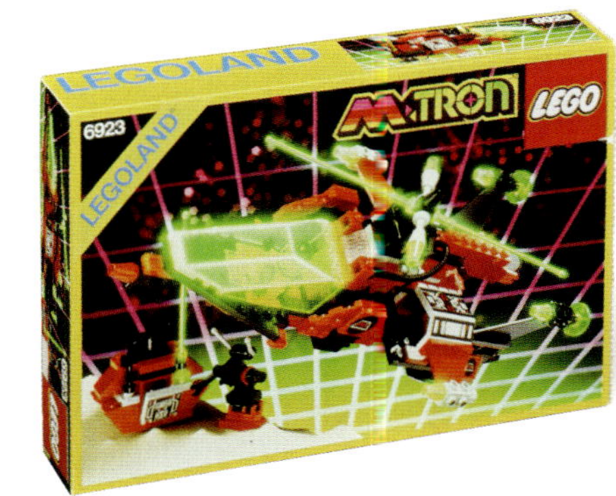

▼ Original mockup of box.

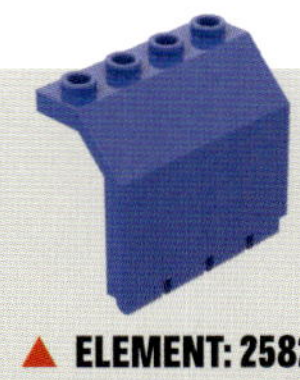

▲ **ELEMENT: 2582**
INTRODUCED: 1990

Hinged panel 2 x 4 x 3⅓ *has a right-angled form with a sloped midsection and a hinge at its base. It often came decorated and appeared in three M:Tron sets with a distinctive electronic pattern matching other printed elements, such as the small container door. Here, it is used as wings. In the same year, it also appeared on a spaceship in the LEGO® Town theme, 1682 Space Shuttle, as the payload doors.*

6989 MEGA CORE MAGNETIZER

Released: 1990
Total elements: 506
Unique elements: 154
Minifigures: 3
Also known as Mobile Recovery Centre in UK retail catalogs.

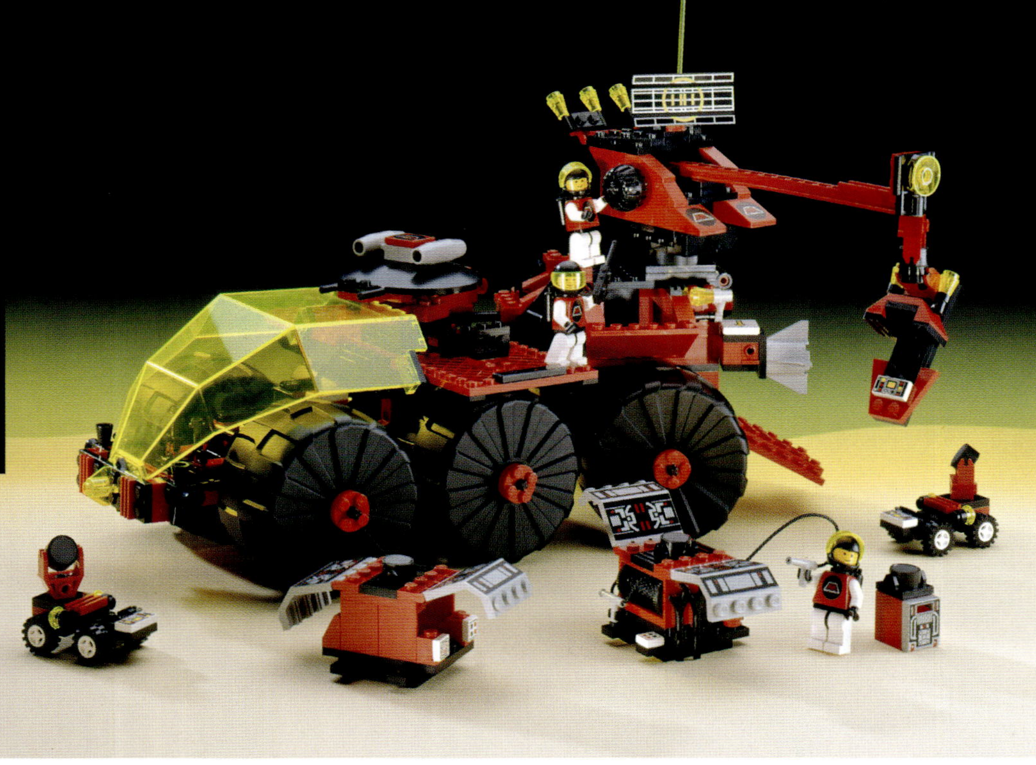

▶ This aerial view highlights the unusual tapered cockpit with room for three minifigures and the vehicle's articulation, with steering controlled by twisting the central dish.

▼ Original mockup of box.

The numerous small containers and secondary vehicles in this set are dwarfed by the bold shapes of the articulated six-wheeled vehicle, emphasized by new large elements such as the hard plastic giant wheel. The driver's cab looks more like a command deck, with tiered seating for three. The rear crane has an extendable jib that can grab various items with its magnet and tilt forward to deposit them on the surface. The six items included are one small container, the two large containers on the projecting platforms, two small buggies nestled in the large inverted sloped elements on each side, and a tiny shuttle stored in the rear cargo. The latter is accessed by swinging the rear thrusters outward to release the ramp. One of the large containers includes nozzles attached to a reel of string.

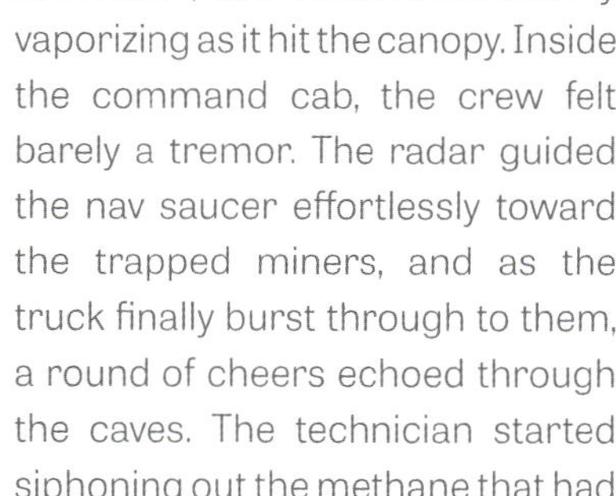

The whole cave shuddered as the magna-drill bored relentlessly through the rubble, each loose rock instantly vaporizing as it hit the canopy. Inside the command cab, the crew felt barely a tremor. The radar guided the nav saucer effortlessly toward the trapped miners, and as the truck finally burst through to them, a round of cheers echoed through the caves. The technician started siphoning out the methane that had been pooling at the roof of the cave as a byproduct of the oxygen pumps while the loadmaster unloaded the cargo. Suddenly, a brief laser beam shot into the cave from the new tunnel. As everyone turned in bewilderment, the cloud of methane exploded with a blinding light. Steam and nitrogen gas enveloped everyone below. The fire officer grabbed the foam induction nozzle while punching a command into the control panel and fired a burst of foam upward that enveloped the gases in a film within seconds.

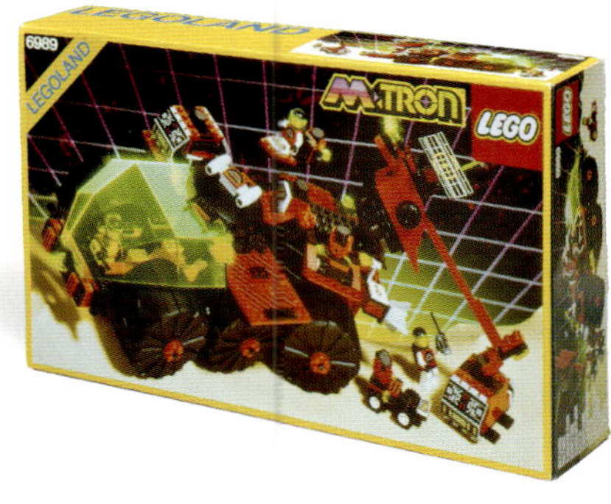

Although it first appeared in a LEGO® Pirates set, the ***modified plate 8 x 8 with grille*** *was initially developed to create platforms on the Mega Core Magnetizer. Its design was an adaptation of a mirror frame only ever used in one set, from the 1980 jewelry range LEGO® Scala. Only minor changes needed to be made to the existing mold, including the addition of four central studs.*

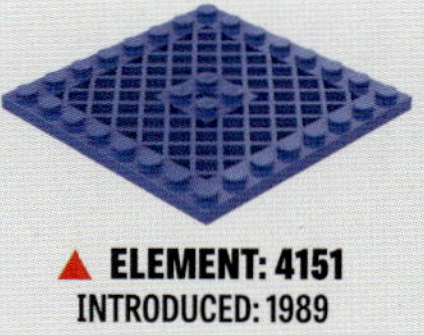

▲ **ELEMENT: 4151**
INTRODUCED: 1989

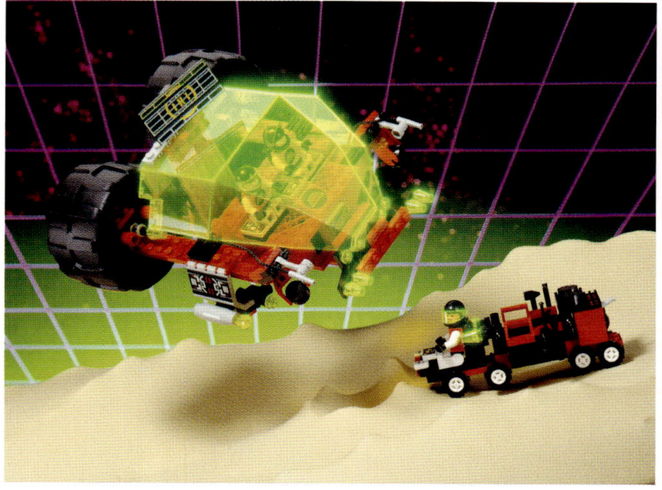

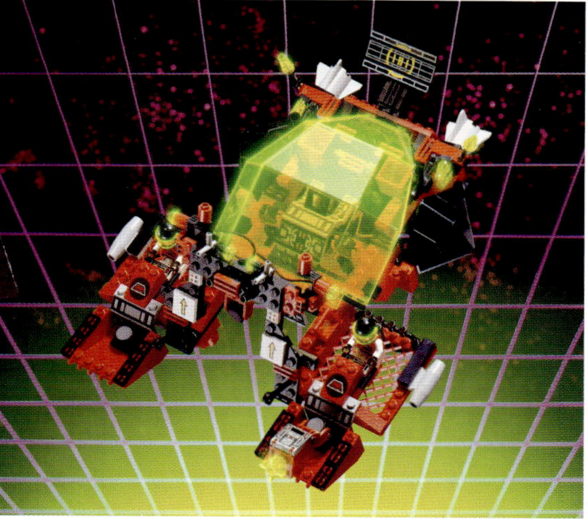

Two highly imaginative alternative models shown on the back of the box: a bubble ship using the large tires as thrusters and a space digger reminiscent of a flying lobster.

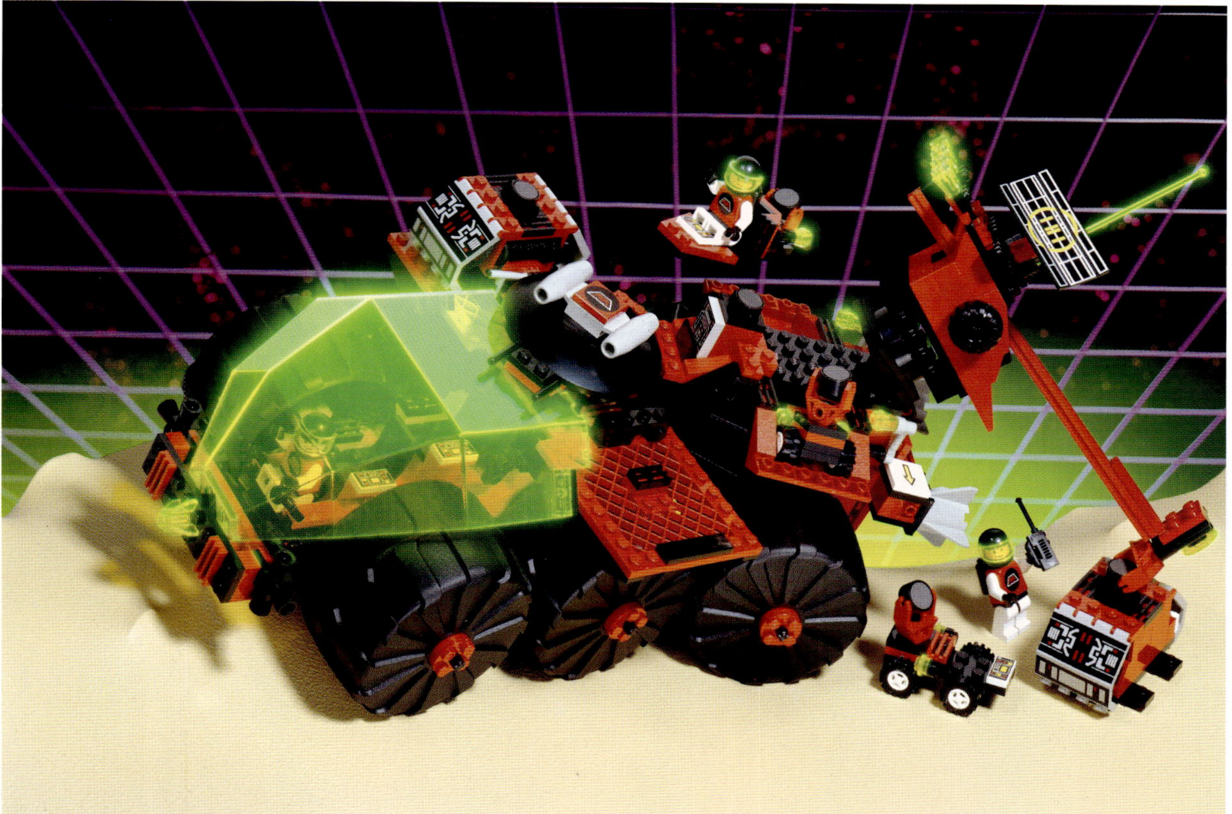

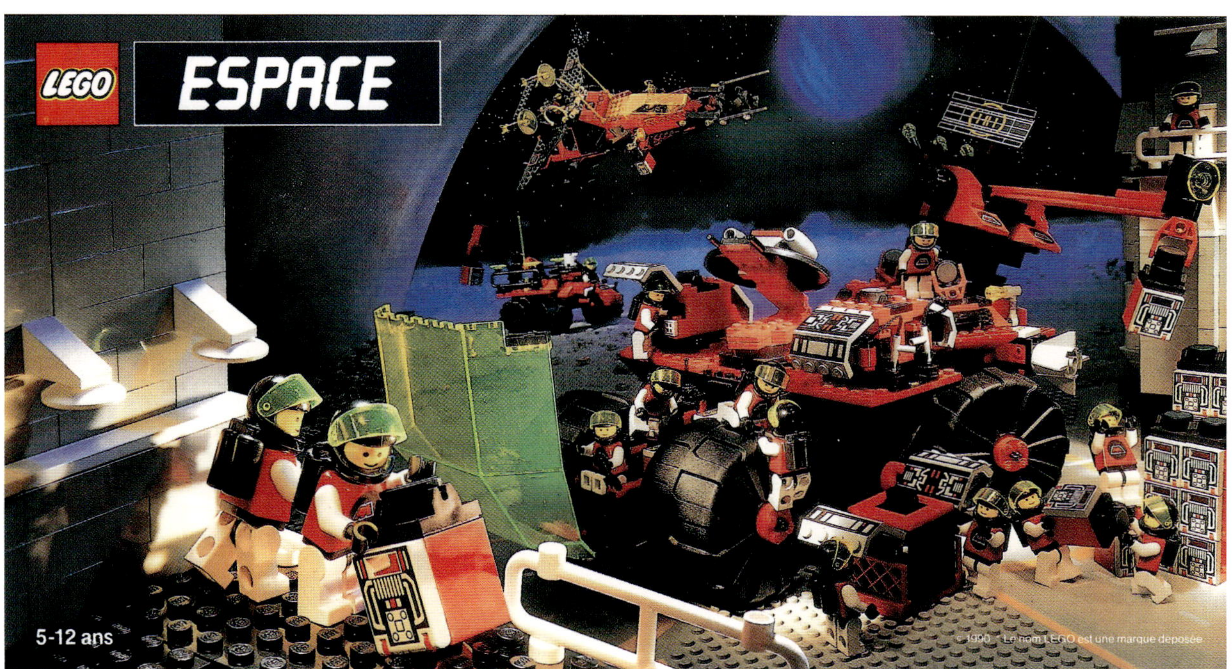

The Mega Core Magnetizer takes center stage in this exciting scene from a French ad in 1990.

6956 STELLAR RECON VOYAGER

Released: 1990
Total elements: 233
Unique elements: 101
Minifigures: 2
Also known as Rescue Star Cruiser in UK retail catalogs.

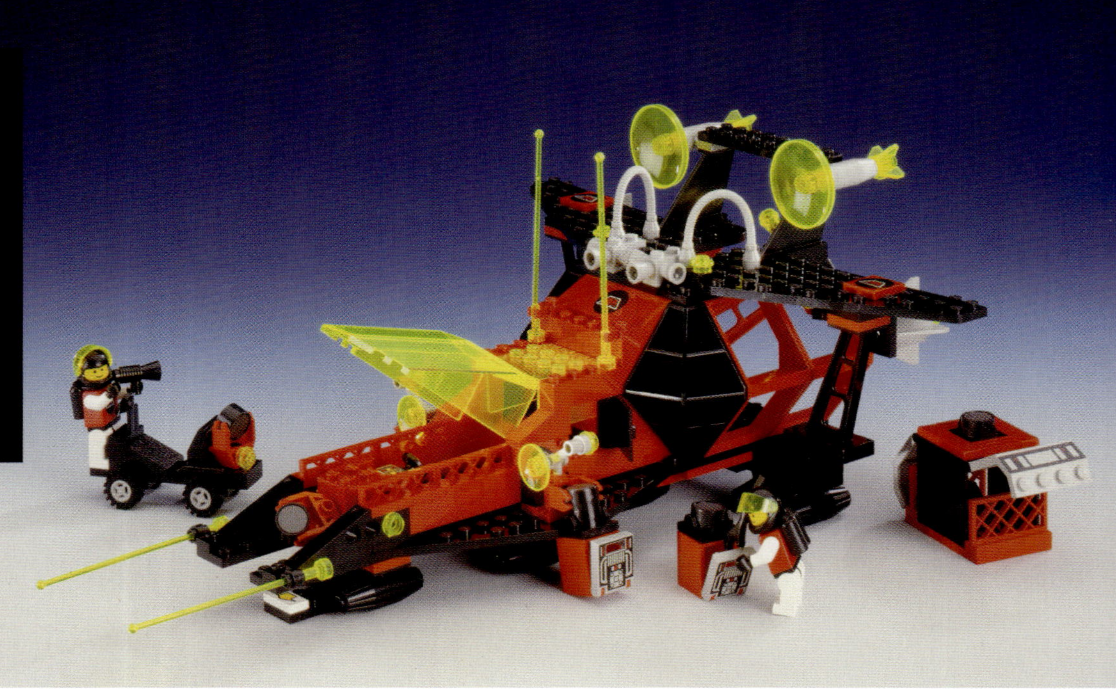

▶ The octagonal container is lifted by the magnetic crane to fit neatly within the rear cargo bay. At the nose, the buggy is attached using magnets.

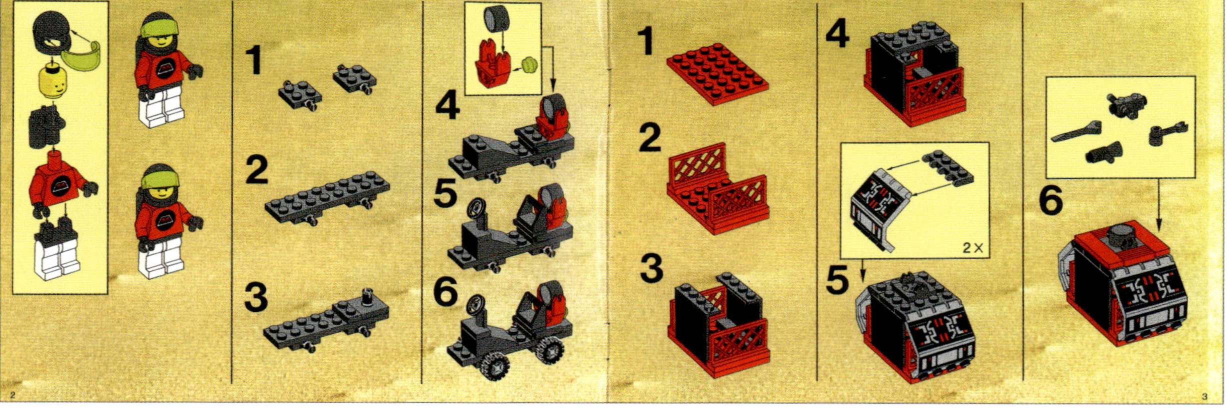

This huge ship has four pairs of magnets to attach the large container at the rear, two small containers on the sides, and most surprisingly, a buggy in its nose! It includes the return of several older elements, such as the steering wheel for the controls, small fences lining the cockpit, gray reaction control system thrusters, and a slope 33° 3 x 6 printed with the subtheme's logo. It also includes some transparent fluorescent green elements not found in other M:Tron sets: the dishes on the empennage and a hinged canopy extender that creates a generously sized cockpit which can be used to seat both minifigures or for storage using the black doors for access.

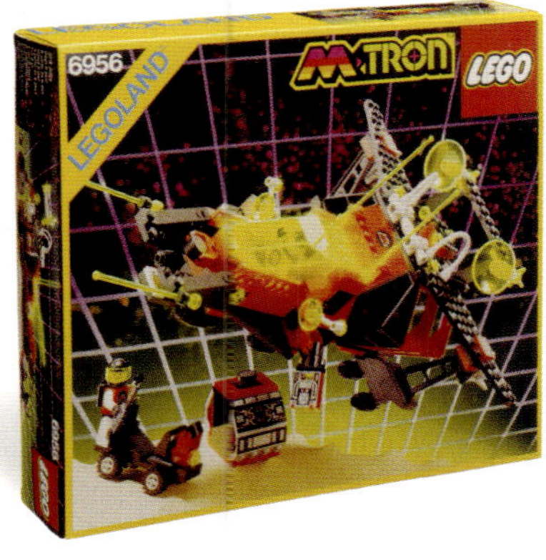

"If we leave the stasis crate behind," shouted the doctor over the din of the evacuation, "we can fit in all the injured." The captain stared at the container, now brimful of ferrocrystals. "Are you crazy?" he replied. "That's enough to harness the field of a neutron star!" The loadmaster stepped in between them. "We can hitch it to one side," she said, "and the buggy to the other, weighted down as best we can." Their cosmocomms units crackled. "Alert: unknown craft approaching system" came the warning. "Let's make this quick!" said the captain. Soon, the cruiser cautiously rose into the rescue shaft, its starboard thruster compensating as best it could for the longitudinal roll. As the captain fired the magna-gimbles in short pulses to clear more rock, the buggy and crate swung perilously and the miners clung to the cargo cage for dear life.

*One of two fastenings introduced by M:Tron, **magnet holder 2 x 3 with clips and pinhole** is designed to allow the magnet to rotate. It takes the form of a 1 x 1 LEGO® Technic brick with angled sides that have projecting arms with clips. After appearing in most LEGO® Space subthemes of the 1990s and 2000s, it was discontinued in 2007 following the introduction of new magnetic elements.*

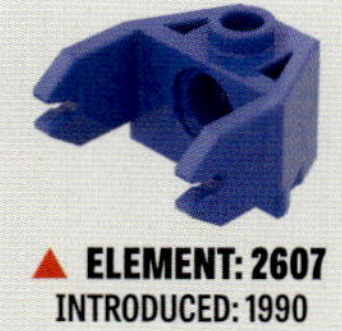

▲ **ELEMENT: 2607**
INTRODUCED: 1990

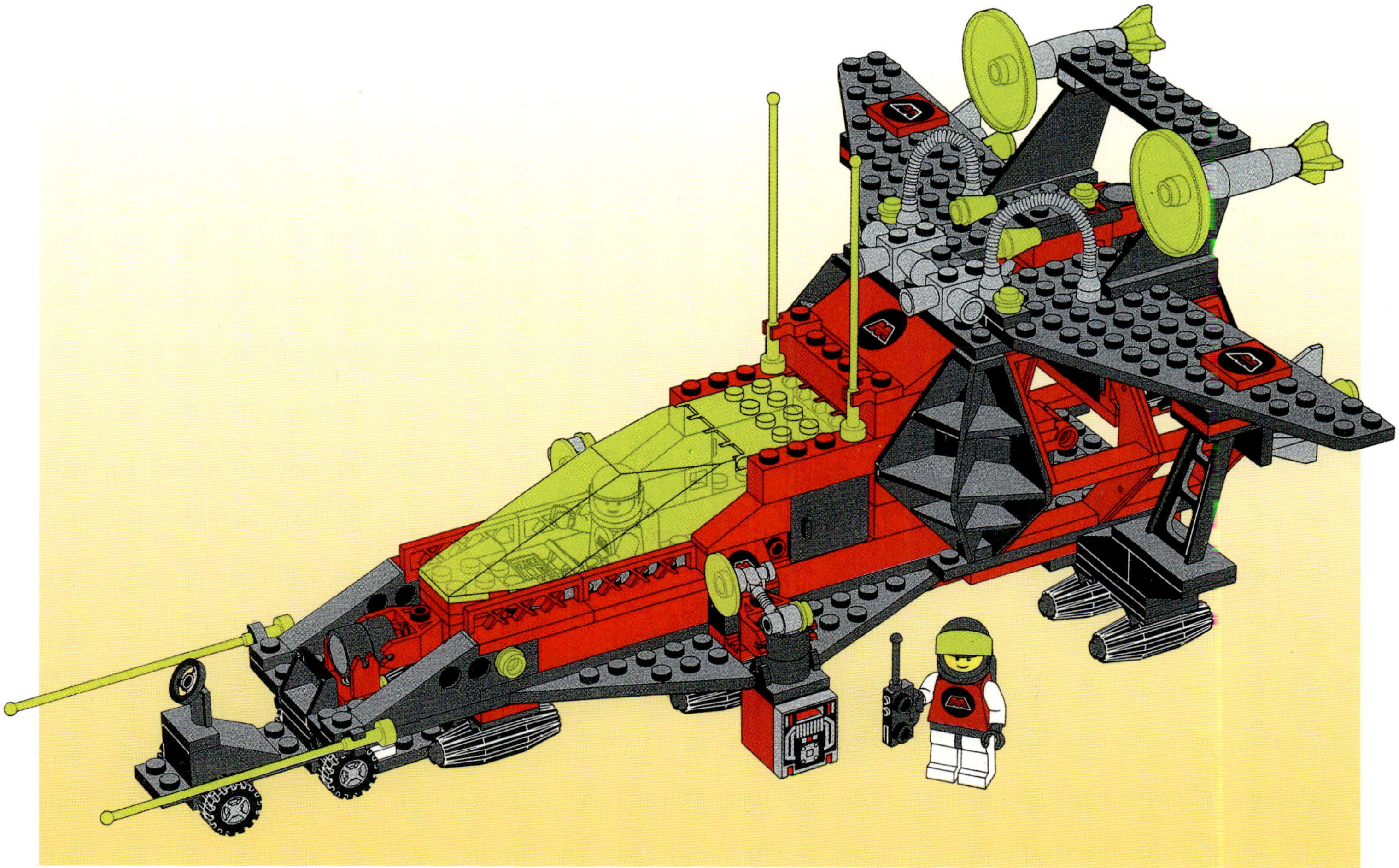

This rover may only consist of a few elements, but it packs in a lot of features. Although small, its chassis is articulated and includes a plate with handlebars behind the hinge to limit rotation. Although the slope at the front is not decorated, the rear features an unusual asymmetrical array of transparent fluorescent green equipment, which the driver can adjust by standing on the raised platform at the rear. The radars and antenna are attached to the vehicle using plates with clip-and-bar connections, allowing a wide range of movement.

The lone scout traversed the jagged rocks of the asteroid, glancing occasionally into the sky to check her bearings. With the cosmocomms system now hacked by an unknown enemy, they were isolated, and she figured her idea was worth a try. Spotting the A-37 gas cluster, she tapped the tablet to stop before detaching it and jumping onto the rear, where she pulled off the ferroantenna and twin magnastrictive transducers. She carefully connected all three together and reprogrammed the aerial to boost ultrasound frequencies. Snapping her new device back onto the vehicle, she angled the aerial toward A-37-K and began recording. "This is the M:Tron mining outpost A-37-K5. Unknown hostile force has severed cosmocomms and sabotaged the mine. I hope they don't intercept this. Send immediate assistance."

1478 MOBILE SATELLITE UP-LINK

Released: 1991
Total elements: 31
Unique elements: 15
Minifigures: 1

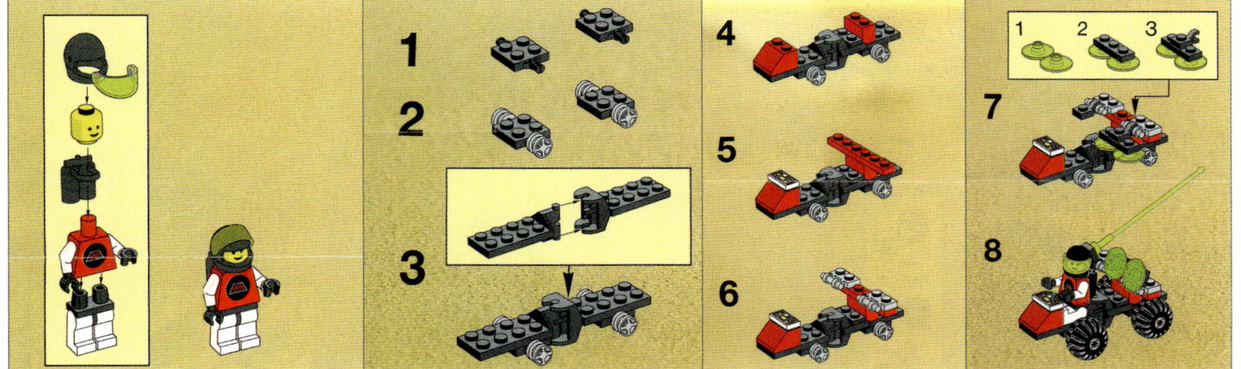

BLACKTRON FUTURE GENERATION

6812 GRID TREKKOR

Released: 1991
Total elements: 25
Unique elements: 12
Minifigures: 1
Also known as Blacktron Spy Buggy in UK retail catalogs.

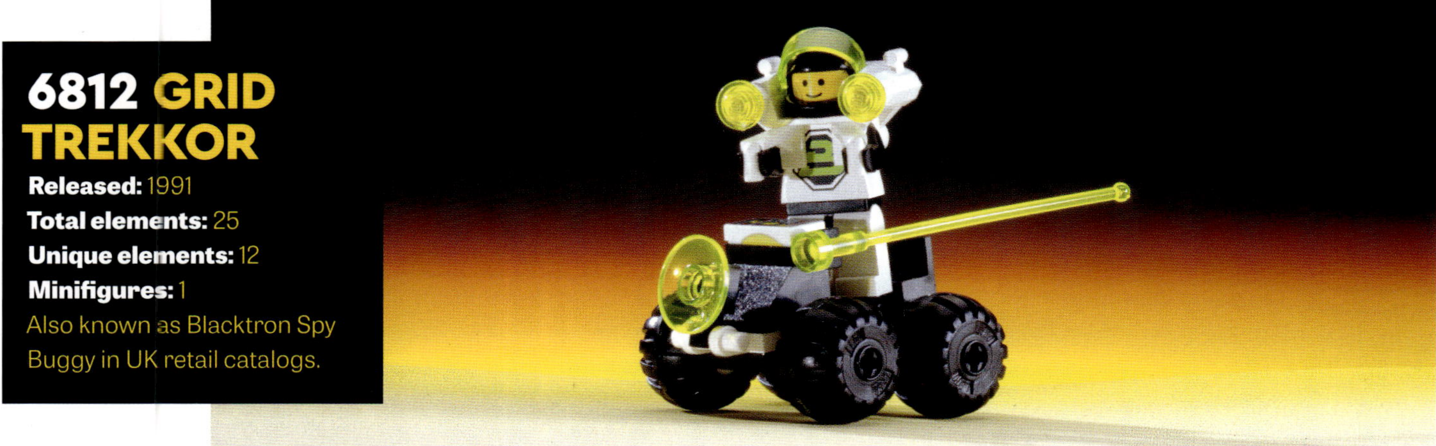

This little rover features an angled chassis and balloon tires. It is controlled by a computer panel, with a horizontal antenna to one side and an adjustable radar at the front. The main focus is on the minifigure, adorned with the exciting new jetpack element which remains relatively rare in sets: apart from Blacktron Future Generation, it only ever appeared in a couple of supplementary packs and one space-themed LEGO® Town set of 1999.

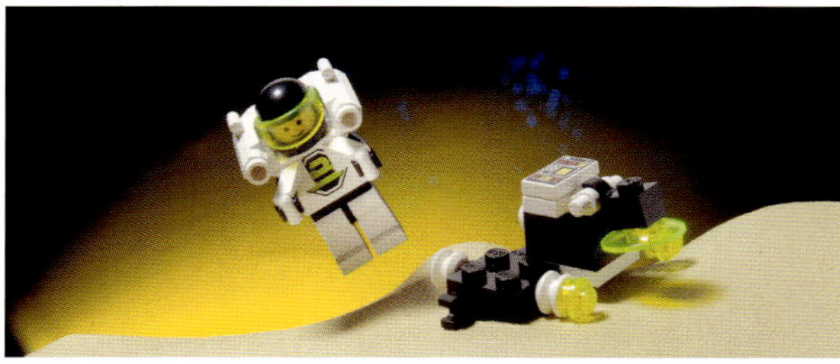

▶ An alternative model featured on the back of the box.

As she entered the tunnel, the vigilist activated her shoulder-mounted gradiometers, but her head-up display showed no signs of life inside the asteroid. Reaching a spacious cave, she tapped the tablet to stop the buggy and control her hoverpack. She lodged the beacon in a concealed nook in the roof of the cave, and just as she was about to update the Future Generation pantheon, her display flashed. Intruders had landed by the tunnel! With no time to blast out an escape route, she paired her suit to the tablet of the buggy waiting below and used the hoverpack controls to steer the buggy remotely. Angling its gravity disk downward, she drove the vehicle straight up the cave wall and roof to her position. With any luck, the intruders would drive right past and not even look up.

*The **minifigure jetpack with twin handles** is worn over the shoulders and has a bulky, U-shaped form that is packed with connections: two studs at the front, four holes at the rear, and two bars on top. These connections could be a means of attaching further technology or simply interpreted as engines or weapons.*

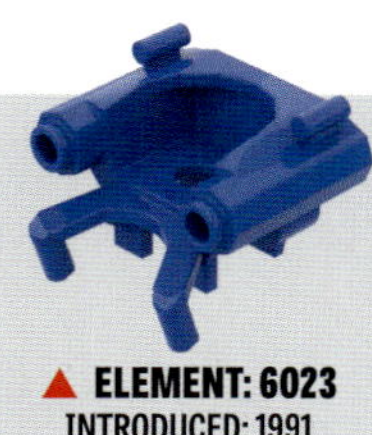

▲ ELEMENT: 6023
INTRODUCED: 1991

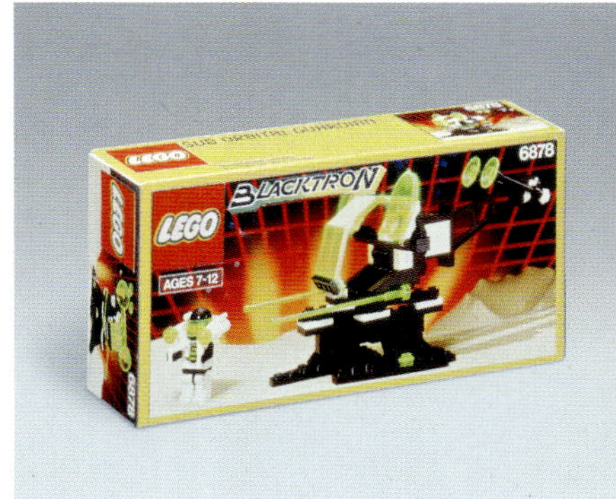

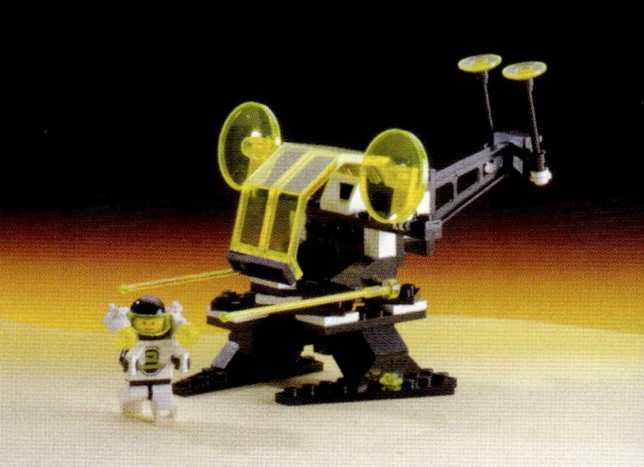

6878 SUB ORBITAL GUARDIAN

Released: 1991
Total elements: 75
Unique elements: 36
Minifigures: 1
Also known as Blacktron Strider in UK retail catalogs.

▲ Original mockup of box.

The legs of this walker use multiple turntables to move in a shuffling motion, like set 6876 Alienator from the first wave of Blacktron. The similarities end there though: the top half emulates the fuselage of a helicopter, adorned with an array of adjustable dishes in transparent fluorescent green. The tail section is made from a container and a stanchion, attached sideways, with hinged antennas at the end.

The custodian remained undetected as she strode across the asteroid in her walker, smashing the intruders' mining equipment. Again, the red-and-black ship passed nearby, closer this time. Flipping open the canopy, she flew up to investigate. At that moment, a huge triangle of light appeared that spanned the landscape around her. The ground began to rumble. She rushed back down to her craft as fast as the hoverpack could go and rotated the grav disks for takeoff. The rocks around her began loosening . . . and tumbled upward into space! Thinking quickly, she reversed the polarity of the grav disks to counteract the unnatural force, delicately maneuvering around the larger chunks of rock as best she could. Escaping the column of debris with minimal damage, she programmed in a suborbital trajectory to avoid detection on her return to the Future Generation outpost.

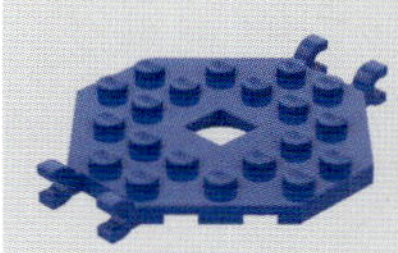

▲ **ELEMENT: 2539**
INTRODUCED: 1989

*Many octagonal elements were introduced by LEGO® Space, but not **octagonal plate 6 x 6 with open center and 4 clips**, which was first used by the LEGO® Pirates theme to create crow's-nests. The central hole is to accommodate sailing ship masts: a feature not required in space.*

6851 TRI-WHEELED TYRAX

Released: 1991
Total elements: 38
Unique elements: 22
Minifigures: 1
Also known as Blacktron Cruncher in UK retail catalogs.

Despite its low element count, this model is still relatively big, thanks to large elements such as the wheels and canopy. The chassis includes several large standard building bricks, and its forward incline gives this vehicle a racing profile. As the wheel elements only require a connection point on one side, the brick situated on the right side of the front wheel is purely aesthetic.

The ATV trike rolled over the rubble left in the wake of the intruders' truck, crunching the debris into a fine powder. The driver pulled up at a vantage point some distance away from the reopened cave and used his visor's enhancement mode to peer inside. The reunited miners and rescuers were so distracted, they had no idea that the Blacktron Duo agent was there. He turned the trike around toward the exit, pointing the rear quantum lasers toward the cave. With two swipes of his tablet, he launched at maximum speed while firing a laser burst backward. The thermal blast from the methane explosion caught up with him almost instantly, but within the protection of his canopy field, the saboteur remained unharmed.

▼ Original mockup of box.

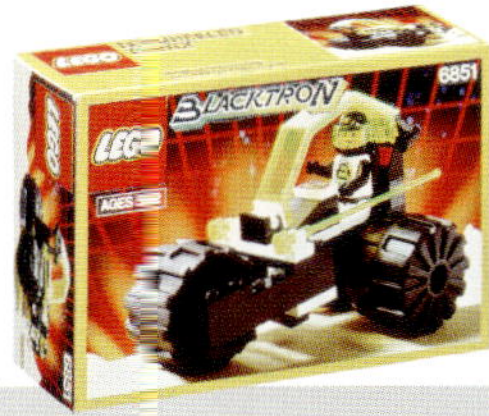

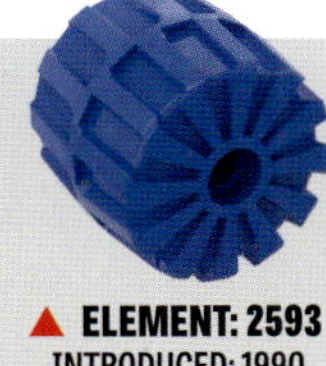

▲ **ELEMENT: 2593**
INTRODUCED: 1990

*The **hard plastic medium wheel** was one of three additional sizes of this wheel design introduced in the years following the large version from 1989, creating a family that is still used in sets to this day. Rather than having an axle hole like some earlier wheels, they attach via a LEGO® Technic pin connection. In 2023, yet another size of wheel was introduced using the same tread pattern: element 3486*

6933 SPECTRAL STARGUIDER

Released: 1991
Total elements: 212
Unique elements: 88
Minifigures: 2
Also known as Blacktron Radar Patrol in UK retail catalogs.

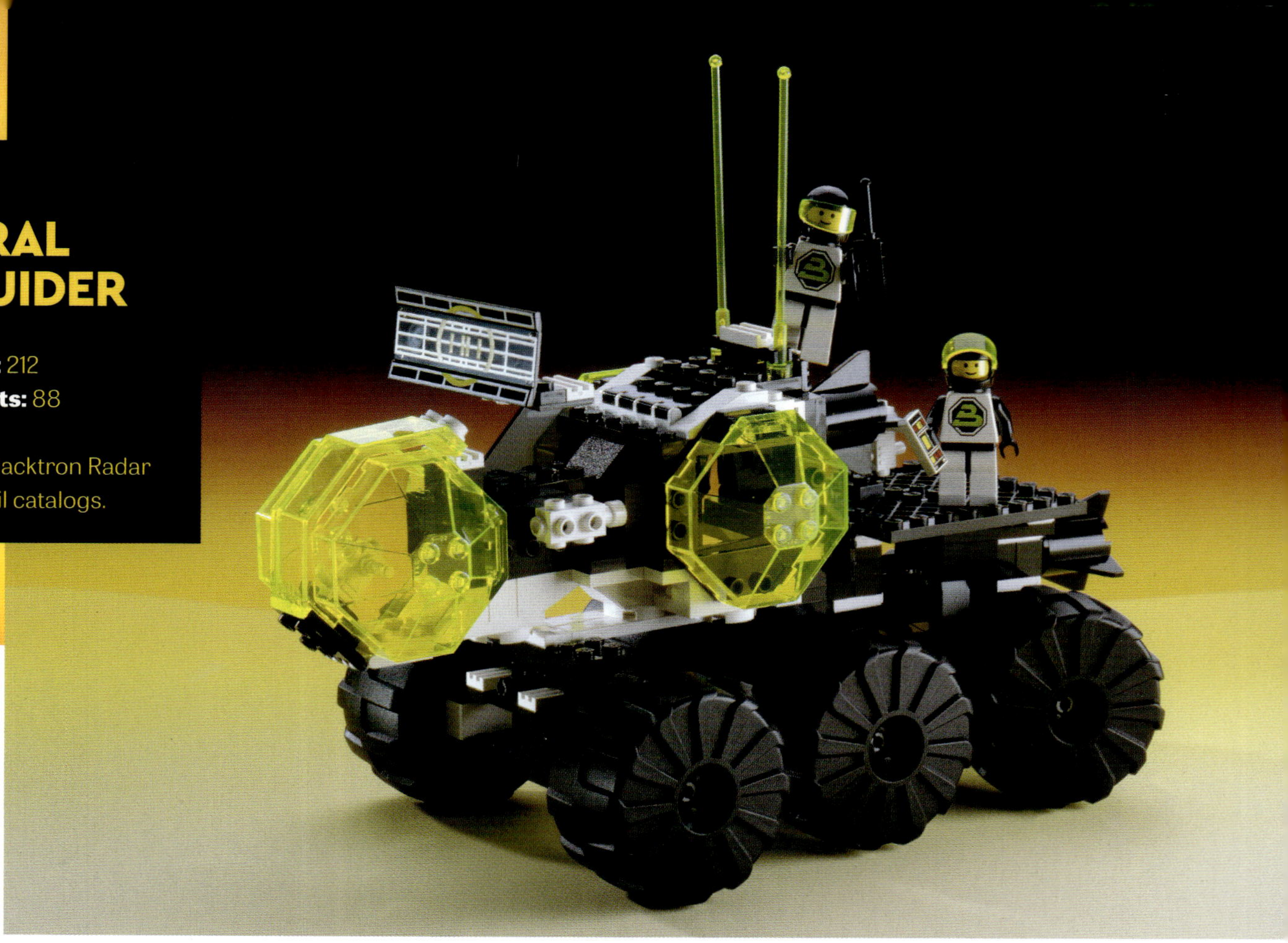

What initially looks like a futuristic truck actually harbors a dual-segment spaceship. The futuristic shape of the completed craft is dominated by its bulbous windows, large thruster, and printed radar made from a snowplow. The remaining articulated chassis has six chunky wheels including four special new bricks containing shock absorbers. Rotating the rear thruster operates the steering mechanism beneath the grille platform, which contains more LEGO® Technic elements than previous LEGO® Space vehicles, including small gears and a rack.

▼ Original mockup of box.

The mercu-rider drove in a race against time as the approaching sunrise of Centaur-F threatened to reach the ATV and scramble the delicate mechanics of the articulator. She combined each twist on the hydraulic arm with a burst of the rear thruster, lurching the vehicle swiftly across the moonscape. The hyperdampers ensured the vehicle rolled over the large, jagged rocks with ease, and she reached the shelter with time to spare. She attached her cab to the studio pod where the helioarchitect was just finalizing the spec. He removed the data tablet from the safe and walked out to the virtual reality platform to connect the tablet to the puppet arm. The mercu-rider flew the ship to the target location directly facing the red giant and paired their command arms and head-up displays. The helio-architect took control from the shelter, activating the ship's magna-helix antennas to grip the star's electro-magnetic field before using the Mercator dish to wrap it in a giant negative-mass grid. As the red giant began to shine in rainbow colors, the mercu-rider picked up her quantum comms. "Centaur-F to Centaur-A. Open wormhole for stellar transport."

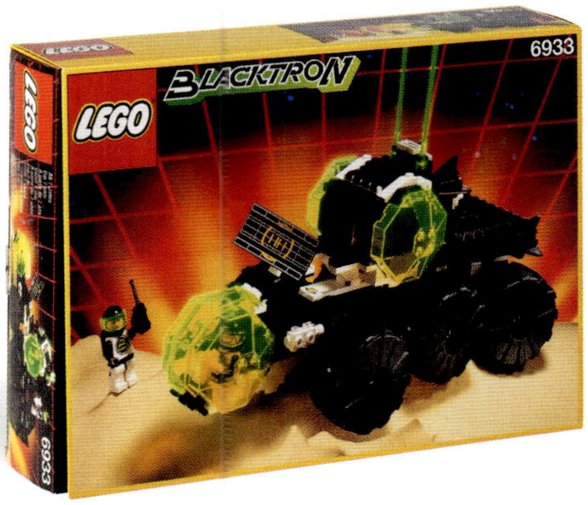

WHAT INITIALLY LOOKS LIKE A FUTURISTIC TRUCK ACTUALLY HARBORS A DUAL-SEGMENT SPACESHIP.

***Brick 1 x 2 x 2⅓ with spring unit** is a tall brick housing a shock absorber assembly with a LEGO Technic pin hole for connecting a wheel. A plate projects outward from the brick to provide stable attachment to the vehicle. This is the only LEGO Space set to include them, and they only appear in seven sets in total.*

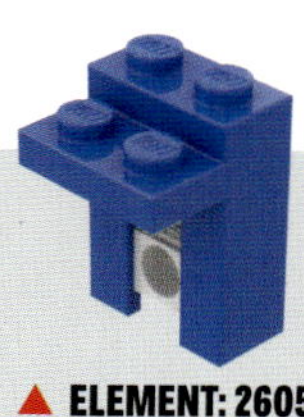

▲ ELEMENT: 2605
INTRODUCED: 1991

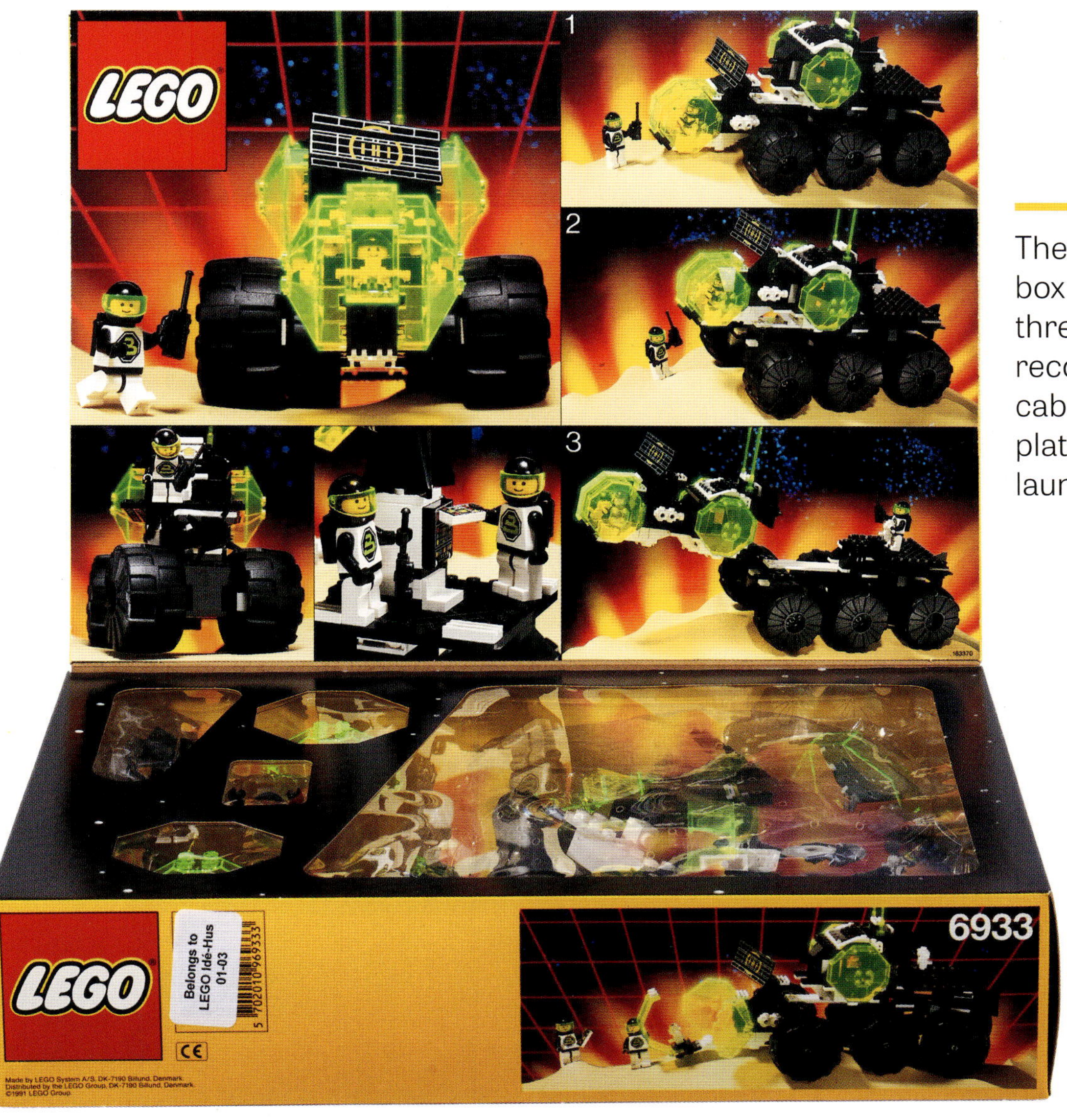

The interior of the box flap showed the three steps required to reconnect the vehicle's cab to the pod on the platform to create and launch the ship.

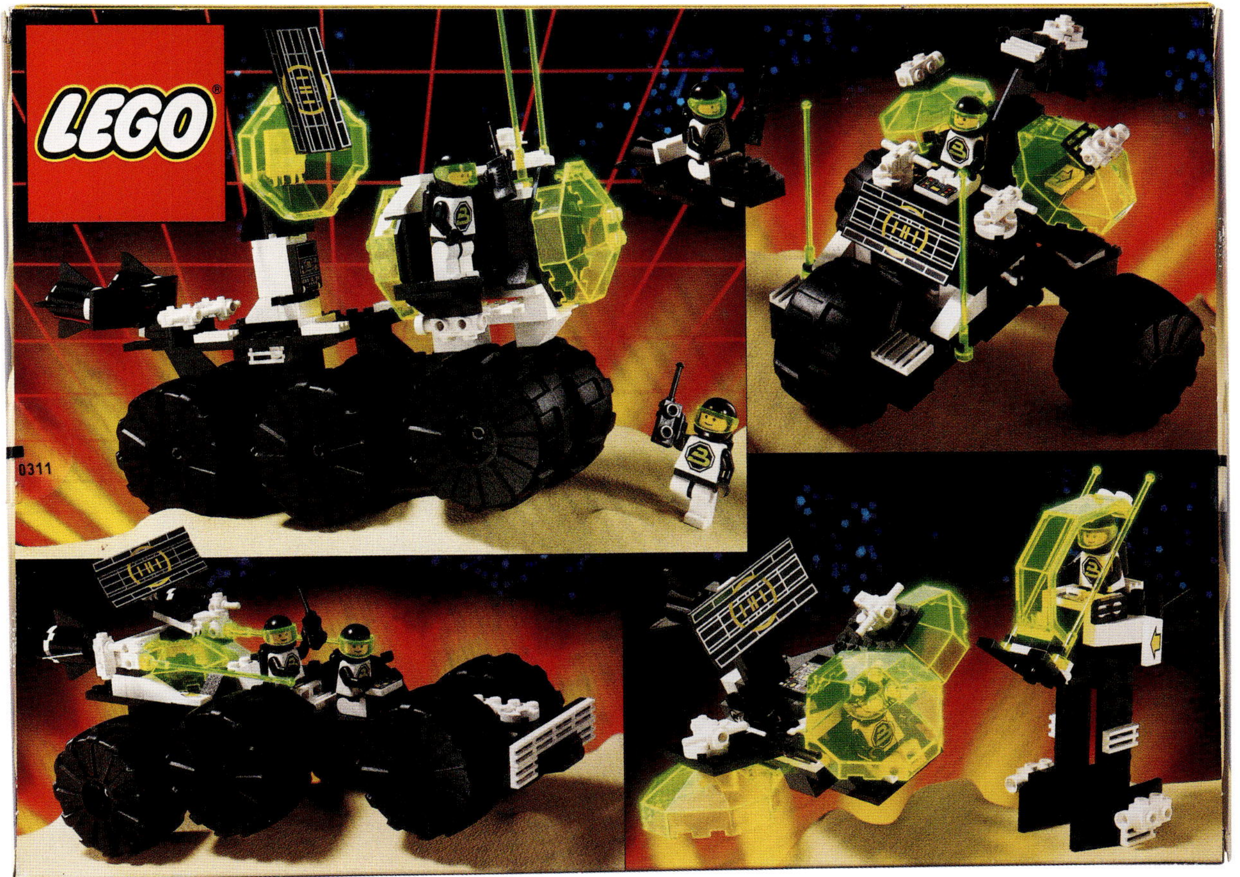

Alternative models shown on the back of the box.

6887 ALLIED AVENGER

Released: 1991
Total elements: 100
Unique elements: 45
Minifigures: 1
Also known as Blacktron Octopod in UK retail catalogs.

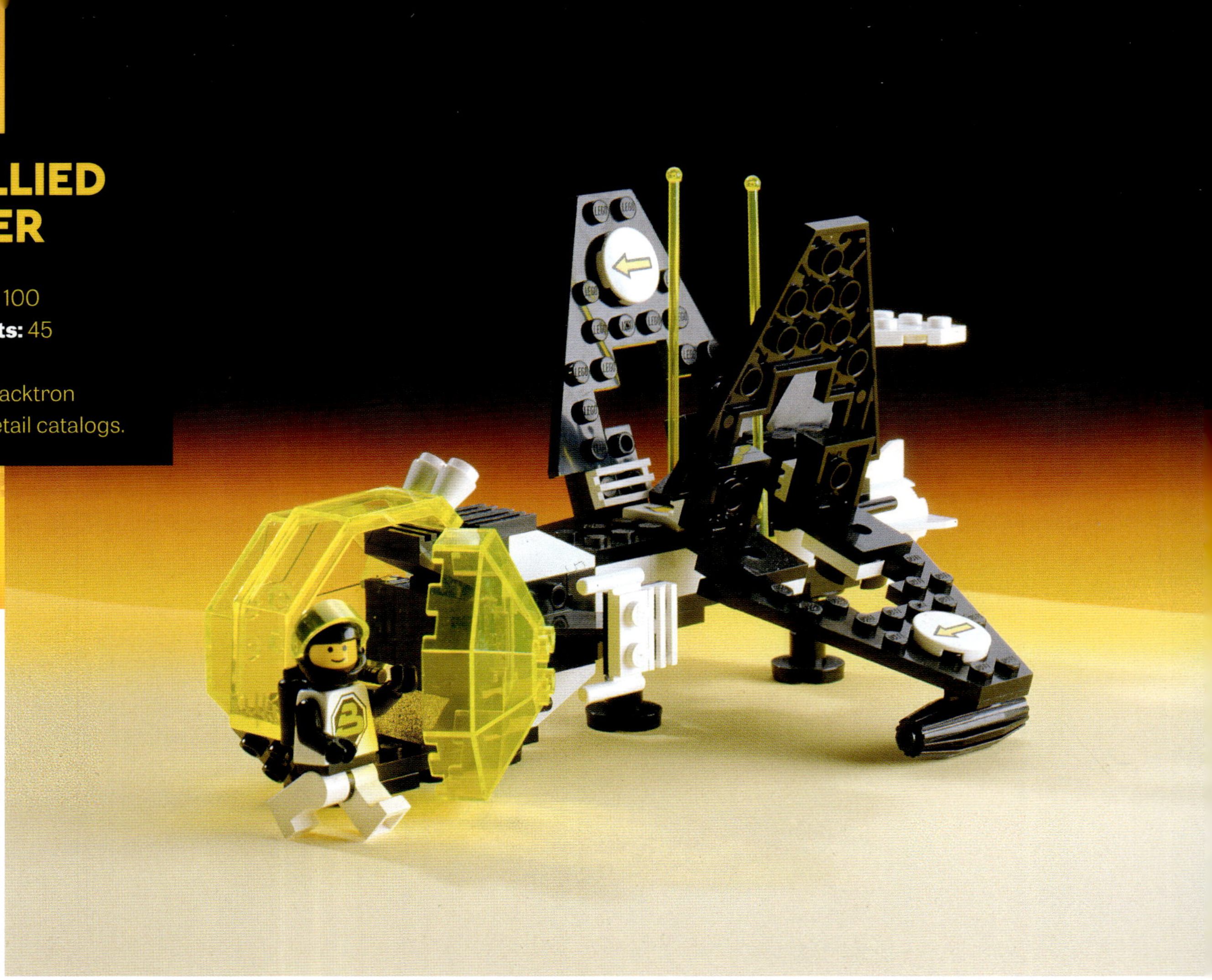

Reminiscent of a fish, an insect, or perhaps even a bomb, this LEGO® Space craft is quite unlike anything that came before it. The teardrop shape of the fuselage tapers backward from the bubble cockpit to a narrow spine, from which four wings project outward in an X formation. The wings have cruciform holes, and each pair is attached with hinges that allow them to rotate slightly for landing. The ship can separate at its midpoint to be interchanged with other models, and like other Blacktron Future Generation ships, the bubble cockpit opens up completely: two octagonal canopies swing back sideways and the central angled panel swings forward.

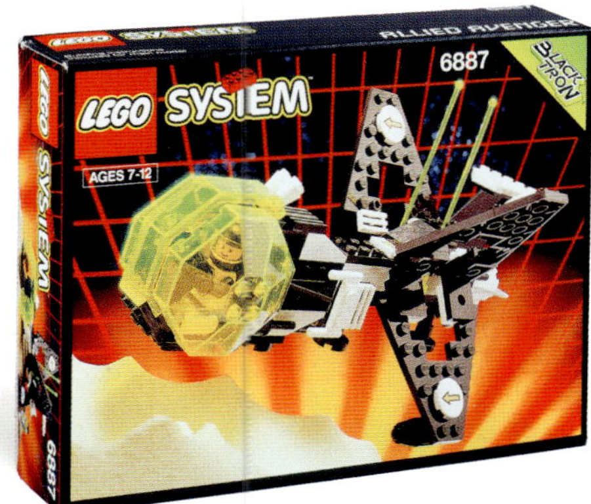

The directive had come at an inconvenient time: the helio-farmer had only recently arrived at his new contract at the giant branch star, and questions would be asked if he didn't collect a full tank of nitrogen. However, a chance to prove his loyalty to the Blacktron Duo was more important than this lousy job. "Disengage stellar drill. Return to harvester," he commanded, and his octagonal pod flew away from the star. He was soon reunited with the scoop field, its four black wings glistening red in the starlight as it gently spun on its axis, mining the nitrogen of every starburst. He reversed the pod to connect with the fusion reactor and requested a systems check on the ramjet. "Plasma accelerator at full power" came the reply. "Nitrogen compression tank at sixty-eight percent." That would have to do.

> **"ONE OF THE COOLEST SPACESHIPS WE HAVE EVER MADE."**
>
> —WILLIAM THOROGOOD, VICE PRESIDENT OF DESIGN, NEW BUSINESS, THE LEGO GROUP, 2004 TO PRESENT

***Brick 2 x 4 x 2 with studs on sides** was introduced in an era when building sideways on models was still uncommon; however, many Blacktron Future Generation sets made use of this new brick to create the signature bubble cockpit. Although many other elements with studs on their sides have been introduced since, this one still finds use in sets occasionally.*

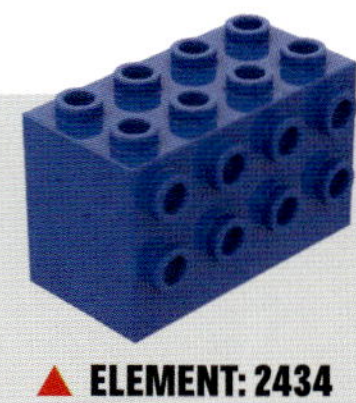

▲ **ELEMENT: 2434**
INTRODUCED: 1990

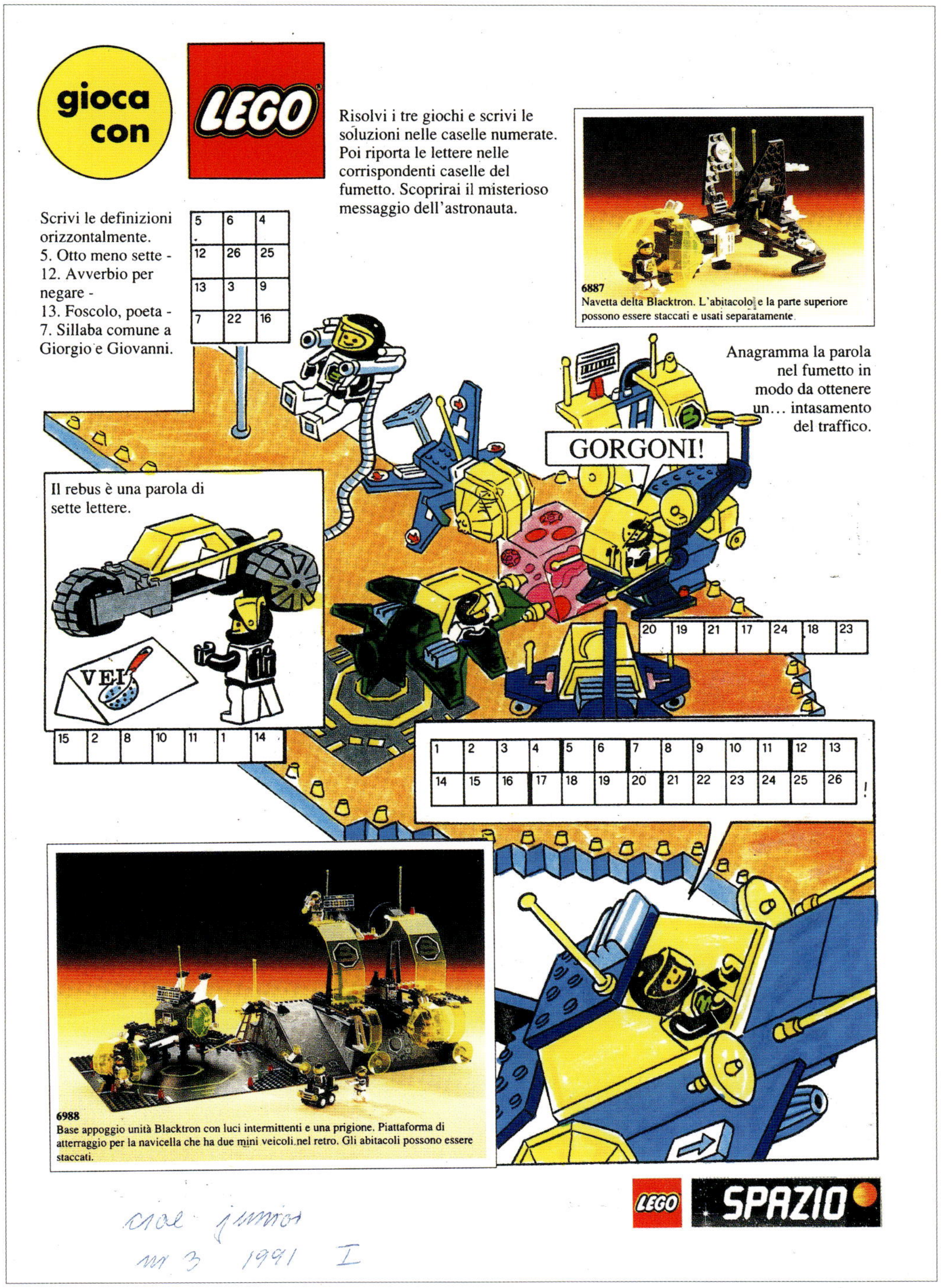

gioca con LEGO

Risolvi i tre giochi e scrivi le soluzioni nelle caselle numerate. Poi riporta le lettere nelle corrispondenti caselle del fumetto. Scoprirai il misterioso messaggio dell'astronauta.

Scrivi le definizioni orizzontalmente.
5. Otto meno sette -
12. Avverbio per negare -
13. Foscolo, poeta -
7. Sillaba comune a Giorgio e Giovanni.

6887
Navetta delta Blacktron. L'abitacolo e la parte superiore possono essere staccati e usati separatamente.

Anagramma la parola nel fumetto in modo da ottenere un... intasamento del traffico.

GORGONI!

Il rebus è una parola di sette lettere.

VEI

6988
Base appoggio unità Blacktron con luci intermittenti e una prigione. Piattaforma di atterraggio per la navicella che ha due mini veicoli nel retro. Gli abitacoli possono essere staccati.

LEGO SPAZIO

The Allied Avenger, known as Blacktron Shuttle in Italy, is among the many products drawn with some creative license in this ad from 1991 featuring word puzzles.

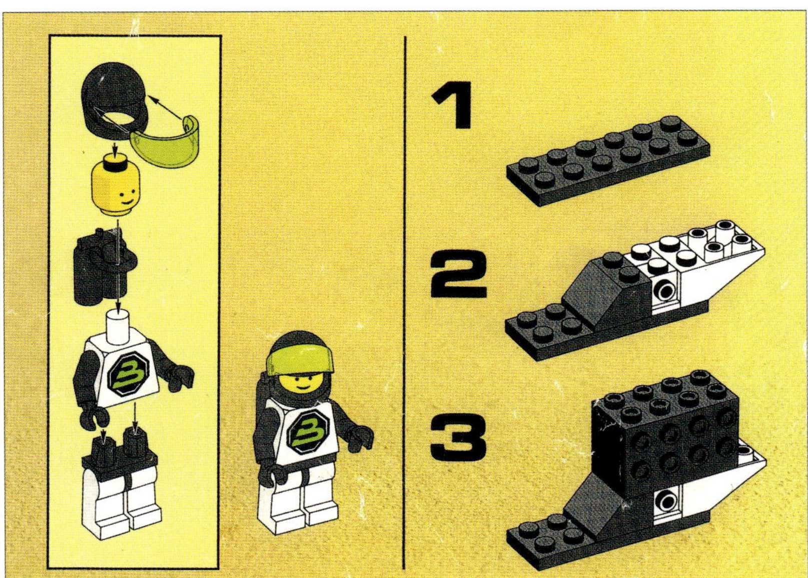

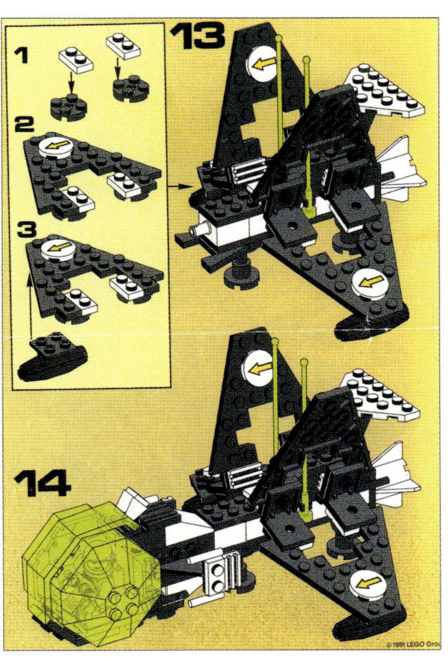

Four large symmetrical wedge plates are joined by pairs of brackets to form the spectacular wings of the Allied Avenger.

6988 ALPHA CENTAURI OUTPOST

Released: 1991
Total elements: 406
Unique elements: 127
Minifigures: 5
Also known as Blacktron Spy Base in UK retail catalogs.

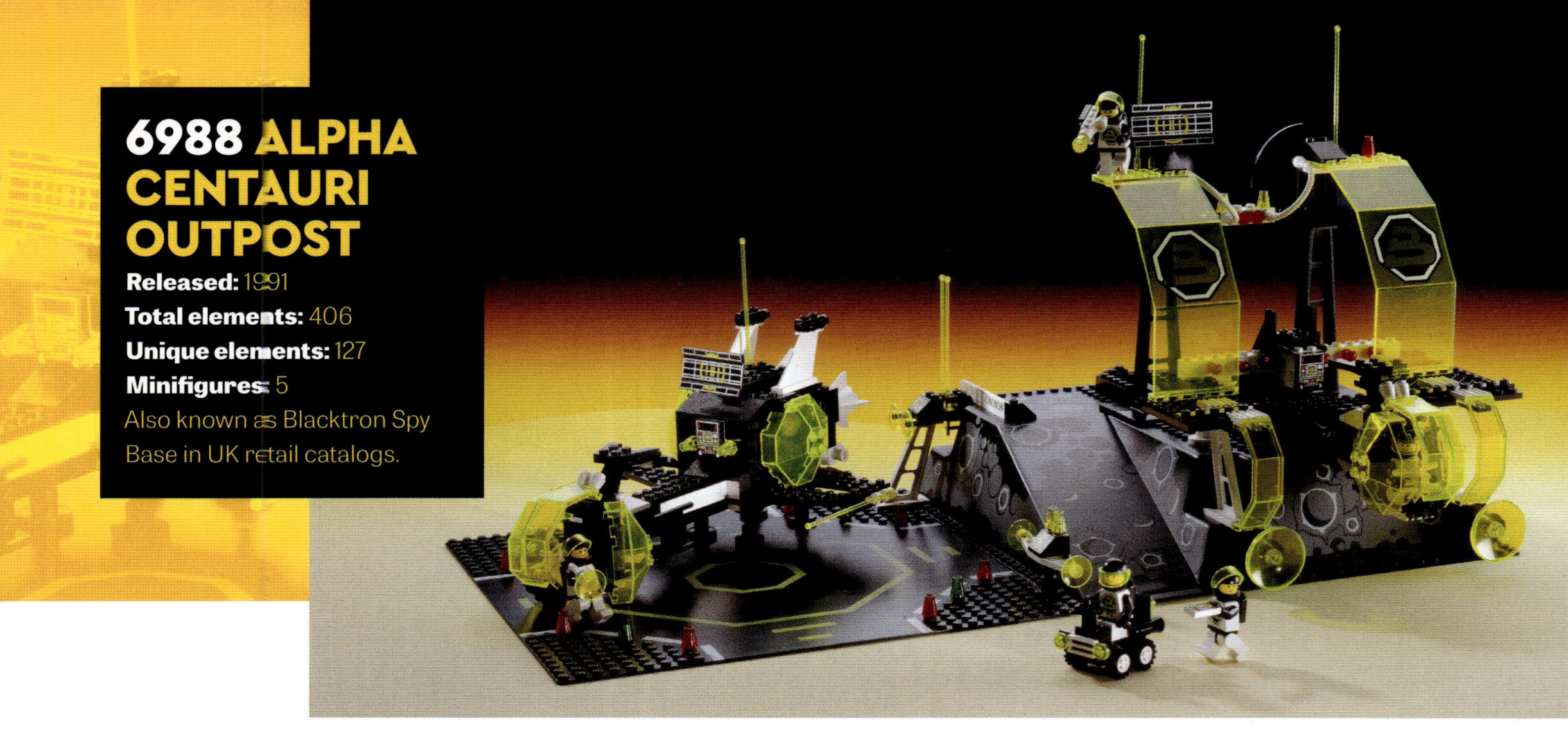

"MY BROTHER HAD ENDLESS ROLE-PLAY STORIES INSPIRED BY THIS SET. WHILE HE TOLD STORIES, I WOULD BUILD MODELS TO SUPPLEMENT THEM AND BUILD UP OUR LEGO SPACE WORLD."

—WILLIAM THOROGOOD, VICE PRESIDENT OF DESIGN, NEW BUSINESS, THE LEGO GROUP, 2004 TO PRESENT

This set includes two new baseplates, providing many open spaces for play. The main craft features a long, prominent spine, with bulk at either end. The design is novel yet also harks back to earlier ships: the thrusters swing sideways to release a ramp and a buggy, and the wings feature rectangular holes like the 6980 Galaxy Commander (see page 56). As with other Blacktron Future Generation sets, the bubble cockpit is modular—and a second is provided, which can be parked at the base of the outpost when not in use. A spare connector is included that combines the two bubble cockpits into one craft. The outpost itself is a sparse framework of panels, stanchions, and dishes, with twin landing pads and a series of lights from the Light & Sound System. The hidden pit within the raised baseplate houses the battery, with stairs and a hatch providing further ideas for narrative play.

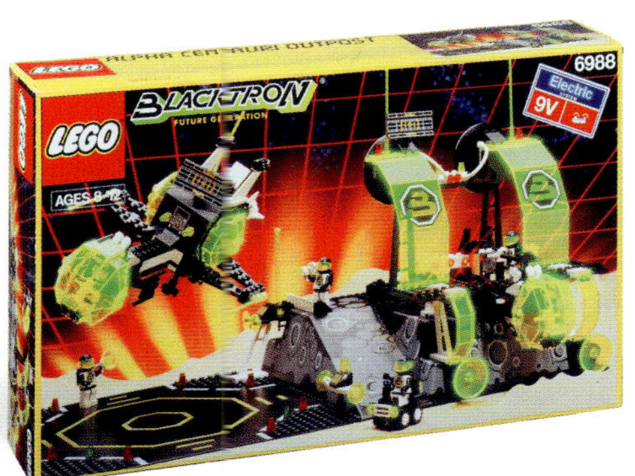

The vast sphere of the negative-mass grid glowed every color of the spectrum as the red giant appeared inside it. From the safety of his craft, the mercu-rider twisted the command arm. "Shipment received, Centaur-F; closing wormhole now. Centaur-A, prepare the electromagnaduct." He switched off the Mercator dish and headed back to the outpost. As he approached, he could see the twin ferroducts glowing brightly as energy from the rapidly dying star behind him was drained into the Future Generation's hyperspace storage tanks. As he got closer, the secondary globeship detached from the fueling station, ready to switch out with him upon arrival, but then he saw something unusual. Two of the crew walked with raised arms up the access ramp as another drove the rover behind, brandishing a negative-mass stapler as though it were some kind of weapon! The mercu-rider watched in astonishment as the prisoners were shepherded down into the hatch of the storage tank farm.

Following the success of the original crater plate, the use of vacuum-formed baseplates in LEGO® Space was expanded—literally—to create ***raised baseplate 32 x 32 with ramp and pit****. Not only could the raised areas be built upon; they could also be decorated. The ink was added to the plastic sheet while it was flat before being pulled up in a vacuum.*

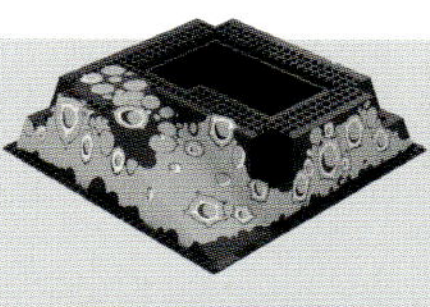

▲ **ELEMENT: 2552**
INTRODUCED: 1991

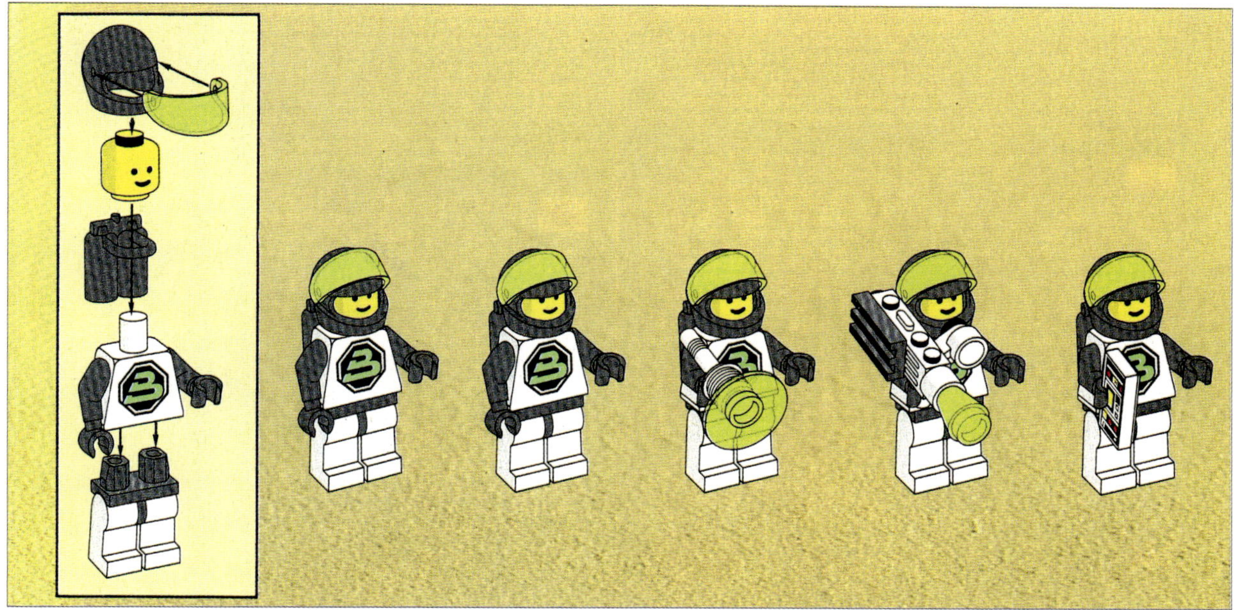

◀ The set came with a team of five identical minifigures and an array of handheld equipment.

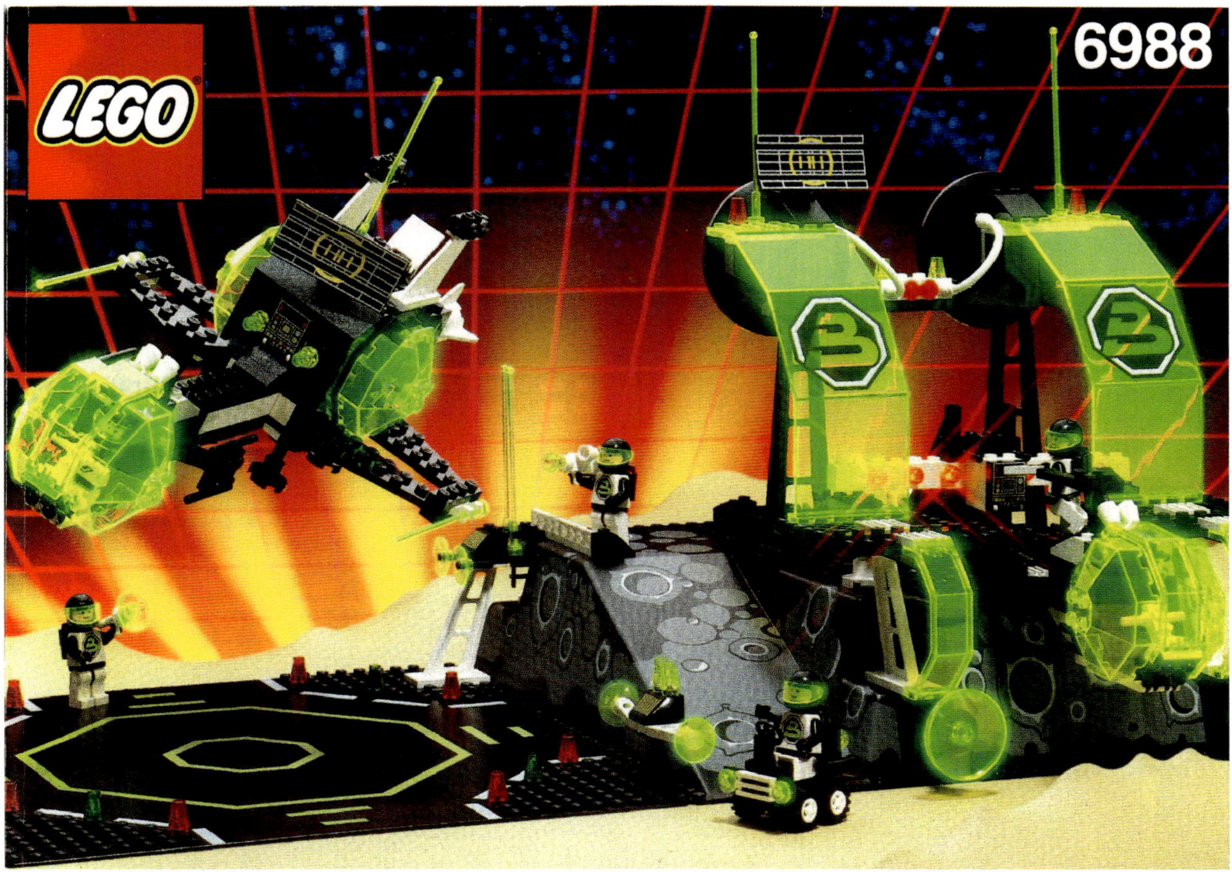

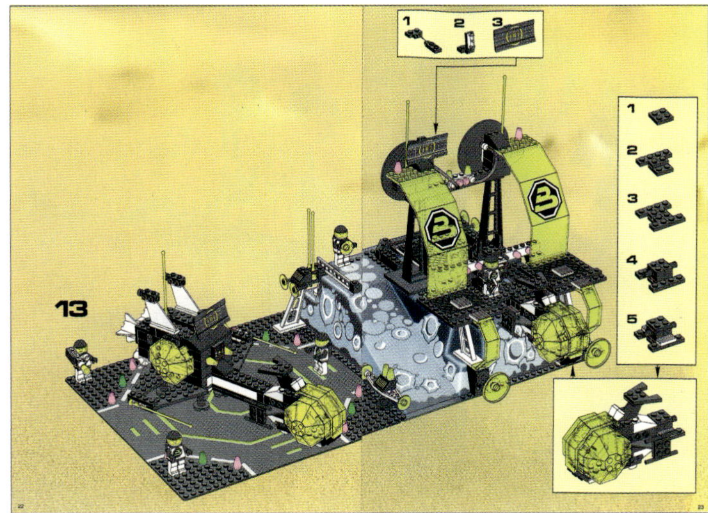

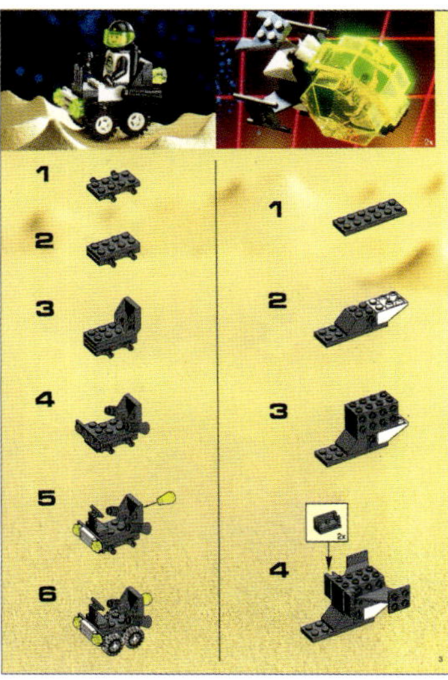

A new feature of building instructions around this time was to make clear which submodel is about to be built by including photographs at their first step.

1479 TWO-PILOT CRAFT

Released: 1991
Total elements: 34
Unique elements: 13
Minifigures: 2

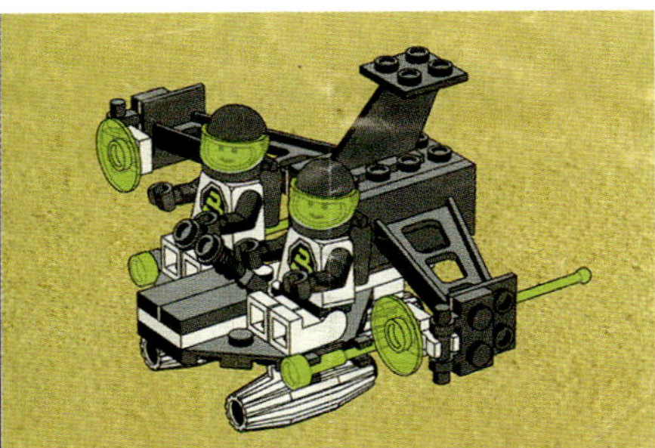

Notably, this small set provides two minifigures and is also packed with exciting building elements, including an octagonal plate with clips, a large brick with studs on the sides, three straked engines, and four transparent fluorescent green elements. Stanchions create dramatic, forward-swept wings, with radars on the tips that can be rotated to the sides. Each pilot uses one joystick to control the craft.

"I'm not so sure now," said the beta-commander. "Having seen these intruders use ferrocrystals . . . ? It's like watching children playing with a new toy." The alpha-commander beside her lifted his half of the signal trigger in readiness. "But they don't understand how the toy works," he countered. "That's kinetic quintessence they're playing with! What if they create a big rip, worse than the Great Milky Way Disaster? We might not be able to contain it to just one galaxy this time." The beta sighed. "You're right. And if they discover that we Future Generation were once the Blacktron, they won't hesitate to wipe us out, like they did our ancestors . . . and use our technology," she said reluctantly, lifting her half of the trigger to prime the quantum laser. They came to a halt and angled their dishes back toward the laboratory, and their hands pressed down in unison.

6832 SUPER NOVA II

Released: 1991
Total elements: 42
Unique elements: 25
Minifigures: 1
Also known as Blacktron Star Rider in UK retail catalogs.

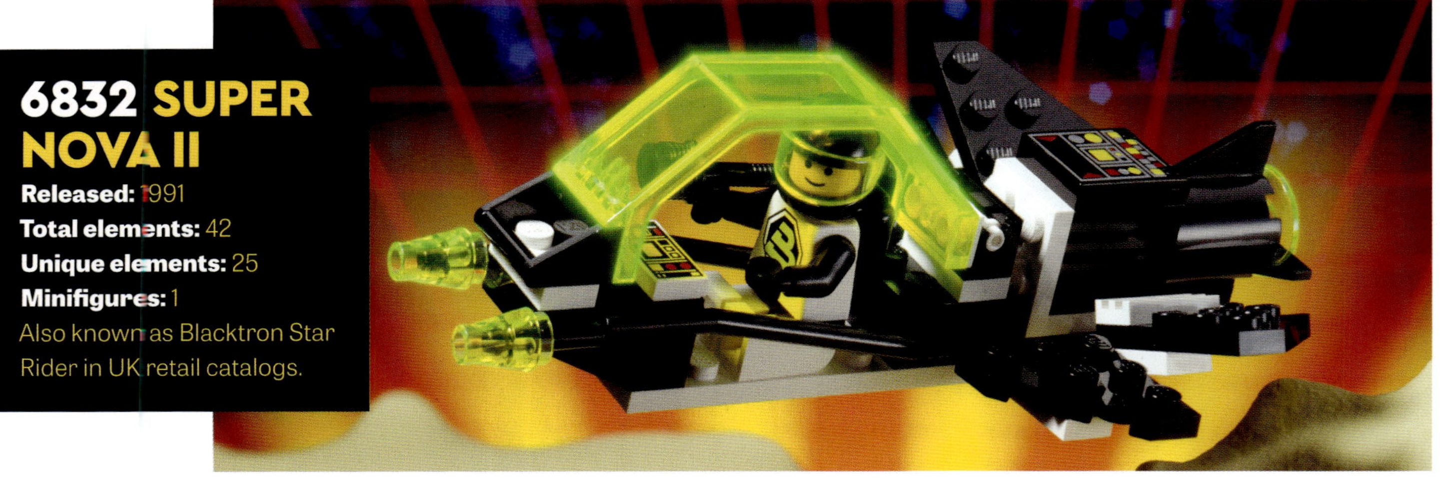

This narrow craft has segmented wings that can be angled, similar to the Blacktron Invader of 1987. The angled canopy is clasped by a handle and swings open backward. A large white bracket mounts both the cargo container and rear thruster sideways, finished with a transparent fluorescent green dish. The cargo is another computer panel tile, the same as the pilot's controls.

▼Original mockup of box.

As the pilot entered the Capitoline Galaxy in the outer reaches of the Future Generation's empire, she disengaged warp speed and the wings slowed to a flutter. Riding from star to star, she used long-range gradiometers to scan for signs of the intruders. A reading soon came in: a small planet on the outskirts of Vespasia. She engaged the quantum comms. "Centaur-A, I'm getting no readings. Heading back to you now." She entered the return course, but before executing, lifted her sabine pistol and fired it at the control tablet. Sliding her hands over the manual drive rails to her sides, she steered down to the planet, where she leapt out and unlocked the rear safe. Inside lay another control tablet. She switched it with the one in the cockpit and opened a virtual channel. "This is Novus-Duo. I have located where the intruders are taking the ferrocrystals . . . some kind of laboratory."

The purpose of LEGO elements, even specialized ones, is always open to interpretation. This set includes the ***horse hitching****, introduced by LEGO® Castle four years earlier as reins to connect wagons to their horses. Here, it creates the angled bars at either side of the pilot; its gentle slope adds flair to the craft's design.*

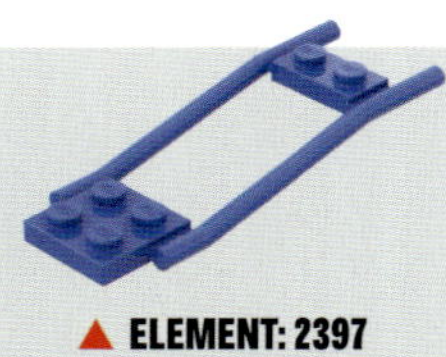

▲ **ELEMENT: 2397**
INTRODUCED: 1987

This little scooter comes with a remarkably large thruster. Its two pairs of wings create a roughly octagonal shape, and the ones with arrow markings can be tilted on their hinge plates. Unusually, the controls are a grille tile.

The speculator swiped the control lattice to adjust their trajectory, and the rear wings angled sharply to fly out of the atmosphere of the rebels' planet. "Yes, they are Future Generation like you and me," they confirmed, "but they follow two self-styled leaders, called the Blacktron Duo, who view the arrival of the intruders as a threat." Now spaceborne, they slid a hand along the thrust groove, speeding even faster away. "Are they responsible for the attacks on the miners?" asked the captain in reply. "Yes, and their intentions are much wider than that," the speculator said gravely. "They say that they seek the revenge of our ancestors . . . they are intent on wiping the intruders out completely."

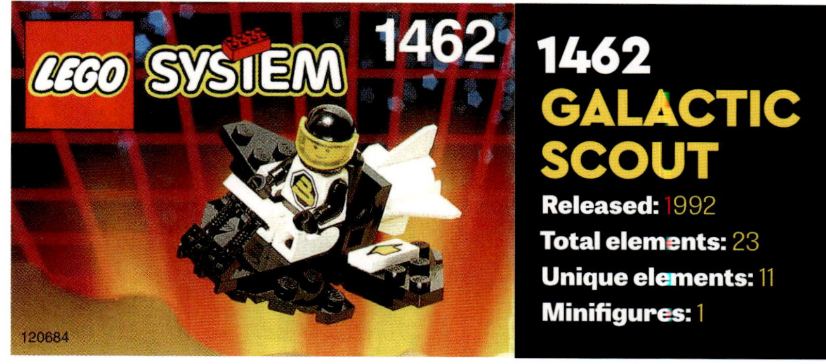

1462 GALACTIC SCOUT

Released: 1992
Total elements: 23
Unique elements: 11
Minifigures: 1

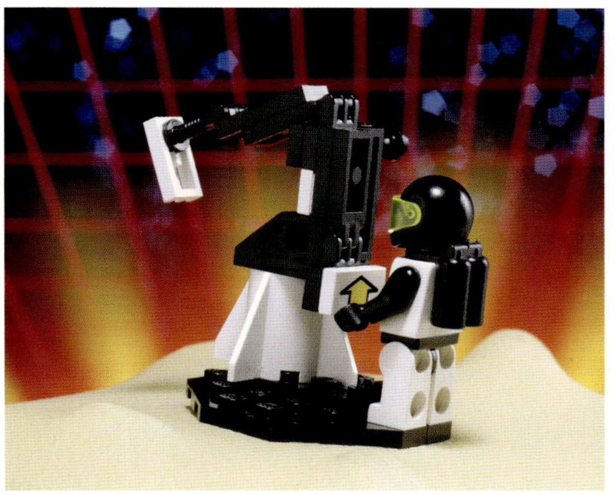

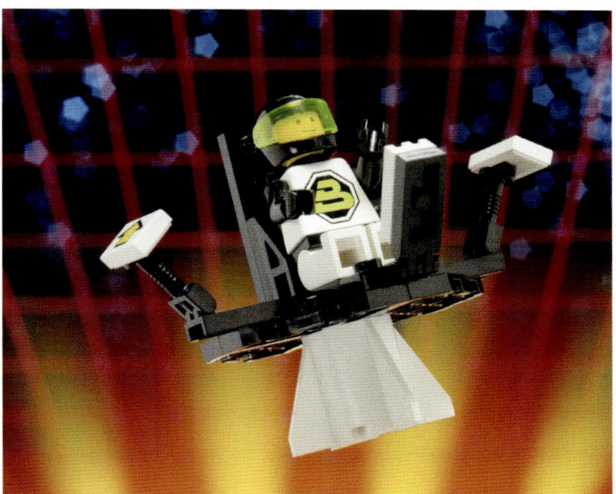

Alternative models featured on the back of the bag.

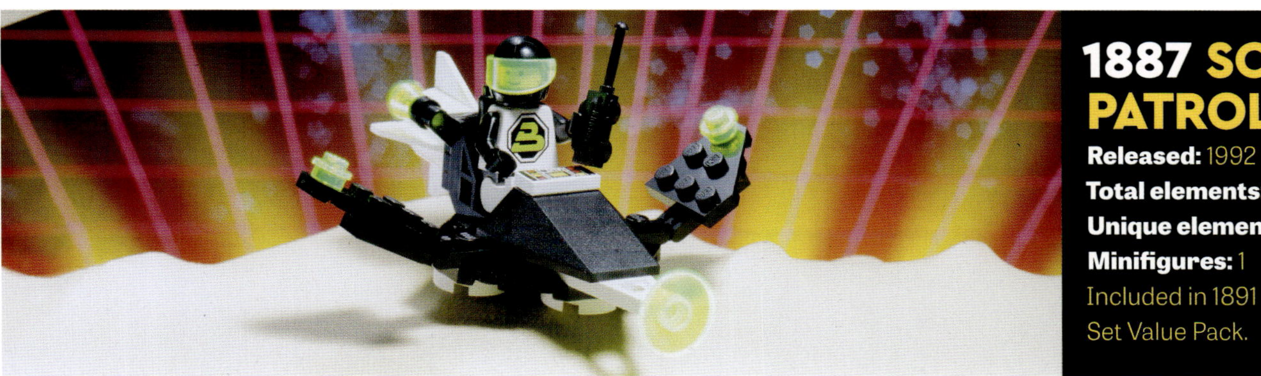

1887 SCOUT PATROL SHIP

Released: 1992
Total elements: 30
Unique elements: 17
Minifigures: 1
Included in 1891 Four Set Value Pack.

This small craft has an almost bird-like appearance, with its octagonal body and hinged wings. There are three thrusters at the rear—a huge white one and two transparent fluorescent green cones—and the four round plates underneath could be another method of propulsion or perhaps landing gear. The printed control panel tile is attached to a new element from the LEGO® Town Race subtheme, the triple wedge slope 4 x 4, and this was the only LEGO® Space set to use it that year.

The agent tore across space, desperate to reach the base before any suspicions were raised. Intercepting this craft had been sheer luck, and the rebel pilot had soon broken under his interrogation. Now all he had to do was infiltrate and get the information back to the Future Generation. "Approaching their planet now," he said into his virtual communications unit. "It's too risky to take any tech with me." With that, he flung the unit into space and fired the front gravity disk to brace for atmospheric entry. The shuttle soared through the sky above the rebel base, and he commenced wing-beat mode to slow down rapidly, landing with an elegant flourish in a free bay of the port. An alarm sounded.

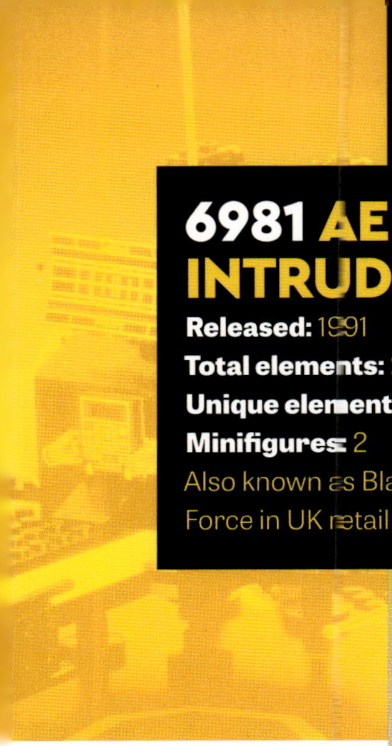

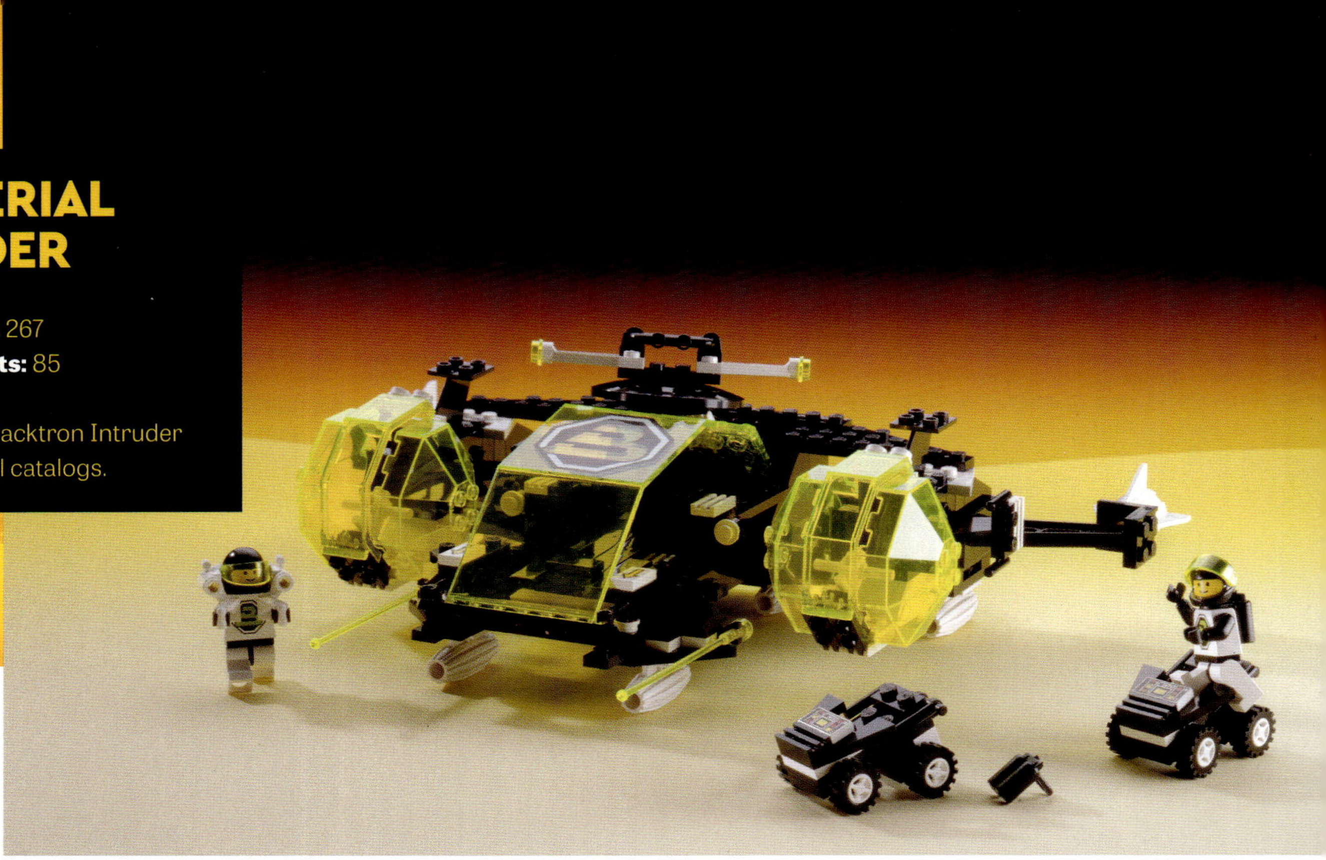

6981 AERIAL INTRUDER

Released: 1991
Total elements: 267
Unique elements: 85
Minifigures: 2
Also known as Blacktron Intruder Force in UK retail catalogs.

▶ The final page of the building instructions reveals that the rear cargo pod is intended for storing the jetpack.

▼ Original mockup of box.

This sleek craft holds many surprises. It has a forked planform with detachable twin pods at the edges, and the rear thrusters are set as far apart as possible, making this ship wider than it is long. Unlike the previous model featuring twin pods, Futuron set 6932 Stardefender 200, these ones actually control the ship. This unusual design frees up the entire central section for cargo: as well as a rear pod housing a jetpack, the front bay has enough space for two small rovers, which are notable for their steeply angled chassis. Unlocking and lifting the large canopy causes a ramp to release at the front, speeding both rovers on their way at once.

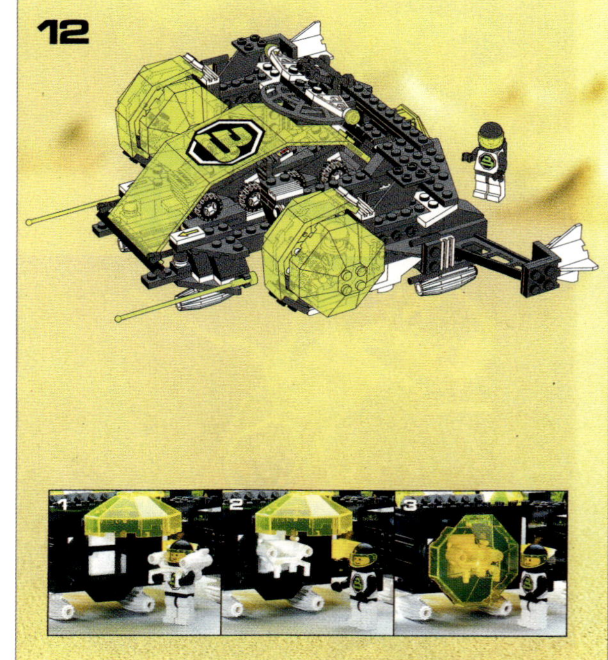

The gradiometer above the cargo bay stopped rotating. "We have pinpointed each of your locations," the Future Generation alpha-captain said to her spies on the planet below. "We'll deliver transport to each of you; then we all attack the Blacktron Duo HQ." As they entered the atmosphere above the rebel outpost, she programmed the deployment sequence as the beta-captain slammed on the command arm to commence a steep descent. With another tug, the beta straightened up just above the surface and the alpha jettisoned the cargo, sending the two buggies bouncing toward her two colleagues hiding near the headquarters. The beta banked to the right and landed close to the arsenal that their third agent was hiding in. "Southeast door," she alerted him as her cockpit sprung open and she rushed around the back to unlock the charge pod. "Leaving you the hoverpack to set the detonators and the craft to rejoin us." The alpha and beta ejected their globeships and sped back to the headquarters to attack.

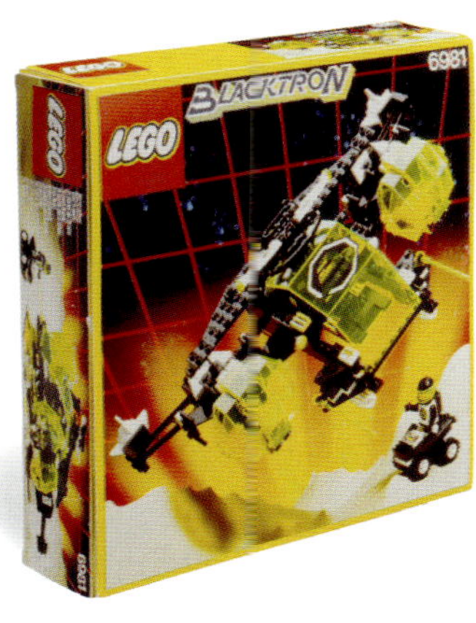

*The fence element on top of this spaceship is attached sideways by using two **modified tiles 1 x 1 with clip**, an extremely common element nowadays. At the time of its introduction by the LEGO® Pirates theme, the existing elements with clips all held bars in a vertical position: this was the first to clutch them horizontally.*

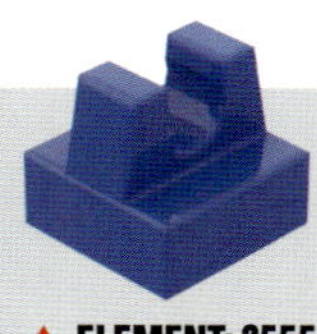

▲ **ELEMENT: 2555**
INTRODUCED: 1989

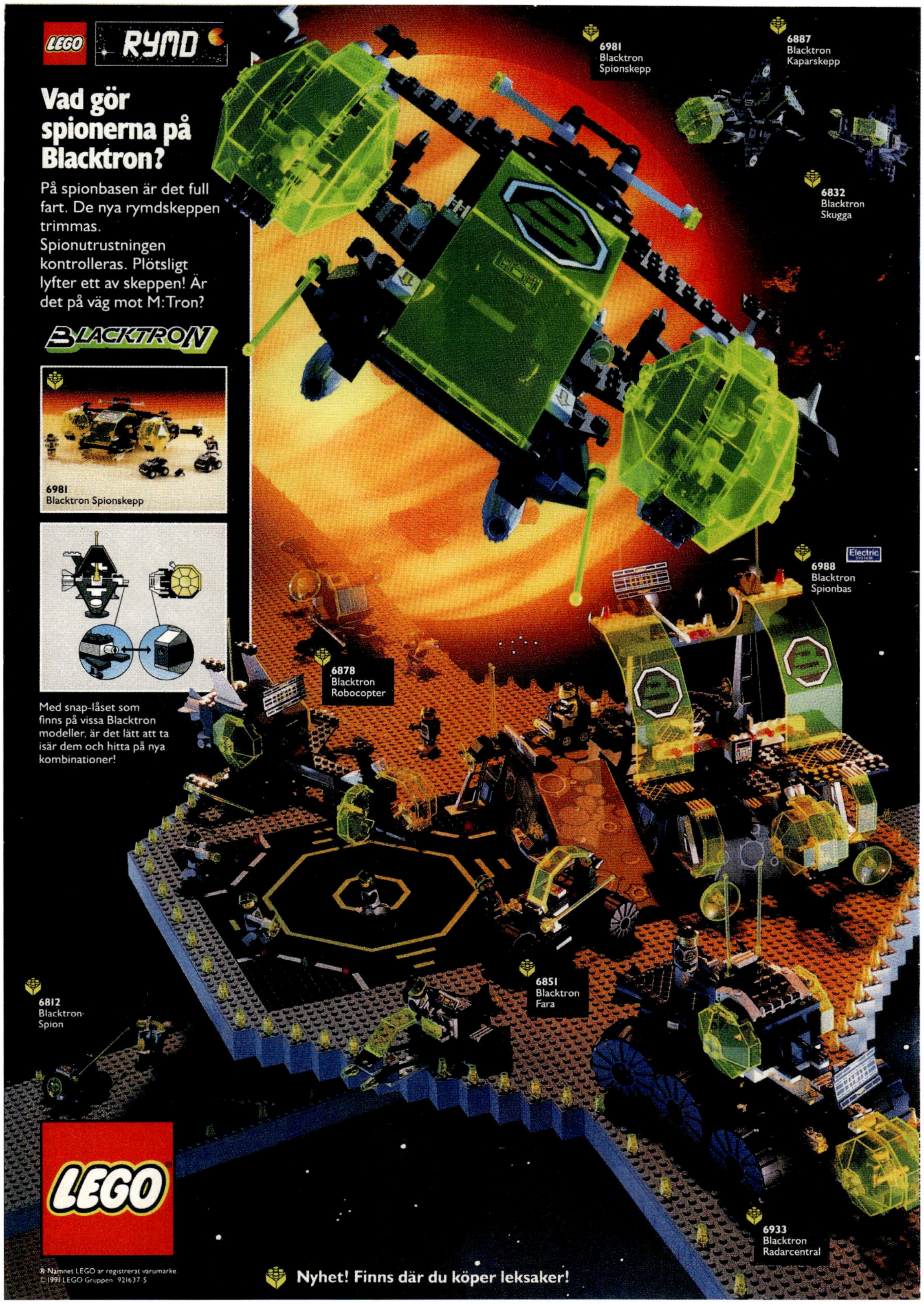

LEGO RYMD

Vad gör spionerna på Blacktron?

På spionbasen är det full fart. De nya rymdskeppen trimmas. Spionutrustningen kontrolleras. Plötsligt lyfter ett av skeppen! Är det på väg mot M:Tron?

BLACKTRON

6981 Blacktron Spionskepp

Med snap-låset som finns på vissa Blacktron modeller, är det lätt att ta isär dem och hitta på nya kombinationer!

6981 Blacktron Spionskepp

6887 Blacktron Kaparskepp

6832 Blacktron Skugga

6988 Blacktron Spionbas

Electric

6878 Blacktron Robocopter

6851 Blacktron Fara

6812 Blacktron Spion

6933 Blacktron Radarcentral

LEGO

® Namnet LEGO är registrerat varumärke
© 1991 LEGO Gruppen 921637-5

Nyhet! Finns där du köper leksaker!

The Aerial Intruder, known as Blacktron Spyship in Sweden, takes pride of place in this 1991 advertisement. An illustration highlights the modular nature of the range.

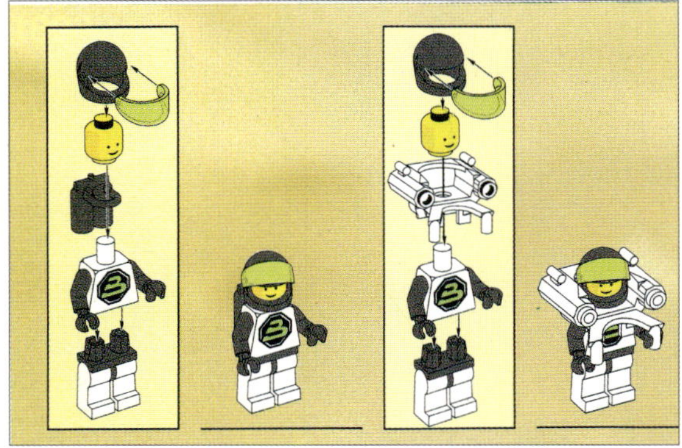

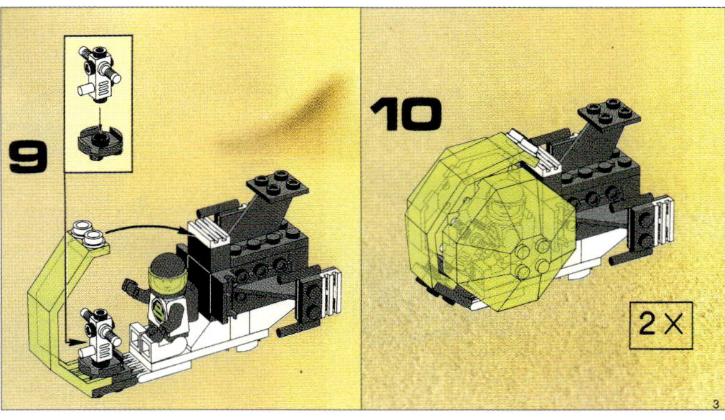

◀ The jetpack element only ever appeared in a handful of sets.

◀ The controls in the twin pods are chainsaw body elements attached to the upside-down windshield.

EPILOGUE

What's next?

Blacktron Future Generation minifigures continued to appear in the LEGO® range as the adversaries of a fresh new wave of Space Police launched in 1992. The revamped subtheme, commonly referred to as Space Police 2, featured police minifigures with a different face print for the first time in LEGO® Space history: eyebrows and a headset were added to the classic smile. In 1993, the mining and research team of the Ice Planet 2002 subtheme introduced three new faces, including Doctor Kelvin, who wears lipstick and hoop earrings.

Faces were changing in the design team as well. Designer Bjarne Panduro Tveskov had moved from the LEGO Space group in 1990 to become a concept designer, working to combine physical LEGO play with digital experiences. He worked on products such as the interactive LEGO® Technic set 8299 Search Sub in 1997 and also investigated how the LEGO Group might make digital products in the future. In 1998 he began working in a freelance capacity and still works with the LEGO Group to this day. "I'm really so lucky that fifty percent of my time is still doing LEGO work," says Tveskov. "It's

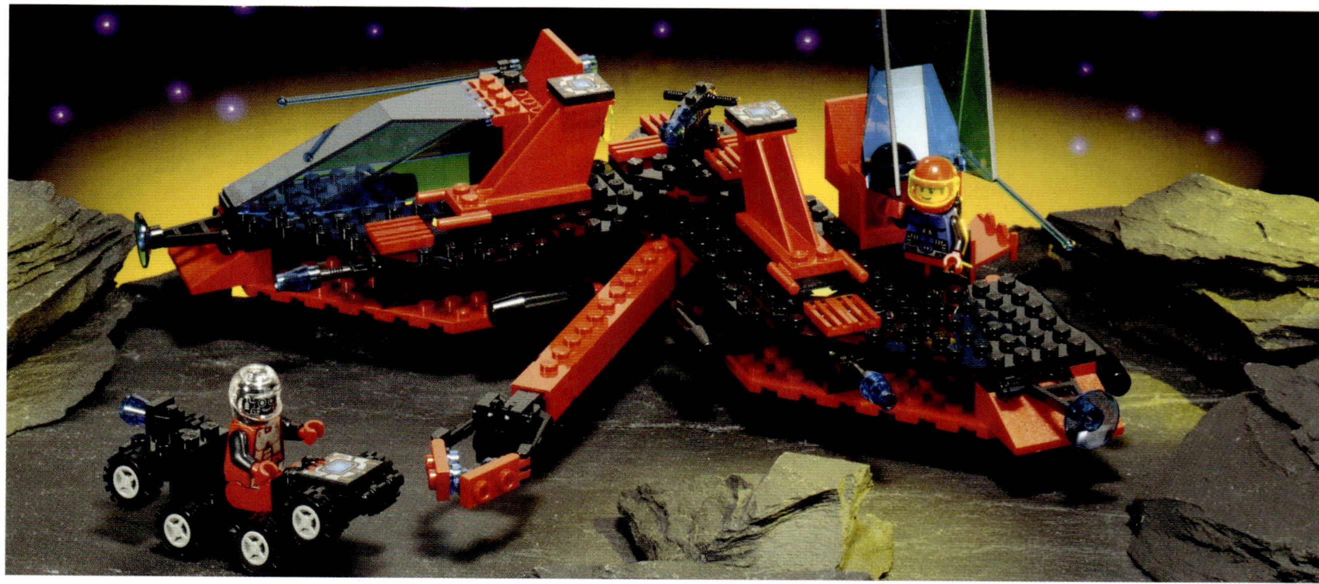

▶ Blacktron may have gone, but "bad guys" remained in LEGO Space play. The Spyrius subtheme of 1994 to 1996 featured technology-thieving spies and their droids. The novel design of set 6939 Saucer Centurion was led by Jørn Thomsen before he moved to LEGO® Trains.

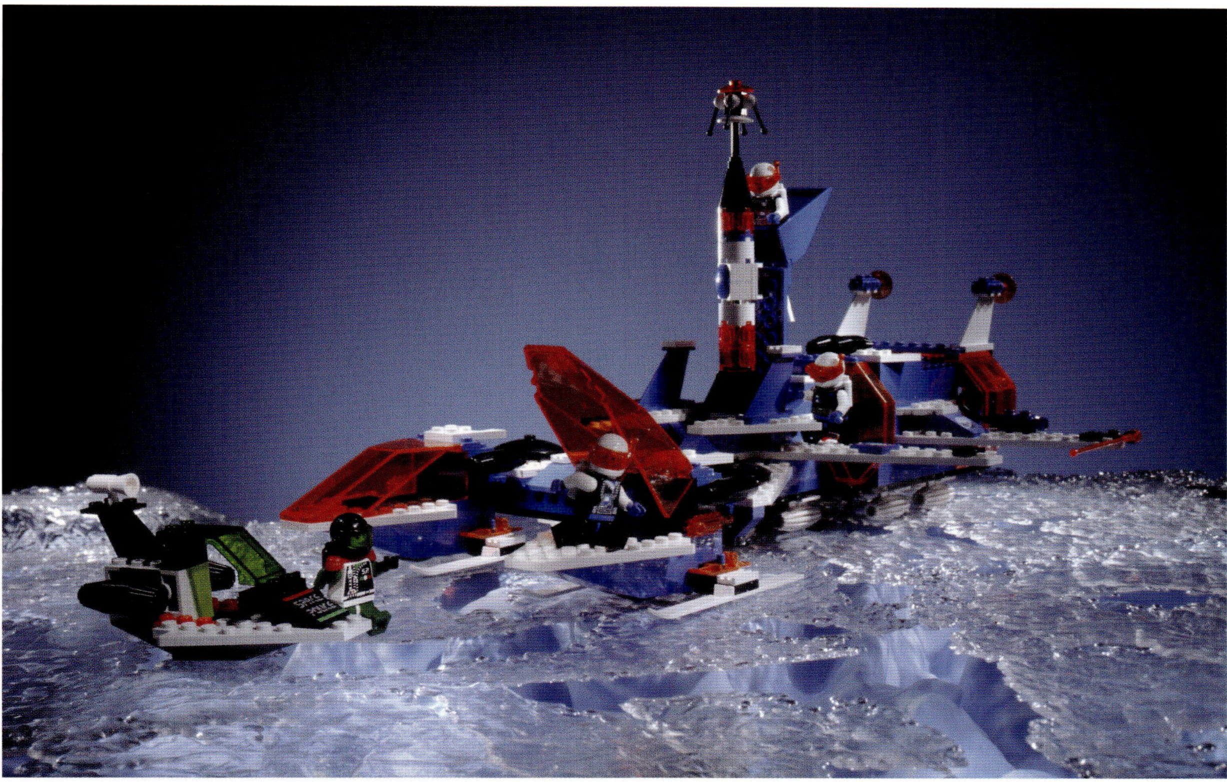

◀ In this stunning promotional image from 1993, a Space Police 2 shuttle visits the mother ship on the frozen world of Ice Planet 2002.

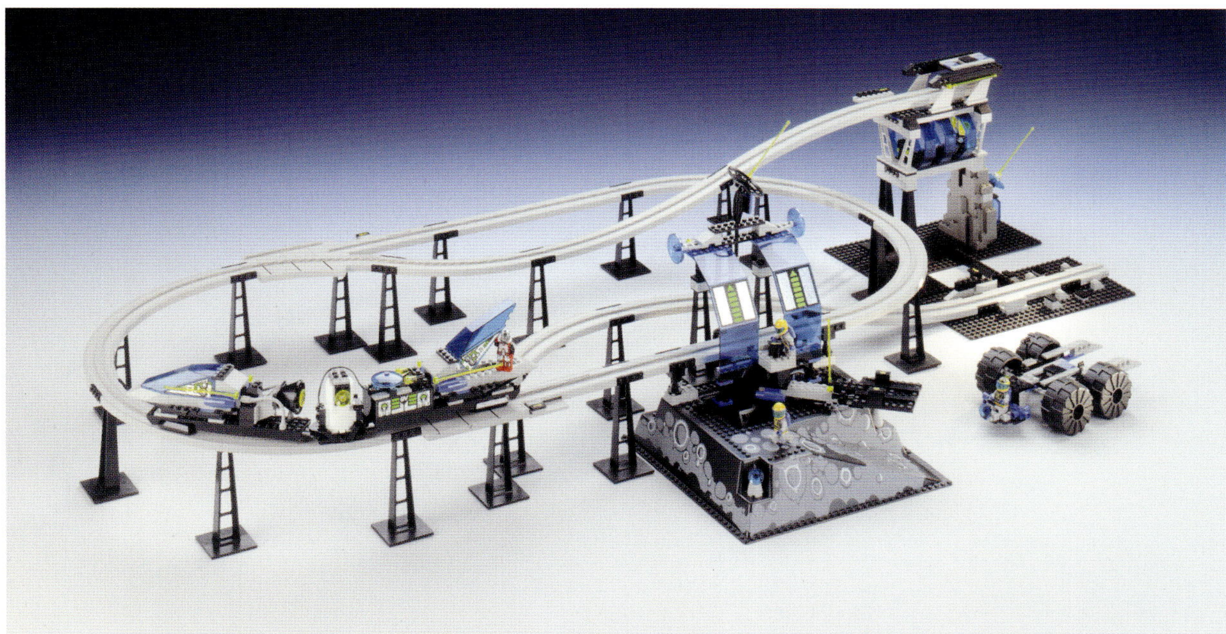

◀ The monorail system returned for the final time in the Unitron set 6991 Monorail Transport Base in 1994, and branching tracks were included.

▼ The first minifigure representing a robot came in the Spyrius subtheme in 1994.

> **"[KNUDSEN] JUST HAD ALL OF THESE SCI-FI IDEAS HE WANTED TO GET OUT."**
>
> —MARK JOHN STAFFORD, SENIOR DESIGNER, THE LEGO GROUP, 2006 TO PRESENT

a great privilege to still walk around and see all the new work being done."

Niels Milan Pedersen was also working freelance for the LEGO Group, having achieved his childhood dream of becoming an archaeologist. "For many years, it was mostly LEGO® Castle and LEGO® Pirates which were my area of expertise," he recalls. "I was mostly making elements, but you were always expected to build some of those alternative models for the boxes, so I was always doing that, throughout all of my years." Pedersen hadn't hung his spacesuit up just yet, though.

Jens Nygaard Knudsen also moved to a different department, where he focused on designing new concepts alongside Daniel August Krentz. The newly appointed chief designer, Søren Vigsø, took the helm of LEGO Space in 1994; however, Knudsen's input was not yet over. Given that LEGO product assortments

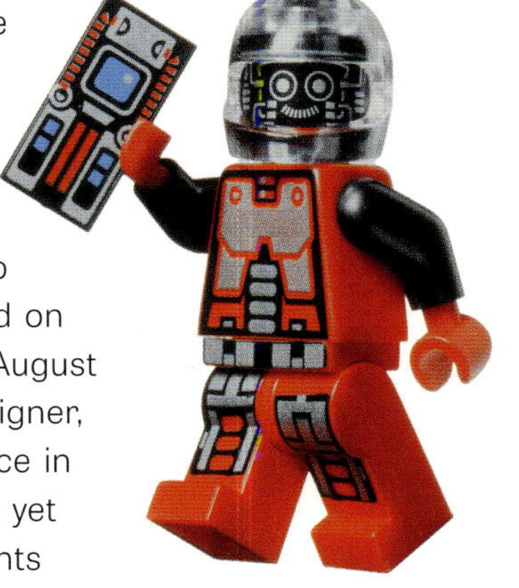

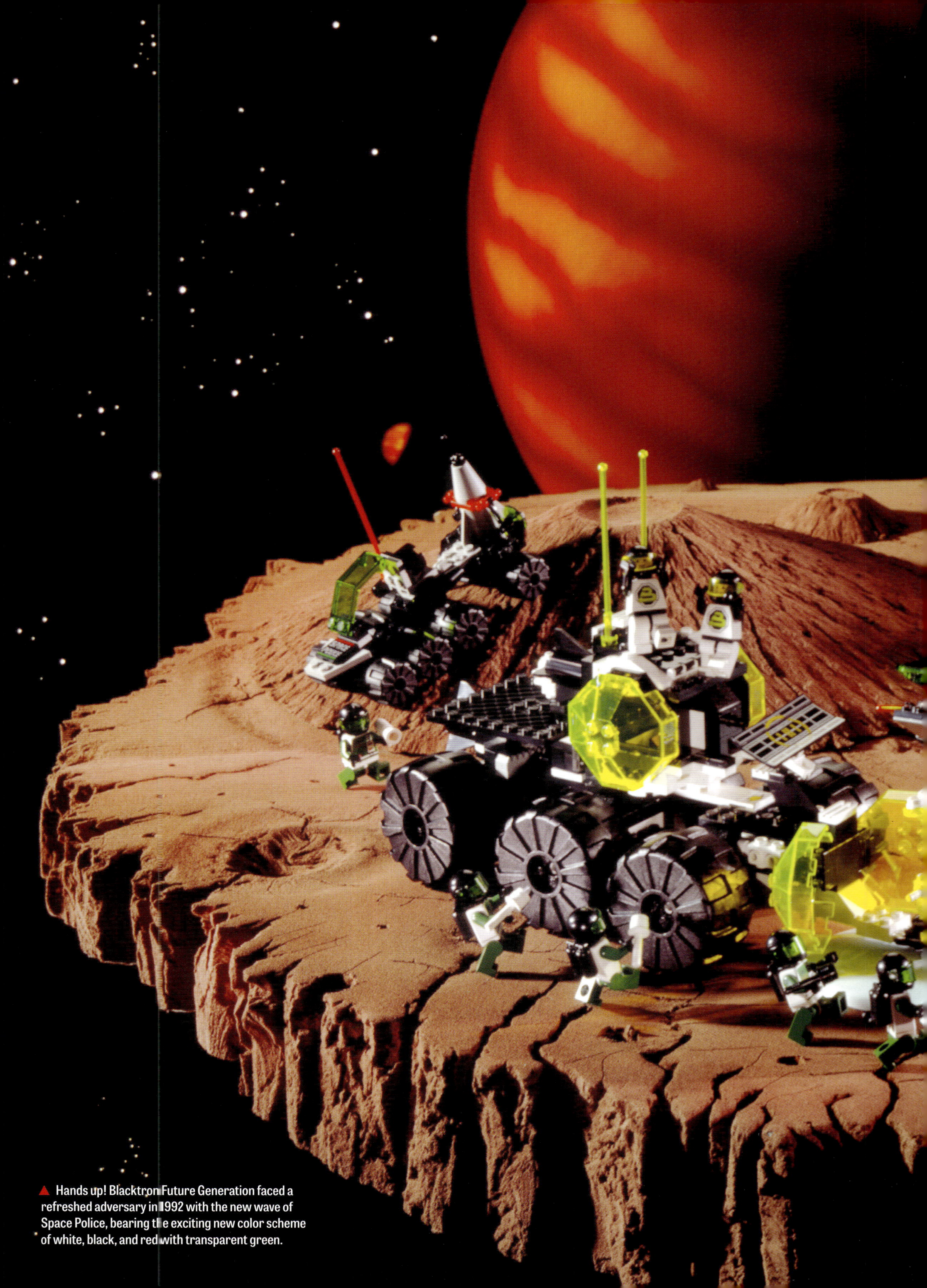

▲ Hands up! Blacktron Future Generation faced a refreshed adversary in 1992 with the new wave of Space Police, bearing the exciting new color scheme of white, black, and red with transparent green.

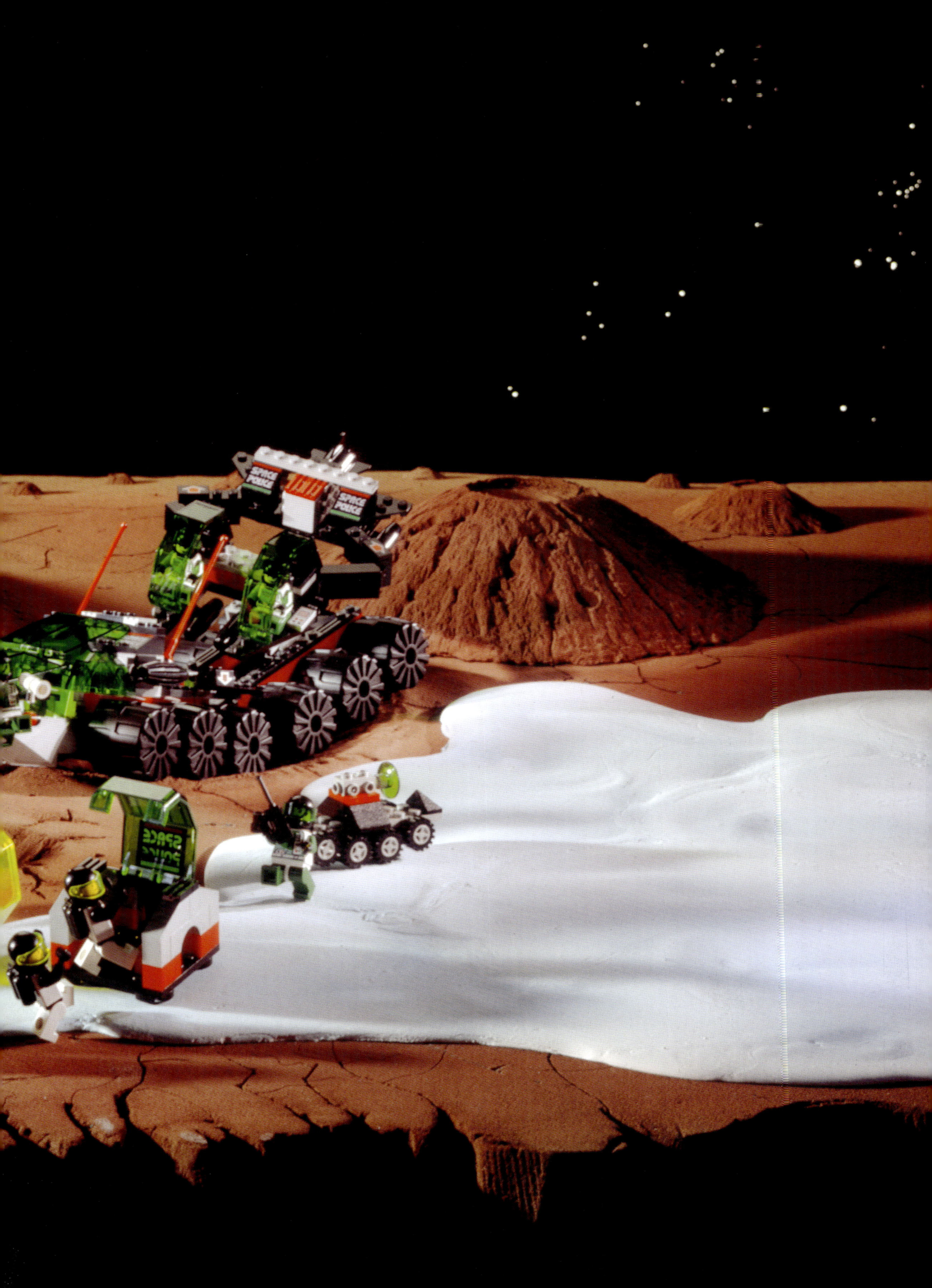
SPACE POLICE
SPACE POLICE

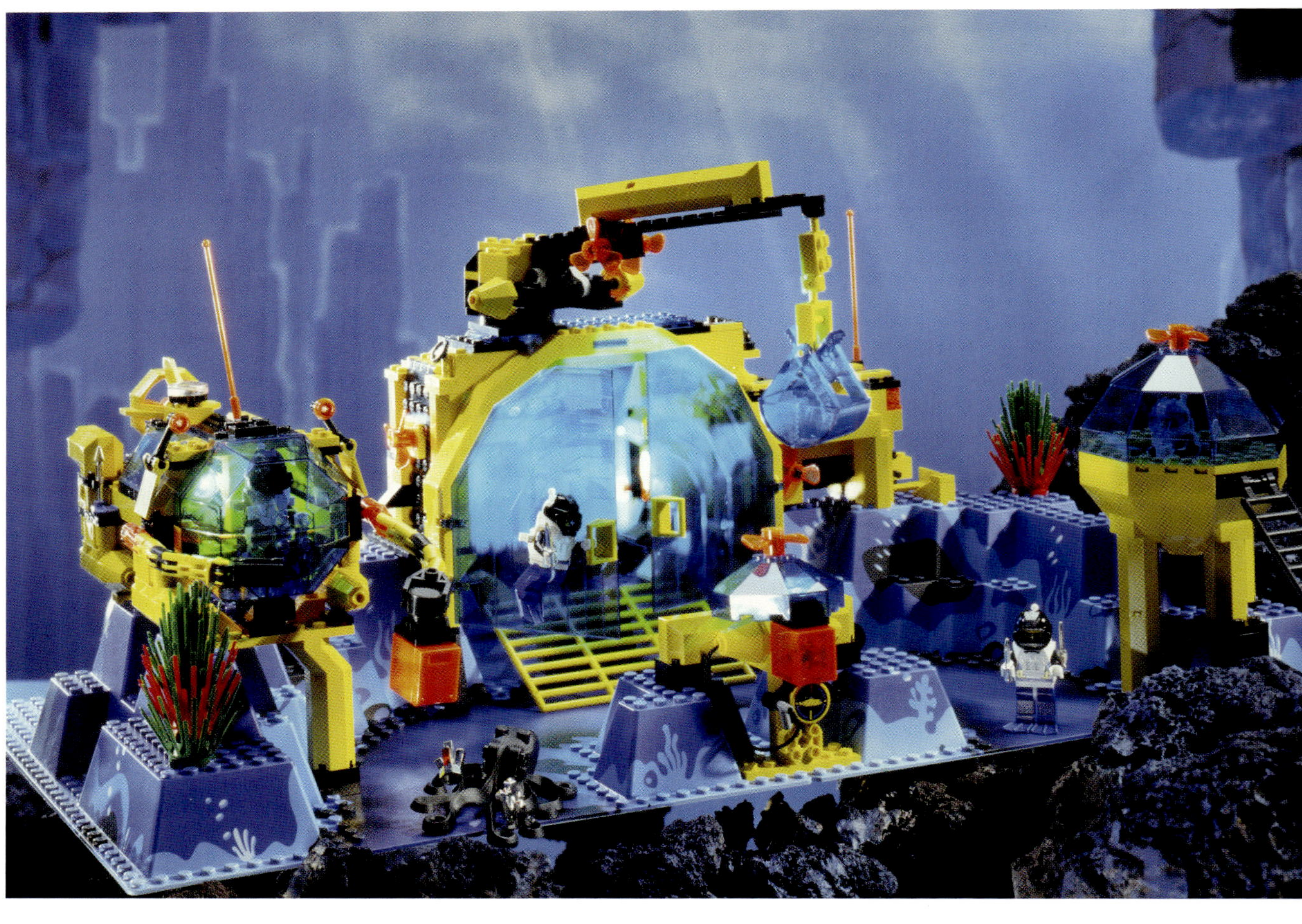

▶ LEGO Aquazone began its development as a LEGO Space concept before becoming its own theme in 1995. The design of set 6195 Neptune Discovery Lab was led by Jørn Thomsen.

take years to reach the shelves, models which Knudsen and his team had developed were released in the years that followed, in the form of new LEGO Space subthemes, such as the heroic Unitron and the seditious Spyrius in 1994 and even a whole new line in 1995: LEGO® Aquazone.

FANTASY WORLDS

If you had been passing by the local stream in Billund one day in the 1970s, you might have witnessed an unusual sight: a grown man plunging LEGO® submarines underwater. It was Knudsen, and he told the story to designer Mark John Stafford three decades later.

"This happened before the LEGO® Space line came out, let alone LEGO® Aquazone," recalls Stafford. "I got the feeling he was almost as excited about sci-fi submarines as he was about space. The idea of him being down there on his hands and knees playing in the stream with submarines, even before the minifigure existed? I can't wrap my head around that, but I love the idea."

The idea that would eventually be realized as LEGO Aquazone began its developmental life in the 1980s as a LEGO Space subtheme named Sea-tron and was ready to be released in 1991 alongside Blacktron Future Generation. Elements had been designed, models were developed, and even the box imagery had been photographed; however, the range was pulled at the last moment.

▼ The 1996 subtheme Exploriens included a holographic sticker of a video screen in set 6899 Nebula Outpost.

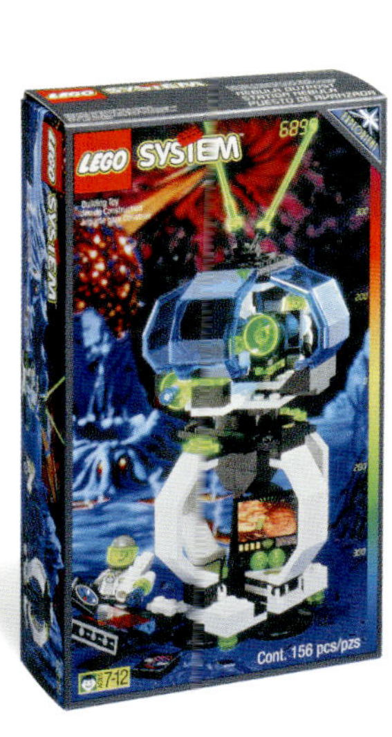

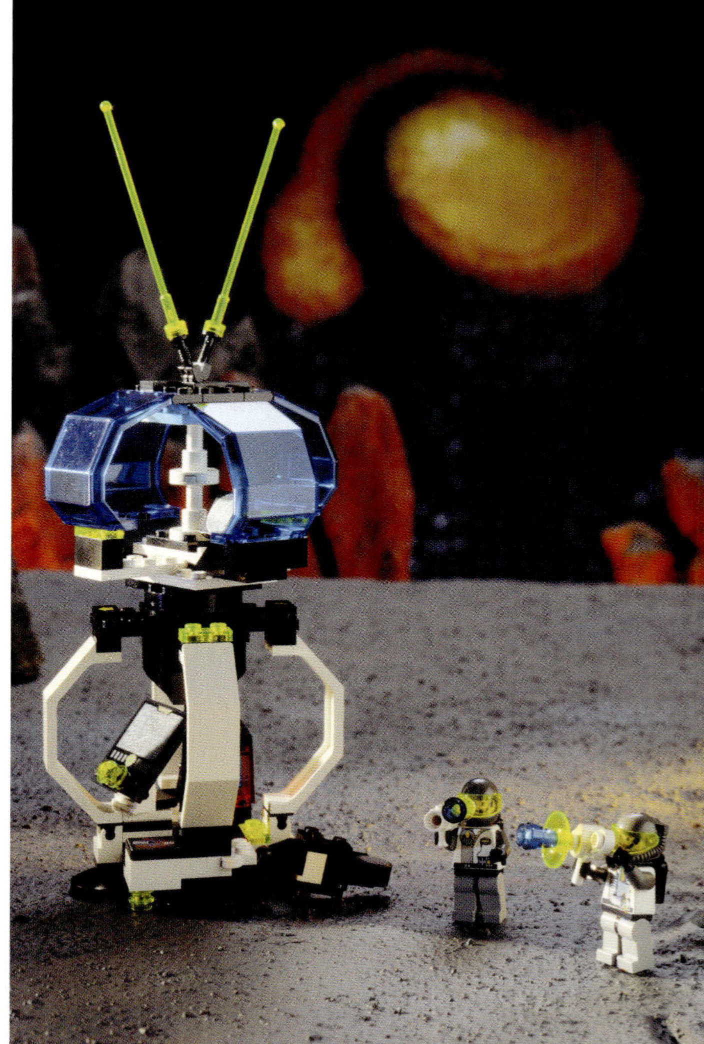

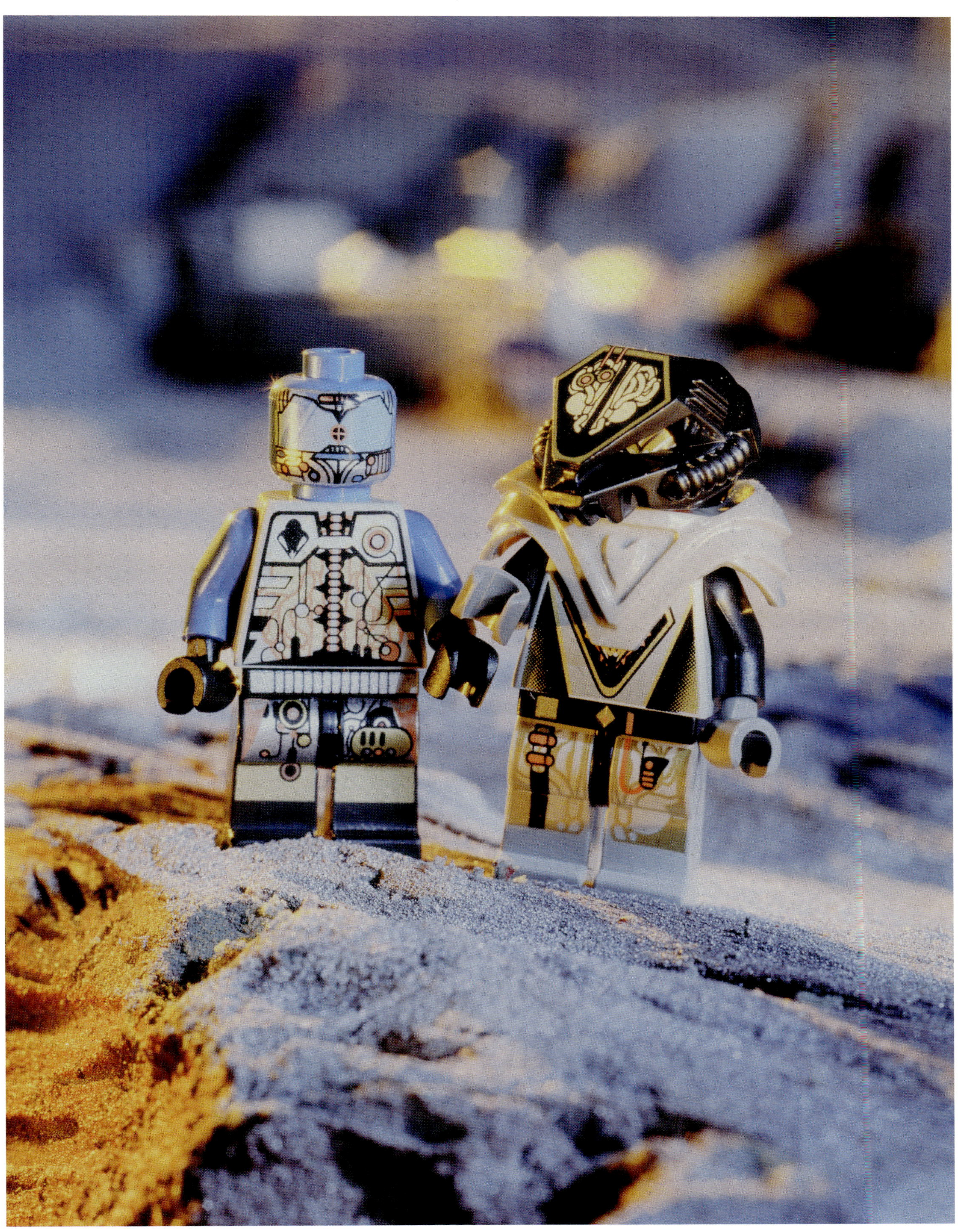

Although Niels Milan Pedersen had proposed alien minifigures in the 1980s, it would not be until 1997's UFO subtheme that they appeared in sets. Here, a Zotaxian Overlord named Alpha Draconis is pictured with a Techdroid.

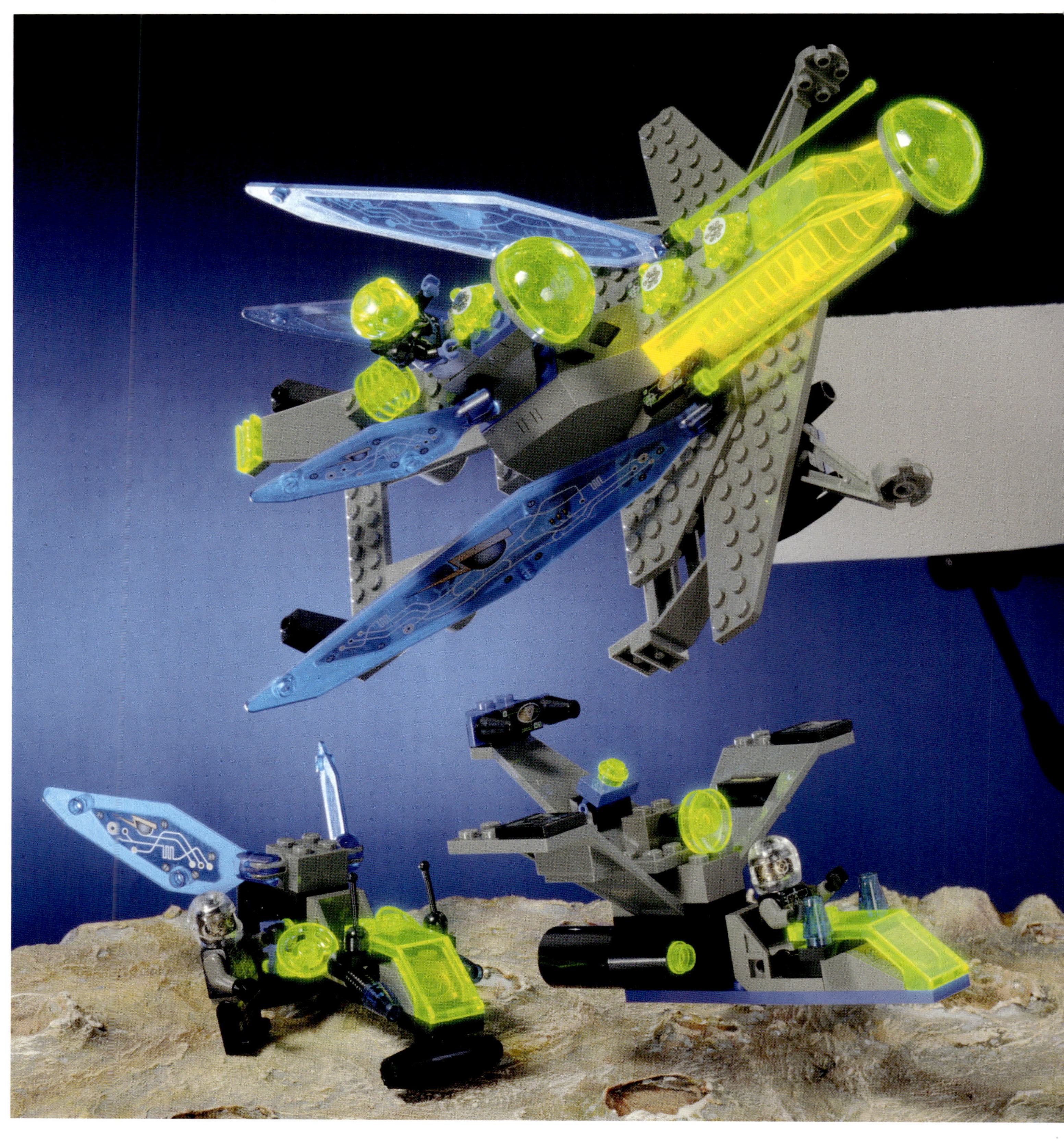

▲ The nose of 1998's Insectoids set 6969 Celestial Stinger is a Light & Sound System element that lights up and makes three different sounds.

Tveskov, who worked on its early development, recalls the reasoning. "It was postponed for a while because there was too much growth in the LEGO Group's pipeline. They had to manage the growth every year in order for the factories to be able to keep up. So that was a very positive problem to have."

"I never thought it would come out," says fellow designer Jørn Thomsen. "By 1995, I had started working with LEGO® Trains, and I was invited over to see the launch presentation of LEGO Aquazone—without changes! Absolutely amazing. Four years waiting on the shelf and still a success."

The LEGO Aquazone theme continued until 1998, paving the way for many other lines that have successfully blended fantasy with science fiction while not falling under the LEGO Space banner. For example, in 1999, LEGO® Rock Raiders featured miners on a planet beset by alien monsters, and the Japanese manga-inspired LEGO® EXO-FORCE™ in 2006 pitched giant rebel robots against their human overlords. Fantasy themes featuring futuristic technology are a core staple of the LEGO product assortment to this day, including LEGO® NINJAGO® and LEGO® Monkie Kid™.

NEW FRONTIERS

During Vigsø's tenure in the 1990s, there were many firsts for the LEGO® Space line.

The proliferation of brick-built droids in the 1980s had dwindled by the start of the 1990s, but robotic companions returned to LEGO Space in 1995 in the Spyrius subtheme. This time, however, they were actual minifigures, and two years later, UFO was the first LEGO Space subtheme to feature alien minifigures. Giant robots also made a return; in fact, in 1997, all four sets of the RoboForce subtheme were large robot-shaped vehicles with detachable spacecraft for heads.

1996's Exploriens subtheme, which depicted paleontologists in space, included a variety of special stickers: some reveal different patterns when viewed through either red or blue transparent elements, while other stickers are prismatic, and one even has a holographic image.

Technological innovation continued within element design as well. One of the products in the UFO line used fiber-optic cables, and 1998's Insectoids subtheme, noted for introducing many unusual elements inspired by insects, included a new Light & Sound System element in the shape of a large alien "stinger" that lights up and makes three different sounds. The Life on Mars subtheme in 2001 included a family of pneumatic elements, with a hand pump powerful enough to shoot Martian minifigures along a system of tubes.

Although the sinister Blacktron had disappeared, conflict retained its newfound position within LEGO Space narrative play. With time, the concept of one subtheme featuring "good guys," such as Unitron, being pitted against "bad guys" from another subtheme, like Spyrius, gave way to both sides being included within the one subtheme. LEGO® Life on Mars was a notable exception to this, with no conflict suggested between the Martians and the human explorers arriving on their world. The subtheme finished in 2002, and LEGO Space did not return until 2007.

A FUTURE GENERATION

Some designers returned to LEGO® Space on occasion in the 2000s. Thomsen worked on Life on Mars, and Pedersen designed elements and models for the Mars Mission subtheme in 2007, where humans went to mine the Red Planet but encountered a hostile alien race.

Pedersen's original colleagues had long since retired—Krentz in 1999 and Knudsen in 2000—and although the LEGO Space line had continued to evolve

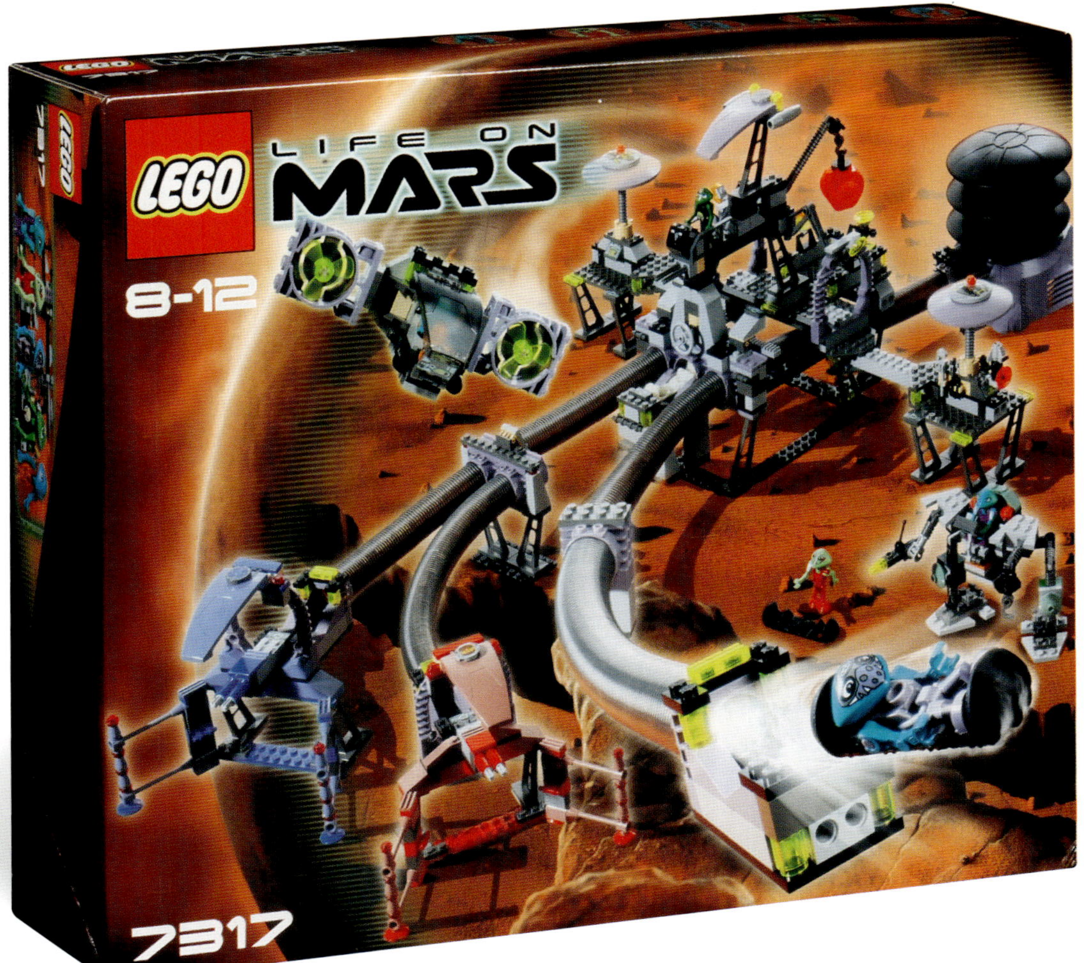

Designer Jørn Thomsen returned to the universe of LEGO Space, working on set 7317 Aero Tube Hanger from the Life on Mars subtheme in 2001. A pump-operated pneumatic system propels Martians in their hypersleds through the tubes.

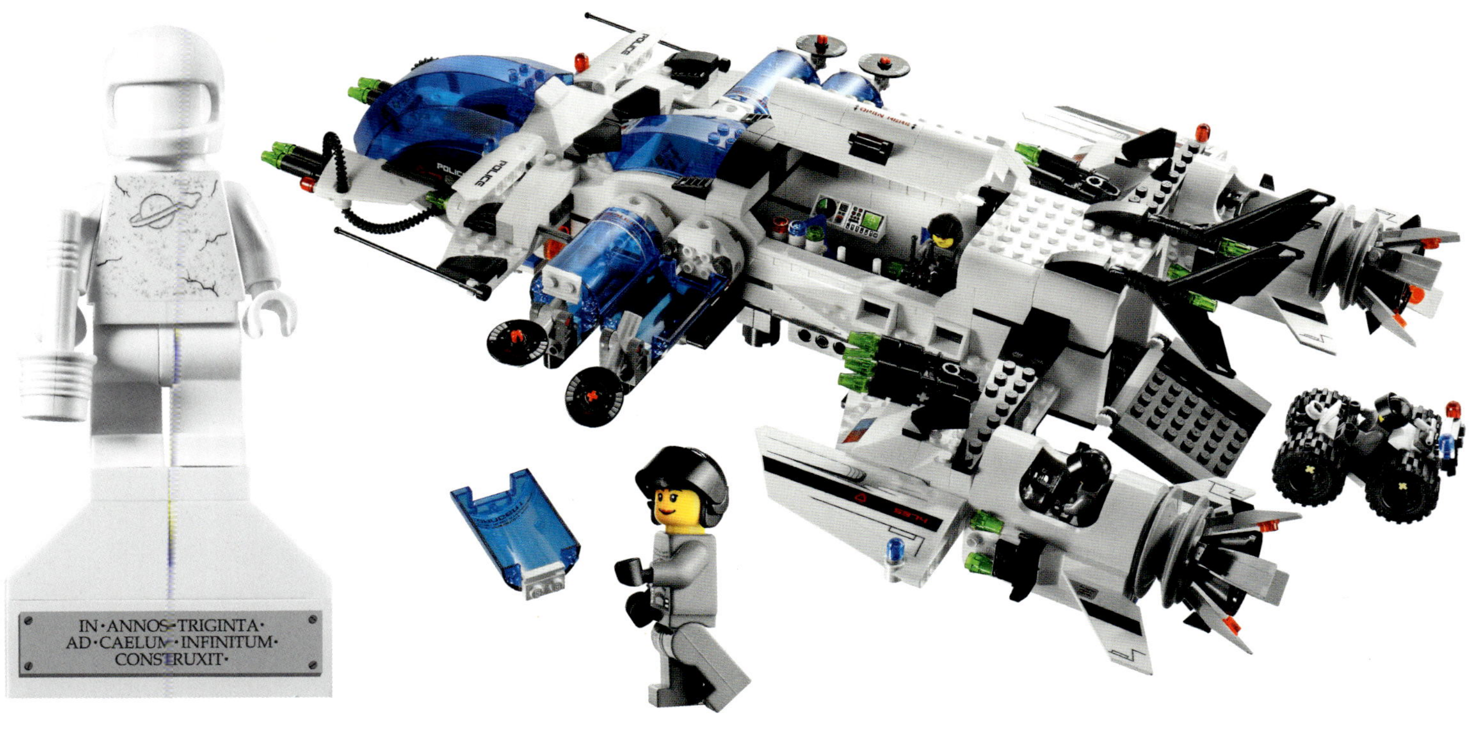

Favorite LEGO Space sets from designer Mark John Stafford's childhood inspired aspects of 2009's Space Police 3 set 5974 Galactic Enforcer. There was even a Classic Space astronaut included in the form of a white statue.

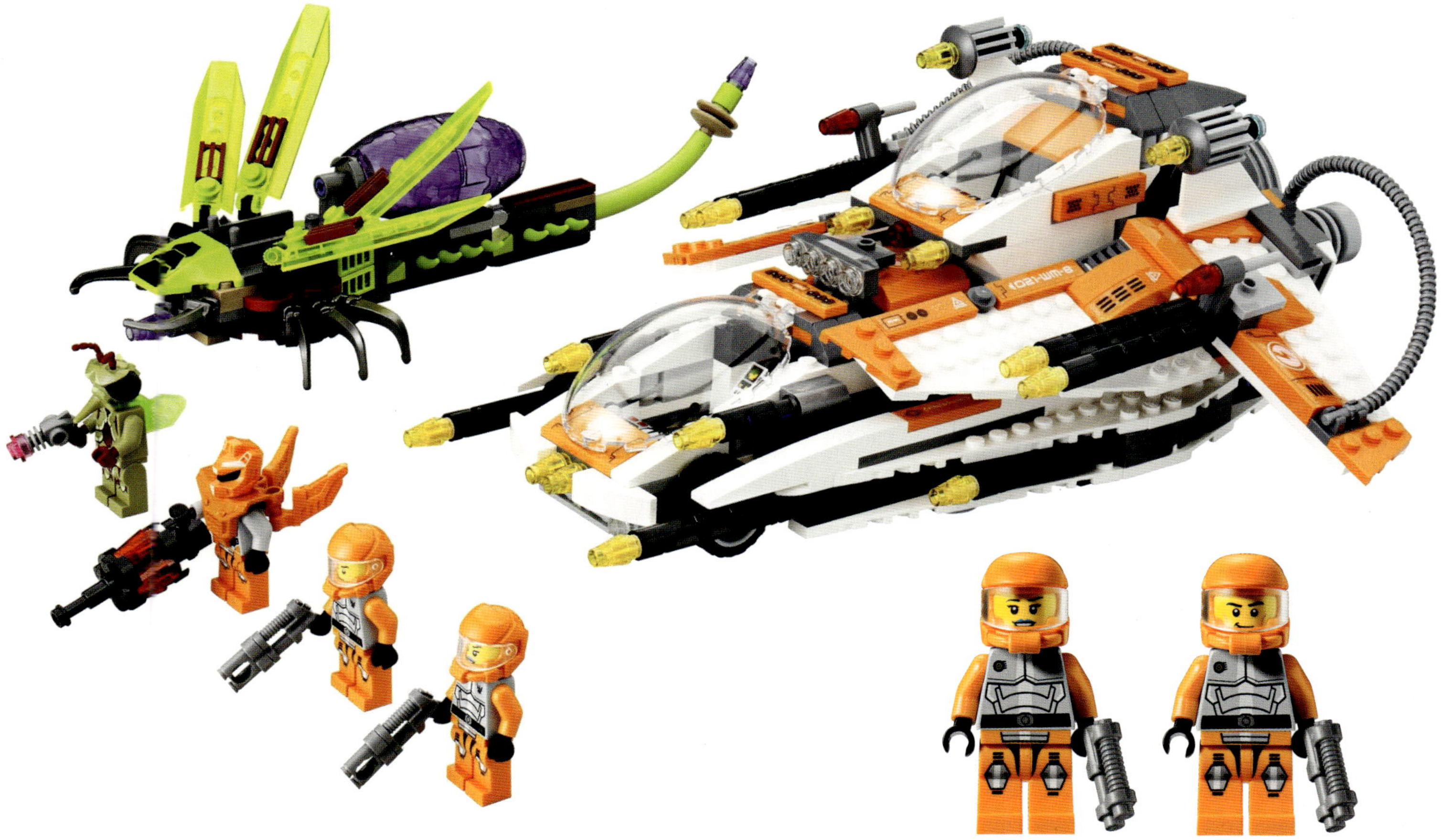

Each of the four squadrons composing Galaxy Squad of 2013 had a different color of suit. Designer Raphael Pretesacque led the design of set 70705 Bug-Obliterator, which features the Orange Team, known for their firepower.

after their era, the past was not forgotten. In fact, quite the opposite was true.

Many children who had grown up playing with LEGO Space products were still creating their own models as adults, and in the 2000s, a handful of them gained coveted design positions at the LEGO Group, including Stafford and William Thorogood. They both became part of the team tasked with bringing Space Police back for a third wave in 2009, along with Pedersen.

"I was a huge fan of Space Police 2 as a child; it was one of the first LEGO® themes I truly fell in love with," recalls Thorogood. "Little did I know that later in my life I would be able to rekindle that passion and lead the design team to create Space Police 3, fusing all of the things I loved about the originals with all of our favorite science-fiction references, plus a humorous LEGO twist."

"JENS HAD A GREAT LIFE, AND A GREAT IMPACT ON SO MANY LIVES."

—NIELS MILAN PEDERSEN, DESIGNER, THE LEGO GROUP, 1980 TO PRESENT

For the design of the large spaceship, Stafford drew inspiration from one of his favorite childhood sets: 6980 Galaxy Commander. "[Set 5974] Galactic Enforcer is a love letter—the front comes off, the back has a base, and I even put a buggy in the back."

The next subtheme, Alien Conquest in 2011, suggested a different kind of nostalgic throwback. Its 1950s B movie aesthetic depicted bug-eyed aliens invading Earth in flying saucers and tripods. Meanwhile, the blue fuselage of humanity's Alien Defence Unit provided many adult fans with welcome recollections of the initial years of LEGO Space.

Spacesuits of differing colors were a key aspect to the aesthetics of the next subtheme, Galaxy Squad, in 2013; the humans were divided into the Blue, Red, Green, and Orange Teams, each with their own specialized skills, vehicles, and robot companions. Together, they fought a swarm of alien bugs who were invading the galaxy, intent on building their hives on every planet they found.

THE PLANET NOSTALGIA

Other lines have also paid tribute to LEGO® Space over the years. The thirtieth anniversary of the minifigure was celebrated across 2008 and 2009, and special Vintage Minifigure Collections were released, including four colors of the Classic Space astronaut and one Blacktron.

In 2010, the very first series of the collectible LEGO® Minifigures range included an astronaut whose detailed chest unit featured the Classic Space logo, and similar nostalgic references have occasionally appeared within the series that have followed. In 2023, a brown Classic Space astronaut was introduced, nurturing their "Spacebaby" clad in a blue spacesuit.

◀ In 2023, the LEGO Minifigures range introduced a "Spacebaby" figure alongside a brown astronaut.

The LEGO® Ideas line, where fans can submit model concepts that have a chance of becoming actual LEGO products, produced a set which embraced the Classic Space aesthetic in 2014. The set 21109 Exo Suit, designed by Peter Reid and developed into a LEGO product by Stafford, even included a new color of astronaut: green.

That same year, nostalgia received a huge boost from THE LEGO® MOVIE™, which included "1980-Something Space Guy" Benny. A blue Classic Space astronaut with a faded chest logo and cracked helmet, he and his spaceships feature in several products, some featuring his "Space Squad" colleagues introduced by the 2019 sequel, including a pink astronaut, Lenny.

To mark the ninetieth anniversary of the LEGO Group, popular classic themes were celebrated with the release of new sets. LEGO Space was among them, in the form of set 10497 Galaxy Explorer in 2022 (see page 200). This reimagined version of the 1979 design by Knudsen was scaled up to be one and a half times the size of the original. In 2023, the celebrations continued with the limited release of set 40580 Blacktron Cruiser: a modern interpretation of 6894 Invader from 1987.

Although these allusions to LEGO Space are only occasional, and no new subthemes have been introduced since Galaxy Squad in 2013, the love for LEGO Space has never waned. Furthermore, the legacy of the line is indelibly present within the LEGO® System in Play.

LEGACY

Krentz passed away in June 2016, and Knudsen less than four years later. Tributes flowed in from their former colleagues and the legions of fans dotted right across planet Earth.

▼ 2011's Alien Conquest subtheme brought the classic blue color back to LEGO Space. The design of set 7067 Jet-Copter Encounter was led by Mark John Stafford.

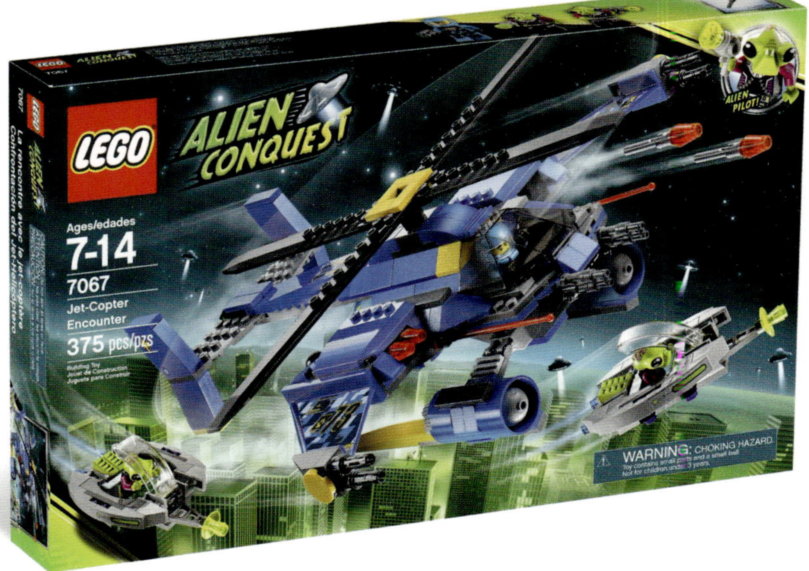

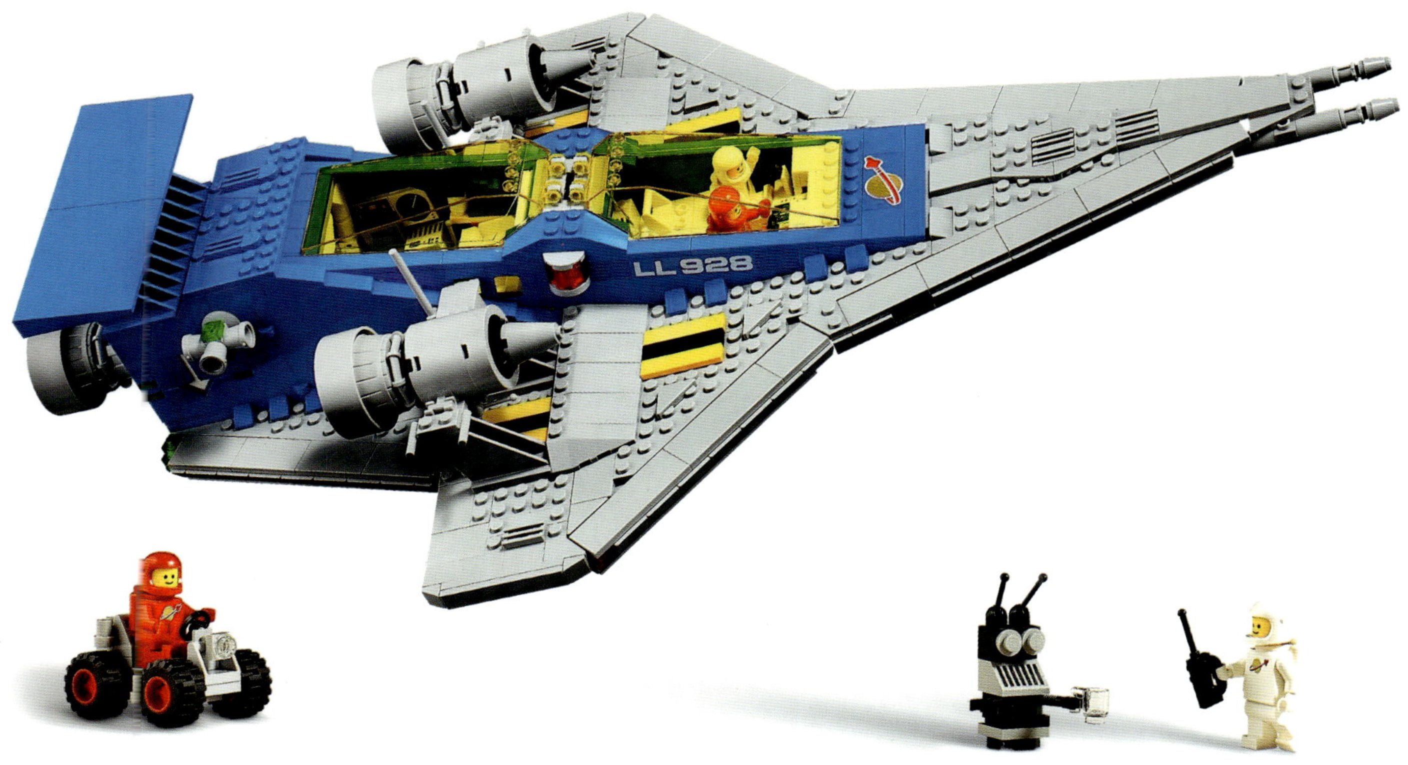

Set 10497 Galaxy Explorer was one of the sets released in 2022 by the LEGO® Icons line to mark the ninetieth anniversary of the LEGO Group. Jens Nygaard Knudsen's 1979 design was reimagined and enlarged, led by designer Michael Psiaki. The droid is identical to the one appearing in 6809 XT-5 and Droid from 1987.

The influence of the early years of LEGO® Space upon the development of LEGO products is far reaching and still in evidence today. "The LEGO Group is always building on the past—literally—and in a way, that's very organic," says Tveskov. "Even some of the elements from back then are around today, including some of those original LEGO Space elements, so I would say that's the most direct influence."

Knudsen's former colleague Connie Bork, who still works with the LEGO Group as a building instructions specialist, is constantly reminded of these elements from day to day. "Some elements can live for so many years, not just the 2 x 4 and 2 x 6 bricks and so on—the special elements are still living too. Also, the design of many elements was developed further from LEGO Space, like the wing elements. They are still here; they're just in a new form.

"I feel LEGO Space did something fundamental to LEGO bricks that is still there: something that made a foundation of how to build. Yes, there are new designers and all sorts of things, but it's still that same way to build a plane or a vehicle and so on. I don't think Jens had ever thought about this, but it is like it just became one part of the 'LEGO DNA.' It's a way that we built upon."

As one of the designers who met Knudsen despite joining after his tenure, Stafford echoes Tveskov and Bork's sentiments. "Whenever we get a really cool new brick that shows a way of building that we haven't seen before, or when things are flowing and everything's working well, there are many designers who say, 'This is like the old days.' That's a very positive thing. It feels like the philosophy of design, and the way we work in teams with design leads, and the fact that nobody's ideas are shot down . . . that feels like his sort of creation. So the influence lives on in that way.

"To some extent, what I'm always trying to do is to 'one-up' Jens: to make sure that what I'm giving to today's children is at least as good." ■

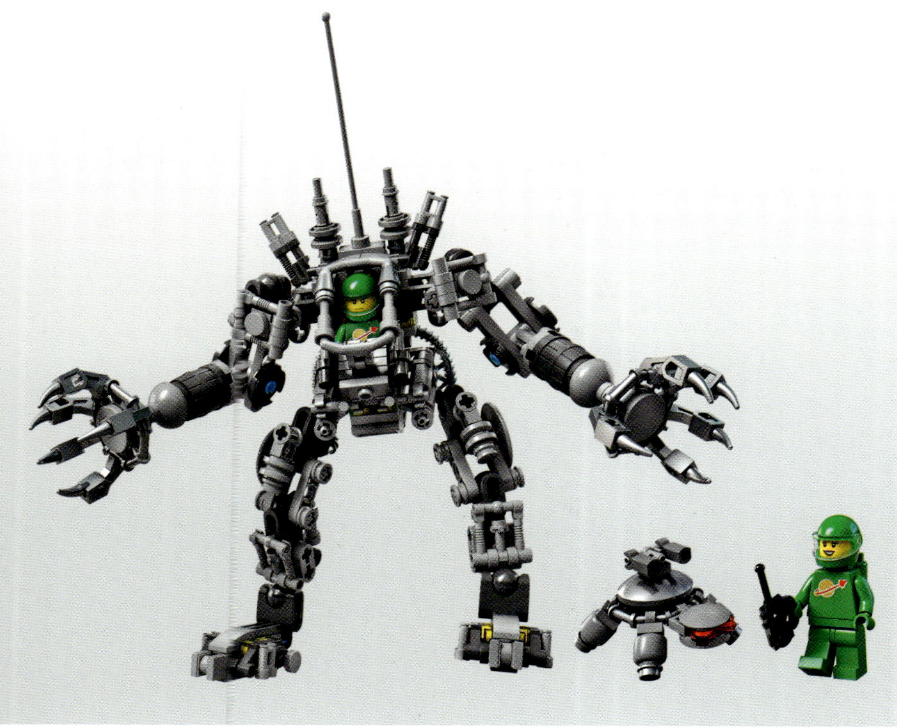

▼ The 2014 LEGO Ideas set 21109 Exo Suit, designed by fan Peter Reid and developed into a LEGO product by designer Mark John Stafford, meshed the past and future with its gray and transparent yellow palette and all-new green astronauts.